Sensor Technology Handbook

Sensor Technology Handbook

Editor-in-Chief
Jon S. Wilson

AMSTERDAM • BOSTON • HEIDELBERG • LONDON
NEW YORK • OXFORD • PARIS • SAN DIEGO
SAN FRANCISCO • SINGAPORE • SYDNEY • TOKYO

Newnes is an imprint of Elsevier

Newnes is an imprint of Elsevier
30 Corporate Drive, Suite 400, Burlington, MA 01803, USA
Linacre House, Jordan Hill, Oxford OX2 8DP, UK

Copyright © 2005, Elsevier Inc. All rights reserved.

No part of this publication may be reproduced, stored in a retrieval system, or transmitted in any form or by any means, electronic, mechanical, photocopying, recording, or otherwise, without the prior written permission of the publisher.

Permissions may be sought directly from Elsevier's Science & Technology Rights Department in Oxford, UK: phone: (+44) 1865 843830, fax: (+44) 1865 853333, e-mail: permissions@elsevier.com.uk. You may also complete your request on-line via the Elsevier homepage (http://elsevier.com), by selecting "Customer Support" and then "Obtaining Permissions."

 Recognizing the importance of preserving what has been written, Elsevier prints its books on acid-free paper whenever possible.

Library of Congress Cataloging-in-Publication Data

(Application submitted.)

British Library Cataloguing-in-Publication Data

A catalogue record for this book is available from the British Library.

ISBN: 0-7506-7729-5

For information on all Newnes publications visit our Web site at: www.books.elsevier.com

Transferred to Digital Printing, 2011

Printed in the United States of America

Contents

Preface ix

CHAPTER 1: Sensor Fundamentals 1
    1.1 Basic Sensor Technology 1
    1.2 Sensor Systems 15

CHAPTER 2: Application Considerations 21
    2.1 Sensor Characteristics 22
    2.2 System Characteristics 22
    2.3 Instrument Selection 23
    2.4 Data Acquisition and Readout 26
    2.5 Installation 26

CHAPTER 3: Measurement Issues and Criteria 29

CHAPTER 4: Sensor Signal Conditioning 31
    4.1 Conditioning Bridge Circuits 31
    4.2 Amplifiers for Signal Conditioning 45
    4.3 Analog to Digital Converters for Signal Conditioning 92
    4.4 Signal Conditioning High Impedance Sensors 108

CHAPTER 5: Acceleration, Shock and Vibration Sensors 137
    5.1 Introduction 137
    5.2 Technology Fundamentals 137
    5.3 Selecting and Specifying Accelerometers 150
    5.4 Applicable Standards 153
    5.5 Interfacing and Designs 155

CHAPTER 6: Biosensors 161
    6.1 Overview: What Is a Biosensor? 161
    6.2 Applications of Biosensors 164
    6.3 Origin of Biosensors 168
    6.4 Bioreceptor Molecules 169
    6.5 Transduction Mechanisms in Biosensors 171
    6.6 Application Range of Biosensors 173
    6.7 Future Prospects 177

Contents

CHAPTER 7: Chemical Sensors ... 181
- 7.1 Technology Fundamentals ... 181
- 7.2 Applications ... 188

CHAPTER 8: Capacitive and Inductive Displacement Sensors ... 193
- 8.1 Introduction ... 193
- 8.2 Capacitive Sensors ... 194
- 8.3 Inductive Sensors ... 196
- 8.4 Capacitive and Inductive Sensor Types ... 198
- 8.5 Selecting and Specifying Capacitive and Inductive Sensors ... 200
- 8.6 Comparing Capacitive and Inductive Sensors ... 203
- 8.7 Applications ... 204
- 8.8 Latest Developments ... 221
- 8.9 Conclusion ... 222

CHAPTER 9: Electromagnetism in Sensing ... 223
- 9.1 Introduction ... 223
- 9.2 Electromagnetism and Inductance ... 223
- 9.3 Sensor Applications ... 226
- 9.4 Magnetic Field Sensors ... 232
- 9.5 Summary ... 235

CHAPTER 10: Flow and Level Sensors ... 237
- 10.1 Methods for Measuring Flow ... 237
- 10.2 Selecting Flow Sensors ... 246
- 10.3 Installation and Maintenance ... 247
- 10.4 Recent Advances in Flow Sensors ... 249
- 10.5 Level Sensors ... 250
- 10.6 Applicable Standards ... 254

CHAPTER 11: Force, Load and Weight Sensors ... 255
- 11.1 Introduction ... 255
- 11.2 Quartz Sensors ... 255
- 11.3 Strain Gage Sensors ... 262

CHAPTER 12: Humidity Sensors ... 271
- 12.1 Humidity ... 271
- 12.2 Sensor Types and Technologies ... 271
- 12.3 Selecting and Specifying Humidity Sensors ... 275
- 12.4 Applicable Standards ... 279
- 12.5 Interfacing and Design Information ... 280

CHAPTER 13: Machinery Vibration Monitoring Sensors ... 285
- 13.1 Introduction ... 285
- 13.2 Technology Fundamentals ... 288
- 13.3 Accelerometer Types ... 291
- 13.4 Selecting Industrial Accelerometers ... 294
- 13.5 Applicable Standards ... 303

Contents

13.6 Latest and Future Developments 304
13.7 Sensor Manufacturers 304
13.8 References and Resources 305

CHAPTER 14: Optical and Radiation Sensors 307
14.1 Photosensors 307
14.2 Thermal Infrared Detectors 317

CHAPTER 15: Position and Motion Sensors 321
15.1 Contact and Non-contact Position Sensors 321
15.2 String Potentiometer and String Encoder Engineering Guide 370
15.3 Linear and Rotary Position and Motion Sensors 379
15.4 Selecting Position and Displacement Transducers 401

CHAPTER 16: Pressure Sensors 411
16.1 Piezoresistive Pressure Sensing 411
16.2 Piezoelectric Pressure Sensors *433*

CHAPTER 17: Sensors for Mechanical Shock 457
17.1 Technology Fundamentals 457
17.2 Sensor Types, Advantages and Disadvantages 459
17.3 Selecting and Specifying 461
17.4 Applicable Standards 473
17.5 Interfacing Information 474
17.6 Design Techniques and Tips, with Examples 478
17.7 Latest and Future Developments 480

CHAPTER 18: Test and Measurement Microphones 481
18.1 Measurement Microphone Characteristics 481
18.3 Traditional Condenser Microphone Design 483
18.4 Prepolarized (or Electret) Microphone Design 484
18.5 Frequency Response 484
18.6 Limitations on Measurement Range 490
18.7 Effect of Environmental Conditions 491
18.8 Microphone Standards 492
18.9 Specialized Microphone Types 494
18.10 Calibration 497
18.11 Major Manufacturers of Test and Measurement Microphones 499

CHAPTER 19: Strain Gages 501
19.1 Introduction to Strain Gages 501
19.2 Strain-Gage Based Measurements 511
19.3 Strain Gage Sensor Installations 522

CHAPTER 20: Temperature Sensors 531
20.1 Sensor Types and Technologies 531
20.2 Selecting and Specifying Temperature Sensors 535

Contents

CHAPTER 21: Nanotechnology-Enabled Sensors *563*
    21.1 Possibilities 564
    21.2 Realities 566
    21.3 Applications 567
    23.4 Summary 571

CHAPTER 22: Wireless Sensor Networks: Principles and Applications *575*
    22.1 Introduction to Wireless Sensor Networks 575
    22.2 Individual Wireless Sensor Node Architecture 576
    22.3 Wireless Sensor Networks Architecture 577
    22.4 Radio Options for the Physical Layer inWireless Sensor Networks 580
    22.5 Power Consideration in Wireless Sensor Networks 583
    22.6 Applications of Wireless Sensor Networks 585
    22.7 Future Developments 588

APPENDIX A: Lifetime Cost of Sensor Ownership *591*

APPENDIX B: Smart Sensors and TEDS FAQ *597*

APPENDIX C: Units and Conversions *601*

APPENDIX D: Physical Constants *607*

APPENDIX E: Dielectric Constants *615*

APPENDIX F: Index of Refraction *617*

APPENDIX G: Engineering Material Properties *619*

APPENDIX H: Emissions Resistivity *625*

APPENDIX I: Physical Properties of Some Typical Liquids *629*

APPENDIX J: Speed of Sound in Various Bulk Media *631*

APPENDIX K: Batteries *633*

APPENDIX L: Temperatures *635*

Contributor's Biographies *637*

Contributing Companies *647*

Sensor Suppliers *655*

Subject Index *683*

Sensor Technology Index *690*

Preface

The first decade of the 21st century has been labeled by some as the "Sensor Decade." With a dramatic increase in sensor R&D and applications over the past 15 years, sensors are certainly poised on the brink of a revolution similar to that experienced in microcomputers in the 1980s. Just in automobiles alone, sensing needs are growing by leaps and bounds, and the sensing technologies used are as varied as the applications. Tremendous advances have been made in sensor technology and many more are on the horizon.

In this volume, we attempted to balance breadth and depth in a single, practical and up-to-date resource. Understanding sensor design and operation typically requires a cross-disciplinary background, as it draws from electrical engineering, mechanical engineering, physics, chemistry, biology, etc. This reference pulls together the most crucial information needed by those who design sensor systems and work with sensors of all types, written by experts from industry and academia. While it would be impossible to cover each and every sensor in use today, we attempted to provide as broad a range of sensor types and applications as possible. The latest technologies, from piezo materials to micro and nano sensors to wireless networks, are discussed, as well as the tried and true methodologies. In addition, information on design, interfacing and signal conditioning is given for each sensor type.

Organized primarily by sensor application, the book is cross-referenced with indices of sensor technology. Manufacturers are listed by sensor type. The other contributors and I have attempted to provide a useful handbook with technical explanations that are clear, simple and thorough. We will also attempt to keep it updated as the technology advances.

Jon S. Wilson
Chandler, Arizona
October, 2004

Where reference is made in the text to the CD-ROM please visit the companion website at
www.elsevierdirect.com/companions/

CHAPTER 1

Sensor Fundamentals

1.1 Basic Sensor Technology

*Dr. Tom Kenny, Department of Mechanical Engineering,
Stanford University*

A *sensor* is a device that converts a physical phenomenon into an electrical signal. As such, sensors represent part of the interface between the physical world and the world of electrical devices, such as computers. The other part of this interface is represented by *actuators*, which convert electrical signals into physical phenomena.

Why do we care so much about this interface? In recent years, enormous capability for information processing has been developed within the electronics industry. The most significant example of this capability is the personal computer. In addition, the availability of inexpensive microprocessors is having a tremendous impact on the design of embedded computing products ranging from automobiles to microwave ovens to toys. In recent years, versions of these products that use microprocessors for control of functionality are becoming widely available. In automobiles, such capability is necessary to achieve compliance with pollution restrictions. In other cases, such capability simply offers an inexpensive performance advantage.

All of these microprocessors need electrical input voltages in order to receive instructions and information. So, along with the availability of inexpensive microprocessors has grown an opportunity for the use of sensors in a wide variety of products. In addition, since the output of the sensor is an electrical signal, sensors tend to be characterized in the same way as electronic devices. The data sheets for many sensors are formatted just like electronic product data sheets.

However, there are many formats in existence, and there is nothing close to an international standard for sensor specifications. The system designer will encounter a variety of interpretations of sensor performance parameters, and it can be confusing. It is important to realize that this confusion is not due to an inability to explain the meaning of the terms—rather it is a result of the fact that different parts of the sensor community have grown comfortable using these terms differently.

Chapter 1

Sensor Data Sheets

It is important to understand the function of the data sheet in order to deal with this variability. The data sheet is primarily a marketing document. It is typically designed to highlight the positive attributes of a particular sensor and emphasize some of the potential uses of the sensor, and might neglect to comment on some of the negative characteristics of the sensor. In many cases, the sensor has been designed to meet a particular performance specification for a specific customer, and the data sheet will concentrate on the performance parameters of greatest interest to this customer. In this case, the vendor and customer might have grown accustomed to unusual definitions for certain sensor performance parameters. Potential new users of such a sensor must recognize this situation and interpret things reasonably. Odd definitions may be encountered here and there, and most sensor data sheets are missing some pieces of information that are of interest to particular applications.

Sensor Performance Characteristics Definitions

The following are some of the more important sensor characteristics:

Transfer Function
> The transfer function shows the functional relationship between physical input signal and electrical output signal. Usually, this relationship is represented as a graph showing the relationship between the input and output signal, and the details of this relationship may constitute a complete description of the sensor characteristics. For expensive sensors that are individually calibrated, this might take the form of the certified calibration curve.

Sensitivity
> The sensitivity is defined in terms of the relationship between input physical signal and output electrical signal. It is generally the ratio between a small change in electrical signal to a small change in physical signal. As such, it may be expressed as the derivative of the transfer function with respect to physical signal. Typical units are volts/kelvin, millivolts/kilopascal, etc.. A thermometer would have "high sensitivity" if a small temperature change resulted in a large voltage change.

Span or Dynamic Range
> The range of input physical signals that may be converted to electrical signals by the sensor is the dynamic range or span. Signals outside of this range are expected to cause unacceptably large inaccuracy. This span or dynamic range is usually specified by the sensor supplier as the range over which other performance characteristics described in the data sheets are expected to apply. Typical units are kelvin, pascal, newtons, etc.

Sensor Fundamentals

Accuracy or Uncertainty

Uncertainty is generally defined as the largest expected error between actual and ideal output signals. Typical units are kelvin. Sometimes this is quoted as a fraction of the full-scale output or a fraction of the reading. For example, a thermometer might be guaranteed accurate to within 5% of FSO (Full Scale Output). "Accuracy" is generally considered by metrologists to be a qualitative term, while "uncertainty" is quantitative. For example one sensor might have better accuracy than another if its uncertainty is 1% compared to the other with an uncertainty of 3%.

Hysteresis

Some sensors do not return to the same output value when the input stimulus is cycled up or down. The width of the expected error in terms of the measured quantity is defined as the hysteresis. Typical units are kelvin or percent of FSO.

Nonlinearity (often called Linearity)

The maximum deviation from a linear transfer function over the specified dynamic range. There are several measures of this error. The most common compares the actual transfer function with the "best straight line," which lies midway between the two parallel lines that encompass the entire transfer function over the specified dynamic range of the device. This choice of comparison method is popular because it makes most sensors look the best. Other reference lines may be used, so the user should be careful to compare using the same reference.

Noise

All sensors produce some output noise in addition to the output signal. In some cases, the noise of the sensor is less than the noise of the next element in the electronics, or less than the fluctuations in the physical signal, in which case it is not important. Many other cases exist in which the noise of the sensor limits the performance of the system based on the sensor. Noise is generally distributed across the frequency spectrum. Many common noise sources produce a white noise distribution, which is to say that the spectral noise density is the same at all frequencies. Johnson noise in a resistor is a good example of such a noise distribution. For white noise, the spectral noise density is characterized in units of volts/Root (Hz). A distribution of this nature adds noise to a measurement with amplitude proportional to the square root of the measurement bandwidth. Since there is an inverse relationship between the bandwidth and measurement time, it can be said that the noise decreases with the square root of the measurement time.

Resolution

The resolution of a sensor is defined as the minimum detectable signal fluctuation. Since fluctuations are temporal phenomena, there is some relationship between the timescale for the fluctuation and the minimum detectable amplitude. Therefore, the definition of resolution must include some information about the nature of the measurement being carried out. Many sensors are limited by noise with a white spectral distribution. In these cases, the resolution may be specified in units of physical signal/root (Hz). Then, the actual resolution for a particular measurement may be obtained by multiplying this quantity by the square root of the measurement bandwidth. Sensor data sheets generally quote resolution in units of signal/root (Hz) or they give a minimum detectable signal for a specific measurement. If the shape of the noise distribution is also specified, it is possible to generalize these results to any measurement.

Bandwidth

All sensors have finite response times to an instantaneous change in physical signal. In addition, many sensors have decay times, which would represent the time after a step change in physical signal for the sensor output to decay to its original value. The reciprocal of these times correspond to the upper and lower cutoff frequencies, respectively. The bandwidth of a sensor is the frequency range between these two frequencies.

Sensor Performance Characteristics of an Example Device

To add substance to these definitions, we will identify the numerical values of these parameters for an off-the-shelf accelerometer, Analog Devices's ADXL150.

Transfer Function

The functional relationship between voltage and acceleration is stated as

$$V(Acc) = 1.5V + \left(Acc \times 167 \frac{mV}{g}\right)$$

This expression may be used to predict the behavior of the sensor, and contains information about the sensitivity and the offset at the output of the sensor.

Sensitivity

The sensitivity of the sensor is given by the derivative of the voltage with respect to acceleration at the initial operating point. For this device, the sensitivity is 167 mV/g.

Dynamic Range

The stated dynamic range for the ADXL322 is ±2g. For signals outside this range, the signal will continue to rise or fall, but the sensitivity is not guaranteed to match 167 mV/g by the manufacturer. The sensor can withstand up to 3500g.

Hysteresis

There is no fundamental source of hysteresis in this device. There is no mention of hysteresis in the data sheets.

Temperature Coefficient

The sensitivity changes with temperature in this sensor, and this change is guaranteed to be less than 0.025%/C. The offset voltage for no acceleration (nominally 1.5 V) also changes by as much as 2 mg/C. Expressed in voltage, this offset change is no larger than 0.3 mV/C.

Linearity

In this case, the linearity is the difference between the actual transfer function and the best straight line over the specified operating range. For this device, this is stated as less than 0.2% of the full-scale output. The data sheets show the expected deviation from linearity.

Noise

Noise is expressed as a noise density and is no more than 300 microg/root Hz. To express this in voltage, we multiply by the sensitivity (167 mV/g) to get 0.5 microV/Rt Hz. Then, in a 10 Hz low-pass-filtered application, we'd have noise of about 1.5 microV RMS, and an acceleration error of about 1 milli g.

Resolution

Resolution is 300 microG/RtHz as stated in the data sheet.

Bandwidth

The bandwidth of this sensor depends on choices of external capacitors and resistors.

Introduction to Sensor Electronics

The electronics that go along with the physical sensor element are often very important to the overall device. The sensor electronics can limit the performance, cost, and range of applicability. If carried out properly, the design of the sensor electronics can allow the optimal extraction of information from a noisy signal.

Chapter 1

Most sensors do not directly produce voltages but rather act like passive devices, such as resistors, whose values change in response to external stimuli. In order to produce voltages suitable for input to microprocessors and their analog-to-digital converters, the resistor must be "biased" and the output signal needs to be "amplified."

Types of Sensors

Resistive sensor circuits

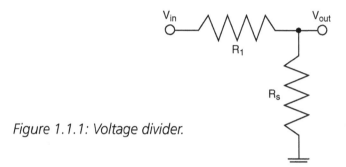

Figure 1.1.1: Voltage divider.

$$V_s = \frac{R_s}{R_1 + R_s} V_{in}$$

$$\text{if } R_1 >> R_s, V_s = \frac{R_s}{R_1} V_{in}$$

Resistive devices obey Ohm's law, which states that the voltage across a resistor is equal to the product of the current flowing through it and the resistance value of the resistor. It is also required that all of the current entering a node in the circuit leave that same node. Taken together, these two rules are called Kirchhoff's Rules for Circuit Analysis, and these may be used to determine the currents and voltages throughout a circuit.

For the example shown in Figure 1.1.1, this analysis is straightforward. First, we recognize that the voltage across the sense resistor is equal to the resistance value times the current. Second, we note that the voltage drop across both resistors (Vin-0) is equal to the sum of the resistances times the current. Taken together, we can solve these two equations for the voltage at the output. This general procedure applies to simple and complicated circuits; for each such circuit, there is an equation for the voltage between each pair of nodes, and another equation that sets the current into a node equal to the current leaving the node. Taken all together, it is always possibly to solve this set of linear equations for all the voltages and currents. So, one way to

Sensor Fundamentals

measure resistance is to force a current to flow and measure the voltage drop. Current sources can be built in number of ways. One of the easiest current sources to build consists of a voltage source and a stable resistor whose resistance is much larger than the one to be measured. The reference resistor is called a load resistor. Analyzing the connected load and sense resistors as shown in Figure 1.1.1, we can see that the current flowing through the circuit is nearly constant, since most of the resistance in the circuit is constant. Therefore, the voltage across the sense resistor is nearly proportional to the resistance of the sense resistor.

As stated, the load resistor must be much larger than the sense resistor for this circuit to offer good linearity. As a result, the output voltage will be much smaller than the input voltage. Therefore, some amplification will be needed.

A Wheatstone bridge circuit is a very common improvement on the simple voltage divider. It consists simply of the same voltage divider in Figure 1.1.1, combined with a second divider composed of fixed resistors only. The point of this additional divider is to make a reference voltage that is the same as the output of the sense voltage divider at some nominal value of the sense resistance. There are many complicated additional features that can be added to bridge circuits to more accurately compensate for particular effects, but for this discussion, we'll concentrate on the simplest designs—the ones with a single sense resistor, and three other bridge resistors that have resistance values that match the sense resistor at some nominal operating point.

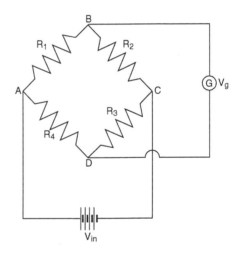

Figure 1.1.2: Wheatstone bridge circuit.

The output of the sense divider and the reference divider are the same when the sense resistance is at its starting value, and changes in the sense resistance lead to small differences between these two voltages. A differential amplifier (such as an instrumentation amplifier) is used to produce the difference between these two voltages and amplify the result. The primary advantages are that there is very little offset voltage at the output of this differential amplifier, and that temperature or other effects that are common to all the resistors are automatically compensated out of the resulting signal. Eliminating the offset means that the small differential signal at the output can be amplified without also amplifying an offset voltage, which makes the design of the rest of the circuit easier.

Chapter 1

Capacitance measuring circuits

Many sensors respond to physical signals by producing a change in capacitance. How is capacitance measured? Essentially, all capacitors have an impedance given by

$$impedance = \frac{1}{i\omega C} = \frac{1}{i2\pi f C}$$

where f is the oscillation frequency in Hz, w is in rad/sec, and C is the capacitance in farads. The i in this equation is the square root of -1, and signifies the phase shift between the current through a capacitor and the voltage across the capacitor.

Now, ideal capacitors cannot pass current at DC, since there is a physical separation between the conductive elements. However, oscillating voltages induce charge oscillations on the plates of the capacitor, which act as if there is physical charge flowing through the circuit. Since the oscillation reverses direction before substantial charges accumulate, there are no problems. The effective resistance of the capacitor is a meaningful characteristic, as long as we are talking about oscillating voltages.

With this in mind, the capacitor looks very much like a resistor. Therefore, we may measure capacitance by building voltage divider circuits as in Figure 1.1.1, and we may use either a resistor or a capacitor as the load resistance. It is generally easiest to use a resistor, since inexpensive resistors are available which have much smaller temperature coefficients than any reference capacitor. Following this analogy, we may build capacitance bridges as well. The only substantial difference is that these circuits must be biased with oscillating voltages. Since the "resistance" of the capacitor depends on the frequency of the AC bias, it is important to select this frequency carefully. By doing so, all of the advantages of bridges for resistance measurement are also available for capacitance measurement.

However, providing an AC bias can be problematic. Moreover, converting the AC signal to a DC signal for a microprocessor interface can be a substantial issue. On the other hand, the availability of a modulated signal creates an opportunity for use of some advanced sampling and processing techniques. Generally speaking, voltage oscillations must be used to bias the sensor. They can also be used to trigger voltage sampling circuits in a way that automatically subtracts the voltages from opposite clock phases. Such a technique is very valuable, because signals that oscillate at the correct frequency are added up, while any noise signals at all other frequencies are subtracted away. One reason these circuits have become popular in recent years is that they can be easily designed and fabricated using ordinary digital VLSI fabrication tools. Clocks and switches are easily made from transistors in CMOS circuits. Therefore, such designs can be included at very small additional cost—remember that the oscillator circuit has to be there to bias the sensor anyway.

Sensor Fundamentals

Capacitance measuring circuits are increasingly implemented as integrated clock/sample circuits of various kinds. Such circuits are capable of good capacitance measurement, but not of very high performance measurement, since the clocked switches inject noise charges into the circuit. These injected charges result in voltage offsets and errors that are very difficult to eliminate entirely. Therefore, very accurate capacitance measurement still requires expensive precision circuitry.

Since most sensor capacitances are relatively small (100 pF is typical), and the measurement frequencies are in the 1–100 kHz range, these capacitors have impedances that are large (> 1 megohm is common). With these high impedances, it is easy for parasitic signals to enter the circuit before the amplifiers and create problems for extracting the measured signal. For capacitive measuring circuits, it is therefore important to minimize the physical separation between the capacitor and the first amplifier. For microsensors made from silicon, this problem can be solved by integrating the measuring circuit and the capacitance element on the same chip, as is done for the ADXL311 mentioned above.

Inductance measurement circuits

Inductances are also essentially resistive elements. The "resistance" of an inductor is given by $X_L = 2\pi f L$, and this resistance may be compared with the resistance of any other passive element in a divider circuit or in a bridge circuit as shown in Figure 1.1.1. Inductive sensors generally require expensive techniques for the fabrication of the sensor mechanical structure, so inexpensive circuits are not generally of much use. In large part, this is because inductors are generally three-dimensional devices, consisting of a wire coiled around a form. As a result, inductive measuring circuits are most often of the traditional variety, relying on resistance divider approaches.

Sensor Limitations

Limitations in resistance measurement

- Lead resistance – The wires leading from the resistive sensor element have a resistance of their own. These resistances may be large enough to add errors to the measurement, and they may have temperature dependencies that are large enough to matter. One useful solution to the problem is the use of the so-called 4-wire resistance approach (Figure 1.1.3). In this case, current (from a current source as in Figure 1.1.1) is passed through the leads and through the sensor element. A second pair of wires is independently attached to the sensor leads, and a voltage reading is made across these two wires alone.

Chapter 1

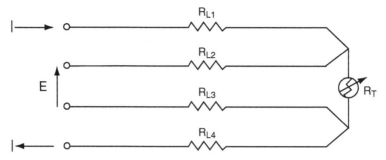

Figure 1.1.3: Lead compensation.

It is assumed that the voltage-measuring instrument does not draw significant current (see next point), so it simply measures the voltage drop across the sensor element alone. Such a 4-wire configuration is especially important when the sensor resistance is small, and the lead resistance is most likely to be a significant problem.

- Output impedance – The measuring network has a characteristic resistance which, simply put, places a lower limit on the value of a resistance which may be connected across the output terminals without changing the output voltage. For example, if the thermistor resistance is 10 kΩ and the load resistor resistance is 1 MΩ, the output impedance of this circuit is approximately 10 kΩ. If a 1 kΩ resistor is connected across the output leads, the output voltage would be reduced by about 90%. This is because the load applied to the circuit (1 kΩ) is much smaller than the output impedance of the circuit (10 kΩ), and the output is "loaded down." So, we must be concerned with the effective resistance of any measuring instrument that we might attach to the output of such a circuit. This is a well-known problem, so measuring instruments are often designed to offer maximum input impedance, so as to minimize loading effects. In our discussions we must be careful to arrange for instrument input impedance to be much greater than sensor output impedance.

Limitations to measurement of capacitance

- Stray capacitance – Any wire in a real-world environment has a finite capacitance with respect to ground. If we have a sensor with an output that looks like a capacitor, we must be careful with the wires that run from the sensor to the rest of the circuit. These stray capacitances appear as additional capacitances in the measuring circuit, and can cause errors. One source of error is the changes in capacitance that result from these wires moving about with respect to ground, causing capacitance fluctuations which might be confused with the signal. Since these effects can be due to acoustic pressure-induced vibrations in the positions of objects, they are often referred to as *microphonics*. An important way to minimize stray capacitances is to minimize the separation between the sensor element and the rest of the circuit. Another way to minimize the effects of stray capacitances is mentioned later—the virtual ground amplifier.

Filters

Electronic filters are important for separating signals from noise in a measurement. The following sections contain descriptions of several simple filters used in sensor-based systems.

- *Low pass* – A low-pass filter (Figure 1.1.4) uses a resistor and a capacitor in a voltage divider configuration. In this case, the "resistance" of the capacitor decreases at high frequency, so the output voltage decreases as the input frequency increases. So, this circuit effectively filters out the high frequencies and "passes" the low frequencies.

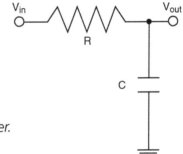

Figure 1.1.4: Low-pass filter.

The mathematical analysis is as follows:

Using the complex notation for the impedance, let

$$Z_1 = R, Z_2 = \frac{1}{i\omega C}$$

Using the voltage divider equation in Figure 1.1.1

$$V_{out} = \frac{Z_2}{Z_1 + Z_2} V_{in}$$

Chapter 1

Substituting for Z_1 and Z_2

$$V_{out} = \frac{\frac{1}{i\omega C}}{R + \frac{1}{i\omega C}} V_{in} = \frac{1}{i\omega RC + 1} V_{in}$$

The magnitude of V_{out} is

$$|V_{out}| = \sqrt{\frac{1}{(\omega RC)^2 + 1}} |V_{in}|$$

and the phase of V_{out} is

$$\phi = \tan^{-1}(-\omega RC)$$

- *High-pass* – The high-pass filter is exactly analogous to the low-pass filter, except that the roles of the resistor and capacitor are reversed. The analysis of a high-pass filter is as follows:

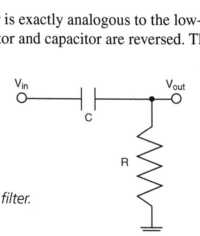

Figure 1.1.5: High-pass filter.

Similar to a low-pass filter,

$$V_{out} = \frac{R}{R + \frac{1}{i\omega C}} V_{in}$$

The magnitude is

$$|V_{out}| = \frac{R}{\sqrt{R^2 + \left(\frac{1}{\omega C}\right)^2}} |V_{in}|$$

12

and the phase is

$$\phi = \tan^{-1}\left(\frac{-1}{\omega RC}\right)$$

- *Bandpass* – By combining low-pass and high-pass filters together, we can create a bandpass filter that allows signals between two preset oscillation frequencies. Its diagram and the derivations are as follows:

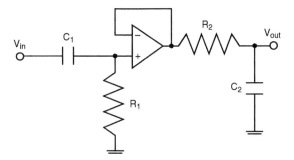

Figure 1.1.6: Band-pass filter.

Let the high-pass filter have the oscillation frequency ω_1 and the low-pass filter have the frequency ω_2 such that

$$\omega_{1co} = \frac{1}{R_1C_1},\ \omega_{2co} = \frac{1}{R_2C_2},\ \omega_1 < \omega_2$$

Then the relation between V_{out} and V_{in} is

$$V_{out} = \left(\frac{1}{i\omega_2 R_2 C_2 + 1}\right)\left(\frac{i\omega_1 R_1 C_1}{i\omega_1 R_1 C_1 + 1}\right)V_{in}$$

The operational amplifier in the middle of the circuit was added in this circuit to isolate the high-pass from the low-pass filter so that they do not effectively load each other. The op-amp simply works as a buffer in this case. In the following section, the role of the op-amps will be discussed more in detail.

Operational amplifiers

Operational amplifiers (op-amps) are electronic devices that are of enormous generic use for signal processing. The use of op-amps can be complicated, but there are a few simple rules and a few simple circuit building blocks which designers need to be familiar with to understand many common sensors and the circuits used with them.

Chapter 1

An op-amp is essentially a simple 2-input, 1-output device. The output voltage is equal to the difference between the non-inverting input and the inverting input multiplied by some extremely large value (10^5). Use of op-amps as simple amplifiers is uncommon.

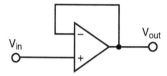

Figure 1.1.7: Non-inverting unity gain amplifier.

Feedback is a particularly valuable concept in op-amp applications. For instance, consider the circuit shown in Figure 1.1.6, called the follower configuration. Notice that the inverting input is tied directly to the output. In this case, if the output is less than the input, the difference between the inputs is a positive quantity, and the output voltage will be increased. This adjustment process continues, until the output is at the same voltage as the non-inverting input. Then, everything stays fixed, and the output will follow the voltage of the non-inverting input. This circuit appears to be useless until you consider that the input impedance of the op-amp can be as high as 10^9 ohms, while the output can be many orders of magnitude smaller. Therefore, this follower circuit is a good way to isolate circuit stages with high output impedance from stages with low input impedance.

This op-amp circuit can be analyzed very easily, using the op-amp golden rules:

1. No current flows into the inputs of the op-amp.
2. When configured for negative feedback, the output will be at whatever value makes the input voltages equal.

Even though these golden rules only apply to ideal operational amplifiers, op-amps can in most cases be treated as ideal. Let's use these rules to analyze more circuits:

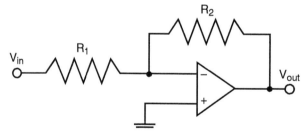

Figure 1.1.8: Inverting amplifier.

Figure 1.1.8 shows an example of an inverting amplifier. We can derive the equation by taking the following steps.

Sensor Fundamentals

1. Point B is ground. Therefore, point A is also ground. (Rule 2)
2. Since the current flowing from V_{in} to V_{out} is constant (Rule 1), $V_{out}/R_2 = -V_{in}/R_1$
3. Therefore, voltage gain = $V_{out}/V_{in} = -R_2/R_1$

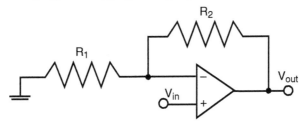

Figure 1.1.9: Non-inverting amplifier.

Figure 1.1.9 illustrates another useful configuration of an op-amp. This is a non-inverting amplifier, which is a slightly different expression than the inverting amplifier. Taking it step-by-step,

1. $V_a = V_{in}$ (Rule 2)
2. Since V_a comes from a voltage divider, $V_a = (R_1/(R_1 + R_2))\, V_{out}$
3. Therefore, $V_{in} = (R_1/(R_1 + R_2))\, V_{out}$
4. $V_{out}/V_{in} = (R_1 + R_2)/R_1 = 1 + R_2/R_1$

The following section provides more details on sensor systems and signal conditioning.

1.2 Sensor Systems

Analog Devices Technical Staff
Walt Kester, Editor

This section deals with sensors and associated signal conditioning circuits. The topic is broad, but the focus here is to concentrate on the sensors with just enough coverage of signal conditioning to introduce it and to at least imply its importance in the overall system.

Strictly speaking, a *sensor* is a device that receives a signal or stimulus and responds with an electrical signal, while a *transducer* is a converter of one type of energy into another. In practice, however, the terms are often used interchangeably.

Sensors and their associated circuits are used to measure various physical properties such as temperature, force, pressure, flow, position, light intensity, etc. These properties act as the stimulus to the sensor, and the sensor output is conditioned and processed to provide the corresponding measurement of the physical property. We will not cover all possible types of sensors here, only the most popular ones, and specifically, those that lend themselves to process control and data acquisition systems.

Excerpted from *Practical Design Techniques for Sensor Signal Conditioning*, Analog Devices, Inc., www.analog.com.

Chapter 1

Sensors do not operate by themselves. They are generally part of a larger system consisting of signal conditioners and various analog or digital signal processing circuits. The *system* could be a measurement system, data acquisition system, or process control system, for example.

Sensors may be classified in a number of ways. From a signal conditioning viewpoint it is useful to classify sensors as either *active* or *passive*. An *active* sensor requires an external source of excitation. Resistor-based sensors such as thermistors, RTDs (Resistance Temperature Detectors), and strain gages are examples of active sensors, because a current must be passed through them and the corresponding voltage measured in order to determine the resistance value. An alternative would be to place the devices in a bridge circuit; however, in either case, an external current or voltage is required.

On the other hand, *passive* (or *self-generating*) sensors generate their own electrical output signal without requiring external voltages or currents. Examples of passive sensors are thermocouples and photodiodes which generate thermoelectric voltages and photocurrents, respectively, which are independent of external circuits. It should be noted that these definitions (*active* vs. *passive*) refer to the need (or lack thereof) of external active circuitry to produce the electrical output signal from the sensor. It would seem equally logical to consider a thermocouple to be active in the sense that it produces an output voltage with no external circuitry. However, the convention in the industry is to classify the sensor with respect to the external circuit requirement as defined above.

SENSORS:

Convert a Signal or Stimulus (Representing a Physical Property) into an Electrical Output

TRANSDUCERS:

Convert One Type of Energy into Another

The Terms are often Interchanged

Active Sensors Require an External Source of Excitation: RTDs, Strain-Gages

Passive (Self-Generating) Sensors do not: Thermocouples, Photodiodes, Piezoelectrics

Figure 1.2.1: Sensor overview.

Sensor Fundamentals

PROPERTY	SENSOR	ACTIVE/PASSIVE	OUTPUT
Temperature	Thermocouple	Passive	Voltage
	Silicon	Active	Voltage/Current
	RTD	Active	Resistance
	Thermistor	Active	Resistance
Force/Pressure	Strain Gage	Active	Resistance
	Piezoelectric	Passive	Voltage
Acceleration	Accelerometer	Active	Capacitance
Position	LVDT	Active	AC Voltage
Light Intensity	Photodiode	Passive	Current

Figure 1.2.2: Typical sensors and their outputs

A logical way to classify sensors—and the method used throughout the remainder of this book—is with respect to the physical property the sensor is designed to measure. Thus, we have temperature sensors, force sensors, pressure sensors, motion sensors, etc. However, sensors which measure different properties may have the same type of electrical output. For instance, a resistance temperature detector (RTD) is a variable resistance, as is a resistive strain gage. Both RTDs and strain gages are often placed in bridge circuits, and the conditioning circuits are therefore quite similar. In fact, bridges and their conditioning circuits deserve a detailed discussion.

The full-scale outputs of most sensors (passive or active) are relatively small voltages, currents, or resistance changes, and therefore their outputs must be properly conditioned before further analog or digital processing can occur. Because of this, an entire class of circuits have evolved, generally referred to as *signal conditioning* circuits. Amplification, level translation, galvanic isolation, impedance transformation, linearization, and filtering are fundamental signal conditioning functions that may be required.

Whatever form the conditioning takes, however, the circuitry and performance will be governed by the electrical character of the sensor and its output. Accurate characterization of the sensor in terms of parameters appropriate to the application, e.g., sensitivity, voltage and current levels, linearity, impedances, gain, offset, drift, time constants, maximum electrical ratings, and stray impedances and other important considerations can spell the difference between substandard and successful application of the device, especially in cases where high resolution and precision, or low-level measurements are involved.

Chapter 1

Higher levels of integration now allow ICs to play a significant role in both analog and digital signal conditioning. ADCs (analog-to-digital converters) specifically designed for measurement applications often contain on-chip programmable-gain amplifiers (PGAs) and other useful circuits, such as current sources for driving RTDs, thereby minimizing the external conditioning circuit requirements.

Most sensor outputs are nonlinear with respect to the stimulus, and their outputs must be linearized in order to yield correct measurements. Analog techniques may be used to perform this function. However, the recent introduction of high performance ADCs now allows linearization to be done much more efficiently and accurately in software and eliminates the need for tedious manual calibration using multiple and sometimes interactive trimpots.

The application of sensors in a typical process control system is shown in Figure 1.2.3. Assume the physical property to be controlled is the temperature. The output of the temperature sensor is conditioned and then digitized by an ADC. The microcontroller or host computer determines if the temperature is above or below the desired value, and outputs a digital word to the digital-to-analog converter (DAC). The DAC output is conditioned and drives the *actuator*, in this case a heater. Notice that the interface between the control center and the remote process is via the industry-standard 4–20mA loop.

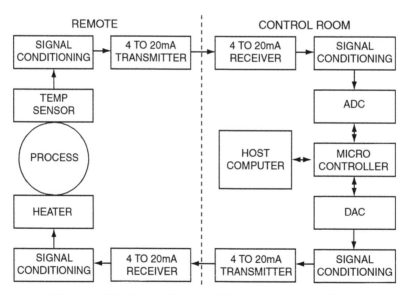

Figure 1.2.3: Typical industrial process control loop.

Sensor Fundamentals

Digital techniques have become increasingly popular in processing sensor outputs in data acquisition, process control, and measurement. Generally, 8-bit microcontrollers (8051-based, for example) have sufficient speed and processing capability for most applications. By including the A/D conversion and the microcontroller programmability on the sensor itself, a "smart sensor" can be implemented with self-contained calibration and linearization features, among others. A smart sensor can then interface directly to an industrial network as shown in Figure 1.2.4.

The basic building blocks of a "smart sensor" are shown in Figure 1.2.5, constructed with multiple ICs. The Analog Devices MicroConverter™-series of products includes on-chip high performance multiplexers, analog-to-digital converters (ADCs) and digital-to-analog converters (DACs), coupled with Flash memory and an industry-standard 8052 microcontroller core, as well as support circuitry and several standard serial port configurations. These are the first integrated circuits which are truly smart sensor data acquisition systems (high-performance data conversion circuits, microcontroller, Flash memory) on a single chip (see Figure 1.2.6).

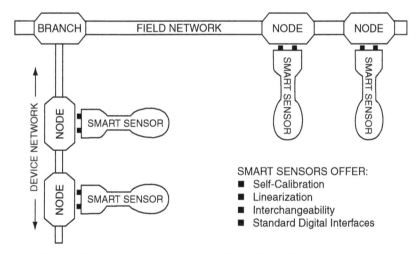

Figure 1.2.4: Standardization at the digital interface using smart sensors.

Chapter 1

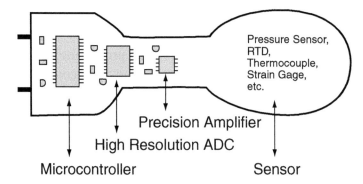

Figure 1.2.5: Basic elements in a smart sensor.

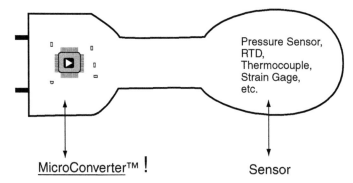

Figure 1.2.6: The even smarter sensor.

CHAPTER 2

Application Considerations

Jon Wilson, Technical Editor

The highest quality, most up-to-date, most accurately calibrated and most carefully selected sensor can still give totally erroneous data if it is not correctly applied. This section will address some of the issues that need to be considered to assure correct application of any sensor.

The following check list is derived from a list originally assembled by Applications Engineering at Endevco® in the late 1970s. It has been sporadically updated as additional issues were encountered. It is generally applicable to all sensor applications, but many of the items mentioned will not apply to any given specific application. However, it provides a reminder of questions that need to be asked and answered during selection and application of any sensor.

Often one of the most difficult tasks facing an instrumentation engineer is the selection of the proper measuring system. Economic realities and the pressing need for safe, properly functioning hardware create an ever-increasing demand to obtain accurate, reliable data on each and every measurement.

On the other hand, each application will have different characteristics from the next and will probably be subjected to different environments with different data requirements. As test or measurement programs progress, data are usually subjected to increasing manipulation, analysis and scrutiny. In this environment, the instrumentation engineer can no longer depend on his general-purpose measurement systems and expect to obtain acceptable data. Indeed, he must carefully analyze every aspect of the test to be performed, the test article, the environmental conditions, and, if available, the analytical predictions. In most cases, this process will indicate a clear choice of acceptable system components. In some cases, this analysis will identify unavoidable compromises or trade-offs and alert the instrumentation engineer and his customer to possible deficiencies in the results.

The intent of this chapter is to assist in the process of selecting an acceptable measuring system. While we hope it will be an aid, we understand it cannot totally address the wide variety of situations likely to arise.

Chapter 2

Let's look at a few hypothetical cases where instrument selection was made with care, but where the tests were failures.

1. A test requires that low g, low-frequency information be measured on the axle bearings of railroad cars to assess the state of the roadbed. After considerable evaluation of the range of conditions to be measured, a high-sensitivity, low-resonance piezoelectric accelerometer is selected. The shocks generated when the wheels hit the gaps between track sections saturate the amplifier, making it impossible to gather any meaningful data.

2. A test article must be exposed to a combined environment of vibration and a rapidly changing temperature. The engineer selects an accelerometer for its high temperature rating without consulting the manufacturer. Thermal transient output swamps the vibration data.

3. Concern over ground loops prompts the selection of an isolated accelerometer. The test structure is made partially from lightweight composites, and the cases of some accelerometers are not referenced to ground. Capacitive coupling of radiated interference to the signal line overwhelms the data.

From these examples, we hope to make the point that, for all measurement systems, it is not adequate to consider only that which we wish to measure. In fact, every physical and electrical phenomenon that is present needs to be considered lest it overwhelm or, perhaps worse, subtly contaminate our data. The user must remember that every measurement system responds to its total environment.

2.1 Sensor Characteristics

The prospective user is generally forced to make a selection based on the characteristics available on the product data sheet. Many performance characteristics are shown on a typical data sheet. Many manufacturers feel that the data sheet should provide as much information as possible. Unfortunately, this abundance of data may create some confusion for a potential user, particularly the new user. Therefore the instrumentation engineer must be sure he or she understands the pertinent characteristics and how they will affect the measurement. If there is any doubt, the manufacturer should be contacted for clarification.

2.2 System Characteristics

The sensor and signal conditioners must be selected to work together as a system. Moreover, the system must be selected to perform well in the intended applications. Overall system accuracy is usually affected most by sensor characteristics such as environmental effects and dynamic characteristics. Amplifier characteristics such as nonlinearity, harmonic distortion and flatness of the frequency response curve are usually negligible when compared to sensor errors.

2.3 Instrument Selection

Selecting a sensor/signal conditioner system for highly accurate measurements requires very skillful and careful measurement engineering. All environmental, mechanical, and measurement conditions must be considered. Installation must be carefully planned and carried out. The following guidelines are offered as an aid to selecting and installing measurement systems for the best possible accuracy.

Sensor

The most important element in a measurement system is the sensor. If the data is distorted or corrupted by the sensor, there is often little that can be done to correct it.

Will the sensor operate satisfactorily in the measurement environment?

    Check:
        Temperature Range
        Maximum Shock and Vibration
        Humidity
        Pressure
        Acoustic Level
        Corrosive Gases
        Magnetic and RF Fields
        Nuclear Radiation
        Salt Spray
        Transient Temperatures
        Strain in the Mounting Surface

Will the sensor characteristics provide the desired data accuracy?

    Check:
        Sensitivity
        Frequency Response
            Resonance Frequency
                Minor Resonances
        Internal Capacitance
        Transverse Sensitivity
        Amplitude Linearity and Hysteresis
        Temperature Deviation
        Weight and size
        Internal Resistance at Maximum Temperature

Chapter 2

  Calibration Accuracy
  Strain Sensitivity
  Damping at Temperature Extremes
  Zero Measurand Output
  Thermal Zero Shift
  Thermal Transient Response

Is the proper mounting being used for this application?
  Check:
    Is Insulating Stud Required?
      Ground Loops
      Calibration Simulation
      Is Adhesive Mounting Required?
      Thread Size, Depth and Class

Cable

Cables and connectors are usually the weakest link in the measurement system chain.

Will the cable operate satisfactorily in the measurement environment?
  Check:
    Temperature Range
    Humidity Conditions

Will the cable characteristics provide the desired data accuracy?
  Check:
    Low Noise
    Size and Weight
    Flexibility
    Is Sealed Connection Required?

Power Supply

Will the power supply operate satisfactorily in the measurement environment?
  Check:
    Temperature Range
    Maximum Shock and Vibration
    Humidity
    Pressure
    Acoustic Level
    Corrosive Gases
    Magnetic and RF Fields
    Nuclear Radiation
    Salt Spray

Application Considerations

Is this the proper power supply for the application?
> Check:
>> Voltage Regulation
>> Current Regulation
> Compliance Voltage
>> Output Voltage Adjustable?
>> Output Current Adjustable?
>> Long Output Lines?
>>> Need for External Sensing
>> Isolation
>> Mode Card, if Required

Will the power supply characteristics provide the desired data accuracy?
> Check:
>> Load Regulation
>> Line Regulation
>> Temperature Stability
>> Time Stability
>> Ripple and Noise
>> Output Impedance
>> Line-Transient Response
>> Noise to Ground
>> DC Isolation

Amplifier

The amplifier must provide gain, impedance matching, output drive current, and other signal processing.

Will the amplifier operate satisfactorily in the measurement environment?
> Check:
>> Temperature Range
>> Maximum Shock and Vibration
>> Humidity
>> Pressure
>> Acoustic Level
>> Corrosive Gases
>> Magnetic and RF Fields
>> Nuclear Radiation
>> Salt Spray

Chapter 2

Is this the proper amplifier for the application?
> Check:
>> Long Input Lines?
>>> Need for Charge Amplifier
>>> Need for Remote Charge Amplifier
>>
>> Long Output Lines
>>> Need for Power Amplifier
>>
>> Airborne
>>> Size, Weight, Power Limitations

Will the amplifier characteristics provide the desired data accuracy?
> Check:
>> Gain and Gain Stability
>> Frequency Response
>> Linearity
>> Stability
>> Phase Shift
>> Output Current and Voltage
>> Residual Noise
>> Input Impedance
>> Transient Response
>> Overload Capability
>> Common Mode Rejection
>> Zero-Temperature Coefficient
>> Gain-Temperature Coefficient

2.4 Data Acquisition and Readout

Does the remainder of the system, including any additional amplifiers, filters, data acquisition and readout devices, introduce any limitation that will tend to degrade the sensor-amplifier characteristics?

> Check: ALL of previous check items, plus Adequate Resolution.

2.5 Installation

Even the most carefully and thoughtfully selected and calibrated system can produce bad data if carelessly or ignorantly installed.

Sensor

Is the unit in good condition and ready to use?
   Check:
      Up-to-Date Calibration
      Physical Condition
         Case
         Mounting Surface
         Connector
         Mounting Hardware
      Inspect for Clean Connector
      Internal Resistance

Is the mounting hardware in good condition and ready to use?
   Check:
      Mounting Surface Condition
      Thread Condition
      Burred End Slots
      Insulated Stud
         Insulation Resistance
         Stud Damage by Over Torquing
      Mounting Surface Clean and Flat
      Sensor Base Surface Clean and Flat
      Hole Drilled and Tapped Deep Enough
      Correct Tap Size
      Hole Properly Aligned Perpendicular to Mounting Surface
      Stud Threads Lubricated
      Sensor Mounted with Recommended Torque

Cement Mounting

   Check:
      Mounting Surface Clean and Flat
      Dental Cement for Uneven Surfaces
      Cement Cured Properly
      Sensor Mounted to Cementing Stud with Recommended Torque

Chapter 2

Cable

Is the cable in good condition and ready for use?
    Check:
        Physical Condition
            Cable Kinked, Crushed
            Connector Threads, Pins
        Inspect for Clean Connectors
        Continuity
        Insulation Resistance
        Capacitance
        All Cable Connections Secure
        Cable Properly restrained
        Excess Cable Coiled and Tied Down
        Drip Loop Provided
        Connectors Sealed and Potted, if Required

Power Supply, Amplifier, and Readout

Are the units in good condition and ready to use?
    Check:
        Up-to-Date Calibration
        Physical Condition
            Connectors
            Case
            Output Cables
        Inspect for Clean Connectors
        Mounted Securely
        All Cable Connections Secure
        Gain Hole Cover Sealed, if Required
        Recommended Grounding in Use

When the above questions have been answered to the user's satisfaction, the measurement system has a high probability of providing accurate data.

CHAPTER 3

Measurement Issues and Criteria

Jon Wilson, Technical Editor

Sensors are most commonly used to make quantifiable measurements, as opposed to qualitative detection or presence sensing. Therefore, it should be obvious that the requirements of the measurement will determine the selection and application of the sensor. How then can we quantify the requirements of the measurement?

First, we must consider what it is we want to measure. Sensors are available to measure almost anything you can think of, and many things you would never think of (but someone has!). Pressure, temperature and flow are probably the most common measurements as they are involved in monitoring and controlling many industrial processes and material transfers. A brief tour of a Sensors Expo exhibition or a quick look at the internet will yield hundreds, if not thousands, of quantities, characteristics or phenomena that can be measured with sensors.

Second, we must consider the environment of the sensor. Environmental effects are perhaps the biggest contributor to measurement errors in most measurement systems. Sensors, and indeed whole measurement systems, respond to their total environment, not just to the measurand. In extreme cases, the response to the combination of environments may be greater than the response to the desired measurand. One of the sensor designer's greatest challenges is to minimize the response to the environment and maximize the response to the desired measurand. Assessing the environment and estimating its effect on the measurement system is an extremely important part of the selection and application process.

The environment includes not only such parameters as temperature, pressure and vibration, but also the mounting or attachment of the sensor, electromagnetic and electrostatic effects, and the rates of change of the various environments. For example, a sensor may be little affected by extreme temperatures, but may produce huge errors in a rapidly changing temperature ("thermal transient sensitivity").

Third, we must consider the requirements for accuracy (uncertainty) of the measurement. Often, we would like to achieve the lowest possible uncertainty, but that may not be economically feasible, or even necessary. How will the information derived from the measurement be used? Will it really make a difference, in the long run, whether the uncertainty is 1% or 1½%? Will highly accurate sensor data be obscured by inaccuracies in the signal conditioning or recording processes? On the other hand, many modern data acquisition systems are capable of much greater accuracy than the sensors making the measurement. A user must not be misled by thinking that high resolution in a data acquisition system will produce high accuracy data from a low accuracy sensor.

Last, but not least, the user must assure that the whole system is calibrated and traceable to a national standards organization (such as National Institute of Standards and Technology [NIST] in the United States). Without documented traceability, the uncertainty of any measurement is unknown. Either each part of the measurement system must be calibrated and an overall uncertainty calculated, or the total system must be calibrated as it will be used ("system calibration" or "end-to-end calibration").

Since most sensors do not have any adjustment capability for conventional "calibration", a characterization or evaluation of sensor parameters is most often required. For the lowest uncertainty in the measurement, the characterization should be done with mounting and environment as similar as possible to the actual measurement conditions.

While this handbook concentrates on sensor technology, a properly selected, calibrated, and applied sensor is necessary but not sufficient to assure accurate measurements. The sensor must be carefully matched with, and integrated into, the total measurement system and its environment.

CHAPTER 4

Sensor Signal Conditioning

Analog Devices Technical Staff
Walt Kester, Editor

Typically a sensor cannot be directly connected to the instruments that record, monitor, or process its signal, because the signal may be incompatible or may be too weak and/or noisy. The signal must be conditioned—i.e., cleaned up, amplified, and put into a compatible format.

The following sections discuss the important aspects of sensor signal conditioning.

4.1 Conditioning Bridge Circuits

Introduction

This section discusses the fundamental concepts of bridge circuits.

Resistive elements are some of the most common sensors. They are inexpensive to manufacture and relatively easy to interface with signal conditioning circuits. Resistive elements can be made sensitive to temperature, strain (by pressure or by flex), and light. Using these basic elements, many complex physical phenomena can be measured, such as fluid or mass flow (by sensing the temperature difference between two calibrated resistances) and dew-point humidity (by measuring two different temperature points), etc. Bridge circuits are often incorporated into force, pressure and acceleration sensors.

Sensor elements' resistances can range from less than 100 Ω to several hundred kΩ, depending on the sensor design and the physical environment to be measured (See Figure 4.1.1). For example, RTDs (resistance temperature devices) are typically 100 Ω or 1000 Ω. Thermistors are typically 3500 Ω or higher.

Sensor	Resistance
■ Strain Gages	120Ω, 350Ω, 3500Ω
■ Weigh-Scale Load Cells	350Ω - 3500Ω
■ Pressure Sensors	350Ω - 3500Ω
■ Relative Humidity	100kΩ - 10MΩ
■ Resistance Temperature Devices (RTDs)	100Ω, 1000Ω
■ Thermistors	100Ω - 10MΩ

Figure 4.1.1: Resistance of popular sensors.

Excerpted from *Practical Design Techniques for Sensor Signal Conditioning*, Analog Devices, Inc., www.analog.com.

Chapter 4

Bridge Circuits

Resistive sensors such as RTDs and strain gages produce small percentage changes in resistance in response to a change in a physical variable such as temperature or force. Platinum RTDs have a temperature coefficient of about 0.385%/°C. Thus, in order to accurately resolve temperature to 1°C, the measurement accuracy must be much better than 0.385 Ω, for a 100 Ω RTD.

Strain gages present a significant measurement challenge because the typical change in resistance over the entire operating range of a strain gage may be less than 1% of the nominal resistance value. Accurately measuring small resistance changes is therefore critical when applying resistive sensors.

One technique for measuring resistance (shown in Figure 4.1.2) is to force a constant current through the resistive sensor and measure the voltage output. This requires both an accurate current source and an accurate means of measuring the voltage. Any change in the current will be interpreted as a resistance change. In addition, the power dissipation in the resistive sensor must be small, in accordance with the manufacturer's recommendations, so that self-heating does not produce errors, therefore the drive current must be small.

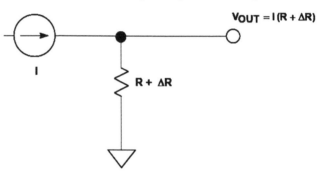

Figure 4.1.2: Measuring resistance indirectly using a constant current source.

Bridges offer an attractive alternative for measuring small resistance changes accurately. The basic Wheatstone bridge (actually developed by S. H. Christie in 1833) is shown in Figure 4.1.3. It consists of four resistors connected to form a quadrilateral, a source of excitation (voltage or current) connected across one of the diagonals, and a voltage detector connected across the other diagonal. The detector measures the difference between the outputs of two voltage dividers connected across the excitation.

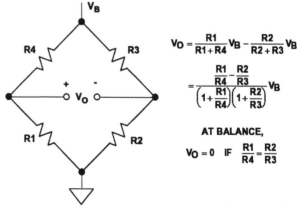

Figure 4.1.3: The Wheatstone bridge.

$$V_O = \frac{R1}{R1+R4} V_B - \frac{R2}{R2+R3} V_B$$

$$= \frac{\frac{R1}{R4} - \frac{R2}{R3}}{\left(1+\frac{R1}{R4}\right)\left(1+\frac{R2}{R3}\right)} V_B$$

AT BALANCE,

$$V_O = 0 \quad \text{IF} \quad \frac{R1}{R4} = \frac{R2}{R3}$$

Sensor Signal Conditioning

A bridge measures resistance indirectly by comparison with a similar resistance. The two principal ways of operating a bridge are as a null detector or as a device that reads a difference directly as voltage.

When R1/R4 = R2/R3, the resistance bridge is at *a null,* regardless of the mode of excitation (current or voltage, AC or DC), the magnitude of excitation, the mode of readout (current or voltage), or the impedance of the detector. Therefore, if the ratio of R2/R3 is fixed at K, a null is achieved when R1 = K·R4. If R1 is unknown and R4 is an accurately determined variable resistance, the magnitude of R1 can be found by adjusting R4 until null is achieved. Conversely, in sensor-type measurements, R4 may be a fixed reference, and a null occurs when the magnitude of the external variable (strain, temperature, etc.) is such that R1 = K·R4.

Null measurements are principally used in feedback systems involving electromechanical and/or human elements. Such systems seek to force the active element (strain gage, RTD, thermistor, etc.) to balance the bridge by influencing the parameter being measured.

For the majority of sensor applications employing bridges, however, the deviation of one or more resistors in a bridge from an initial value is measured as an indication of the magnitude (or a change) in the measured variable. In this case, the output voltage change is an indication of the resistance change. Because very small resistance changes are common, the output voltage change may be as small as tens of millivolts, even with V_B = 10 V (a typical excitation voltage for a load cell application).

In many bridge applications, there may be two, or even four, elements that vary. Figure 4.1.4 shows the four commonly used bridges suitable for sensor applications and the corresponding equations which relate the bridge output voltage to the excitation voltage and the bridge resistance values. In this case, we assume a constant voltage drive, VB. Note that since the bridge output is directly proportional to VB, the measurement accuracy can be no better than that of the accuracy of the excitation voltage.

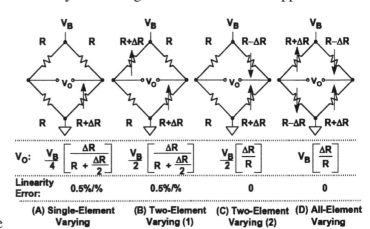

Figure 4.1.4: Output voltage and linearity error for constant voltage drive bridge configurations.

Chapter 4

In each case, the value of the fixed bridge resistor, R, is chosen to be equal to the nominal value of the variable resistor(s). The deviation of the variable resistor(s) about the nominal value is proportional to the quantity being measured, such as strain (in the case of a strain gage) or temperature (in the case of an RTD).

The *sensitivity* of a bridge is the ratio of the maximum expected change in the output voltage to the excitation voltage. For instance, if $V_B = 10$ V, and the full-scale bridge output is 10 mV, then the sensitivity is 1 mV/V.

The *single-element varying* bridge is most suited for temperature sensing using RTDs or thermistors. This configuration is also used with a single resistive strain gage. All the resistances are nominally equal, but one of them (the sensor) is variable by an amount ΔR. As the equation indicates, the relationship between the bridge output and ΔR is not linear. For example, if $R = 100 \, \Omega$, and $\Delta R = 0.152$, (0.1% change in resistance), the output of the bridge is 2.49875 mV for $V_B = 10$ V. The error is 2.50000 mV $-$ 2.49875 mV, or 0.00125 mV. Converting this to a percent of full scale by dividing by 2.5 mV yields an end-point linearity error in percent of approximately 0.05%. (Bridge end-point linearity error is calculated as the worst error in % FS from a straight line which connects the origin and the end point at FS, i.e. the FS gain error is not included). If $\Delta R = 1 \, \Omega$ (1% change in resistance), the output of the bridge is 24.8756 mV, representing an end-point linearity error of approximately 0.5%. The end-point linearity error of the single-element bridge can be expressed in equation form:

Single-Element Varying Bridge End-Point Linearity Error ≈ % Change in Resistance ÷ 2

It should be noted that the above nonlinearity refers to the nonlinearity of the bridge itself and not the sensor. In practice, most sensors exhibit a certain amount of their own nonlinearity which must be accounted for in the final measurement.

In some applications, the bridge nonlinearity may be acceptable, but there are various methods available to linearize bridges. Since there is a fixed relationship between the bridge resistance change and its output (shown in the equations), software can be used to remove the linearity error in digital systems. Circuit techniques can also be used to linearize the bridge output directly, and these will be discussed shortly.

There are two possibilities to consider in the case of the two-element varying bridge. In the first, Case (1), both elements change in the same direction, such as two identical strain gages mounted adjacent to each other with their axes in parallel.

The nonlinearity is the same as that of the single-element varying bridge, however the gain is twice that of the single-element varying bridge. The two-element varying bridge is commonly found in pressure sensors and flow meter systems.

Sensor Signal Conditioning

A second configuration of the two-element varying bridge, Case (2), requires two identical elements that vary in *opposite* directions. This could correspond to two identical strain gages: one mounted on top of a flexing surface, and one on the bottom. Note that this configuration is linear, and like two-element Case (1), has twice the gain of the single-element configuration. Another way to view this configuration is to consider the terms R + ΔR and R − ΔR as comprising the two sections of a center-tapped potentiometer.

The *all-element varying* bridge produces the most signal for a given resistance change and is inherently linear. It is an industry-standard configuration for load cells which are constructed from four identical strain gages.

Bridges may also be driven from constant current sources as shown in Figure 4.1.5. Current drive, although not as popular as voltage drive, has an advantage when the bridge is located remotely from the source of excitation because the wiring resistance does not introduce errors in the measurement. Note also that with constant current excitation, all configurations are linear with the exception of the single-element varying case.

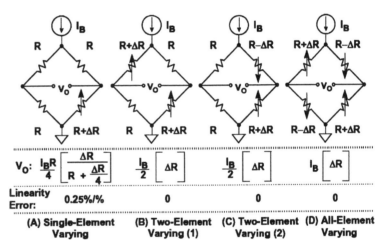

Figure 4.1.5: Output voltage and linearity error for constant current drive bridge configurations.

In summary, there are many design issues relating to bridge circuits. After selecting the basic configuration, the excitation method must be determined. The value of the excitation voltage or current must first be determined. Recall that the full scale bridge output is directly proportional to the excitation voltage (or current). Typical bridge sensitivities are 1 mV/V to 10 mV/V. Although large excitation voltages yield proportionally larger full scale output voltages, they also result in higher power dissipation and the possibility of sensor resistor self-heating errors. On the other hand, low values of excitation voltage require more gain in the conditioning circuits and increase the sensitivity to noise.

Chapter 4

Regardless of its value, the stability of the excitation voltage or current directly affects the overall accuracy of the bridge output. Stable references and/or ratiometric techniques are required to maintain desired accuracy.

Amplifying and Linearizing Bridge Outputs

The output of a single-element varying bridge may be amplified by a single precision op-amp connected in the inverting mode as shown in Figure 4.1.7. This circuit, although simple, has poor gain accuracy and also unbalances the bridge due to loading from RF and the op amp bias current. The RF resistors must be carefully chosen and matched to maximize the common mode rejection (CMR). Also it is difficult to maximize the CMR while at the same time allowing different gain options. In addition, the output is nonlinear. The key redeeming feature of the circuit is that it is capable of single supply operation and requires a single op amp. Note that the RF resistor connected to the non-inverting input is returned to $V_S/2$ (rather than ground) so that both positive and negative values of ΔR can be accommodated, and the op amp output is referenced to $V_S/2$.

- Selecting Configuration (1, 2, 4 - Element Varying)
- Selection of Voltage or Current Excitation
- Stability of Excitation Voltage or Current
- Bridge Sensitivity: FS Output / Excitation Voltage 1mV / V to 10mV / V Typical
- Fullscale Bridge Outputs: 10mV - 100mV Typical
- Precision, Low Noise Amplification / Conditioning Techniques Required
- Linearization Techniques May Be Required
- Remote Sensors Present Challenges

Figure 4.1.6: Bridge considerations.

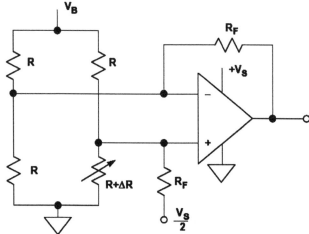

Figure 4.1.7: Using a single op amp as a bridge amplifier for a single-element varying bridge.

A much better approach is to use an instrumentation amplifier (in-amp) as shown in Figure 4.1.8. This efficient circuit provides better gain accuracy (usually set with a single resistor, RG) and does not unbalance the bridge. Excellent common mode rejection can be achieved with modern in-amps. Due to the bridge's intrinsic characteristics, the output is nonlinear, but this can be corrected in the software (assuming that the in-amp output is digitized using an analog-to-digital converter and followed by a microcontroller or microprocessor).

Various techniques are available to linearize bridges, but it is important to distinguish between the linearity of the bridge equation and the linearity of the sensor response to the phenomenon being sensed. For example, if the active element is an RTD, the bridge used to implement the measurement might have perfectly adequate linearity; yet the output could still be nonlinear due to the RTD's nonlinearity. Manufacturers of sensors employing bridges

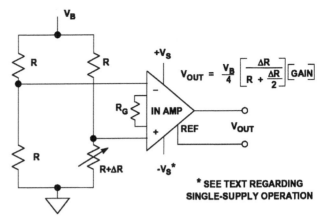

Figure 4.1.8: Using an instrumentation amplifier with a single-element varying bridge.

address the nonlinearity issue in a variety of ways, including keeping the resistive swings in the bridge small, shaping complementary nonlinear response into the active elements of the bridge, using resistive trims for first-order corrections, and others.

Figure 4.1.9 shows a single-element varying active bridge in which an op amp produces a forced null, by adding a voltage in series with the variable arm. That voltage is equal in magnitude and opposite in polarity to the incremental voltage across the varying element and is linear with ΔR. Since it is an op amp output, it can be used as a low impedance output point for the bridge measurement. This active bridge has a gain of two over the standard single-element varying bridge, and the output is linear, even for large values of ΔR. Because of the small output signal, this bridge must usually be followed by a second amplifier.

The amplifier used in this circuit requires dual supplies because its output must go negative.

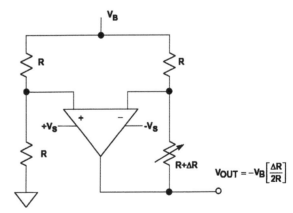

Figure 4.1.9: Linearizing a single-element varying bridge method 1.

Chapter 4

Another circuit for linearizing a single-element varying bridge is shown in Figure 4.1.10. The bottom of the bridge is driven by an op amp, which maintains a constant current in the varying resistance element. The output signal is taken from the right hand leg of the bridge and amplified by a non-inverting op amp. The output is linear, but the circuit requires two op amps which must operate on dual supplies. In addition, R1 and R2 must be matched for accurate gain.

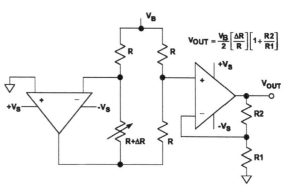

Figure 4.1.10: Linearizing a single-element varying bridge method 2.

A circuit for linearizing a voltage-driven two-element varying bridge is shown in Figure 4.1.11. This circuit is similar to Figure 4.1.9 and has twice the sensitivity. A dual supply op amp is required. Additional gain may be necessary.

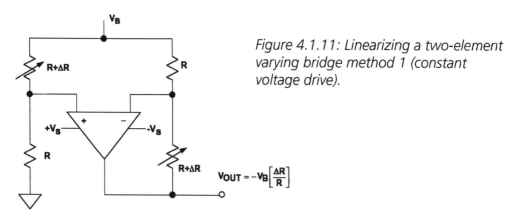

Figure 4.1.11: Linearizing a two-element varying bridge method 1 (constant voltage drive).

The two-element varying bridge circuit in Figure 4.1.12 uses an op amp, a sense resistor, and a voltage reference to maintain a constant current through the bridge

$$(I_B = V_{REF}/R_{SENSE}).$$

The current through each leg of the bridge remains constant ($I_B/2$) as the resistances change; therefore the output is a linear function of ΔR. An instrumentation amplifier provides the additional gain. This circuit can be operated on a single supply with the proper choice of amplifiers and signal levels.

Sensor Signal Conditioning

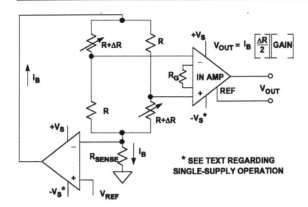

Figure 4.1.12: Linearizing a two-element varying bridge method 2 (constant voltage drive).

* SEE TEXT REGARDING SINGLE-SUPPLY OPERATION

Driving Bridges

Wiring resistance and noise pickup are the biggest problems associated with remotely located bridges. Figure 4.1.13 shows a 350 Ω strain gage which is connected to the rest of the bridge circuit by 100 feet of 30 gage twisted pair copper wire. The resistance of the wire at 25°C is 0.105 Ω/ft, or 10.5 Ω for 100ft. The total lead resistance in series with the 350 Ω strain gage is therefore 21 Ω. The temperature coefficient of the copper wire is 0.385%/°C. Now we will calculate the gain and offset error in the bridge output due to a +10°C temperature rise in the cable. These calculations are easy to make, because the bridge output voltage is simply the difference between the output of two voltage dividers, each driven from a +10 V source.

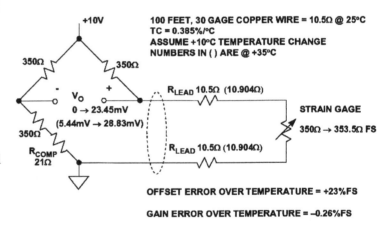

Figure 4.1.13: Errors produced by wiring resistance for remote resistive bridge sensor.

The full-scale variation of the strain gage resistance (with flex) above its nominal 350 Ω value is +1% (+3.5 Ω), corresponding to a full-scale strain gage resistance of 353.5 Ω, which causes a bridge output voltage of +23.45 mV. Notice that the addition of the 21 Ω R_{COMP} resistor compensates for the wiring resistance and balances the bridge when the strain gage resistance is 350 Ω. Without R_{COMP}, the bridge would have an output offset voltage of 145.63 mV for a nominal strain gage resistance of 350 Ω. This offset could be compensated for in software just as easily, but for this example, we chose to do it with R_{COMP}.

Assume that the cable temperature increases +10°C above nominal room temperature. This results in a total lead resistance increase of +0.404 Ω (10.5 Ω × 0.00385/°C × 10°C) in each lead. *Note: The values in parentheses in the diagram indicate the values at +35°C.* The total additional lead resistance (of the two leads) is +0.808 Ω. With no strain, this additional lead resistance produces an offset of +5.44 mV in the bridge output. Full-scale strain produces a bridge output of +28.83 mV (a change of +23.39 mV from no strain). Thus the increase in temperature produces an offset voltage error of +5.44 mV (+23% full scale) and a gain error of −0.06 mV (23.39 mV − 23.45 mV), or −0.26% full scale. Note that these errors are produced solely by the 30 gage wire, and do not include any temperature coefficient errors in the strain gage itself.

The effects of wiring resistance on the bridge output can be minimized by the 3-wire connection shown in Figure 4.1.14. We assume that the bridge output voltage is measured by a high impedance device, therefore there is no current in the sense lead. Note that the sense lead measures the voltage output of a divider: the top half is the bridge resistor plus the lead resistance, and the bottom half is strain gage resistance plus the lead resistance. The nominal sense voltage is therefore independent of the lead resistance. When the strain gage resistance increases to full scale (353.5 Ω,), the bridge output increases to +24.15 mV.

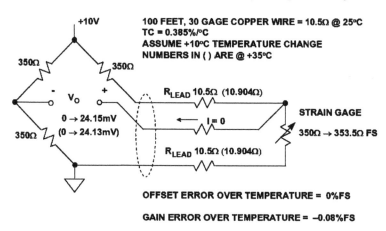

Figure 4.1.14: 3-wire connection to remote bridge element (single-element varying).

Increasing the temperature to +35°C increases the lead resistance by +0.404 Ω in each half of the divider. The full scale bridge output voltage decreases to +24.13 mV because of the small loss in sensitivity, but there is no offset error. The gain error due to the temperature increase of +10°C is therefore only −0.02 mV, or −0.08% of full scale. Compare this to the +23% full scale offset error and the −0.26% gain error for the two-wire connection shown in Figure 4.1.13.

Sensor Signal Conditioning

The three-wire method works well for remotely located resistive elements which make up one leg of a single-element varying bridge. However, all-element varying bridges generally are housed in a complete assembly, as in the case of a load cell. When these bridges are remotely located from the conditioning electronics, special techniques must be used to maintain accuracy.

Of particular concern is maintaining the accuracy and stability of the bridge excitation voltage. The bridge output is directly proportional to the excitation voltage, and any drift in the excitation voltage produces a corresponding drift in the output voltage.

For this reason, most all-element varying bridges (such as load cells) are six-lead assemblies: two leads for the bridge output, two leads for the bridge excitation, and two *sense* leads. This method (called Kelvin or 4-wire sensing) is shown in Figure 4.1.15. The sense lines go to high impedance op amp inputs, so there is minimal error due to the bias current induced voltage drop across their lead resistance. The op amps maintain the required excitation voltage to make the voltage measured between the sense leads always equal to V_B. Although Kelvin sensing eliminates errors due to voltage drops in the wiring resistance, the drive voltages must still be highly stable since they directly affect the bridge output voltage. In addition, the op amps must have low offset, low drift, and low noise.

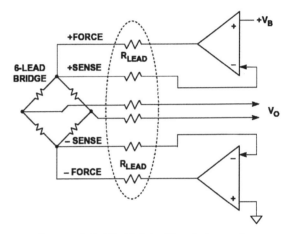

Figure 4.1.15: Kelvin (4-wire) sensing minimizes errors due to lead resistance.

The constant current excitation method shown in Figure 4.1.16 is another method for minimizing the effects of wiring resistance on the measurement accuracy. However, the accuracy of the reference, the sense resistor, and the op amp all influence the overall accuracy.

A very powerful *ratiometric* technique which includes Kelvin sensing to minimize errors due to wiring resistance and also eliminates the need for an accurate excitation voltage is shown in Figure 4.1.17. The AD7730 measurement ADC can be driven from a single supply voltage which is also used to excite the remote bridge. Both the analog input and the reference input to the ADC are high impedance and fully differential. By using the + and – SENSE outputs from the bridge as the differential reference to the ADC, there is no loss in measurement accuracy if the actual bridge

Chapter 4

excitation voltage varies. The AD7730 is one of a family of sigma-delta ADCs with high resolution (24 bits) and internal programmable gain amplifiers (PGAs) and is ideally suited for bridge applications. These ADCs have self- and system calibration features which allow offset and gain errors due to the ADC to be minimized. For instance, the AD7730 has an offset drift of 5 nV/°C and a gain drift of 2 ppm/°C. Offset and gain errors can be reduced to a few microvolts using the system calibration feature.

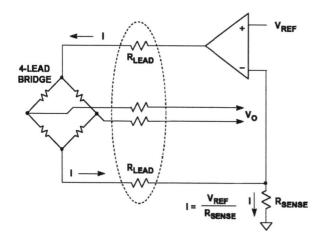

Figure 4.1.16: Constant current excitation minimizes wiring resistance errors.

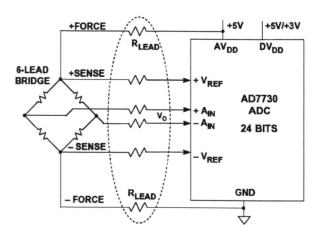

Figure 4.1.17: Driving remote bridge using Kelvin (4-wire) sensing and ratiometric connection to ADC.

Sensor Signal Conditioning

Maintaining an accuracy of 0.1% or better with a full-scale bridge output voltage of 20 mV requires that the sum of all offset errors be less than 20 µV. Figure 4.1.18 shows some typical sources of offset error that are inevitable in a system. Parasitic thermocouples whose junctions are at different temperatures can generate voltages between a few and tens of microvolts for a 1°C temperature differential. The diagram shows a typical parasitic junction formed between the copper printed circuit board traces and the kovar pins of the IC amplifier. This thermocouple voltage is about 35 µV/°C temperature differential. The thermocouple voltage is significantly less when using a plastic package with a copper lead frame.

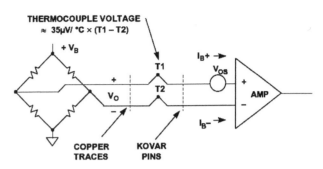

Figure 4.1.18: Typical sources of offset voltage.

The amplifier offset voltage and bias current are other sources of offset error. The amplifier bias current must flow through the source impedance. Any unbalance in either the source resistances or the bias currents produce offset errors. In addition, the offset voltage and bias currents are a function of temperature. High performance low offset, low offset drift, low bias current, and low noise precision amplifiers are required. In some cases, chopper-stabilized amplifiers may be the only solution.

AC bridge excitation as shown in Figure 4.1.19 can effectively remove offset voltages in series with the bridge output. The concept is simple. The net bridge output voltage is measured under two conditions as shown. The first measurement yields a measurement V_A, where V_A is the sum of the desired bridge output voltage V_O and the net offset error voltage E_{OS}. The polarity of the bridge excitation is reversed, and a second measurement V_B is made. Subtracting V_B from V_A yields $2V_O$, and the offset error term E_{OS} cancels as shown.

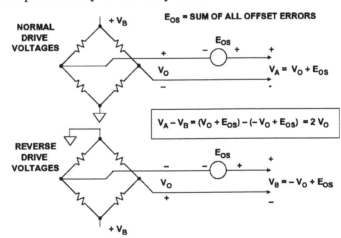

Figure 4.1.19: AC excitation minimizes offset errors.

Chapter 4

Obviously, this technique requires a highly accurate measurement ADC (such as the AD7730) as well as a microcontroller to perform the subtraction. If a ratiometric reference is desired, the ADC must also accommodate the changing polarity of the reference voltage. Again, the AD7730 includes this capability.

P-Channel and N-Channel MOSFETs can be configured as an AC bridge driver as shown in Figure 4.1.20. Dedicated bridge driver chips are also available, such as the Micrel MIC4427. Note that because of the on-resistance of the MOSFETs, Kelvin sensing must be used in these applications. It is also important that the drive signals be non-overlapping to prevent excessive MOSFET switching currents. The AD7730 ADC has on chip circuitry to generate the required non-overlapping drive signals for AC excitation.

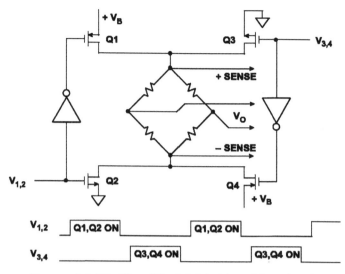

Figure 4.1.20: Simplified AC bridge drive circuit.

References

1. Ramon Pallas-Areny and John G. Webster, Sensors and Signal Conditioning, John Wiley, New York, 1991.
2. Dan Sheingold, Editor, Transducer Interfacing Handbook, Analog Devices, Inc., 1980.
3. Walt Kester, Editor, 1992 Amplifier Applications Guide, Section 2, 3, Analog Devices, Inc., 1992.
4. Walt Kester, Editor, System Applications Guide, Section 1, 6, Analog Devices, Inc., 1993.
5. AD7730 Data Sheet, Analog Devices, available at http://www.analog.com.

4.2 Amplifiers for Signal Conditioning

Introduction

This section examines the critical parameters of amplifiers for use in precision signal conditioning applications. Offset voltages for precision IC op amps can be as low as 10 µV with corresponding temperature drifts of 0.1 µV/°C. Chopper stabilized op amps provide offsets and offset voltage drifts which cannot be distinguished from noise. Open loop gains greater than 1 million are common, along with common mode and power supply rejection ratios of the same magnitude. Applying these precision amplifiers while maintaining the amplifier performance can present significant challenges to a design engineer, i.e., external passive component selection and PC board layout.

It is important to understand that DC open-loop gain, offset voltage, power supply rejection (PSR), and common mode rejection (CMR) should not be the only considerations in selecting precision amplifiers. The AC performance of the amplifier is also important, even at "low" frequencies. Open-loop gain, PSR, and CMR all have relatively low corner frequencies, and therefore what may be considered "low" frequency may actually fall above these corner frequencies, increasing errors above the value predicted solely by the DC parameters. For example, an amplifier having a DC open-loop gain of 10 million and a unity-gain crossover frequency of 1 MHz has a corresponding corner frequency of 0.1 Hz! One must therefore consider the open loop gain at the actual *signal* frequency. The relationship between the single-pole unity-gain crossover frequency, f_u, the signal frequency, f_{sig}, and the open-loop gain $A_{VOL}(f_{sig})$ (measured at the signal frequency) is given by:

$$A_{VOL}(f_{sig}) = \frac{f_u}{f_{sig}} \qquad \text{Eq. 4.2.1}$$

In the example above, the open loop gain is 10 at 100 kHz, and 100,000 at 10 Hz. Loss of open loop gain at the frequency of interest can introduce distortion, especially at audio frequencies. Loss of CMR or PSR at the line frequency or harmonics thereof can also introduce errors.

The challenge of selecting the right amplifier for a particular signal conditioning application has been complicated by the sheer proliferation of various types of amplifiers in various processes (Bipolar, Complementary Bipolar, BiFET, CMOS, BiCMOS, etc.) and architectures (traditional op amps, instrumentation amplifiers, chopper amplifiers, isolation amplifiers, etc.) In addition, a wide selection of precision amplifiers are now available which operate on single supply voltages, which

Chapter 4

- Input Offset Voltage   <100µV
- Input Offset Voltage Drift   <1µV/°C
- Input Bias Current   <2nA
- Input Offset Current   <2nA
- DC Open Loop Gain   >1,000,000
- Unity Gain Bandwidth Product, f_u 500kHz - 5MHz
- Always Check Open Loop Gain at Signal Frequency!
- 1/f (0.1Hz to 10Hz) Noise   <1µV p-p
- Wideband Noise   <10nV/√Hz
- CMR, PSR   >100dB
    - Single Supply Operation
    - Power Dissipation

Figure 4.2.1: Amplifiers for signal conditioning.

complicates the design process even further because of the reduced signal swings and voltage input and output restrictions. Offset voltage and noise are now a more significant portion of the input signal. Selection guides and parametric search engines which can simplify this process somewhat are available on the Internet (http://www.analog.com) as well as on CD-ROM. Other manufacturers have similar information available.

In this section, we will first look at some key performance specifications for precision op amps. Other amplifiers will then be examined such as instrumentation amplifiers, chopper amplifiers, and isolation amplifiers. The implications of single supply operation will be discussed in detail because of their significance in today's designs, which often operate from batteries or other low power sources.

Precision Op Amp Characteristics

Input Offset Voltage

Input offset voltage error is usually one of the largest error sources for precision amplifier circuit designs. However, it is a systemic error and can usually be dealt with by using a manual offset null trim or by system calibration techniques using a microcontroller or microprocessor. Both solutions carry a cost penalty, and today's precision op amps offer initial offset voltages as low as 10 µV for bipolar devices, and far less for chopper stabilized amplifiers. With low offset amplifiers, it is possible to eliminate the need for manual trims or system calibration routines.

Measuring input offset voltages of a few microvolts requires that the test circuit does not introduce more error than the offset voltage itself. Figure 4.2.2 shows a circuit for measuring offset voltage. The circuit amplifies the input offset voltage by the noise gain (1001). The measurement is made at the amplifier output using an accurate digital voltmeter. The offset re-

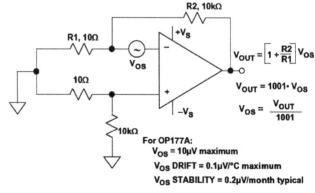

Figure 4.2.2: Measuring input offset voltage.

ferred to the input (RTI) is calculated by dividing the output voltage by the noise gain. The small source resistance seen at R1∥R2 results in negligible bias current contribution to the measured offset voltage. For example, 2 nA bias current flowing through the 10 Ω resistor produces a 0.02 µV error referred to the input.

As simple as it looks, this circuit may give inaccurate results. The largest potential source of error comes from parasitic thermocouple junctions formed where two different metals are joined. The thermocouple voltage formed by temperature difference between two junctions can range from 2 µV/°C to more than 40 µV/°C. Note that in the circuit additional resistors have been added to the non-inverting input in order to exactly match the thermocouple junctions in the inverting input path.

The accuracy of the measurement depends on the mechanical layout of the components and how they are placed on the PC board. Keep in mind that the two connections of a component such as a resistor create two equal, but opposite polarity thermoelectric voltages (assuming they are connected to the same metal, such as the copper trace on a PC board) which cancel each other *assuming both are at exactly the same temperature*. Clean connections and short lead lengths help to minimize temperature gradients and increase the accuracy of the measurement.

Airflow should be minimal so that all the thermocouple junctions stabilize at the same temperature. In some cases, the circuit should be placed in a small closed container to eliminate the effects of external air currents. The circuit should be placed flat on a surface so that convection currents flow up and off the top of the board, not across the components as would be the case if the board was mounted vertically.

Measuring the offset voltage shift over temperature is an even more demanding challenge. Placing the printed circuit board containing the amplifier being tested in a small box or plastic bag with foam insulation prevents the temperature chamber air current from causing thermal gradients across the parasitic thermocouples. If cold testing is required, a dry nitrogen purge is recommended. Localized temperature cycling of the amplifier itself using a Thermostream-type heater/cooler may be an alternative. However, these units tend to generate quite a bit of airflow, which can be troublesome.

In addition to temperature related drift, the offset voltage of an amplifier changes as time passes. This aging effect is generally specified as *long-term stability* in µV/month, or µV/1000 hours, but this is misleading. Since aging is a "drunkard's walk" phenomenon, it is proportional to the *square root* of the elapsed time. An aging rate of 1 µV/1000 hours becomes about 3 µV/year, not 9 µV/year. Long-term stability of the OP177 and the AD707 is approximately 0.3 µV/month. This refers to a time period *after* the first 30 days of operation. Excluding the initial hour of operation, changes in the offset voltage of these devices during the first 30 days of operation are typically less than 2 µV.

As a general rule of thumb, it is prudent to control amplifier offset voltage by device selection whenever possible, bus sometimes trim may be desired. Many precision op amps have pins available for optional offset null. Generally, two pins are joined by a potentiometer, and the wiper goes to one of the supplies through a resistor as shown in Figure 4.2.3. If the wiper is connected to the wrong supply, the op amp will probably be destroyed, so the data sheet instructions must be carefully observed! The range of offset adjustment in a precision op amp should be no more than two or three times the maximum offset voltage of the lowest grade device, in order to minimize the sensitivity of these pins. The voltage gain of an op amp between its offset adjustment pins and its output may actually be greater than the gain at its signal inputs! It is therefore very important to keep these pins free of noise. It is inadvisable to have long leads from an op amp to a remote potentiometer. To minimize any offset error due to supply current, connect R1 directly to the pertinent device supply pin, such as pin 7 shown in the diagram.

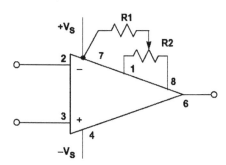

- R1 = 10kΩ, R2 = 2kΩ, OFFSET ADJUST RANGE = 200µV
- R1 = 0, R1 = 20kΩ, OFFSET ADJUST RANGE = 3mV

Figure 4.2.3: OP177/AD707 offset adjustment pins.

It is important to note that the offset drift of an op amp with temperature will vary with the setting of its offset adjustment. In most cases a bipolar op amp will have minimum drift at minimum offset. The offset adjustment pins should therefore be used only to adjust the op amp's own offset, not to correct any system offset errors, since this would be at the expense of increased temperature drift. The drift penalty for a JFET input op amp is much worse than for a bipolar input and is in the order of 4 µV/°C for each millivolt of nulled offset voltage. It is generally better to control the offset voltage by proper selection of devices and device grades. Dual, triple, quad, and single op amps in small packages do not generally have null capability because of pin count limitations, and offset adjustments must be done elsewhere in the system when using these devices. This can be accomplished with minimal impact on drift by a universal trim, which sums a small voltage into the input.

Sensor Signal Conditioning

Input Offset Voltage and Input Bias Current Models

Thus far, we have considered only the op amp input offset voltage. However, the input bias currents also contribute to offset error as shown in the generalized model of Figure 4.2.4. It is useful to refer all offsets to the op amp input (RTI) so that they can be easily compared with the input signal. The equations in the diagram are given for the total offset voltage referred to input (RTI) and referred to output (RTO).

For a precision op amp having a standard bipolar input stage using either PNPs or NPNs, the input bias currents are typically 50 nA to 400 nA and are well matched. By making R3 equal to the parallel combination of R1 and R2, their effect on the net RTI and RTO offset voltage is approximately canceled, thus leaving the offset current, i.e., the difference between the input currents as an error. This current is usually an order of magnitude lower than the bias current specification. This scheme, however, does not work for bias-current compensated bipolar op amps (such as the OP177 and the AD707) as shown in Figure 4.2.5. Bias-current compensated input stages have most of the good features of the simple bipolar input stage: low offset and drift, and low voltage noise. Their bias current is low and fairly stable over temperature. The additional current sources reduce the net bias currents typically to between 0.5 nA and 10 nA. However, the signs of the + and − input bias currents may or may not be the same, and they are not well matched, but are very low. Typically, the specification for the *offset current* (the difference between the + and − input bias currents) in bias-current compensated op amps is generally about the same as the individual bias currents. In the case of the standard bipolar differential

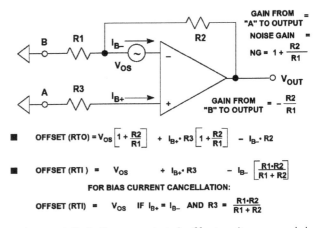

Figure 4.2.4: Op amp total offset voltage model.

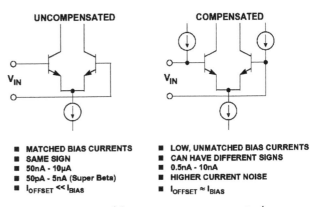

Figure 4.2.5: Input bias current compensated op amps.

Chapter 4

pair with no bias-current compensation, the offset current specification is typically five to ten times lower than the bias current specification.

DC Open Loop Gain Nonlinearity

It is well understood that in order to maintain accuracy, a precision amplifier's DC open loop gain, A_{VOL}, should be high. This can be seen by examining the equation for the closed loop gain:

$$\text{Closed Loop Gain} = A_{VCL} = \frac{NG}{1 + \frac{NG}{A_{VOL}}} \qquad \text{Eq. 4.2.2}$$

Noise gain (NG) is simply the gain seen by a small voltage source in series with the op amp input and is also the amplifier signal gain in the noninverting mode. If A_{VOL} in the above equation is infinite, the closed loop gain is exactly equal to the noise gain. However, for finite values of A_{VOL}, there is a closed loop gain error given by the equation:

$$\% \text{ Gain Error} = \frac{NG}{NG + A_{VOL}} \times 100\% \approx \frac{NG}{A_{VOL}} \times 100\%, \text{ for } NG \ll A_{VOL}$$

$$\text{Eq. 4.2.3}$$

Notice from the equation that the percent gain error is directly proportional to the noise gain, therefore the effects of finite A_{VOL} are less for low gain. The first example in Figure 4.2.6 where the noise gain is 1000 shows that for an open loop gain of 2 million, there is a gain error of about 0.05%. If the open loop gain stays constant over temperature and for various output loads and voltages, the gain error can be calibrated out of the measurement, and there is then no overall system gain error.

- "IDEAL" CLOSED LOOP GAIN = NOISE GAIN = NG
- ACTUAL CLOSED LOOP GAIN = $\frac{NG}{1 + \frac{NG}{A_{VOL}}}$
- % CLOSED LOOP GAIN ERROR = $\frac{NG}{NG + A_{VOL}} \times 100\% \approx \frac{NG}{A_{VOL}} \times 100\%$
  - Assume A_{VOL} = 2,000,000, NG = 1,000
    %GAIN ERROR ≈ 0.05%
  - Assume A_{VOL} Drops to 300,000
    %GAIN ERROR ≈ 0.33%
  - CLOSED LOOP GAIN UNCERTAINTY
    = 0.33% – 0.05% = 0.28%

Figure 4.2.6: Changes in DC open loop gain cause closed loop gain uncertainty.

Sensor Signal Conditioning

If, however, the open loop gain *changes*, the closed loop gain will also change, thereby introducing a *gain uncertainty*. In the second example in the figure, an A_{VOL} decrease to 300,000 produces a gain error of 0.33%, introducing a *gain uncertainty* of 0.28% in the closed loop gain. In most applications, using the proper amplifier, the resistors around the circuit will be the largest source of gain error.

Changes in the output voltage level and the output loading are the most common causes of changes in the open loop gain of op amps. A change in open loop gain with signal level produces *nonlinearity* in the closed loop gain transfer function which cannot be removed during system calibration. Most op amps have fixed loads, so A_{VOL} changes with load are not generally important. However, the sensitivity of A_{VOL} to output signal level may increase for higher load currents.

The severity of the nonlinearity varies widely from device type to device type, and is generally not specified on the data sheet. The minimum A_{VOL} is always specified, and choosing an op amp with a high A_{VOL} will minimize the probability of gain nonlinearity errors. Gain nonlinearity can come from many sources, depending on the design of the op amp. One common source is thermal feedback. If temperature shift is the sole cause of the nonlinearity error, it can be assumed that minimizing the output loading will help. To verify this, the nonlinearity is measured with no load and then compared to the loaded condition.

An oscilloscope X-Y display test circuit for measuring DC open loop gain nonlinearity is shown in Figure 4.2.7. The same precautions previously discussed relating to the offset voltage test circuit must be observed in this circuit. The amplifier is configured for a signal gain of –1. The open loop gain is defined as the change in output voltage divided by the change in the input offset voltage. However, for large values of A_{VOL}, the offset may change only a few microvolts over the entire output voltage swing. Therefore the divider consisting of the 10 Ω resistor and R_G (1 MΩ) forces the voltage V_Y to be :

$$V_Y = \left[1 + \frac{R_G}{10\Omega}\right] V_{OS} = 100,001 \bullet V_{OS} \qquad \text{Eq. 4.2.4}$$

The value of R_G is chosen to give measurable voltages at V_Y depending on the expected values of V_{OS}.

The ±10 V ramp generator output is multiplied by the signal gain, –1, and forces the op amp output voltage V_X to swing from +10 V to –10 V. Because of the gain factor applied to the offset voltage, the offset adjust potentiometer is added to allow the initial output offset to be set to zero. The resistor values chosen will null an input offset voltage of up to ±10 mV. Stable 10 V voltage references should be used at each end

Chapter 4

of the potentiometer to prevent output drift. *Also, the frequency of the ramp generator must be quite low, probably no more than a fraction of 1 Hz because of the low corner frequency of the open loop gain (0.1 Hz for the OP177).*

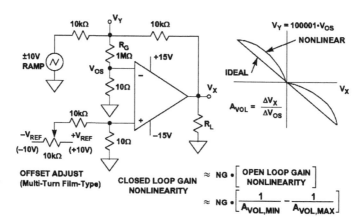

Figure 4.2.7: Circuit measures open loop gain nonlinearity.

The plot on the right-hand side of Figure 4.2.7 shows V_Y plotted against V_X. If there is no gain nonlinearity the graph will have a constant slope, and A_{VOL} is calculated as follows:

$$A_{VOL} = \frac{\Delta V_X}{\Delta V_{OS}} = \left[1 + \frac{R_G}{10\Omega}\right]\left[\frac{\Delta V_X}{\Delta V_Y}\right] = 100,001 \bullet \left[\frac{\Delta V_X}{\Delta V_Y}\right] \qquad \text{Eq. 4.2.5}$$

If there is nonlinearity, A_{VOL} will vary as the output signal changes. The approximate open loop gain nonlinearity is calculated based on the maximum and minimum values of A_{VOL} over the output voltage range:

$$\text{Open Loop Gain Nonlinearity} = \frac{1}{A_{VOL,MIN}} - \frac{1}{A_{VOL,MAX}} \qquad \text{Eq. 4.2.6}$$

The closed loop gain nonlinearity is obtained by multiplying the open loop gain nonlinearity by the noise gain, NG:

$$\text{Closed Loop Gain Nonlinearity} \approx NG \bullet \left[\frac{1}{A_{VOL,MIN}} - \frac{1}{A_{VOL,MAX}}\right] \qquad \text{Eq. 4.2.7}$$

In the ideal case, the plot of V_{OS} versus V_X would have a constant slope, and the reciprocal of the slope is the open loop gain, A_{VOL}. A horizontal line with zero slope would indicate infinite open loop gain. In an actual op amp, the slope may change across the output range because of nonlinearity, thermal feedback, etc. In fact, the slope can even change sign.

Sensor Signal Conditioning

Figure 4.2.8 shows the V_Y (and V_{OS}) versus V_X plot for the OP177 precision op amp. The plot is shown for two different loads, 2 kΩ and 10 kΩ. The reciprocal of the slope is calculated based on the end points, and the average A_{VOL} is about 8 million. The maximum and minimum values of A_{VOL} across the output voltage range are measured to be approximately 9.1 million, and 5.7 million, respectively. This corresponds to an open loop gain nonlinearity of about 0.07ppm. Thus, for a noise gain of 100, the corresponding closed loop gain nonlinearity is about 7ppm.

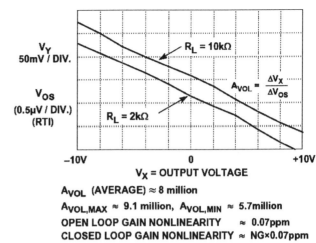

Figure 4.2.8: OP177 gain nonlinearity.

Op Amp Noise

The three noise sources in an op amp circuit are the voltage noise of the op amp, the current noise of the op amp (there are two uncorrelated sources, one in each input), and the Johnson noise of the resistances in the circuit. Op amp noise has two components, "white" noise at medium frequencies and low frequency "1/f" noise, whose spectral density is inversely proportional to the square root of the frequency. It should be noted that, though both the voltage and the current noise may have the same characteristic behavior, in a particular amplifier the 1/f corner frequency is not necessarily the same for voltage and current noise (it is usually specified for the voltage noise as shown in Figure 4.2.9.

The low frequency noise is generally known as 1/f noise (the noise power obeys a 1/f law—the noise voltage or noise current is proportional

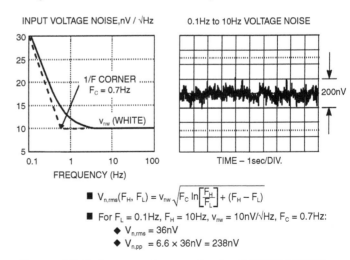

Figure 4.2.9: Input voltage noise for OP177/AD707.

to $1/\sqrt{f}$). The frequency at which the 1/f noise spectral density equals the white noise is known as the *1/f corner frequency*, F_C, and is a figure of merit for an op amp, with low corner frequencies indicating better performance. Values of 1/f corner frequency vary from less than 1 Hz high accuracy bipolar op amps like the OP177/AD707, several hundred Hz for the AD743/745 FET-input op amps, to several thousands of Hz for some high speed op amps where process compromises favor high speed rather than low frequency noise.

For the OP177/AD707 shown in Figure 4.2.9, the 1/f corner frequency is 0.7 Hz, and the white noise is 10 nV/√Hz. The low frequency 1/f noise is often expressed as the peak-to-peak noise in the bandwidth 0.1 Hz to 10 Hz as shown in the scope photo in Figure 4.2.9. Note that this noise ultimately limits the resolution of a precision measurement system because the bandwidth up to 10 Hz is usually the bandwidth of most interest. The equation for the total rms noise, $V_{n,rms}$, in the bandwidth F_L to F_H is given by the equation:

$$V_{n,rms}(F_H, F_L) = v_{nw}\sqrt{F_C \ln\left[\frac{F_H}{F_L}\right] + (F_H - F_L)} \qquad \text{Eq. 4.2.8}$$

where v_{nw} is the noise spectral density in the "white noise" region (usually specified at a frequency of 1 kHz), F_C is the 1/f corner frequency, and F_L and F_H is the measurement bandwidth of interest. In the example shown, the 0.1 Hz to 10 Hz noise is calculated to be 36nV rms, or approximately 238 nV peak-to-peak, which closely agrees with the scope photo on the right (a factor of 6.6 is generally used to convert rms values to peak-to-peak values).

It should be noted that at higher frequencies, the term in the equation containing the natural logarithm becomes insignificant, and the expression for the rms noise becomes:

$$V_{n,rms}(F_H, F_L) \approx v_{nw}\sqrt{F_H - F_L}$$

And, if $F_H \gg F_L$,

$$V_{n,rms}(F_H) \approx v_{nw}\sqrt{F_H} \qquad \text{Eq. 4.2.9}$$

However, some op amps (such as the OP07 and OP27) have voltage noise characteristics that increase slightly at high frequencies. The voltage noise versus frequency curve for op amps should therefore be examined carefully for flatness when calculating high frequency noise using this approximation.

Sensor Signal Conditioning

At very low frequencies when operating exclusively in the 1/f region, $F_C \gg (F_H - F_L)$, and the expression for the rms noise reduces to:

$$V_{n,rms}(F_H, F_L) \approx v_{nw}\sqrt{F_C \ln\left[\frac{F_H}{F_L}\right]} \qquad \text{Eq. 4.2.10}$$

Note that there is no way of reducing this 1/f noise by filtering if operation extends to DC. Making $F_H = 0.1$ Hz and $F_L = 0.001$ still yields an rms 1/f noise of about 18 nV rms, or 119 nV peak-to-peak.

The point is that averaging the results of a large number of measurements taken over a long period of time has practically no effect on the error produced by 1/f noise. The only method of reducing it further is to use a chopper stabilized op amp which does not pass the low frequency noise components.

A generalized noise model for an op amp is shown in Figure 4.2.10. All uncorrelated noise sources add as a root-sum-of-squares manner, i.e., noise voltages V1, V2, and V3 give a result of:

$$\sqrt{V1^2 + V2^2 + V3^2} \qquad \text{Eq. 4.2.11}$$

Thus, any noise voltage which is more than four or five times any of the others is dominant, and the others may generally be ignored. This simplifies noise analysis.

In this diagram, the total noise of all sources is shown referred to the input (RTI). The RTI noise is useful because it can be compared directly to the input signal level. The total noise referred to the output (RTO) is obtained by simply multiplying the RTI noise by the noise gain.

The diagram assumes that the feedback network is purely resistive. If it contains reactive elements (usually capacitors), the noise gain is not constant over the bandwidth of interest, and more complex techniques must be used to calculate the total noise (see in particular, Reference 12). However, for precision applications where the feedback network is most likely to be resistive, the equations are valid.

Notice that the Johnson noise voltage associated with the three resistors has been included. All resistors have a Johnson noise of 4 kTBR, where k is Boltzmann's Constant (1.38×10^{-23} J/K), T is the absolute temperature, B is the bandwidth in Hz, and R is the resistance in Ω. A simple relationship which is easy to remember is that *a 1000 Ω resistor generates a Johnson noise of 4nV/$\sqrt{Hz}$ at 25°C.*

Chapter 4

The voltage noise of various op amps may vary from under 1nV/√Hz to 20nV/√Hz, or even more. Bipolar input op amps tend to have lower voltage noise than JFET input ones, although it is possible to make JFET input op amps with low voltage noise (such as the AD743/AD745), at the cost of large input devices and hence large (~20pF) input capacitance. Current noise can vary much more widely, from around 0.1fA/√Hz (in JFET input electrometer op amps) to several pA/√Hz (in high speed bipolar op amps). For bipolar or JFET input devices where all the bias current flows into the input junction, the current noise is simply the Schottky (or shot) noise of the bias current. The shot noise spectral density is simply $2I_Bq$ amps/√Hz, where I_B is the bias current (in amps) and q is the charge on an electron (1.6×10^{-19} C). It cannot be calculated for bias-compensated or current feedback op amps where the external bias current is the difference between two internal current sources.

Current noise is only important when it flows through an impedance and in turn generates a noise voltage. The equation shown in Figure 4.2.10 shows how the current noise flowing in the resistors contribute to the total noise. The choice of a low noise op amp therefore depends on the impedances around it. Consider an OP27, a bias compensated op amp with low voltage noise (3 nV/√Hz), but quite high current noise (1 pA/√Hz) as shown in the schematic of Figure 4.2.11. With zero source impedance, the voltage noise dominates. With a source resistance of 3 kΩ, the current noise (1 pA/√Hz) flowing in 3 kΩ will equal the voltage noise, but the Johnson noise of the 3 kΩ resistor is 7 nV/√Hz and so is dominant. With a source resistance of 300 kΩ, the effect of the current noise increases a hundredfold to 300 nV/√Hz, while the voltage noise is unchanged, and the Johnson noise (which is proportional to the *square root* of the resistance) increases tenfold. Here, the current noise dominates.

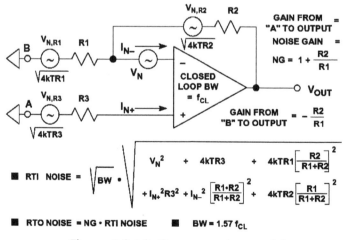

Figure 4.2.10: Op amp noise model.

Sensor Signal Conditioning

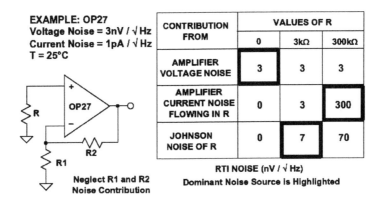

Figure 4.2.11: Different noise sources dominate at different source impedance.

The previous example shows that the choice of a low noise op amp depends on the source impedance of the input signal, and at high impedances, current noise always dominates. This is shown in Figure 4.2.12 for several bipolar (OP07, OP27, 741) and JFET (AD645, AD743, AD744) op amps.

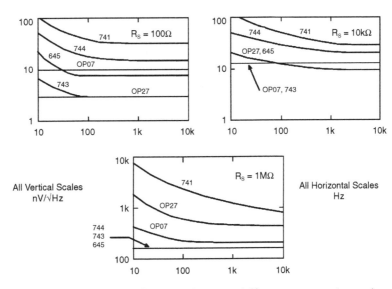

Figure 4.2.12: Different amplifiers are best at different source impedance levels.

For low impedance circuitry (generally < 1 kΩ), amplifiers with low voltage noise, such as the OP27 will be the obvious choice, and their comparatively large current noise will not affect the application. At medium resistances, the Johnson noise of resistors is dominant, while at very high resistances, we must choose an op amp with the smallest possible current noise, such as the AD549 or AD645.

Chapter 4

Until recently, BiFET amplifiers (with JFET inputs) tended to have comparatively high voltage noise (though very low current noise), and thus were more suitable for low noise applications in high rather than low impedance circuitry. The AD645, AD743, and AD745 have very low values of both voltage and current noise. The AD645 specifications at 10 kHz are 10nV/√Hz and 0.6fA/√Hz, and the AD743/AD745 specifications at 10 kHz are 2.0 nV/√Hz and 6.9 fA/√Hz. These make possible the design of low noise amplifier circuits which have low noise over a wide range of source impedances.

Common Mode Rejection and Power Supply Rejection

If a signal is applied equally to both inputs of an op amp so that the differential input voltage is unaffected, the output should not be affected. In practice, changes in common mode voltage will produce changes in the output. The *common mode rejection ratio* or CMRR is the ratio of the common mode gain to the differential- mode gain of an op amp. For example, if a differential input change of Y volts will produce a change of 1 V at the output, and a common mode change of X volts produces a similar change of 1 V, then the CMRR is X/Y. It is normally expressed in dB, and typical LF values are between 70 and 120 dB. When expressed in dB, it is generally referred to as *common mode rejection* (CMR). At higher frequencies, CMR deteriorates—many op amp data sheets show a plot of CMR versus frequency as shown in Figure 4.2.13 for the OP177/AD707 precision op amps.

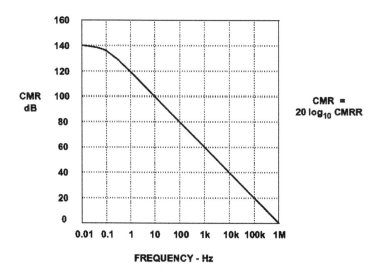

Figure 4.2.13: OP177/AD707 common mode rejection (CMR).

Sensor Signal Conditioning

CMRR produces a corresponding output offset voltage error in op amps configured in the non-inverting mode as shown in Figure 4.2.14. Op amps configured in the inverting mode have no CMRR output error because both inputs are at ground or virtual ground, so there is no common mode voltage, only the offset voltage of the amplifier if un-nulled.

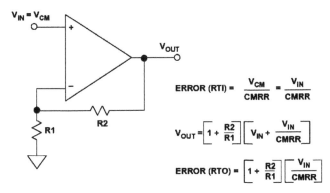

Figure 4.2.14: Calculating offset error due to common mode rejection ratio (CMRR).

If the supply of an op amp changes, its output should not, but it will. The specification of *power supply rejection ratio* or PSRR is defined similarly to the definition of CMRR. If a change of X volts in the supply produces the same output change as a differential input change of Y volts, then the PSRR on that supply is X/Y. When the ratio is expressed in dB, it is generally referred to as *power supply rejection*, or PSR. The definition of PSRR assumes that both supplies are altered equally in opposite directions—otherwise the change will introduce a common mode change as well as a supply change, and the analysis becomes considerably more complex. It is this effect which causes apparent differences in PSRR between the positive and negative supplies. In the case of single supply op amps, PSR is generally defined with respect to the change in the positive supply. Many single supply op amps have separate PSR specifications for the positive and negative supplies. The PSR of the OP177/AD707 is shown in Figure 4.2.15.

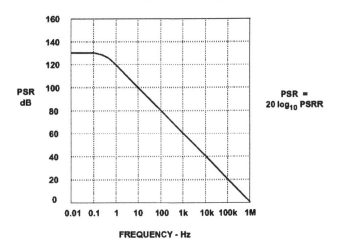

Figure 4.2.15: OP177/AD707 power supply rejection (PSR).

Chapter 4

The PSRR of op amps is frequency dependent, therefore power supplies must be well decoupled as shown in Figure 4.2.16. At low frequencies, several devices may share a 10–50 µF capacitor on each supply, provided it is no more than 10cm (PC track distance) from any of them. At high frequencies, each IC must have every supply decoupled by a low inductance capacitor (0.1 µF or so) with short leads and PC tracks. These capacitors must also provide a return path for HF currents in the op amp load. Decoupling capacitors should be connected to a low impedance large area ground plane with minimum lead lengths. Surface mount capacitors minimize lead inductance and are a good choice.

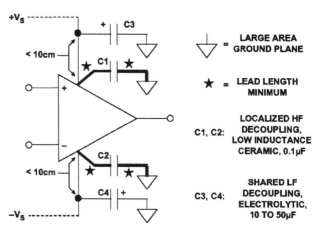

Figure 4.2.16: Proper low and high-frequency decoupling techniques for op amps.

Amplifier DC Error Budget Analysis

A room temperature error budget analysis for the OP177A op amp is shown in Figure 4.2.17. The amplifier is connected in the inverting mode with a signal gain of 100. The key data sheet specifications are also shown in the diagram. We assume an input signal of 100 mV full scale which corresponds to an output signal of 10 V. The various error sources are normalized to full scale and expressed in parts per million (ppm). Note: parts per million (ppm) error = fractional error $\times 10^6$ = % error $\times 10^4$.

Note that the offset errors due to V_{OS} and I_{OS} and the gain error due to finite A_{VOL} can be removed with a system calibration. However, the error due to open loop gain nonlinearity cannot be removed with calibration and produces a relative accuracy error, often called *resolution error*.

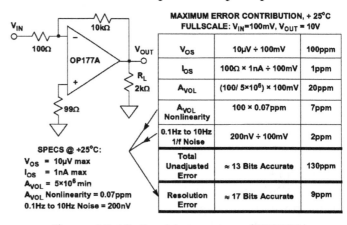

Figure 4.2.17: Precision op amp (OP177A) DC error budget.

The second contributor to resolution error is the 1/f noise. This noise is always present and adds to the uncertainty of the measurement. The overall relative accuracy of the circuit at room temperature is 9 ppm which is equivalent to approximately 17 bits of resolution.

Single Supply Op Amps

Over the last several years, single-supply operation has become an increasingly important requirement because of market requirements. Automotive, set-top box, camera/camcorder, PC, and laptop computer applications are demanding IC vendors to supply an array of linear devices that operate on a single supply rail, with the same performance of dual supply parts. Power consumption is now a key parameter for line or battery operated systems, and in some instances, more important than cost. This makes low-voltage/low supply current operation critical; at the same time, however, accuracy and precision requirements have forced IC manufacturers to meet the challenge of "doing more with less" in their amplifier designs.

- Single Supply Offers:
  - Lower Power
  - Battery Operated Portable Equipment
  - Requires Only One Voltage

- Design Tradeoffs:
  - Reduced Signal Swing Increases Sensitivity to Errors Caused by Offset Voltage, Bias Current, Finite Open-Loop Gain, Noise, etc.
  - Must Usually Share Noisy Digital Supply
  - Rail-to-Rail Input and Output Needed to Increase Signal Swing
  - Precision Less than the best Dual Supply Op Amps but not Required for All Applications
  - Many Op Amps Specified for Single Supply, but do not have Rail-to-Rail Inputs or Outputs

Figure 4.2.18: Single supply amplifiers.

In a single-supply application, the most immediate effect on the performance of an amplifier is the reduced input and output signal range. As a result of these lower input and output signal excursions, amplifier circuits become more sensitive to internal and external error sources. Precision amplifier offset voltages on the order of 0.1 mV are less than a 0.04 LSB error source in a 12-bit, 10 V full-scale system. In a single-supply system, however, a "rail-to-rail" precision amplifier with an offset voltage of 1 mV represents a 0.8 LSB error in a 5 V full-scale system, and 1.6 LSB error in a 2.5 V full-scale system.

To keep battery current drain low, larger resistors are usually used around the op amp. Since the bias current flows through these larger resistors, they can generate offset errors equal to or greater than the amplifier's own offset voltage.

Gain accuracy in some low voltage single-supply devices is also reduced, so device selection needs careful consideration. Many amplifiers having open-loop gains in the millions typically operate on dual supplies: for example, the OP07 family types. However, many single-supply/rail-to-rail amplifiers for precision applications typically have open-loop gains between 25,000 and 30,000 under light loading (>10 kΩ). Selected devices, like the OP113/213/413 family, do have high open-loop gains (i.e., > 1M).

Chapter 4

Many trade-offs are possible in the design of a single-supply amplifier circuit: speed versus power, noise versus power, precision versus speed and power, etc. Even if the noise floor remains constant (highly unlikely), the signal-to-noise ratio will drop as the signal amplitude decreases.

Besides these limitations, many other design considerations that are otherwise minor issues in dual-supply amplifiers now become important. For example, signal-to-noise (SNR) performance degrades as a result of reduced signal swing. "Ground reference" is no longer a simple choice, as one reference voltage may work for some devices, but not others. Amplifier voltage noise increases as operating supply current drops, and bandwidth decreases. Achieving adequate bandwidth and required precision with a somewhat limited selection of amplifiers presents significant system design challenges in single-supply, low-power applications.

Most circuit designers take "ground" reference for granted. Many analog circuits scale their input and output ranges about a ground reference. In dual-supply applications, a reference that splits the supplies (0 V) is very convenient, as there is equal supply headroom in each direction, and 0 V is generally the voltage on the low impedance ground plane.

In single-supply/rail-to-rail circuits, however, the ground reference can be chosen anywhere within the supply range of the circuit, since there is no standard to follow. The choice of ground reference depends on the type of signals processed and the amplifier characteristics. For example, choosing the negative rail as the ground reference may optimize the dynamic range of an op amp whose output is designed to swing to 0 V. On the other hand, the signal may require level shifting in order to be compatible with the input of other devices (such as ADCs) that are not designed to operate at 0 V input.

Early single-supply "zero-in, zero-out" amplifiers were designed on bipolar processes which optimized the performance of the NPN transistors. The PNP transistors were either lateral or substrate PNPs with much less bandwidth than the NPNs. Fully complementary processes are now required for the new-breed of single-supply/rail-to-rail operational amplifiers. These new amplifier designs do not use lateral or substrate PNP transistors within the signal path, but incorporate parallel NPN and PNP input stages to accommodate input signal swings from ground to the positive supply rail. Furthermore, rail-to-rail output stages are designed with bipolar NPN and PNP common-emitter, or N-channel/P-channel common-source amplifiers whose collector-emitter saturation voltage or drain-source channel on-resistance determine output signal swing as a function of the load current.

The characteristics of a single-supply amplifier input stage (common mode rejection, input offset voltage and its temperature coefficient, and noise) are critical in precision, low-voltage applications. Rail-to-rail input operational amplifiers must resolve small signals, whether their inputs are at ground, or in some cases near the amplifier's positive supply. Amplifiers having a minimum of 60 dB common mode rejection over the entire input common mode voltage range from 0 V to the positive supply are good candidates. It is not necessary that amplifiers maintain common mode rejection for signals beyond the supply voltages: *what is required is that they do not self-destruct for momentary overvoltage conditions*. Furthermore, amplifiers that have offset voltages less than 1 mV and offset voltage drifts less than 2 µV/°C are also very good candidates for precision applications. Since *input* signal dynamic range and SNR are equally if not more important than *output* dynamic range and SNR, precision single-supply/rail-to-rail operational amplifiers should have noise levels referred-to-input (RTI) less than 5 µVp-p in the 0.1 Hz to 10 Hz band.

The need for rail-to-rail amplifier output stages is driven by the need to maintain wide dynamic range in low-supply voltage applications. A single-supply/rail-to-rail amplifier should have output voltage swings which are within at least 100 mV of either supply rail (under a nominal load). The output voltage swing is very dependent on output stage topology and load current. The voltage swing of a good output stage should maintain its rated swing for loads down to 10 kΩ. The smaller the V_{OL} and the larger the V_{OH}, the better. System parameters, such as "zero-scale" or "full-scale" output voltage, should be determined by an amplifier's V_{OL} (for zero-scale) and V_{OH} (for full-scale).

Since the majority of single-supply data acquisition systems require at least 12- to 14-bit performance, amplifiers which exhibit an open-loop gain greater than 30,000 for all loading conditions are good choices in precision applications.

Single Supply Op Amp Input Stages

There is some demand for op amps whose input common mode voltage includes *both* supply rails. Such a feature is undoubtedly useful in some applications, but engineers should recognize that there are relatively few applications where it is absolutely essential. These should be carefully distinguished from the many applications where common mode range *close* to the supplies or one that includes *one* of the supplies is necessary, but input rail-to-rail operation is not.

In many single-supply applications, it is required that the input go to only one of the supply rails (usually ground). High-side or low-side sensing applications are good examples of this. Amplifiers which will handle zero-volt inputs are relatively easily designed using PNP differential pairs (or N-channel JFET pairs) as shown in Figure 4.2.19. The input common mode range of such an op amp extends from about 200 mV below the negative supply to within about 1 V of the positive supply.

The input stage could also be designed with NPN transistors (or P-channel JFETs), in which case the input common mode range would include the positive rail and to within about 1 V of the negative rail. This requirement typically occurs in applications such as high-side current sensing, a low-frequency measurement application. The OP282/OP482 input stage uses the P-channel JFET input pair whose input common mode range includes the positive rail. Other circuit topologies for high-side sensing (such as the AD626) use the precision resistors to attenuate the common mode voltage.

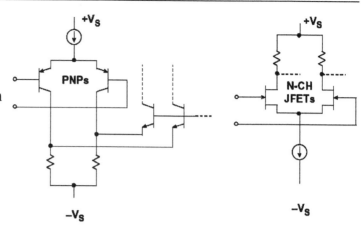

Figure 4.2.19: PNP or N-channel JFET stages allow input signal to go to the negative rail.

True rail-to-rail input stages require two long-tailed pairs (see Figure 4.2.20), one of NPN bipolar transistors (or N-channel JFETs), the other of PNP transistors (or P-channel JFETs). These two pairs exhibit *different* offsets and bias currents, so when the applied input common mode voltage changes, the amplifier input offset voltage and input bias current does also. In fact, when both current sources remain active throughout the entire input common mode range, amplifier input offset voltage is the *average* offset voltage of the NPN pair and the PNP pair. In those designs where the current sources are alternatively switched off at some point along the input common mode voltage, amplifier input offset voltage is dominated by the PNP pair offset voltage for signals near the negative supply, and by the NPN pair

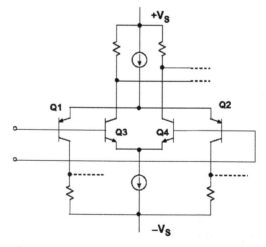

Figure 4.2.20: True rail-to-rail input stage.

offset voltage for signals near the positive supply. It should be noted that true rail-to-rail input stages can also be constructed from CMOS transistors as in the case of the OP250/450 and the AD8531/8532/8534.

Amplifier input bias current, a function of transistor current gain, is also a function of the applied input common mode voltage. The result is relatively poor common mode rejection (CMR), and a changing common mode input impedance over the common mode input voltage range, compared to familiar dual-supply devices. These specifications should be considered carefully when choosing a rail-rail input op amp, especially for a non-inverting configuration. Input offset voltage, input bias current, and even CMR may be quite good over *part* of the common mode range, but much worse in the region where operation shifts between the NPN and PNP devices and vice versa.

True rail-to-rail amplifier input stage designs must transition from one differential pair to the other differential pair somewhere along the input common mode voltage range. Some devices like the OP191/291/491 family and the OP279 have a common mode crossover threshold at approximately 1 V below the positive supply. The PNP differential input stage is active from about 200 mV below the negative supply to within about 1 V of the positive supply. Over this common mode range, amplifier input offset voltage, input bias current, CMR, input noise voltage/current are primarily determined by the characteristics of the PNP differential pair. At the crossover threshold, however, amplifier input offset voltage becomes the average offset voltage of the NPN/PNP pairs and can change rapidly. Also, amplifier bias currents, dominated by the PNP differential pair over most of the input common mode range, change polarity and magnitude at the crossover threshold when the NPN differential pair becomes active.

Op amps like the OP184/284/484 utilize a rail-to-rail input stage design where both NPN and PNP transistor pairs are active throughout the entire input common mode voltage range, and there is no common mode crossover threshold. Amplifier input offset voltage is the average offset voltage of the NPN and the PNP stages. Amplifier input offset voltage exhibits a smooth transition throughout the entire input common mode range because of careful laser trimming of the resistors in the input stage. In the same manner, through careful input stage current balancing and input transistor design, amplifier input bias currents also exhibit a smooth transition throughout the entire common mode input voltage range. The exception occurs at the extremes of the input common mode range, where amplifier offset voltages and bias currents increase sharply due to the slight forward-biasing of parasitic p-n junctions. This occurs for input voltages within approximately 1 V of either supply rail.

When *both* differential pairs are active throughout the entire input common mode range, amplifier transient response is faster through the middle of the common mode range by as much as a factor of 2 for bipolar input stages and by a factor of √2 for JFET input stages. Input stage transconductance determines the slew rate and the unity-gain crossover frequency of the amplifier, hence response time degrades slightly at the extremes of the input common mode range when either the PNP stage (signals approaching the positive supply rail) or the NPN stage (signals approaching the negative supply rail) are forced into cutoff. The thresholds at which the transconductance changes occur are approximately within 1 V of either supply rail, and the behavior is similar to that of the input bias currents.

Applications which require true rail-rail inputs should therefore be carefully evaluated, and the amplifier chosen to ensure that its input offset voltage, input bias current, common mode rejection, and noise (voltage and current) are suitable.

Single Supply Op Amp Output Stages

The earliest IC op amp output stages were NPN emitter followers with NPN current sources or resistive pull-downs, as shown in the left-hand diagram of Figure 4.2.21. Naturally, the slew rates were greater for positive-going than for negative-going signals. While all modern op amps have push-pull output stages of some sort, many are still asymmetrical, and have a greater slew rate in one direction than the other. Asymmetry tends to introduce distortion on AC signals and generally results from the use of IC processes with faster NPN than PNP transistors. It may also result in the ability of the output to approach one supply more closely than the other.

In many applications, the output is required to swing only to one rail, usually the negative rail (i.e., ground in single-supply systems). A pull-down resistor to the negative rail will allow the output to approach that rail (provided the load impedance is high enough, or is also grounded to that rail), but only slowly. Using an FET current source instead of a resistor can speed things up, but this adds complexity.

With new complementary bipolar processes (CB), well matched high speed PNP and NPN transistors are available. The complementary emitter follower output stage shown in the right-hand diagram of Figure 4.2.21 has many advantages including low output impedance. However, the output can only swing within about one V_{BE} drop of either supply rail. An output swing of +1 V to +4 V is typical of such stages when operated on a single +5 V supply.

Sensor Signal Conditioning

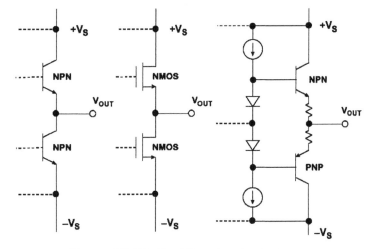

Figure 4.2.21: Traditional output stages.

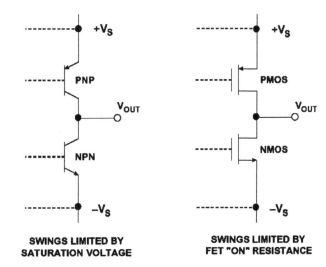

Figure 4.2.22: "Almost" rail-to-rail output structures.

The complementary common-emitter/common-source output stages shown in Figure 4.2.22 allow the output voltage to swing much closer to the output rails, but these stages have higher open loop output impedance than the emitter follower- based stages. In practice, however, the amplifier's open loop gain and local feedback produce an apparent low output impedance, particularly at frequencies below 10 Hz.

Chapter 4

The complementary common emitter output stage using BJTs (left-hand diagram in Figure 4.2.22) cannot swing completely to the rails, but only to within the transistor saturation voltage (V_{CESAT}) of the rails. For small amounts of load current (less than 100 µA), the saturation voltage may be as low as 5 to 10 mV, but for higher load currents, the saturation voltage can increase to several hundred mV (for example, 500 mV at 50 mA).

On the other hand, an output stage constructed of CMOS FETs can provide nearly true rail-to-rail performance, but only under no-load conditions. If the output must source or sink current, the output swing is reduced by the voltage dropped across the FETs internal "on" resistance (typically, 100 Ω for precision amplifiers, but can be less than 10 Ω for high current drive CMOS amplifiers).

For these reasons, it is apparent that there is no such thing as a true rail-to-rail output stage, hence the title of Figure 4.2.22 ("almost" rail-to-rail output stages).

Figure 4.2.23 summarizes the performance characteristics of a number of single-supply op amps suitable for some precision applications. The devices are listed in order of increasing supply current. Single, dual, and quad versions of each op amp are available, so the supply current is the normalized I_{SY}/amplifier for comparison. The input and output voltage ranges (V_S = +5 V) are also supplied in the table. The "0, 4 V" inputs are PNP pairs, with the exception of the AD820/822/824 which use N-Channel JFETs. Output stages having voltage ranges designated "5 mV, 4 V" are NPN emitter-followers with current source pull-downs (OP193/293/493, OP113/213/413). Output stages designated "R/R" use CMOS common source stages (OP181/281/481)

**LISTED IN ORDER OF INCREASING SUPPLY CURRENT

**PART NO.	V_{OS} max	V_{OS} TC	A_{VOL} min	NOISE (1kHz)	INPUT	OUTPUT	I_{SY}/AMP
OP181/281/481	1500µV	10µV/°C	5M	70nV/√Hz	0, 4V	"R/R"	4µA
OP193/293/493	75µV	0.2µV/°C	200k	65nV/√Hz	0, 4V	5mV, 4V	15µA
OP196/296/496	300µV	1.5µV/°C	150k	26nV/√Hz	R/R	"R/R"	50µA
OP191/291/491	700µV	1.1µV/°C	25k	35nV/√Hz	R/R	"R/R"	400µA
*AD820/822/824	400µV	2µV/°C	500k	16nV/√Hz	0, 4V	"R/R"	800µA
OP184/284/484	65µV	0.2µV/°C	50k	3.9nV/√Hz	R/R	"R/R"	1250µA
OP113/213/413	125µV	0.2µV/°C	2M	4.7nV/√Hz	0, 4V	5mV, 4V	1750µA

*JFET INPUT

NOTE: Unless Otherwise Stated Specifications are Typical @ +25°C
V_S = +5V

Figure 4.2.23: Precision single-supply op amp performance characteristics.

Sensor Signal Conditioning

or CB common emitter stages (OP196/296/496, OP191/291/491, AD820/822/824, OP184/284/484).

In summary, the following points should be considered when selecting amplifiers for single-supply/rail-to-rail applications:

First, input offset voltage and input bias currents are a function of the applied input common mode voltage (for true rail-to-rail input op amps). Circuits using this class of amplifiers should be designed to minimize resulting errors. An inverting amplifier configuration with a false ground reference at the non-inverting input prevents these errors by holding the input common mode voltage constant. If the inverting amplifier configuration cannot be used, then amplifiers like the OP184/284/OP484 which do not exhibit any common mode crossover thresholds should be used.

Second, since input bias currents are not always small and can exhibit different polarities, source impedance levels should be carefully matched to minimize additional input bias current-induced offset voltages and increased distortion. Again, consider using amplifiers that exhibit a smooth input bias current transition throughout the applied input common mode voltage.

Third, rail-to-rail amplifier output stages exhibit load-dependent gain which affects amplifier open-loop gain, and hence closed-loop gain accuracy. Amplifiers with open-loop gains greater than 30,000 for resistive loads less than 10 kΩ are good choices in precision applications. For applications not requiring full rail-rail swings, device families like the OP113/213/413 and OP193/293/493 offer DC gains of 200,000 or more.

Lastly, no matter what claims are made, rail-to-rail output voltage swings are functions of the amplifier's output stage devices and load current. The saturation voltage (V_{CESAT}), saturation resistance (R_{SAT}) for bipolar output stages, and FET on-resistance for CMOS output stages, as well as load current all affect the amplifier output voltage swing.

Op Amp Process Technologies

The wide variety of processes used to make op amps are shown in Figure 4.2.24. The earliest op amps were made using standard NPN-based bipolar processes. The PNP transistors available on these processes were extremely slow and were used primarily for current sources and level shifting.

The ability to produce matching high speed PNP transistors on a bipolar process added great flexibility to op amp circuit designs. These complementary bipolar (CB) processes are widely used in today's precision op amps, as well as those requiring

Chapter 4

wide bandwidths. The high-speed PNP transistors have f_ts which are greater than one-half the f_ts of the NPNs.

The addition of JFETs to the complementary bipolar process (CBFET) allow high input impedance op amps to be designed suitable for such applications as photodiode or electrometer preamplifiers.

- **BIPOLAR (NPN-BASED):** This is Where it All Started!!
- **COMPLEMENTARY BIPOLAR (CB):** Rail-to-Rail, Precision, High Speed
- **BIPOLAR + JFET (BiFET):** High Input Impedance, High Speed
- **COMPLEMENTARY BIPOLAR + JFET (CBFET):** High Input Impedance, Rail-to-Rail Output, High Speed
- **COMPLEMENTARY MOSFET (CMOS):** Low Cost, Non-Critical Op Amps
- **BIPOLAR + CMOS (BiCMOS):** Bipolar Input Stage adds Linearity, Low Power, Rail-to-Rail Output
- **COMPLEMENTARY BIPOLAR + CMOS (CBCMOS):** Rail-to-Rail Inputs, Rail-to-Rail Outputs, Good Linearity, Low Power

Figure 4.2.24: Op amp process technology summary.

CMOS op amps, with a few exceptions, generally have relatively poor offset voltage, drift, and voltage noise. However, the input bias current is very low. They offer low power and cost, however, and improved performance can be achieved with BiFET or CBFET processes.

The addition of bipolar or complementary devices to a CMOS process (BiMOS or CBCMOS) adds great flexibility, better linearity, and low power. The bipolar devices are typically used for the input stage to provide good gain and linearity, and CMOS devices for the rail-to-rail output stage.

In summary, there is no single IC process which is optimum for all op amps. Process selection and the resulting op amp design depends on the targeted applications and ultimately should be transparent to the customer.

Instrumentation Amplifiers (In-Amps)

An instrumentation amplifier is a closed-loop gain block which has a differential input and an output which is single-ended with respect to a reference terminal (see Figure 4.2.25). The input impedances are balanced and have high values, typically $10^9\ \Omega$ or higher. Unlike an op amp, which has its closed-loop gain determined by external resistors connected between its inverting input and its output, an in-amp employs an internal feedback resistor network which is isolated from its signal input terminals. With the input signal applied across the two differential inputs, gain is either preset internally or is user-set by an internal (via pins) or external gain resistor, which is also isolated from the signal inputs. Typical in-amp gain settings range from 1 to 10,000.

In order to be effective, an in-amp needs to be able to amplify microvolt-level signals, while simultaneously rejecting volts of common mode signal at its inputs. This requires that in-amps have very high common mode rejection (CMR): typical values of CMR are 70 dB to over 100 dB, with CMR usually improving at higher gains.

Sensor Signal Conditioning

It is important to note that a CMR specification for DC inputs alone is not sufficient in most practical applications. In industrial applications, the most common cause of external interference is pickup from the 50/60 Hz AC power mains. Harmonics of the power mains frequency can also be troublesome. In differential measurements, this type of interference tends to be induced equally onto both in-amp inputs. The interfering signal therefore appears as a common mode signal to the in-amp. Specifying CMR over frequency is more important than specifying its DC value. Imbalance in the source impedance can degrade the CMR of some in-amps. Analog Devices fully specifies in-amp CMR at 50/60 Hz with a source impedance imbalance of 1 kΩ.

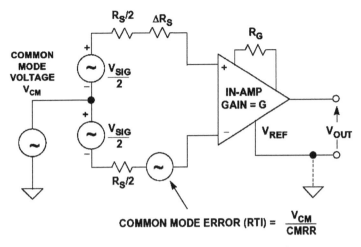

Figure 4.2.25: Instrumentation amplifier.

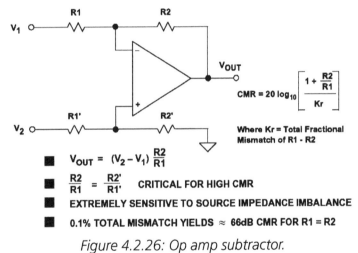

Figure 4.2.26: Op amp subtractor.

Low-frequency CMR of op amps, connected as subtractors as shown in Figure 4.2.26, generally is a function of the resistors around the circuit, not the op amp. A mismatch of only 0.1% in the resistor ratios will reduce the DC CMR to approximately 66dB. Another problem with the simple op amp subtractor is that the input impedances are relatively low and are unbalanced between the two sides. The input impedance seen by V_1 is R_1, but the input impedance seen by V_2 is $R1' + R2'$. This configuration can be quite problematic in terms of CMR, since even a small source impedance imbalance (~10 Ω) will degrade the workable CMR.

Chapter 4

Instrumentation Amplifier Configurations

Instrumentation amplifier configurations are based on op amps, but the simple subtractor circuit described above lacks the performance required for precision applications. An in-amp architecture which overcomes some of the weaknesses of the subtractor circuit uses two op amps as shown in Figure 4.2.27. This circuit is typically referred to as the *two op amp in-amp*. Dual IC op amps are used in most cases for good matching. The circuit gain may be trimmed with an external resistor, R_G. The input impedance is high, permitting the impedance of the signal sources to be high and unbalanced. The DC common mode rejection is limited by the matching of R1/R2 to R1'/R2'. If there is a mismatch in any of the four resistors, the DC common mode rejection is limited to:

$$CMR \leq 20 \log \left[\frac{GAIN \times 100}{\% MISMATCH} \right] \qquad \text{Eq. 4.2.12}$$

There is an implicit advantage to this configuration due to the gain executed on the signal. This raises the CMR in proportion.

Integrated instrumentation amplifiers are particularly well suited to meeting the combined needs of ratio matching and temperature tracking of the gain-setting resistors. While thin film resistors fabricated on silicon have an initial tolerance of up to ±20%, laser trimming during production allows the ratio error between the resistors to be reduced to 0.01% (100 ppm). Furthermore, the tracking between the temperature coefficients of the thin film resistors is inherently low and is typically less than 3 ppm/°C (0.0003%/°C).

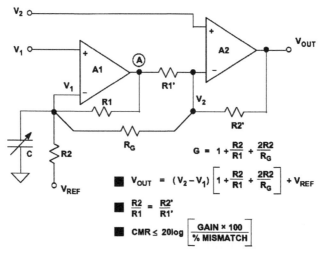

Figure 4.2.27: Two op amp instrumentation amplifier.

Sensor Signal Conditioning

When dual supplies are used, V_{REF} is normally connected directly to ground. In single supply applications, V_{REF} is usually connected to a low impedance voltage source equal to one-half the supply voltage. The gain from V_{REF} to node "A" is R1/R2, and the gain from node "A" to the output is R2'/R1'. This makes the gain from V_{REF} to the output equal to unity, assuming perfect ratio matching. Note that it is critical that the source impedance seen by V_{REF} be low, otherwise CMR will be degraded.

One major disadvantage of this design is that common mode voltage input range must be traded off against gain. The amplifier A1 must amplify the signal at V_1 by

$$1 + \frac{R1}{R2} \qquad \text{Eq. 4.2.13}$$

If R1 >> R2 (low gain in Figure 4.2.27), A1 will saturate if the common mode signal is too high, leaving no headroom to amplify the wanted differential signal. For high gains (R1<< R2), there is correspondingly more headroom at node "A" allowing larger common mode input voltages.

The AC common mode rejection of this configuration is generally poor because the signal from V_1 to V_{OUT} has the additional phase shift of A1. In addition, the two amplifiers are operating at different closed-loop gains (and thus at different bandwidths). The use of a small trim capacitor "C" as shown in the diagram can improve the AC CMR somewhat.

A low gain (G = 2) single supply two op amp in-amp configuration results when R_G is not used, and is shown in Figure 4.2.28. The input common mode and differential signals must be limited to values which prevent saturation of either A1 or A2. In the

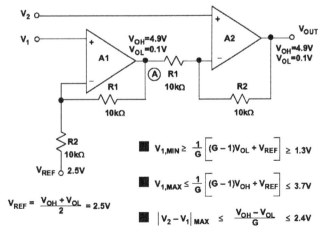

Figure 4.2.28: Single supply restrictions: V_S = +5 V, G = 2.

Chapter 4

example, the op amps remain linear to within 0.1 V of the supply rails, and their upper and lower output limits are designated V_{OH} and V_{OL}, respectively. Using the equations shown in the diagram, the voltage at V_1 must fall between 1.3 V and 2.4 V to prevent A1 from saturating. Notice that V_{REF} is connected to the average of V_{OH} and V_{OL} (2.5 V). This allows for bipolar differential input signals with V_{OUT} referenced to +2.5 V.

A high gain (G = 100) single supply two op amp in-amp configuration is shown in Figure 4.2.29. Using the same equations, note that the voltage at V_1 can now swing between 0.124 V and 4.876 V. Again, V_{REF} is connected to 2.5 V to allow for bipolar differential input and output signals.

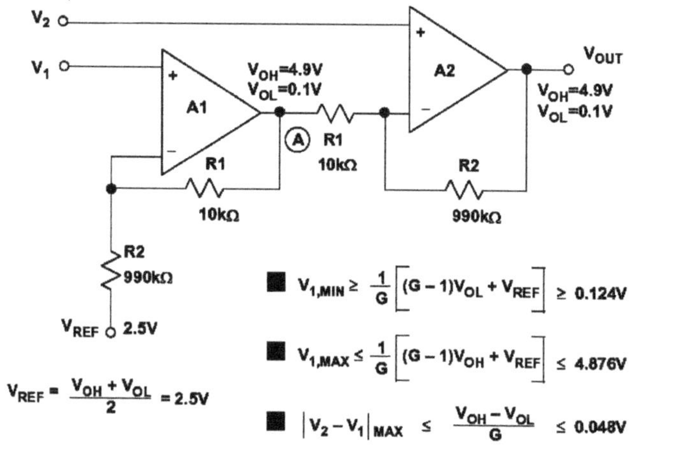

The above discussion shows that regardless of gain, the basic two op amp in-amp does not allow for zero-volt common mode input voltages when operated on a single supply. This limitation can be overcome using the circuit shown in Figure 4.2.30 which is implemented in the AD627 in-amp. Each op amp is composed of a PNP common emitter input stage and a gain stage, designated Q1/A1 and Q2/A2, respectively. The PNP transistors not only provide gain but also level shift the input signal positive by about 0.5 V, thereby allowing the common mode input voltage to go to 0.1 V below the negative supply rail. The maximum positive input voltage allowed is 1 V less than the positive supply rail.

Figure 4.2.29: Single supply restrictions: V_S = +5 V, G = 100.

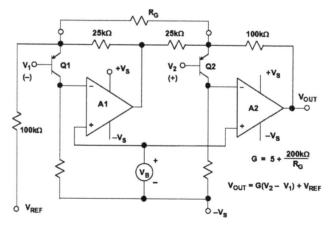

Figure 4.2.30: AD627 in-amp architecture.

Sensor Signal Conditioning

The AD627 in-amp delivers rail-to-rail output swing and operates over a wide supply voltage range (+2.7 V to ±18 V). Without R_G, the external gain setting resistor, the in-amp gain is 5. Gains up to 1000 can be set with a single external resistor. Common mode rejection of the AD627B at 60 Hz with a 1 kΩ source imbalance is 85dB when operating on a single +3 V supply and G = 5. Even though the AD627 is a two op amp in-amp, a patented circuit keeps the CMR flat out to a much higher frequency than would be achievable with a conventional discrete two op amp in-amp. The AD627 data sheet (available at http://www.analog.com) has a detailed discussion of allowable input/output voltage ranges as a function of gain and power supply voltages. Key specifications for the AD627 are summarized in Figure 4.2.31.

- Wide Supply Range : +2.7V to ±18V
- Input Voltage Range: $-V_S - 0.1V$ to $+V_S - 1V$
- 85µA Supply Current
- Gain Range: 5 to 1000
- 75µV Maximum Input Offset Volage (AD627B)
- 10ppm/°C Maximum Offset Voltage TC (AD627B)
- 10ppm Gain Nonlinearity
- 85dB CMR @ 60Hz, 1kΩ Source Imbalance (G = 5)
- 3µV p-p 0.1Hz to 10Hz Input Voltage Noise (G = 5)

Figure 4.2.31: AD627 in-amp key specifications.

For true balanced high impedance inputs, three op amps may be connected to form the in-amp shown in Figure 4.2.32. This circuit is typically referred to as the *three op amp in-amp*. The gain of the amplifier is set by the resistor, R_G, which may be internal, external, or (software or pin-strap) programmable. In this configuration, CMR depends upon the ratio matching of R3/R2 to R3'/R2'. Furthermore, common mode signals are only amplified by a factor of 1 regardless of gain (no common mode voltage will appear across R_G, hence, no common mode current will flow in it because the input terminals of an op amp will have no significant potential difference between them). Thus, CMR will theoretically increase in direct proportion to gain. Large common mode signals (within the A1-A2 op amp headroom limits) may be handled at all gains. Finally,

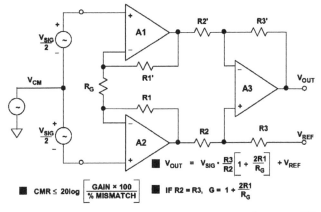

Figure 4.2.32: Three op amp instrumentation amplifier.

because of the symmetry of this configuration, common mode errors in the input amplifiers, if they track, tend to be canceled out by the subtractor output stage. These features explain the popularity of the three op amp in-amp configuration.

The classic three op amp configuration has been used in a number of monolithic IC instrumentation amplifiers. Besides offering excellent matching between the three internal op amps, thin film laser trimmed resistors provide excellent ratio matching and gain accuracy at much lower cost than using discrete op amps and resistor networks. The AD620 is an excellent example of monolithic in-amp technology, and a simplified schematic is shown in Figure 4.2.33.

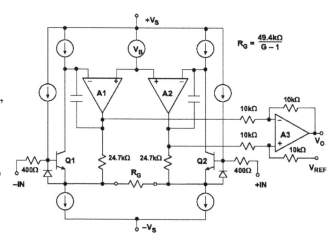

Figure 4.2.33: AD620 in-amp simplified schematic.

The AD620 is a highly popular in-amp and is specified for power supply voltages from ±2.3 V to ±18 V. Input voltage noise is only 9 nV/√Hz @ 1 kHz. Maximum input bias current is only 1 nA maximum because of the Superbeta input stage.

Overvoltage protection is provided by the internal 400 Ω thin-film current-limit resistors in conjunction with the diodes which are connected from the emitter-to-base of Q1 and Q2. The gain is set with a single external R_G resistor. The appropriate internal resistors are trimmed so that standard 1% or 0.1% resistors can be used to set the AD620 gain to popular gain values.

As in the case of the two op amp in-amp configuration, single supply operation of the three op amp in-amp requires an understanding of the internal node voltages. Figure 4.2.34 shows a generalized diagram of the in-amp operating on a single +5 V supply. The maximum and minimum

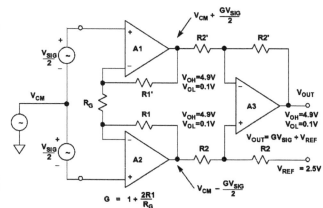

Figure 4.2.34: Three op amp in-amp single +5 V supply restrictions.

Sensor Signal Conditioning

allowable output voltages of the individual op amps are designated V_{OH} (maximum high output) and V_{OL} (minimum low output) respectively. Note that the gain from the common mode voltage to the outputs of A1 and A2 is unity, and that *the sum of the common mode voltage and the signal voltage at these outputs must fall within the amplifier output voltage range.* It is obvious that this configuration cannot handle input common mode voltages of either zero volts or +5 V because of saturation of A1 and A2. As in the case of the two op amp in-amp, the output reference is positioned halfway between V_{OH} and V_{OL} in order to allow for bipolar differential input signals.

This chapter has emphasized the operation of high performance linear circuits from a single, low-voltage supply (5 V or less) is a common requirement. While there are many precision single supply operational amplifiers, such as the OP213, the OP291, and the OP284, and some good single-supply instrumentation amplifiers, the highest performance instrumentation amplifiers are still specified for dual-supply operation.

One way to achieve both high precision and single-supply operation takes advantage of the fact that several popular sensors (e.g., strain gages) provide an output signal centered around the (approximate) mid-point of the supply voltage (or the reference voltage), where the inputs of the signal conditioning amplifier need not operate near "ground" or the positive supply voltage.

Under these conditions, a dual-supply instrumentation amplifier referenced to the supply mid-point followed by a "rail-to-rail" operational amplifier gain stage provides very high DC precision. Figure 4.2.35 illustrates one such high-performance instrumentation amplifier operating on a single, +5 V supply. This circuit uses an AD620 low-cost precision instrumentation amplifier for the input stage, and an AD822 JFET-input dual rail-to-rail output operational amplifier for the output stage.

In this circuit, R3 and R4 form a voltage divider which splits the supply voltage in half to +2.5 V, with fine adjustment provided by a trimming potentiometer, P1. This voltage is applied to the input of A1, an AD822 which buffers it and provides a low-impedance source needed to drive the AD620's reference pin. The AD620's Reference pin has a 10 kΩ input resistance and an input signal current of up to 200µA. The

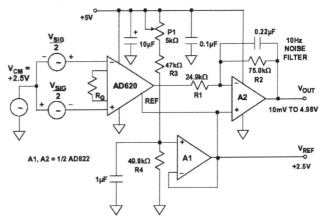

Figure 4.2.35: A precision single-supply composite in-amp with rail-to-rail output.

Chapter 4

other half of the AD822 is connected as a gain-of-3 inverter, so that it can output ±2.5 V, "rail-to-rail," with only ±0.83 V required of the AD620. This output voltage level of the AD620 is well within the AD620's capability, thus ensuring high linearity for the "dual-supply" front end. *Note that the final output voltage must be measured with respect to the +2.5 V reference, and not to GND.*

The general gain expression for this composite instrumentation amplifier is the product of the AD620 and the inverting amplifier gains:

$$GAIN = \left(\frac{49.4k\Omega}{R_G} + 1\right)\left(\frac{R2}{R1}\right) \qquad \text{Eq. 4.2.14}$$

For this example, an overall gain of 10 is realized with $R_G = 21.5\ k\Omega$ (closest standard value). The table (Figure 4.2.36) summarizes various R_G/gain values and performance.

In this application, the allowable input voltage on either input to the AD620 must lie between +2 V and +3.5 V in order to maintain linearity. For example, at an overall circuit gain of 10, the common mode input voltage range spans 2.25 V to 3.25 V, allowing room for the ±0.25 V full-scale differential input voltage required to drive the output ±2.5 V about V_{REF}.

CIRCUIT GAIN	R_G (Ω)	V_{OS}, RTI (µV)	TC V_{OS}, RTI (µV/°C)	NONLINEARITY (ppm) *	BANDWIDTH (kHz)**
10	21.5k	1000	1000	< 50	600
30	5.49k	430	430	< 50	600
100	1.53k	215	215	< 50	300
300	499	150	150	< 50	120
1000	149	150	150	< 50	30

* Nonlinearity Measured Over Output Range: $0.1V < V_{OUT} < 4.90V$
** Without 10Hz Noise Filter

Figure 4.2.36: Performance summary of the +5 V single-supply AD620/AD822 composite in-amp.

The inverting configuration was chosen for the output buffer to facilitate system output offset voltage adjustment by summing currents into the A2 stage buffer's feedback summing node. These offset currents can be provided by an external DAC, or from a resistor connected to a reference voltage.

The AD822 rail-to-rail output stage exhibits a very clean transient response (not shown) and a small-signal bandwidth over 100 kHz for gain configurations up to 300. Note that excellent linearity is maintained over 0.1 V to 4.9 V V_{OUT}. To reduce the effects of unwanted noise pickup, a capacitor is recommended across A2's feedback resistance to limit the circuit bandwidth to the frequencies of interest.

In cases where zero-volt inputs are required, the AD623 single supply in-amp configuration shown in Figure 4.2.37 offers an attractive solution. The PNP emitter follower level shifters, Q1/Q2, allow the input signal to go 150 mV below the negative supply

Sensor Signal Conditioning

and to within 1.5 V of the positive supply. The AD623 is fully specified for single power supplies between +3 V and +12 V and dual supplies between ±2.5 V and ±6 V (see Figure 4.2.38). The AD623 data sheet (available at http://www.analog.com) contains an excellent discussion of allowable input/output voltage ranges as a function of gain and power supply voltages.

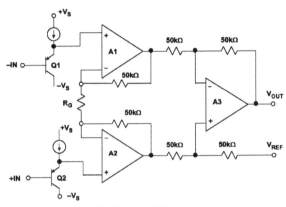

Figure 4.2.37: AD623 single-supply in-amp architecture.

- Wide Supply Range: +3V to ±6V
- Input Voltage Range: $-V_S - 0.15V$ to $+V_S - 1.5V$
- 575µA Maximum Supply Current
- Gain Range: 1 to 1000
- 100µV Maximum Input Offset Voltage (AD623B)
- 1µV/°C Maximum Offset Voltage TC (AD623B)
- 50ppm Gain Nonlinearity
- 105dB CMR @ 60Hz, 1kΩ Source Imbalance, G ≥ 100
- 3µV p-p 0.1Hz to 10Hz Input Voltage Noise (G = 1)

Figure 4.2.38: AD623 in-amp key specifications.

Instrumentation Amplifier DC Error Sources

The DC and noise specifications for instrumentation amplifiers differ slightly from conventional op amps, so some discussion is required in order to fully understand the error sources.

The gain of an in-amp is usually set by a single resistor. If the resistor is external to the in-amp, its value is either calculated from a formula or chosen from a table on the data sheet, depending on the desired gain.

Absolute value laser wafer trimming allows the user to program gain accurately with this single resistor. The absolute accuracy and temperature coefficient of this resistor directly affects the in-amp gain accuracy and drift. Since the external resistor will never exactly match the internal thin film resistor tempcos, a low TC (<25 ppm/°C) metal film resistor should be chosen, preferably with a 0.1% or better accuracy.

Often specified as having a gain range of 1 to 1000, or 1 to 10,000, many in-amps will work at higher gains, but the manufacturer will not guarantee a specific level of performance at these high gains. In practice, as the gain-setting resistor becomes smaller, any errors due to the resistance of the metal runs and bond wires become significant. These errors, along with an increase in noise and drift, may make higher single-stage gains impractical. In addition, input offset voltages can become quite sizable when

Chapter 4

reflected to output at high gains. For instance, a 0.5 mV input offset voltage becomes 5 V at the output for a gain of 10,000. For high gains, the best practice is to use an instrumentation amplifier as a preamplifier then use a post amplifier for further amplification.

In a pin-programmable gain in-amp such as the AD621, the gain setting resistors are internal, well matched, and the gain accuracy and gain drift specifications include their effects. The AD621 is otherwise generally similar to the externally gain-programmed AD620.

The *gain error* specification is the maximum deviation from the gain equation. Monolithic in-amps such as the AD624C have very low factory trimmed gain errors, with its maximum error of 0.02% at G = 1 and 0.25% at G = 500 being typical for this high quality in-amp. Notice that the gain error increases with increasing gain. Although externally connected gain networks allow the user to set the gain exactly, the temperature coefficients of the external resistors and the temperature differences between individual resistors within the network all contribute to the overall gain error. If the data is eventually digitized and presented to a digital processor, it may be possible to correct for gain errors by measuring a known reference voltage and then multiplying by a constant.

Nonlinearity is defined as the maximum deviation from a straight line on the plot of output versus input. The straight line is drawn between the end-points of the actual transfer function. Gain nonlinearity in a high quality in-amp is usually 0.01% (100 ppm) or less, and is relatively insensitive to gain over the recommended gain range.

The total input offset voltage of an in-amp consists of two components (see Figure 4.2.39). Input offset voltage, V_{OSI}, is that component of input offset which is reflected to the output of the in-amp by the gain G. Output offset voltage, V_{OSO}, is independent of gain. At low gains, output offset voltage is dominant, while at high gains input offset dominates. The output offset voltage drift is normally specified as drift at G = 1 (where input effects are insignificant), while input offset voltage drift is given by a drift specification at a high gain (where output offset effects are negligible). The total output offset error, referred to the input

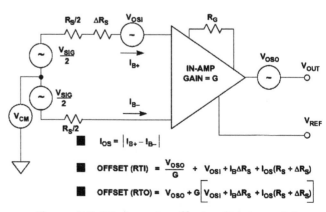

Figure 4.2.39: In-amp offset voltage model.

Sensor Signal Conditioning

(RTI), is equal to $V_{OSI} + V_{OSO}/G$. In-amp data sheets may specify V_{OSI} and V_{OSO} separately or give the total RTI input offset voltage for different values of gain.

Input bias currents may also produce offset errors in in-amp circuits (see Figure 4.2.39). If the source resistance, R_S, is unbalanced by an amount, ΔR_S, (often the case in bridge circuits), then there is an additional input offset voltage error due to the bias current, equal to $I_B \Delta R_S$ (assuming that $I_{B+} \approx I_{B-} = I_B$). This error is reflected to the output, scaled by the gain G. The input offset current, I_{OS}, creates an input offset voltage error across the source resistance, $R_S + \Delta R_S$, equal to $I_{OS}(R_S + \Delta R_S)$, which is also reflected to the output by the gain, G.

In-amp common mode error is a function of both gain and frequency. Analog Devices specifies in-amp CMR for a 1 kΩ source impedance unbalance at a frequency of 60 Hz. The RTI common mode error is obtained by dividing the common mode voltage, V_{CM}, by the common mode rejection ratio, CMRR.

Power supply rejection (PSR) is also a function of gain and frequency. For in-amps, it is customary to specify the sensitivity to each power supply separately. Now that all DC error sources have been accounted for, a worst case DC error budget can be calculated by reflecting all the sources to the in-amp input (Figure 4.2.40).

ERROR SOURCE	RTI VALUE
Gain Accuracy (ppm)	Gain Accuracy × FS Input
Gain Nonlinearity (ppm)	Gain Nonlinearity × FS Input
Input Offset Voltage, V_{OSI}	V_{OSI}
Output Offset Voltage, V_{OSO}	$V_{OSO} \div G$
Input Bias Current, I_B, Flowing in ΔR_S	$I_B \Delta R_S$
Input Offset Current, I_{OS}, Flowing in R_S	$I_{OS}(R_S + \Delta R_S)$
Common Mode Input Voltage, V_{CM}	$V_{CM} \div$ CMRR
Power Supply Variation, ΔV_S	$\Delta V_S \div$ PSRR

Figure 4.2.40: Instrumentation amplifier DC errors referred to the input (RTI).

Instrumentation Amplifier Noise Sources

Since in-amps are primarily used to amplify small precision signals, it is important to understand the effects of all the associated noise sources. The in-amp noise model is shown in Figure 4.2.41. There are two sources of input voltage noise. The first is represented as a noise source, V_{NI}, in series with the input, as in a conventional op amp circuit. This noise is reflected to the output by the in-amp gain, G. The second noise source is the output noise, V_{NO}, represented as a noise voltage in series with the in-amp output. The output noise, shown here referred to V_{OUT}, can be referred to the input by dividing by the gain, G.

There are two noise sources associated with the input noise currents I_{N+} and I_{N-}. Even though I_{N+} and I_{N-} are usually equal ($I_{N+} \approx I_{N-} = I_N$), they are uncorrelated, and therefore, the noise they each create must be summed in a root-sum-squares (RSS) fashion. I_{N+} flows through one half of R_S, and I_{N-} the other half. This generates two

Chapter 4

noise voltages, each having an amplitude, $I_N R_S/2$. Each of these two noise sources is reflected to the output by the in-amp gain, G.

The total output noise is calculated by combining all four noise sources in an RSS manner:

In-amp data sheets often present the total voltage noise RTI as a function of gain. This noise spectral density includes both the input (V_{NI}) and output (V_{NO}) noise contributions. The input

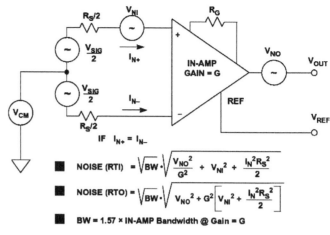

Figure 4.2.41: In-amp noise model.

current noise spectral density is specified separately. As in the case of op amps, the total noise RTI must be integrated over the in-amp closed-loop bandwidth to compute the RMS value. The bandwidth may be determined from data sheet curves which show frequency response as a function of gain.

In-Amp Bridge Amplifier Error Budget Analysis

It is important to understand in-amp error sources in a typical application. Figure 4.2.42 shows a 350 Ω load cell which has a full scale output of 100 mV when excited with a 10 V source. The AD620 is configured for a gain of 100 using the external 499 Ω gain-setting resistor. The table shows how each error source contributes to the total unadjusted error of 2145ppm. The gain, offset, and CMR errors can be removed with a system calibration. The remaining errors—gain nonlinearity and 0.1 Hz to 10 Hz noise—cannot be removed with calibration and limit the system resolution to 42.8 ppm (approximately 14-bit accuracy).

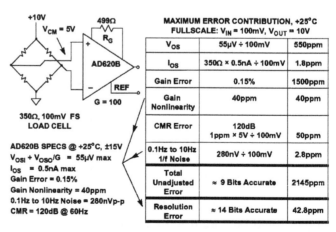

Figure 4.2.42: AD620B bridge amplifier DC error budget.

Sensor Signal Conditioning

In-Amp Performance Tables

Figure 4.2.43 shows a selection of precision in-amps designed primarily for operation on dual supplies. It should be noted that the AD620 is capable of single +5 V supply operation (see Figure 4.2.35), but neither its input nor its output are capable of rail-to-rail swings.

	Gain Accuracy *	Gain Nonlinearity	V_{OS} Max	V_{OS} TC	CMR Min	0.1Hz to 10Hz p-p Noise
AD524C	0.5% / P	100ppm	50µV	0.5µV/°C	120dB	0.3µV
AD620B	0.5% / R	40ppm	50µV	0.6µV/°C	120dB	0.28µV
AD621B[1]	0.05% / P	10ppm	50µV	1.6µV/°C	100dB	0.28µV
AD622	0.5% / R	40ppm	125µV	1µV/°C	103dB	0.3µV
AD624C[2]	0.25% / R	50ppm	25µV	0.25µV/°C	130dB	0.2µV
AD625C	0.02% / R	50ppm	25µV	0.25µV/°C	125dB	0.2µV
AMP01A	0.6% / R	50ppm	50µV	0.3µV/°C	125dB	0.12µV
AMP02E	0.5% / R	60ppm	100µV	2µV/°C	115dB	0.4µV

* / P = Pin Programmable
* / R = Resistor Programmable
[1] G = 100
[2] G = 500

Figure 4.2.43: Precision in-amps: data for $V_S = \pm15$ V, G = 1000.

Instrumentation amplifiers specifically designed for single supply operation are shown in Figure 4.2.44. It should be noted that although the specifications in the figure are given for a single +5 V supply, all of the amplifiers are also capable of dual supply operation and are specified for both dual and single supply operation on their data sheets. In addition, the AD623 and AD627 will operate on a single +3 V supply.

	Gain Accuracy *	Gain Nonlinearity	V_{OS} Max	V_{OS} TC	CMR Min	0.1Hz to 10Hz p-p Noise	Supply Current
AD623B	0.5% / R	50ppm	100µV	1µV/°C	105dB	1.5µV	575µA
AD627B	0.35% / R	10ppm	75µV	1µV/°C	85dB	1.5µV	85µA
AMP04E	0.4% / R	250ppm	150µV	3µV/°C	90dB	0.7µV	290µA
AD626B[1]	0.6% / P	200ppm	2.5mV	6µV/°C	80dB	2µV	700µA

* / P = Pin Programmable
* / R = Resistor Programmable
[1] Differential Amplifier, G = 100

Figure 4.2.44: Single supply in-amps: data for $V_S = \pm5$ V, G = 1000.

The AD626 is not a true in-amp but is a differential amplifier with a thin-film input attenuator which allows the common mode voltage to exceed the supply voltages. This device is designed primarily for high and low-side current-sensing applications. It will also operate on a single +3 V supply.

Chapter 4

In-Amp Input Overvoltage Protection

As interface amplifiers for data acquisition systems, instrumentation amplifiers are often subjected to input overloads, i.e., voltage levels in excess of the full scale for the selected gain range. The manufacturer's "absolute maximum" input ratings for the device should be closely observed. As with op amps, many in-amps have absolute maximum input voltage specifications equal to $\pm V_S$. External series resistors (for current limiting) and Schottky diode clamps may be used to prevent overload, if necessary. Some instrumentation amplifiers have built-in overload protection circuits in the form of series resistors (thin film) or series-protection FETs. In-amps such as the AMP-02 and the AD524 utilize series-protection FETs, because they act as a low impedance during normal operation, and a high impedance during fault conditions.

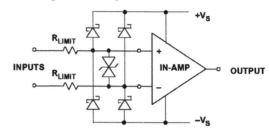

- Always Observe Absolute Maximum Data Sheet Specs!
- Schottky Diode Clamps to the Supply Rails Will Limit Input to Approximately $\pm V_S \pm 0.3V$, TVSs Limit Differential Voltage
- External Resistors (or Internal Thin-Film Resistors) Can Limit Input Current, but will Increase Noise
- Some In-Amps Have Series-Protection Input FETs for Lower Noise and Higher Input Over-Voltages (up to $\pm 60V$, Depending on Device)

Figure 4.2.45: Instrumentation amplifier input overvoltage considerations.

An additional Transient Voltage Suppresser (TVS) may be required across the input pins to limit the maximum differential input voltage. This is especially applicable to three op amp in-amps operating at high gain with low values of R_G.

Chopper Stabilized Amplifiers

For the lowest offset and drift performance, chopper-stabilized amplifiers may be the only solution. The best bipolar amplifiers offer offset voltages of 10 µV and 0.1 µV/°C drift. Offset voltages less than 5 µV with practically no measurable offset drift are obtainable with choppers, albeit with some penalties.

The basic chopper amplifier circuit is shown in Figure 4.2.46. When the switches are in the "Z" (auto-zero) position, capacitors C2 and C3 are charged to the amplifier input and output offset voltage, respectively. When the switches are in the "S"

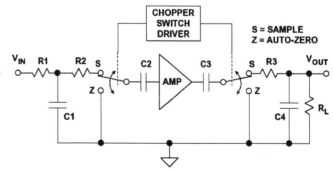

Figure 4.2.46: Classic chopper amplifier.

Sensor Signal Conditioning

(sample) position, V_{IN} is connected to V_{OUT} through the path comprised of R1, R2, C2, the amplifier, C3, and R3. The chopping frequency is usually between a few hundred Hz and several kHz, and it should be noted that because this is a sampling system, the input frequency must be much less than one-half the chopping frequency in order to prevent errors due to aliasing. The R1/C1 combination serves as an antialiasing filter. It is also assumed that after a steady state condition is reached, there is only a minimal amount of charge transferred during the switching cycles. The output capacitor, C4, and the load, R_L, must be chosen such that there is minimal V_{OUT} droop during the auto-zero cycle.

The basic chopper amplifier of Figure 4.2.46 can pass only very low frequencies because of the input filtering required to prevent aliasing. The *chopper-stabilized* architecture shown in Figure 4.2.47 is most often used in chopper amplifier implementations. In this circuit, A1 is the *main* amplifier, and A2 is the *nulling* amplifier. In the sample mode (switches in "S" position), the nulling amplifier, A2, monitors the input offset voltage of A1 and drives its output to zero by applying a suitable correcting voltage at A1's null pin. Note, however, that A2 also has an input offset voltage, so it must correct its own error before attempting to null A1's offset. This is achieved in the auto-zero mode (switches in "Z" position) by momentarily disconnecting A2 from A1, shorting its inputs together, and coupling its output to its own null pin. During the auto-zero mode, the correction voltage for A1 is momentarily held by C1. Similarly, C2 holds the correction voltage for A2 during the sample mode. In modern IC chopper-stabilized op amps, the storage capacitors C1 and C2 are on-chip.

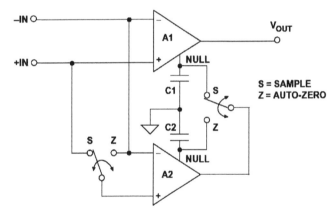

Figure 4.2.47: Chopper stabalized amplifier.

Note in this architecture that the input signal is always connected to the output through A1. The bandwidth of A1 thus determines the overall signal bandwidth, and the input signal is not limited to less than one-half the chopping frequency as in the case of the traditional chopper amplifier architecture. However, the switching action does produce small transients at the chopping frequency which can mix with the input signal frequency and produce in-band distortion.

Chapter 4

It is interesting to consider the effects of a chopper amplifier on low frequency 1/f noise. If the chopping frequency is considerably higher than the 1/f corner frequency of the input noise, the chopper-stabilized amplifier continuously nulls out the 1/f noise on a sample-by-sample basis. Theoretically, a chopper op amp therefore has no 1/f noise. However, the chopping action produces wideband noise which is generally much worse than that of a precision bipolar op amp.

Figure 4.2.48 shows the noise of a precision bipolar amplifier (OP177/AD707) versus that of the AD8551/52/54 chopper-stabilized op amp. The peak-to-peak noise in various bandwidths is calculated for each in the table below the graphs. Note that as the frequency is lowered, the chopper amplifier noise continues to drop, while the bipolar amplifier noise approaches a limit determined by the 1/f corner frequency and its white noise (see Figure 4.2.9). At a very low frequency, the noise performance of the chopper is superior to that of the bipolar op amp.

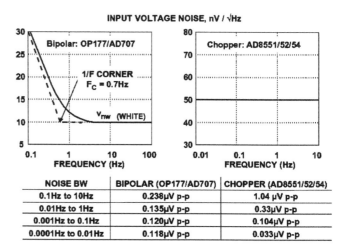

Figure 4.2.48: Noise: bipolar vs. chopper amplifier.

NOISE BW	BIPOLAR (OP177/AD707)	CHOPPER (AD8551/52/54)
0.1Hz to 10Hz	0.238µV p-p	1.04 µV p-p
0.01Hz to 1Hz	0.135µV p-p	0.33µV p-p
0.001Hz to 0.1Hz	0.120µV p-p	0.104µV p-p
0.0001Hz to 0.01Hz	0.118µV p-p	0.033µV p-p

The AD8551/8552/8554 family of chopper-stabilized op amps offers rail-to-rail input and output single supply operation, low offset voltage, and low offset drift. The storage capacitors are internal to the IC, and no external capacitors other than standard decoupling capacitors are required. Key specifications for the devices are given in Figure 4.2.49.

- Single Supply: +3V to +5V
- 5µV Max. Input Offset Voltage
- 0.04µV/°C Input Offset Voltage Drift
- 120dB CMR, PSR
- 800µA Supply Current / Op Amp
- 100µs Overload Recovery Time
- 50nV/√Hz Input Voltage Noise
- 1.5MHz Gain-Bandwidth Product
- Single (AD8551), Dual (AD8552) and Quad (AD8554)

Figure 4.2.49: AD8551/52/54 chopper stabilized rail-to-rail input/output amplifiers.

It should be noted that extreme care must be taken when applying these devices to avoid parasitic thermocouple effects in order to fully realize the offset and drift performance.

Sensor Signal Conditioning

Isolation Amplifiers

There are many applications where it is desirable, or even essential, for a sensor to have no direct ("galvanic") electrical connection with the system to which it is supplying data, either in order to avoid the possibility of dangerous voltages or currents from one half of the system doing damage in the other, or to break an intractable ground loop. Such a system is said to be "isolated," and the arrangement which passes a signal without galvanic connections is known as an "isolation barrier."

The protection of an isolation barrier works in both directions, and may be needed in either, or even in both. The obvious application is where a sensor may accidentally encounter high voltages, and the system it is driving must be protected. Or a sensor may need to be isolated from accidental high voltages arising downstream, in order to protect its environment: examples include the need to prevent the ignition of explosive gases by sparks at sensors and the protection from electric shock of patients whose ECG, EEG or EMG is being monitored. The ECG case is interesting, as protection may be required in *both* directions: the patient must be protected from accidental electric shock, but if the patient's heart should stop, the ECG machine must be protected from the very high voltages (>7.5 kV) applied to the patient by the defibrillator which will be used to attempt to restart it.

- Sensor is at a High Potential Relative to Other Circuitry (or may become so under Fault Conditions)
- Sensor May Not Carry Dangerous Voltages, Irrespective of Faults in Other Circuitry (e.g. Patient Monitoring and Intrinsically Safe Equipment for use with Explosive Gases)
- To Break Ground Loops

Figure 4.2.50: Applications for isolation amplifiers.

Just as interference, or *unwanted* information, may be coupled by electric or magnetic fields, or by electromagnetic radiation, these phenomena may be used for the transmission of *wanted* information in the design of isolated systems. The most common isolation amplifiers use transformers, which exploit magnetic fields, and another common type uses small high voltage capacitors, exploiting electric fields. Opto-isolators, which consist of an LED and a photocell, provide isolation by using light, a form of electromagnetic radiation. Different isolators have differing performance: some are sufficiently linear to pass high accuracy analog signals across an isolation barrier, with others the signal may need to be converted to digital form before transmission, if accuracy is to be maintained, a common application for V/F converters.

Transformers are capable of analog accuracy of 12–16 bits and bandwidths up to several hundred kHz, but their maximum voltage rating rarely exceeds 10 kV, and is often much lower. Capacitively coupled isolation amplifiers have lower accuracy, perhaps 12-bits maximum, lower bandwidth, and lower voltage ratings—but they are cheap. Optical isolators are fast and cheap, and can be made with very high voltage ratings (4–7 kV is one of the more common ratings), but they have poor analog domain linearity, and are not usually suitable for direct coupling of precision analog signals.

Linearity and isolation voltage are not the only issues to be considered in the choice of isolation systems. Power is essential. Both the input and the output circuitry must be powered, and unless there is a battery on the isolated side of the isolation barrier (which is possible, but rarely convenient), some form of isolated power must be provided. Systems using transformer isolation can easily use a transformer (either the signal transformer or another one) to provide isolated power, but it is impractical to transmit useful amounts of power by capacitive or optical means. Systems using these forms of isolation must make other arrangements to obtain isolated power supplies—this is a powerful consideration in favor of choosing transformer isolated isolation amplifiers: they almost invariably include an isolated power supply.

The isolation amplifier has an input circuit that is galvanically isolated from the power supply and the output circuit. In addition, there is minimal capacitance between the input and the rest of the device. Therefore, there is no possibility for DC current flow, and minimum AC coupling. Isolation amplifiers are intended for applications requiring safe, accurate measurement of low frequency voltage or current (up to about 100 kHz) in the presence of high common-mode voltage (to thousands of volts) with high common mode rejection. They are also useful for line-receiving of signals transmitted at high impedance in noisy environments, and for safety in general-purpose measurements, where DC and line-frequency leakage must be maintained at levels well below certain mandated minimums. Principal applications are in electrical environments of the kind associated with medical equipment, conventional and nuclear power plants, automatic test equipment, and industrial process control systems.

In the basic two-port form, the output and power circuits are not isolated from one another. In the three-port isolator shown in Figure 4.2.51, the input circuits, output circuits, and power source are all isolated from one another. The figure shows the circuit architecture of a self-contained isolator, the AD210. An isolator of this type requires power from a two-terminal DC power supply. An internal oscillator (50 kHz) converts the DC power to AC, which is transformer-coupled to the shielded input section, then converted to DC for the input stage and the auxiliary power output. The AC carrier is also modulated by the amplifier output, transformer-coupled to the output stage, demodulated by a phase-sensitive demodulator (using the carrier as the reference), filtered, and buffered using isolated DC power derived from the carrier. The

Sensor Signal Conditioning

AD210 allows the user to select gains from 1 to 100 using an external resistor. Bandwidth is 20 kHz, and voltage isolation is 2500 V RMS (continuous) and ±3500 V peak (continuous).

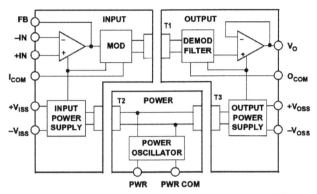

Figure 4.2.51: AD210 3-port isolation amplifier.

The AD210 is a 3-port isolation amplifier: the power circuitry is isolated from both the input and the output stages and may therefore be connected to either—or to neither. It uses transformer isolation to achieve 3500 V isolation with 12-bit accuracy. Key specifications for the AD210 are summarized in Figure 4.2.52.

- **Transformer Coupled**
- **High Common Mode Voltage Isolation:**
  - ◆ 2500V RMS Continuous
  - ◆ ±3500V Peak Continuous
- **Wide Bandwidth: 20kHz (Full Power)**
- **0.012% Maximum Linearity Error**
- **Input Amplifier: Gain 1 to 100**
- **Isolated Input and Output Power Supplies, ±15V, ±5mA**

Figure 4.2.52: AD210 isolation amplifier key features.

A typical isolation amplifier application using the AD210 is shown in Figure 4.2.53. The AD210 is used with an AD620 instrumentation amplifier in a current-sensing system for motor control. The input of the AD210, being isolated, can be connected to a 110 or 230 V power line without any protection, and the isolated ±15 V powers the AD620, which senses the voltage drop in a small current sensing resistor. The 110 or 230 V RMS common-mode voltage is ignored by the isolated system. The AD620 is used to improve system accuracy: the V_{OS} of the AD210 is 15 mV, while the AD620 has V_{OS} of 30 μV and correspondingly lower drift. If higher DC offset and drift are acceptable, the AD620 may be omitted, and the AD210 used directly at a closed loop gain of 100.

Chapter 4

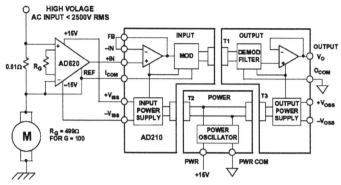

Figure 4.2.53: Motor control currrent sensing.

References

1. Walt Jung, Ed., **Op Amp Applications Handbook,** 2005, Newnes, ISBN: 0-672-22453-4.
3. **Amplifier Applications Guide**, Analog Devices, Inc., 1992.
4. **System Applications Guide**, Analog Devices, Inc., 1994.
5. **Linear Design Seminar, Analog Devices**, Inc., 1995.
6. **Practical Analog Design Techniques**, Analog Devices, Inc., 1995.
7. **High Speed Design Techniques**, Analog Devices, Inc., 1996.
8. James L. Melsa and Donald G. Schultz, **Linear Control Systems**, McGraw-Hill, 1969, pp. 196-220.
9. Thomas M. Fredrickson, **Intuitive Operational Amplifiers**, McGraw-Hill, 1988.
10. Paul R. Gray and Robert G. Meyer, **Analysis and Design of Analog Integrated Circuits, Second Edition**, John Wiley, 1984.
11. J. K. Roberge, **Operational Amplifiers-Theory and Practice**, John Wiley, 1975.
12. Lewis Smith and Dan Sheingold, *Noise and Operational Amplifier Circuits*, **Analog Dialogue 25th Anniversary Issue**, pp. 19-31, 1991. (Also AN358)
13. D. Stout, M. Kaufman, **Handbook of Operational Amplifier Circuit Design**, New York, McGraw-Hill, 1976.
14. Joe Buxton, *Careful Design Tames High-Speed Op Amps*, **Electronic Design**, April 11, 1991.
15. J. Dostal, **Operational Amplifiers**, Elsevier Scientific Publishing, New York, 1981.
16. Sergio Franco, **Design with Operational Amplifiers and Analog Integrated Circuits**, Second Edition, McGraw-Hill, 1998.

17. Charles Kitchin and Lew Counts, **Instrumentation Amplifier Application Guide**, Analog Devices, 1991.
18. AD623 and AD627 Instrumentation Amplifier Data Sheets, Analog Devices, http://www.analog.com
19. Eamon Nash, *A Practical Review of Common Mode and Instrumentation Amplifiers*, **Sensors Magazine**, July 1998, pp. 26–33.
20. Eamon Nash, *Errors and Error Budget Analysis in Instrumentation Amplifiers*, **Application Note AN-539**, Analog Devices.

4.3 Analog to Digital Converters for Signal Conditioning

The trend in ADCs and DACs is toward higher speeds and higher resolutions at reduced power levels. Modern data converters generally operate on ±5 V (dual supply) or +5 V (single supply). In fact, many new converters operate on a single +3 V supply. This trend has created a number of design and applications problems which were much less important in earlier data converters, where ±15 V supplies and ±10 V input ranges were the standard.

Lower supply voltages imply smaller input voltage ranges, and hence more susceptibility to noise from all potential sources: power supplies, references, digital signals, EMI/RFI, and probably most important, improper layout, grounding, and decoupling techniques. Single-supply ADCs often have an input range which is not referenced to ground. Finding compatible single-supply drive amplifiers and dealing with level shifting of the input signal in direct-coupled applications also becomes a challenge.

In spite of these issues, components are now available which allow extremely high resolutions at low supply voltages and low power. This section discusses the applications problems associated with such components and shows techniques for successfully designing them into systems.

The most popular precision signal conditioning ADCs are based on two fundamental architectures: *successive approximation* and *sigma-delta*. The *tracking* ADC architecture is particularly suited for resolver-to-digital converters, but it is rarely used in other precision signal conditioning applications. The *flash* converter and the *subranging (or pipelined)* converter architectures are widely used where sampling frequencies extend into the megahertz and hundreds of megahertz region, but are overkills in both speed and cost for low frequency precision signal conditioning applications.

- Typical Supply Voltages: ±5V, +5V, +5/+3V, +3V
- Lower Signal Swings Increase Sensitivity to all Types of Noise (Device, Power Supply, Logic, etc.)
- Device Noise Increases at Low Currents
- Common Mode Input Voltage Restrictions
- Input Buffer Amplifier Selection Critical
- Auto-Calibration Modes Desirable at High Resolutions

Figure 4.3.1: Low power, low voltage ADC design issues.

Sensor Signal Conditioning

- **Successive Approximation**
  - Resolutions to 16-bits
  - Minimal Throughput Delay Time
  - Used in Multiplexed Data Acquisition Systems
- **Sigma-Delta**
  - Resolutions to 24-bits
  - Excellent Differential Linearity
  - Internal Digital Filter, Excellent AC Line Rejection
  - Long Throughput Delay Time
  - Difficult to Multiplex Inputs Due to Digital Filter Settling Time
- **High Speed Architectures:**
  - Flash Converter
  - Subranging or Pipelined

Figure 4.3.2: ADCs for signal conditioning.

Successive Approximation ADCs

The successive approximation ADC has been the mainstay of signal conditioning for many years. Recent design improvements have extended the sampling frequency of these ADCs into the megahertz region. The use of internal switched capacitor techniques along with auto calibration techniques extend the resolution of these ADCs to 16-bits on standard CMOS processes without the need for expensive thin-film laser trimming.

The basic successive approximation ADC is shown in Figure 4.3.3. It performs conversions on command. On the assertion of the CONVERT START command, the sample-and-hold (SHA) is placed in the *hold* mode, and all the bits of the successive approximation register (SAR) are reset to "0" except the MSB which is set to "1". The SAR output drives the internal DAC. If the DAC output is greater than the analog input, this bit in the SAR is reset, otherwise it is left set. The next most significant bit is then set to "1". If the DAC output is greater than the analog input, this bit in the SAR is reset, otherwise it is left set. The process is repeated with each bit in turn. When all the bits have been set, tested, and reset or not as appropriate, the contents of the SAR correspond to the value of the analog input, and the conversion is complete.

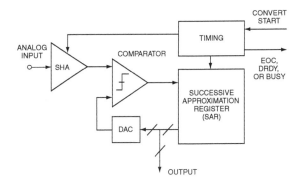

Figure 4.3.3: Successive approximation ADC.

Chapter 4

The end of conversion is generally indicated by an end-of-convert (EOC), data-ready (DRDY), or a busy signal (actually, *not*-BUSY indicates end of conversion). The polarities and name of this signal may be different for different SAR ADCs, but the fundamental concept is the same. At the beginning of the conversion interval, the signal goes high (or low) and remains in that state until the conversion is completed, at which time it goes low (or high). The trailing edge is generally an indication of valid output data.

An N-bit conversion takes N steps. It would seem on superficial examination that a 16-bit converter would have twice the conversion time of an 8-bit one, but this is not the case. In an 8-bit converter, the DAC must settle to 8-bit accuracy before the bit decision is made, whereas in a 16-bit converter, it must settle to 16-bit accuracy, which takes a lot longer. In practice, 8-bit successive approximation ADCs can convert in a few hundred nanoseconds, while 16-bit ones will generally take several microseconds.

Notice that the overall accuracy and linearity of the SAR ADC is determined primarily by the internal DAC. Until recently, most precision SAR ADCs used laser-trimmed thin-film DACs to achieve the desired accuracy and linearity. The thin-film resistor trimming process adds cost, and the thin-film resistor values may be affected when subjected to the mechanical stresses of packaging.

For these reasons, switched capacitor (or charge-redistribution) DACs have become popular in newer SAR ADCs. The advantage of the switched capacitor DAC is that the accuracy and linearity is primarily determined by photolithography, which in turn controls the capacitor plate area and the capacitance as well as matching. In addition, small capacitors can be placed in parallel with the main capacitors which can be switched in and out under control of autocalibration routines to achieve high accuracy and linearity without the need for thin-film laser trimming. Temperature tracking between the switched capacitors can be better than 1 ppm/°C, thereby offering a high degree of temperature stability.

A simple 3-bit capacitor DAC is shown in Figure 4.3.4. The switches are shown in the *track*, or *sample* mode where the analog input voltage, A_{IN}, is constantly charging and discharging the parallel combination of all the capacitors. The *hold* mode is initiated by opening S_{IN}, leaving the sampled analog input volt-

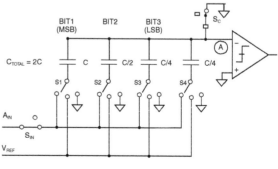

Figure 4.3.4: 3-bit switched capacitor DAC.

Sensor Signal Conditioning

age on the capacitor array. Switch S_C is then opened allowing the voltage at node A to move as the bit switches are manipulated. If S1, S2, S3, and S4 are all connected to ground, a voltage equal to $-A_{IN}$ appears at node A. Connecting S1 to V_{REF} adds a voltage equal to $V_{REF}/2$ to $-A_{IN}$. The comparator then makes the MSB bit decision, and the SAR either leaves S1 connected to V_{REF} or connects it to ground depending on the comparator output (which is high or low depending on whether the voltage at node A is negative or positive, respectively). A similar process is followed for the remaining two bits. At the end of the conversion interval, S1, S2, S3, S4, and S_{IN} are connected to A_{IN}, S_C is connected to ground, and the converter is ready for another cycle.

Note that the extra LSB capacitor (C/4 in the case of the 3-bit DAC) is required to make the total value of the capacitor array equal to 2C so that binary division is accomplished when the individual bit capacitors are manipulated.

The operation of the capacitor DAC (cap DAC) is similar to an R/2R resistive DAC. When a particular bit capacitor is switched to VREF, the voltage divider created by the bit capacitor and the total array capacitance (2C) adds a voltage to node A equal to the weight of that bit. When the bit capacitor is switched to ground, the same voltage is subtracted from node A.

Because of their popularity, successive approximation ADCs are available in a wide variety of resolutions, sampling rates, input and output options, package styles, and costs. It would be impossible to attempt to list all types, but Figure 4.3.5 shows a number of recent Analog Devices' SAR ADCs which are representative. Note that many devices are complete data acquisition systems with input multiplexers which allow a single ADC core to process multiple analog channels.

	RESOLUTION	SAMPLING RATE	POWER	CHANNELS
AD7472	12-BITS	1.5 MSPS	9 mW	1
AD7891	12-BITS	500 kSPS	85 mW	8
AD7858/59	12-BITS	200 kSPS	20 mW	8
AD7887/88	12-BITS	125 kSPS	3.5 mW	8
AD7856/57	14-BITS	285 kSPS	60 mW	8
AD974	16-BITS	200 kSPS	120 mW	4
AD7670	16-BITS	1 MSPS	250 mW	1

Figure 4.3.5: Resolution/conversion time comparison for representative single-supply SAR ADCs.

Chapter 4

While there are some variations, the fundamental timing of most SAR ADCs is similar and relatively straightforward (see Figure 4.3.6). The conversion process is initiated by asserting a CONVERT START signal. The CONVST signal is a negative-going pulse whose positive-going edge actually initiates the conversion. The internal sample-and-hold (SHA) amplifier is placed in the hold mode on this edge, and the various bits are determined using the SAR algorithm. The negative-going edge of the CONVST pulse causes the EOC or BUSY line to go high. When the conversion is complete, the BUSY line goes low, indicating the completion of the conversion process. In most cases the trailing edge of the BUSY line can be used as an indication that the output data is valid and can be used to strobe the output data into an external register. However, because of the many variations in terminology and design, the individual data sheet should always be consulted when using with a specific ADC.

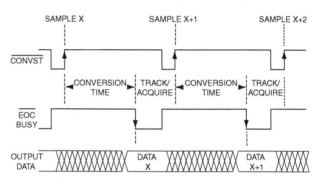

It should also be noted that some SAR ADCs require an external high frequency clock in addition to the CONVERT START command. In most cases, there is no need to synchronize the two. The frequency of the external clock, if required, generally falls in the

Figure 4.3.6: Typical SAR ADC timing.

range of 1 MHz to 30 MHz depending on the conversion time and resolution of the ADC. Other SAR ADCs have an internal oscillator which is used to perform the conversions and only require the CONVERT START command. Because of their architecture, SAR ADCs allow single-shot conversion at any repetition rate from DC to the converter's maximum conversion rate.

In a SAR ADC, the output data for a particular cycle is valid at the end of the conversion interval. In other ADC architectures, such as sigma-delta or the two- stage subranging architecture shown in Figure 4.3.7, this is not the case. The subranging ADC shown in the figure is a two-stage *pipelined* or subranging 12-bit converter. The first conversion is done by the 6-bit ADC which drives a 6-bit DAC. The output of the 6-bit DAC represents a 6-bit approximation to the analog input. Note that SHA2 delays the analog signal while the 6-bit ADC makes its decision and the 6-bit DAC settles. The DAC approximation is then subtracted from the analog signal from SHA2, amplified, and digitized by a 7-bit ADC. The outputs of the two conversions are combined, and the extra bit used to correct errors made in the first conversion. The typical timing associated with this type of converter is shown in Figure 4.3.8.

Sensor Signal Conditioning

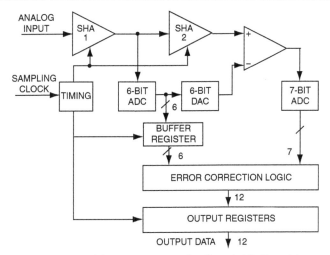

Figure 4.3.7: 12-bit two-stage pipelined ADC architecture.

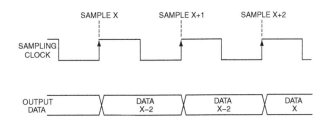

ABOVE SHOWS TWO CLOCK-CYCLES PIPELINE DELAY

Figure 4.3.8: Typical pipelined ADC timing.

Note that the output data presented immediately after sample X actually corresponds to sample X–2, i.e., there is a two clock-cycle "pipeline" delay. The pipelined ADC architecture is generally associated with high speed ADCs, and in most cases the pipeline delay, or *latency*, is not a major system problem in most applications where this type of converter is used.

Pipelined ADCs may have more than two clock-cycles latency depending on the particular architecture. For instance, the conversion could be done in three, or four, or perhaps even more pipelined stages causing additional latency in the output data.

Therefore, if the ADC is to be used in an event-triggered (or single-shot) mode where there must be a one-to-one time correspondence between each sample and the corresponding data, then the pipeline delay can be troublesome, and the SAR architecture is advantageous. Pipeline delay or latency can also be a problem in high speed

Chapter 4

servo-loop control systems or multiplexed applications. In addition, some pipelined converters have a *minimum* allowable conversion rate and must be kept running to prevent saturation of internal nodes.

Switched capacitor SAR ADCs generally have unbuffered input circuits similar to the circuit shown in Figure 4.3.9 for the AD7858/59 ADC. During the acquisition time, the analog input must charge the 20 pF equivalent input capacitance to the correct value. If the input is a DC signal, then the source resistance, R_S, in series with the 125 Ω internal switch resistance creates a time constant. In order to settle to 12-bit accuracy, approximately 9 time constants must be allowed for settling, and this defines the minimum allowable acquisition time. (Settling to 14-bits requires about 10 time constants, and 16-bits requires about 11).

$$T_{ACQ} > 9 \times (R_S + 125)\Omega \times 20 \text{ pF}.$$

For example, if $R_S = 50\ \Omega$, the acquisition time per the above formula must be at least 310 ns.

For AC applications, a low impedance source should be used to prevent distortion due to the non-linear ADC input circuit. In a single supply application, a fast settling rail-to-rail op amp such as the AD820 should be used. Fast settling allows the op amp to settle quickly from the transient currents induced on its input by the internal ADC switches. In Figure 4.3.9, the AD820 drives a lowpass filter consisting of the 50 Ω series resistor and the 10 nF capacitor (cutoff frequency approximately 320 kHz). This filter removes high frequency components which could result in aliasing and increased noise.

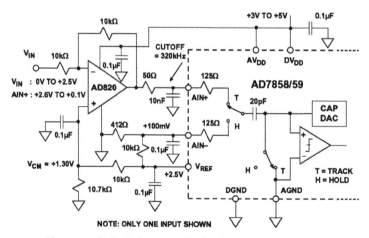

Figure 4.3.9: Driving switched capacitor inputs of AD7858/59 12-bit, 200 kSPS ADC.

Sensor Signal Conditioning

Using a single supply op amp in this application requires special consideration of signal levels. The AD820 is connected in the inverting mode and has a signal gain of −1. The noninverting input is biased at a common mode voltage of +1.3 V with the 10.7 kΩ/10 kΩ divider, resulting in an output voltage of +2.6 V for V_{IN} = 0 V, and +0.1 V for V_{IN} = +2.5 V. This offset is provided because the AD820 output cannot go all the way to ground, but is limited to the VCESAT of the output stage NPN transistor, which under these loading conditions is about 50 mV. The input range of the ADC is also offset by +100 mV by applying the +100mV offset from the 412 Ω/10 kΩ divider to the AIN− input.

The AD789X-family of single supply SAR ADCs (as well as the AD974, AD976, and AD977) includes a thin film resistive attenuator and level shifter on the analog input to allow a variety of input range options, both bipolar and unipolar. A simplified diagram of the input circuit of the AD7890-10 12-bit, 8-channel ADC is shown in Figure 4.3.10. This arrangement allows the converter to digitize a ±10V input while operating on a single +5 V supply. The R1/R2/R3 thin film network provides the attenuation and level shifting to convert the ±10 V input to a 0 V to +2.5 V signal which is digitized by the internal ADC. This type of input requires no special drive circuitry because R1 isolates the input from the actual converter circuitry. Nevertheless, the source resistance, R_S, should be kept reasonably low to prevent gain errors caused by the R_S/R1 divider.

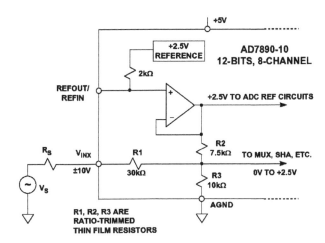

Figure 4.3.10: Driving single-supply ADCs with scaled inputs.

Chapter 4

SAR ADCs with Multiplexed Inputs

Multiplexing is a fundamental part of many data acquisition systems, and a fundamental understanding of multiplexers is required to design a data acquisition system. Switches for data acquisition systems, especially when integrated into the IC, generally are CMOS-types shown in Figure 4.3.11. Utilizing the P-Channel and N-Channel MOSFET switches in parallel minimizes the change of on-resistance (R_{ON}) as a function of signal voltage. On-resistance can vary from less than 5 Ω to several hundred ohms depending upon the device. Variation in on-resistance as a function of signal level (often called R_{ON}-modulation) can cause distortion if the multiplexer must drive a load, and therefore R_{ON} *flatness* is also an important specification.

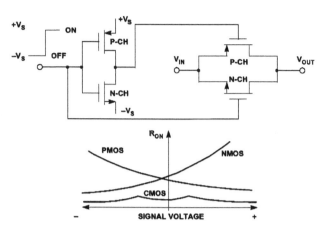

Figure 4.3.11: Basic CMOS analog switch.

Because of non-zero R_{ON} and R_{ON}-modulation, multiplexer outputs should be isolated from the load with a suitable buffer amplifier. A separate buffer is not required if the multiplexer drives a high input impedance, such as a PGA, SHA or ADC—but beware! Some SHAs and ADCs draw high frequency pulse current at their sampling rate and cannot tolerate being driven by an unbuffered multiplexer.

The key multiplexer specifications are *switching time*, *on-resistance*, *on-resistance flatness*, and *off-channel isolation, and crosstalk*. Multiplexer switching time ranges from less than 20 ns to over 1μs, RON from less than 5 Ω to several hundred ohms, and off-channel isolation from 50 to 90 dB.

A number of CMOS switches can be connected to form a multiplexer as shown in Figure 4.3.12. The number of input channels typically ranges from 4 to 16, and some multiplexers have in-

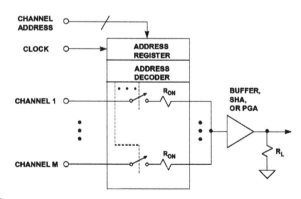

Figure 4.3.12: Simplified diagram of a typical analog multiplexer.

Sensor Signal Conditioning

ternal channel-address decoding logic and registers, while with others, these functions must be performed externally. Unused multiplexer inputs *must* be grounded or severe loss of system accuracy may result.

Complete Data Acquisition Systems on a Chip

VLSI mixed-signal processing allows the integration of large and complex data acquisition circuits on a single chip. Most signal conditioning circuits including multiplexers, PGAs, and SHAs, can now be manufactured on the same chip as the ADC. This high level of integration permits data acquisition systems (DASs) to be specified and tested as a single complex function.

Such functionality relieves the designer of most of the burden of testing and calculating error budgets. The DC and AC characteristics of a complete data acquisition system are specified as a complete function, which removes the necessity of calculating performance from a collection of individual worst case device specifications. A complete monolithic system should achieve a higher performance at much lower cost than would be possible with a system built up from discrete functions. Furthermore, system calibration is easier, and in fact many monolithic DASs are self calibrating, offering both internal and system calibration functions.

The AD7858 is an example of a highly integrated IC DAS (see Figure 4.3.13). The device operates on a single supply voltage of +3 V to +5.5 V and dissipates only 15 mW. The resolution is 12-bits, and the maximum sampling frequency is 200 kSPS. The input multiplexer can be configured either as eight single-ended inputs or four pseudo-differential inputs. The AD7858 requires an external 4 MHz clock and initiates the conversion on the positive-going edge of the CONVST pulse which does not need to be synchronized to the high frequency clock. Conversion can also be initiated via software by setting a bit in the proper control register.

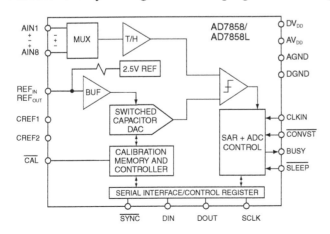

Figure 4.3.13: AD7858 12-bit, 200 kSPS 8-channel single-supply ADC.

Chapter 4

The AD7858 contains an on-chip 2.5 V reference (which can be overridden with an external one), and the fullscale input voltage range is 0 V to VREF. The internal DAC is a switched capacitor type, and the ADC contains a self-calibration and system calibration option to ensure accurate operation over time and temperature. The input/output port is a serial one and is SPI, QSPI, 8051, and µP compatible.

The AD7858L is a lower power (5.5 mW) version of the AD7858 which operates at a maximum sampling rate of 100 kSPS.

Sigma-Delta ($\Sigma\Delta$) Measurement ADCs

Sigma-delta analog-digital converters ($\Sigma\Delta$ ADCs) have been known for nearly thirty years, but only recently has the technology (high-density digital VLSI) existed to manufacture them as inexpensive monolithic integrated circuits. They are now used in many applications where a low-cost, low-bandwidth, low-power, high-resolution ADC is required.

There have been innumerable descriptions of the architecture and theory of $\Sigma\Delta$ ADCs, but most commence with a maze of integrals and deteriorate from there. In the Applications Department at Analog Devices, we frequently encounter engineers who do not understand the theory of operation of $\Sigma\Delta$ ADCs and are convinced, from study of a typical published article, that it is too complex to comprehend easily.

There is nothing particularly difficult to understand about $\Sigma\Delta$ ADCs, as long as you avoid the detailed mathematics, and this section has been written in an attempt to clarify the subject. A $\Sigma\Delta$ ADC contains very simple analog electronics (a comparator, a switch, and one or more integrators and analog summing circuits), and quite complex digital computational circuitry. This circuitry consists of a digital signal processor (DSP) which acts as a filter (generally, but not invariably, a low pass filter). It is not necessary to know precisely how the filter works to appreciate what it does. To understand how a $\Sigma\Delta$ ADC works, familiarity with the concepts of *over-sampling, quantization noise shaping, digital filtering,* and *decimation* is required.

Let us consider the technique of over-sampling with an analysis in the frequency domain. Where a DC conversion has a *quantization error* of up to ½ LSB, a

- Low Cost, High Resolution (to 24-bits) Excellent DNL,
- Low Power, but Limited Bandwidth
- Key Concepts are Simple, but Math is Complex
    - Oversampling
    - Quantization Noise Shaping
    - Digital Filtering
    - Decimation
- Ideal for Sensor Signal Conditioning
    - High Resolution
    - Self, System, and Auto Calibration Modes

Figure 4.3.14: Sigma-delta ADCs.

sampled data system has *quantization noise*. A perfect classical N-bit sampling ADC has an RMS quantization noise of q/√12 uniformly distributed within the Nyquist band of DC to $f_s/2$ (where q is the value of an LSB and f_s is the sampling rate) as shown in Figure 4.3.15A. Therefore, its SNR with a full-scale sinewave input will be (6.02N + 1.76) dB. If the ADC is less than perfect, and its noise is greater than its theoretical minimum quantization noise, then its *effective* resolution will be less than N-bits. Its actual resolution (often known as its effective number of bits or ENOB) will be defined by

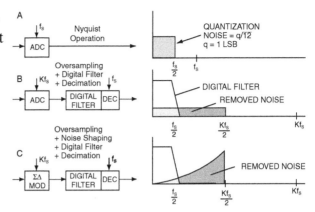

Figure 4.3.15: Oversampling, digital filtering, noise shaping, and decimation.

$$ENOB = \frac{SNR - 1.76\ dB}{6.02\ dB}$$

If we choose a much higher sampling rate, Kf_s (see Figure 4.3.15B), the quantization noise is distributed over a wider bandwidth DC to $Kf_s/2$. If we then apply a digital low pass filter (LPF) to the output, we remove much of the quantization noise, but do not affect the wanted signal—so the ENOB is improved. We have accomplished a high resolution A/D conversion with a low resolution ADC. The factor K is generally referred to as the *oversampling ratio*.

Since the bandwidth is reduced by the digital output filter, the output data rate may be lower than the original sampling rate (Kfs) and still satisfy the Nyquist criterion. This may be achieved by passing every Mth result to the output and discarding the remainder. The process is known as "decimation" by a factor of M. Despite the origins of the term (*decem* is Latin for ten), M can have any integer value, provided that the output data rate is more than twice the signal bandwidth. Decimation does not cause any loss of information (see Figure 4.3.15B).

If we simply use over-sampling to improve resolution, we must over-sample by a factor of 2^{2N} to obtain an N-bit increase in resolution. The ΣΔ converter does not need such a high over-sampling ratio because it not only limits the signal passband, but also shapes the quantization noise so that most of it falls outside this passband as shown in Figure 4.3.15C.

Chapter 4

If we take a 1-bit ADC (generally known as a comparator), drive it with the output of an integrator, and feed the integrator with an input signal summed with the output of a 1-bit DAC fed from the ADC output, we have a first-order ΣΔ modulator as shown in Figure 4.3.16. Add a digital low pass filter (LPF) and decimator at the digital output, and we have a ΣΔ ADC: the ΣΔ modulator shapes the quantization noise so that it lies above the passband of the digital output filter, and the ENOB is therefore much larger than would otherwise be expected from the over-sampling ratio.

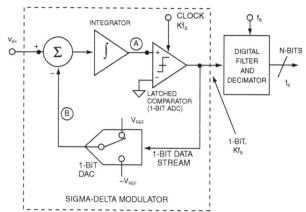

Figure 4.3.16: First order sigma-delta ADC.

Intuitively, a ΣΔ ADC operates as follows. Assume a DC input at V_{IN}. The integrator is constantly ramping up or down at node A. The output of the comparator is fed back through a 1-bit DAC to the summing input at node B. The negative feedback loop from the comparator output through the 1-bit DAC back to the summing point will force the average DC voltage at node B to be equal to V_{IN}. This implies that the average DAC output voltage must equal to the input voltage V_{IN}. The average DAC output voltage is controlled by the *ones-density* in the 1-bit data stream from the comparator output. As the input signal increases towards $+V_{REF}$, the number of "ones" in the serial bit stream increases, and the number of "zeros" decreases. Similarly, as the signal goes negative towards $-V_{REF}$, the number of "ones" in the serial bit stream decreases, and the number of "zeros" increases. From a very simplistic standpoint, this analysis shows that the average value of the input voltage is contained in the serial bit stream out of the comparator. The digital filter and decimator process the serial bit stream and produce the final output data.

The concept of noise shaping is best explained in the frequency domain by considering the simple ΣΔ modulator model in Figure 4.3.17.

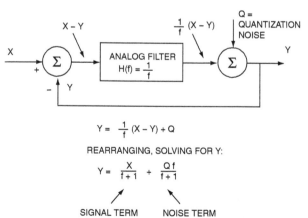

Figure 4.3.17: Simplified frequency domain linearized model of a sigma-delta modulator.

104

Sensor Signal Conditioning

The integrator in the modulator is represented as an analog lowpass filter with a transfer function equal to H(f) = 1/f. This transfer function has an amplitude response which is inversely proportional to the input frequency. The 1-bit quantizer generates quantization noise, Q, which is injected into the output summing block. If we let the input signal be X, and the output Y, the signal coming out of the input summer must be X – Y. This is multiplied by the filter transfer function, 1/f, and the result goes to one input to the output summer. By inspection, we can then write the expression for the output voltage Y as:

$$Y = \frac{1}{f}(X - Y) + Q$$

This expression can easily be rearranged and solved for Y in terms of X, f, and Q:

$$Y = \frac{X}{f+1} + \frac{Q \cdot f}{f+1}$$

Note that as the frequency f approaches zero, the output voltage Y approaches X with no noise component. At higher frequencies, the amplitude of the signal component decreases, and the noise component increases. At high frequency, the output consists primarily of quantization noise. In essence, the analog filter has a lowpass effect on the signal, and a highpass effect on the quantization noise. Thus the analog filter performs the noise shaping function in the ΣΔ modulator model.

For a given input frequency, higher order analog filters offer more attenuation. The same is true of ΣΔ modulators, provided certain precautions are taken.

By using more than one integration and summing stage in the ΣΔ modulator, we can achieve higher orders of quantization noise shaping and even better ENOB for a given over-sampling ratio as is shown in Figure 4.3.18 for both a first and second-order ΣΔ modulator. The block diagram for the second-order ΣΔ modulator is shown in Figure 4.3.19. Third, and higher, order ΣΔ ADCs were once thought to be potentially unstable at some values of input. Recent analyses using *finite* rather than infinite gains in the comparator have shown that this is not necessarily so, but even if instability does start to occur, it is not important, since the DSP in the digital filter and decimator can

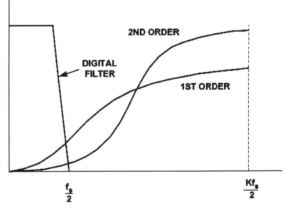

Figure 4.3.18: Sigma-delta modulators shape quantization noise.

105

be made to recognize incipient instability and react to prevent it.

Figure 4.3.20 shows the relationship between the order of the ΣΔ modulator and the amount of over-sampling necessary to achieve a particular SNR. For instance, if the oversampling ratio is 64, an ideal second-order system is capable of providing an SNR of about 80dB. This implies approximately 13 effective number of bits (ENOB). Although the filtering done by the digital filter and decimator can be done to any degree of precision desirable, it would be pointless to carry more than 13 binary bits to the outside world. Additional bits would carry no useful signal information, and would be buried in the quantization noise unless post-filtering techniques were employed.

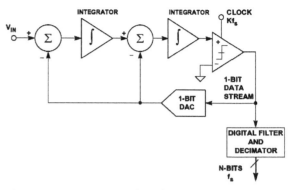

Figure 4.3.19: Second-order sigma-delta ADC.

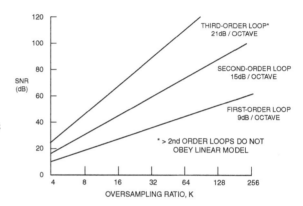

Figure 4.3.20: SNR versus oversampling ratio for first, second, and third-order loops.

The ΣΔ ADCs that we have described so far contain integrators, which are low pass filters, whose passband extends from DC. Thus, their quantization noise is pushed up in frequency. At present, most commercially available ΣΔ ADCs are of this type (although some which are intended for use in audio or telecommunications applications contain bandpass rather than lowpass digital filters to eliminate any system DC offsets). Sigma-delta ADCs are available with resolutions up to 24-bits for DC measurement applications (AD77XX-family), and with resolutions of 18-bits for high quality digital audio applications (AD1879).

But there is no particular reason why the filters of the ΣΔ modulator should be LPFs, except that traditionally ADCs have been thought of as being baseband devices, and that integrators are somewhat easier to construct than bandpass filters. If we replace the integrators in a ΣΔ ADC with bandpass filters (BPFs), the quantization noise is moved up and down in frequency to leave a virtually noise-free region in the passband (see Reference 1). If the digital filter is then programmed to have its pass-band in this region, we have a ΣΔ ADC with a bandpass, rather than a lowpass characteristic. Although studies of this architecture are in their infancy, such ADCs would seem to be ideally suited for use in digital radio receivers, medical ultrasound, and a number of other applications.

A $\Sigma\Delta$ ADC works by over-sampling, where simple analog filters in the $\Sigma\Delta$ modulator shape the quantization noise so that the SNR *in the bandwidth of interest* is much greater than would otherwise be the case, and by using high performance digital filters and decimation to eliminate noise outside the required passband. Because the analog circuitry is so simple and undemanding, it may be built with the same digital VLSI process that is used to fabricate the DSP circuitry of the digital filter. Because the basic ADC is 1-bit (a comparator), the technique is inherently linear.

Although the detailed analysis of $\Sigma\Delta$ ADCs involves quite complex mathematics, their basic design can be understood without the necessity of any mathematics at all. For further discussion on $\Sigma\Delta$ ADCs, refer to References 2 and 3.

High Resolution, Low-Frequency Sigma-Delta Measurement ADCs

The AD7710, AD7711, AD7712, AD7713, and AD7714, AD7730, and AD7731 are members of a family of sigma-delta converters designed for high accuracy, low frequency measurements. They have no missing codes to 24-bits, and their effective resolutions extend to 22.5 bits depending upon the device, update rate, programmed filter bandwidth, PGA gain, post-filtering, etc. They all use similar sigma-delta cores, and their main differences are in their analog inputs, which are optimized for different transducers. Newer members of the family, such as the AD7714, AD7730/7730L, and the AD7731/7731L are designed and specified for single supply operation.

There are also similar 16-bit devices available (AD7705, AD7706, AD7715) which also operate on single supplies.

The AD1555/AD1556 is a 24-bit two-chip $\Sigma\Delta$ modulator/filter specifically designed for seismic data acquisition systems. This combination yields a dynamic range of 120 dB. The AD1555 contains a PGA and a 4th-order $\Sigma\Delta$ modulator. The AD1555 outputs a serial 1-bit data stream to the AD1556 which contains the digital filter and decimator.

Chapter 4

4.4 Signal Conditioning High Impedance Sensors

Many popular sensors have output impedances greater than several MΩ, and the associated signal conditioning circuitry must be carefully designed to meet the challenges of low bias current, low noise, and high gain. A large portion of this section is devoted to the analysis of a photodiode preamplifier. This application points out many of the problems associated with high impedance sensor signal conditioning circuits and offers practical solutions which can be applied to practically all such sensors. Other examples of high impedance sensors discussed are piezoelectric sensors, charge output sensors, and charge coupled devices (CCDs).

- Photodiode Preamplifiers
- Piezoelectric Sensors
  - Accelerometers
  - Hydrophones
- Humidity Monitors
- pH Monitors
- Chemical Sensors
- Smoke Detectors
- Charge Coupled Devices and Contact Image Sensors for Imaging

Figure 4.4.1: High impedance sensors.

Photodiode Preamplifier Design

Photodiodes generate a small current which is proportional to the level of illumination. They have many applications ranging from precision light meters to high-speed fiber optic receivers.

- Optical: Light Meters, Auto-Focus, Flash Controls
- Medical: CAT Scanners (X-Ray Detection), Blood Particle Analyzers
- Automotive: Headlight Dimmers, Twilight Detectors
- Communications: Fiber Optic Receivers
- Industrial: Bar Code Scanners, Position Sensors, Laser Printers

Figure 4.4.2: Photodiode applications.

The equivalent circuit for a photodiode is shown in Figure 4.4.3. One of the standard methods for specifying the sensitivity of a photodiode is to state its short circuit photocurrent (I_{SC}) at a given light level from a well defined light source. The most commonly used source is an incandescent tungsten lamp running at a color temperature of 2850K. At 100 fc (footcandles) illumination (approximately the light level on an overcast day), the short circuit current is usually in the picoamps to hundreds of microamps range for small area (less than 1mm^2) diodes.

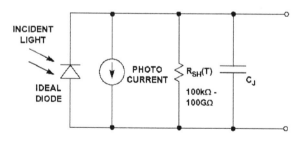

Figure 4.4.3: Photodiode equivalent circuit.

Sensor Signal Conditioning

The short circuit current is very linear over 6 to 9 decades of light intensity, and is therefore often used as a measure of absolute light levels. The open circuit forward voltage drop across the photodiode varies logarithmically with light level, but, because of its large temperature coefficient, the diode voltage is seldom used as an accurate measure of light intensity.

The shunt resistance R_{SH} is usually in the order of 1000 MΩ at room temperature, and decreases by a factor of two for every 10°C rise in temperature. Diode capacitance C_J is a function of junction area and the diode bias voltage. A value of 50 pF at zero bias is typical for small area diodes.

Photodiodes may either be operated with zero bias (*photovoltaic* mode, left) or reverse bias (*photoconductive* mode, right) as shown in Figure 4.4.4. The most precise linear operation is obtained in the photovoltaic mode, while higher switching speeds are realizable when the diode is operated in the photoconductive mode at the expense of linearity. Under these reverse bias conditions, a small amount of current called *dark current* will flow even when there is no illumination. There is no dark current in the photovoltaic mode. In the photovoltaic mode, the diode noise is basically the thermal noise generated by the shunt resistance. In the photoconductive mode, shot noise due to conduction is an additional source of noise. Photodiodes are usually optimized during the design process for use in either the photovoltaic mode or the photoconductive mode, but not both. Figure 4.4.5 shows the photosensitivity for a small photodiode (Silicon Detector Part Number SD-020-12-001), and specifications for the diode are summarized in Figure 4.4.6. This diode was chosen for the design example to follow.

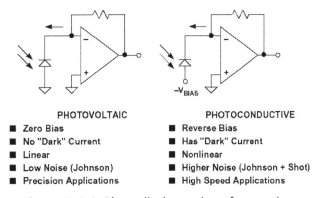

Figure 4.4.4: Photodiode modes of operation.

- Area: 0.2mm²
- Capacitance: 50pF
- Shunt Resistance @ 25°C: 1000MΩ
- Maximum Linear Output Current: 40µA
- Response Time: 12ns
- Photosensitivity: 0.03µA / foot candle (fc)

Figure 4.4.5: Photodiode specifications: silicon detector part number SD-020-12-001.

Chapter 4

A convenient way to convert the photodiode current into a usable voltage is to use an op amp as a current-to-voltage converter as shown in Figure 4.4.7. The diode bias is maintained at zero volts by the virtual ground of the op amp, and the short circuit current is converted into a voltage. At maximum sensitivity, the amplifier must be able to detect a diode current of 30 pA. This implies that the feedback resistor must be very large, and the amplifier bias current very small. For example, 1000 MΩ will yield a corresponding voltage of 30 mV for this amount of current. Larger resistor values are impractical, so we will use 1000 MΩ for the most sensitive range. This will give an output voltage range of 10 mV for 10pA

ENVIRONMENT	ILLUMINATION (fc)	SHORT CIRCUIT CURRENT
Direct Sunlight	1000	30μA
Overcast Day	100	3μA
Twilight	1	0.03μA
Full Moonlit Night	0.1	3000pA
Clear Night / No Moon	0.001	30pA

Figure 4.4.6: Short circuit current versus light intensity for photodiode (photovoltaic mode).

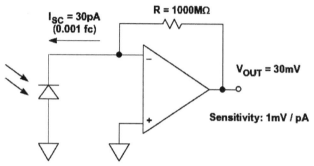

Figure 4.4.7: Current-to-voltage converter (simplified).

of diode current and 10 V for 10 nA of diode current. This yields a range of 60 dB. For higher values of light intensity, the gain of the circuit must be reduced by using a smaller feedback resistor. For this range of maximum sensitivity, we should be able to easily distinguish between the light intensity on a clear moonless night (0.001fc) and that of a full moon (0.1fc)!

Notice that we have chosen to get as much gain as possible from one stage, rather than cascading two stages. This is in order to maximize the signal-to-noise ratio (SNR). If we halve the feedback resistor value, the signal level decreases by a factor of 2, while the noise due to the feedback resistor (4 kTR · Bandwidth) decreases by only 2. This reduces the SNR by 3 dB, assuming the closed loop bandwidth remains constant. Later in the analysis, we will see that the resistors are one of the largest contributors to the overall output noise.

To accurately measure photodiode currents in the tens of picoamps range, the bias current of the op amp should be no more than a few picoamps. This narrows the choice considerably. The industry-standard OP07 is an ultra-low offset voltage (10 μV) bipolar op amp, but its bias current is 4 nA (4000 pA!). Even super-beta bipolar op amps with bias current compensation (such as the OP97) have bias currents on the order of 100 pA at room temperature, but may be suitable for very high temperature applications, as these currents do not double every 10°C rise like FETs. A FET-input electrometer-grade op amp is chosen for our photodiode preamp, since it must operate only over a limited temperature range. Figure 4.4.8 summarizes the performance of several popular "electrometer grade" FET input op amps. These devices are fabricated on a BiFET process and use P-Channel JFETs as the input stage (see Figure 4.4.9). The rest of the op amp circuit is designed using bipolar devices. The BiFET op amps are laser trimmed at the wafer level to minimize offset voltage and offset voltage drift. The offset voltage drift is minimized by first trimming the input stage for equal currents in the two JFETs which comprise the differential pair. A second trim of the JFET source resistors minimizes the input offset voltage. The AD795 was selected for the photodiode preamplifier, and its key specifications are summarized in Figure 4.4.10.

PART #	V_{OS}, MAX*	TC V_{OS}, MAX	I_B, MAX*	0.1Hz TO 10Hz NOISE	PACKAGE
AD549	250µV	5µV/°C	100fA	4µV p-p	TO-99
AD645	250µV	1µV/°C	1.5pA	2µV p-p	TO-99, DIP
AD795	250µV	3µV/°C	1pA	2.5µV p-p	SOIC, DIP

* 25°C SPECIFICATION

Figure 4.4.8: Low bias current precision BiFET op amps (electrometer grade).

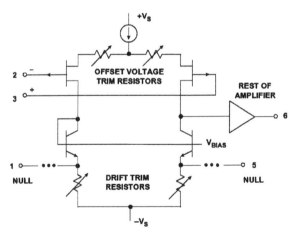

- Offset Voltage: 250µV Max. @ 25°C (K Grade)
- Offset Voltage Drift: 3µV / °C Max (K Grade)
- Input Bias Current: 1pA Max @ 25°C (K Grade)
- 0.1Hz to 10Hz Voltage Noise: 2.5µV p-p
- 1/f Corner Frequency: 12Hz
- Voltage Noise: 10nV / √Hz @ 100Hz
- Current Noise: 0.6fA / √Hz @ 100Hz
- 40mW Power Dissipation @ ±15V
- 1MHz Gain Bandwidth Product

Figure 4.4.9: BiFET op amp input stage.

Figure 4.4.10: AD795 BiFET op amp key specifications.

Since the diode current is measured in terms of picoamperes, extreme attention must be given to potential leakage paths in the actual circuit. Two parallel conductor stripes on a high-quality well-cleaned epoxy-glass PC board 0.05 inches apart running parallel for 1 inch have a leakage resistance of approximately 10^{11} ohms at +125°C. If there is 15 volts between these runs, there will be a current flow of 150 pA.

Chapter 4

The critical leakage paths for the photodiode circuit are enclosed by the dotted lines in Figure 4.4.11. The feedback resistor should be thin film on ceramic or glass with glass insulation. The compensation capacitor across the feedback resistor should have a polypropylene or polystyrene dielectric. All connections to the summing junction should be kept short. If a cable is used to connect the photodiode to the preamp, it should be kept as short as possible and have Teflon insulation.

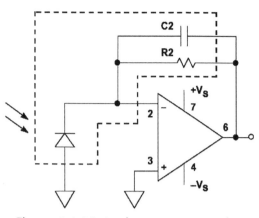

Figure 4.4.11: Leakage current paths.

Guarding techniques can be used to reduce parasitic leakage currents by isolating the amplifier's input from large voltage gradients across the PC board. Physically, a guard is a low impedance conductor that surrounds an input line and is raised to the line's voltage. It serves to buffer leakage by diverting it away from the sensitive nodes.

The technique for guarding depends on the mode of operation, i.e., inverting or non-inverting. Figure 4.4.12 shows a PC board layout for guarding the inputs of the AD795 op amp in the DIP ("N") package. Note that the pin spacing allows a trace to pass between the pins of this package. In the inverting mode, the guard traces surround the inverting input (pin 2) and run parallel to the input trace. In the follower mode, the guard voltage is the feedback voltage to pin 2, the inverting input. In both modes, the guard traces should be located on both sides of the PC board if at all possible and connected together.

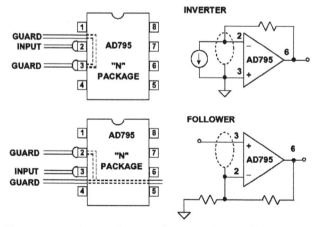

Figure 4.4.12: PCB layout for guarding DIP package.

Sensor Signal Conditioning

Things are slightly more complicated when using guarding techniques with the SOIC surface mount ("R") package because the pin spacing does not allow for PC board traces between the pins. Figure 4.4.13 shows the preferred method. In the SOIC "R" package, pins 1, 5, and 8 are "no connect" pins and can be used to route signal traces as shown. In the case of the follower, the guard trace must be routed around the $-V_S$ pin.

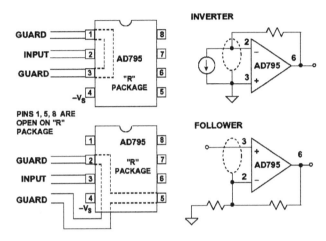

Figure 4.4.13: PCB layout for guarding SOIC package.

For extremely low bias current applications (such as using the AD549 with an input bias current of 100 fA), all connections to the input of the op amp should be made to a virgin Teflon standoff insulator ("Virgin" Teflon is a solid piece of new Teflon material which has been machined to shape and has not been welded together from powder or grains). If mechanical and manufacturing considerations allow, the inverting input pin of the op amp should be soldered directly to the Teflon standoff (see Figure 4.4.14) rather than going through a hole in the PC board. The PC board itself must be cleaned carefully and then sealed against humidity and dirt using a high quality conformal coating material.

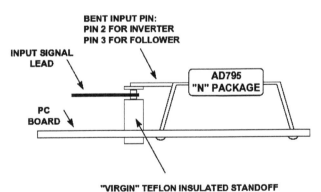

Figure 4.4.14: Input pin connected to "virgin" Teflon insulated standoff.

Chapter 4

In addition to minimizing leakage currents, the entire circuit should be well shielded with a grounded metal shield to prevent stray signal pickup.

Preamplifier Offset Voltage and Drift Analysis

An offset voltage and bias current model for the photodiode preamp is shown in Figure 4.4.15. There are two important considerations in this circuit. First, the diode shunt resistance (R1) is a function of temperature—it halves every time the temperature increases by 10°C. At room temperature (+25°C), R1 = 1000 MΩ, but at +70°C it decreases to 43 MΩ. This has a drastic impact on the circuit DC noise gain and hence the output offset voltage. In the example, at +25°C the DC noise gain is 2, but at +70°C it increases to 24.

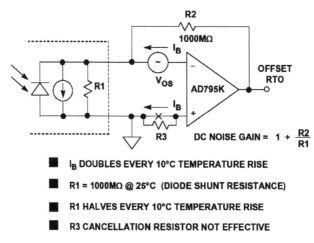

Figure 4.4.15: AD795 preamplifier DC offset errors.

The second difficulty with the circuit is that the input bias current doubles every 10°C rise in temperature. The bias current produces an output offset error equal to $I_B R2$. At +70°C the bias current increases to 24 pA compared to its room temperature value of 1 pA. Normally, the addition of a resistor (R3) between the non-inverting input of the op amp and ground having a value of R1∥R2 would yield a first-order cancellation of this effect. However, because R1 changes with temperature, this method is not effective. In addition, the bias current develops a voltage across the R3 cancellation resistor, which in turn is applied to the photodiode, thereby causing the diode response to become nonlinear.

Sensor Signal Conditioning

The total referred to output (RTO) offset voltage errors are summarized in Figure 4.4.16. Notice that at +70°C the total error is 33.24 mV. This error is acceptable for the design under consideration. The primary contributor to the error at high temperature is of course the bias current. Operating the amplifier at reduced supply voltages, minimizing output drive requirements, and heat sinking are some ways to reduce this error source. The addition of an external offset nulling circuit would minimize the error due to the initial input offset voltage.

	0°C	25°C	50°C	70°C
V_{OS}	0.325mV	0.250mV	0.325mV	0.385mV
Noise Gain	1.1	2	7	24
V_{OS} Error RTO	0.358mV	0.500mV	2.28mV	9.24mV
I_B	0.2pA	1.0pA	6.0pA	24pA
I_B Error RTO	0.2mV	1mV	6.0mV	24mV
Total Error RTO	0.558mV	1.50mV	8.28mV	33.24mV

Figure 4.4.16: AD795K preamplifier total output offset error.

Thermoelectric Voltages as Sources of Input Offset Voltage

Thermoelectric potentials are generated by electrical connections which are made between different metals at different temperatures. For example, the copper PC board electrical contacts to the kovar input pins of a TO-99 IC package can create an offset voltage of 40 µV/°C when the two metals are at different temperatures. Common lead-tin solder, when used with copper, creates a thermoelectric voltage of 1 to 3 µV/°C. Special cadmium-tin solders are available that reduce this to 0.3 µV/°C.

The solution to this problem is to ensure that the connections to the inverting and non-inverting input pins of the IC are made with the same material and that the PC board thermal layout is such that these two pins remain at the same temperature. In the case where a Teflon standoff is used as an insulated connection point for the inverting input (as in the case of the photodiode preamp), prudence dictates that connections to the non-inverting inputs be made in a similar manner to minimize possible thermoelectric effects.

Preamplifier AC Design, Bandwidth, and Stability

The key to the preamplifier AC design is an understanding of the circuit noise gain as a function of frequency. Plotting gain versus frequency on a log-log scale makes the analysis relatively simple (see Figure 4.4.17). This type of plot is also referred to as a Bode plot. The noise gain is the gain seen by a small voltage source in series with the op amp input terminals. It is also the same as the non-inverting signal gain (the gain from "A" to the output). In the photodiode preamplifier, the signal current from the photodiode passes through the C2/R2 network. It is important to distinguish between the signal gain and the noise gain, because it is the noise gain characteristic which determines stability regardless of where the actual signal is applied.

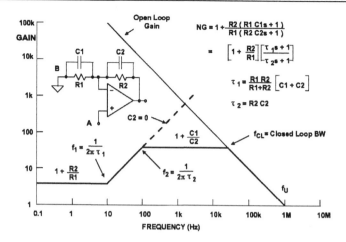

Figure 4.4.17: Generalized noise gain (NG) Bode plot.

Stability of the system is determined by the net slope of the noise gain and the open loop gain where they intersect. For unconditional stability, the noise gain curve must intersect the open loop response with a net slope of less than 12 dB/octave (20 dB per decade). The dotted line shows a noise gain which intersects the open loop gain at a net slope of 12dB/octave, indicating an unstable condition. This is what would occur in our photodiode circuit if there were no feedback capacitor (i.e., C2 = 0).

The general equations for determining the break points and gain values in the Bode plot are also given in Figure 4.4.17. A zero in the noise gain transfer function occurs at a frequency of $1/2\pi\tau_1$, where $\tau_1 = R1\|R2(C1 + C2)$. The pole of the transfer function occurs at a corner frequency of $1/2\pi\tau_2$, where $\tau_2 = R2C2$ which is also equal to the signal bandwidth if the signal is applied at point "B". At low frequencies, the noise gain is 1 + R2/R1. At high frequencies, it is 1 + C1/C2. Plotting the curve on the log- log graph is a simple matter of connecting the breakpoints with a line having a slope of 45°. The point at which the noise gain intersects the op amp open loop gain is called the *closed loop bandwidth*. Notice that the *signal bandwidth* for a signal applied at point "B" is much less, and is $1/2\pi R2C2$.

Figure 4.4.18 shows the noise gain plot for the photodiode preamplifier using the actual circuit values. The choice of C2 determines the actual signal bandwidth and also the phase margin. In the example, a signal bandwidth of 16 Hz was chosen. Notice that a smaller value of C2 would result in a higher signal bandwidth and a corresponding reduction in phase margin. It is also interesting to note that although the signal bandwidth is only 16 Hz, the closed loop bandwidth is 167 kHz. This will have important implications with respect to the output noise voltage analysis to follow.

Sensor Signal Conditioning

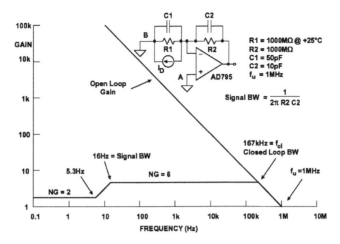

Figure 4.4.18: Noise gain of AD795 preamplifier at 25°C.

It is important to note that temperature changes do not significantly affect the stability of the circuit. Changes in R1 (the photodiode shunt resistance) only affect the low frequency noise gain and the frequency at which the zero in the noise gain response occurs. The high frequency noise gain is determined by the C1/C2 ratio.

Photodiode Preamplifier Noise Analysis

To begin the analysis, we consider the AD795 input voltage and current noise spectral densities shown in Figure 4.4.19. The AD795 performance is truly impressive for a JFET input op amp: 2.5 µV p-p 0.1 Hz to 10 Hz noise, and a 1/f corner frequency of 12 Hz, comparing favorably with all but the best bipolar op amps. As shown in the figure, the current noise is much lower than bipolar op amps, making it an ideal choice for high impedance applications.

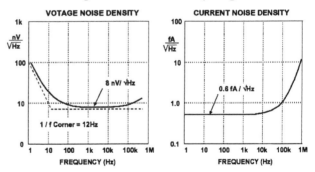

Figure 4.4.19: Voltage and current noise of AD795.

The complete noise model for an op amp is shown in Figure 4.4.20. This model includes the reactive elements C1 and C2. Each individual output noise contributor is calculated by integrating the square of its spectral density over the appropriate frequency bandwidth and then taking the square root:

$$\text{RMS Output Noise Due to } V_1 = \sqrt{\int V_1(f)^2 \, df} \qquad \text{Eq. 4.4.1}$$

Chapter 4

In most cases, this integration can be done by inspection of the graph of the individual spectral densities superimposed on a graph of the noise gain. The total output noise is then obtained by combining the individual components in a root-sum-squares manner. The table below the diagram in Figure 4.4.20 shows how each individual source is reflected to the output and the corresponding bandwidth for inte-

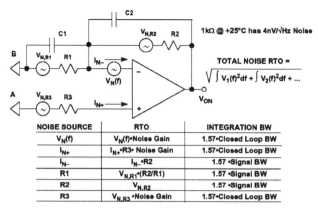

Figure 4.4.20: Amplifier noise model.

gration. The factor of 1.57 ($\pi/2$) is required to convert the single pole bandwidth into its equivalent noise bandwidth. The resistor Johnson noise spectral density is given by:

$$V_R = \sqrt{4kTR} \qquad \text{Eq. 4.4.2}$$

where k is Boltzmann's constant (1.38×10^{-23} J/K) and T is the absolute temperature in K. A simple way to compute this is to remember that the noise spectral density of a 1 kΩ resistor is 4 nV/$\sqrt{Hz}$ at +25°C. The Johnson noise of another resistor value can be found by multiplying by the square root of the ratio of the resistor value to 1000 Ω. Johnson noise is broadband, and its spectral density is constant with frequency.

Input Voltage Noise

In order to obtain the output voltage noise spectral density plot due to the input voltage noise, the input voltage noise spectral density plot is multiplied by the noise gain plot. This is easily accomplished using the Bode plot on a log-log scale. The total RMS output voltage noise due to the input voltage noise is then obtained by integrating the square of the output voltage noise spectral density plot and then taking the square root. In most cases, this integration may be approximated. A lower frequency limit of 0.01 Hz in the 1/f region is normally used. If the bandwidth of integration for the input voltage noise is greater than a few hundred Hz, the input voltage noise spectral density may be assumed to be constant. Usually, the value of the input voltage noise spectral density at 1 kHz will provide sufficient accuracy.

It is important to note that the input voltage noise contribution must be integrated over the entire closed loop bandwidth of the circuit (the closed loop bandwidth, f_{cl}, is the frequency at which the noise gain intersects the op amp open loop response). This is also true of the other noise contributors which are reflected to the output by the noise gain (namely, the non-inverting input current noise and the non-inverting input resistor noise).

Sensor Signal Conditioning

The inverting input noise current flows through the feedback network to produce a noise voltage contribution at the output The input noise current is approximately constant with frequency, therefore, the integration is accomplished by multiplying the noise current spectral density (measured at 1 kHz) by the noise bandwidth which is 1.57 times the signal bandwidth (1/2πR2C2). The factor of 1.57 (π/2) arises when single-pole 3 dB bandwidth is converted to equivalent noise bandwidth.

High Impedance Sensors

Johnson Noise Due to Feedforward Resistor R1

The noise current produced by the feedforward resistor R1 also flows through the feedback network to produce a contribution at the output. The noise bandwidth for integration is also 1.57 times the signal bandwidth.

Non-Inverting Input Current Noise

The non-inverting input current noise, I_{N+}, develops a voltage noise across R3 which is reflected to the output by the noise gain of the circuit. The bandwidth for integration is therefore the closed loop bandwidth of the circuit. However, there is no contribution at the output if R3 = 0 or if R3 is bypassed with a large capacitor which is usually desirable when operating the op amp in the inverting mode.

Johnson Noise Due to Resistor in Non-Inverting Input

The Johnson voltage noise due to R3 is also reflected to the output by the noise gain of the circuit. If R3 is bypassed sufficiently, it makes no significant contribution to the output noise.

Summary of Photodiode Circuit Noise Performance

Figure 4.4.21 shows the output noise spectral densities for each of the contributors at +25°C. Note that there is no contribution due to I_{N+} or R3 since the non-inverting input of the op amp is grounded.

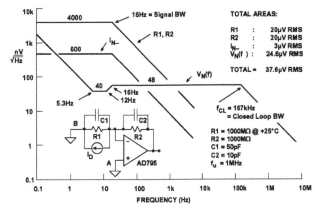

Figure 4.4.21: Ouput voltage noise components spectral densities (nV/√Hz) at +25°C.

Chapter 4

Noise Reduction Using Output Filtering

From the above analysis, the largest contributor to the output noise voltage at +25°C is the input voltage noise of the op amp reflected to the output by the noise gain. This contributor is large primarily because the noise gain over which the integration is performed extends to a bandwidth of 167 kHz (the intersection of the noise gain curve with the open-loop response of the op amp). If the op amp output is filtered by a single pole filter (as shown in Figure 4.4.22) with a 20 Hz cutoff frequency (R = 80 MΩ, C = 0.1 µF), this contribution is reduced to less than 1 µV rms. Notice that the same results would not be achieved simply by increasing the feedback capacitor, C2. Increasing C2 lowers the high frequency noise gain, but the integration bandwidth becomes proportionally higher. Larger values of C2 may also decrease the signal bandwidth to unacceptable levels. The addition of the simple filter reduces the output noise to 28.5 µV rms; approximately 75% of its former value. After inserting the filter, the resistor noise and current noise are now the largest contributors to the output noise.

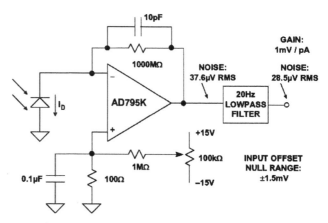

Figure 4.4.22: AD795 photodiode preamp with offset null adjustment.

Summary of Circuit Performance

The diagram for the final optimized design of the photodiode circuit is shown in Figure 4.4.22. Performance characteristics are summarized in Figure 4.4.23. The total output voltage drift over 0 to +70°C is 33 mV. This corresponds to 33 pA of diode current, or approximately 0.001 foot-candles. (The level of illumination on a clear moonless night). The offset nulling circuit shown on the non-inverting input can be used to null out the room temperature offset. Note that this method is better than using the offset null pins because using the offset null pins will increase the offset voltage TC by about 3 µV/°C for each millivolt nulled. In addition, the AD795 SOIC package does not have offset nulling pins.

- Output Offset Error (0°C to +70°C) : 33mV
- Output Sensitivity: 1mV / pA
- Output Photosensitivity: 30V / foot-candle
- Total Output Noise @ +25°C : 28.5µV RMS
- Total Noise RTI @ +25°C : 44fA RMS, or 26.4pA p-p
- Range with R2 = 1000MΩ : 0.001 to 0.33 foot-candles
- Bandwidth: 16Hz

Figure 4.4.23: AD795 photodiode circuit performance summary.

The input sensitivity based on a total output voltage noise of 44 µV is obtained by dividing the output voltage noise by the value of the feedback resistor R2. This yields a minimum detectable diode current of 44 fA. If a 12-bit ADC is used to digitize the 10 V full scale output, the weight of the least significant bit (LSB) is 2.5 mV. The output noise level is much less than this.

Photodiode Circuit Tradeoffs

There are many tradeoffs which could be made in the basic photodiode circuit design we have described. More signal bandwidth can be achieved in exchange for a larger output noise level. Reducing the feedback capacitor C2 to 1 pF increases the signal bandwidth to approximately 160 Hz. Further reductions in C2 are not practical because the parasitic capacitance is probably in the order of 1 to 2 pF. A small amount of feedback capacitance is also required to maintain stability.

If the circuit is to be operated at higher levels of illumination (greater than approximately 0.3 fc), the value of the feedback resistor can be reduced thereby resulting in further increases in circuit bandwidth and less resistor noise. If gain-ranging is to be used to measure the higher light levels, extreme care must be taken in the design and layout of the additional switching networks to minimize leakage paths.

Compensation of a High Speed Photodiode I/V Converter

A classical I/V converter is shown in Figure 4.4.24. Note that it is the same as the photodiode preamplifier if we assume that R1 >> R2. The total input capacitance, C1, is the sum of the diode capacitance and the op amp input capacitance. This is a classical second-order system, and the following guidelines can be applied in order to determine the proper compensation.

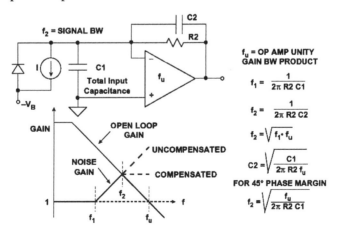

Figure 4.4.24: Compensating for input capacitance in a current-to-voltage converter.

The net input capacitance, C1, forms a zero at a frequency f_1 in the noise gain transfer function as shown in the Bode plot.

$$f_1 = \frac{1}{2\pi R2 C1} \qquad \text{Eq. 4.4.3}$$

Note that we are neglecting the effects of the compensation capacitor C2 and are assuming that it is small relative to C1 and will not significantly affect the zero frequency f_1 when it is added to the circuit. In most cases, this approximation yields results which are close enough, considering the other variables in the circuit.

If left uncompensated, the phase shift at the frequency of intersection, f_2, will cause instability and oscillation. Introducing a pole at f_2 by adding the feedback capacitor C2 stabilizes the circuit and yields a phase margin of about 45 degrees.

$$f_2 = \frac{1}{2\pi R2 C2} \qquad \text{Eq. 4.4.4}$$

Since f_2 is the geometric mean of f_1 and the unity-gain bandwidth frequency of the op amp, f_u,

$$f_2 = \sqrt{f_1 \cdot f_u} \qquad \text{Eq. 4.4.5}$$

These equations can be combined and solved for C2:

$$C_2 = \sqrt{\frac{C1}{2\pi R2 \cdot f_u}} \qquad \text{Eq. 4.4.6}$$

This value of C2 will yield a phase margin of about 45 degrees. Increasing the capacitor by a factor of 2 increases the phase margin to about 65 degrees.

In practice, the optimum value of C2 should be determined experimentally by varying it slightly to optimize the output pulse response.

Selection of the Op Amp for Wideband Photodiode I/V Converters

The op amp in the high speed photodiode I/V converter should be a wideband FET-input one in order to minimize the effects of input bias current and allow low values of photocurrents to be detected. In addition, if the equation for the 3 dB bandwidth, f_2, is rearranged in terms of f_u, R2, and C1, then

$$f_2 = \sqrt{\frac{f_u}{2\pi R2 C1}} \qquad \text{Eq. 4.4.7}$$

where C1 is the sum of the diode capacitance, C_D, and the op amp input capacitance, C_{IN}. In a high speed application, the diode capacitance will be much smaller than that of the low frequency preamplifier design previously discussed—perhaps as low as a few pF.

Sensor Signal Conditioning

By inspection of this equation, it is clear that in order to maximize f_2, the FET-input op amp should have both a high unity gain-bandwidth product, f_u, and a low input capacitance, C_{IN}. In fact, the ratio of f_u to C_{IN} is a good figure-of-merit when evaluating different op amps for this application.

Figure 4.4.25 compares a number of FET-input op amps suitable for photodiode preamps. By inspection, the AD823 op amp has the highest ratio of unity gain-bandwidth product to input capacitance, in addition to relatively low input bias current. For these reasons, it was chosen for the wideband photodiode preamp design.

	Unity GBW Product f_u (MHz)	Input Capacitance C_{IN} (pF)	f_u/C_{IN} (MHz/pF)	Input Bias Current I_B (pA)	Voltage Noise @ 10kHz (nV/√Hz)
AD823	16	1.8	8.9	3	16
AD843	34	6	5.7	600	19
AD744	13	5.5	2.4	100	16
AD845	16	8	2	500	18
OP42	10	6	1.6	100	12
AD745*	20	20	1	250	2.9
AD795	1	1	1	1	8
AD820	1.9	2.8	0.7	2	13
AD743	4.5	20	0.2	250	2.9

*Stable for Noise Gains ≥ 5, Usually the Case, Since High Frequency Noise Gain = 1 + C1/C2, and C1 Usually ≥ 4C2

Figure 4.4.25: FET-input op amp comparison table for wide bandwidth photodiode preamps.

High Speed Photodiode Preamp Design

The HP 5082-4204 PIN Photodiode will be used as an example for our discussion. Its characteristics are given in Figure 4.4.26. It is typical of many commercially available PIN photodiodes. As in most high-speed photodiode applications, the diode is operated in the reverse-biased or *photoconductive* mode. This greatly lowers the diode junction capacitance, but causes a small amount of *dark current* to flow even when the diode is not illuminated (we will show a circuit which compensates for the dark current error later in the section).

- Sensitivity: 350µA @ 1mW, 900nm
- Maximum Linear Output Current: 100µA
- Area: 0.002cm² (0.2mm²)
- Capacitance: 4pF @ 10V Reverse Bias
- Shunt Resistance: $10^{11}\Omega$
- Risetime: 10ns
- Dark Current: 600pA @ 10V Reverse Bias

Figure 4.4.26: HP 5082-4204 photodiode.

This photodiode is linear with illumination up to approximately 50 to 100 µA of output current. The dynamic range is limited by the total circuit noise and the diode dark current (assuming no dark current compensation).

Using the circuit shown in Figure 4.4.27, assume that we wish to have a full scale output of 10V for a diode current of 100 µA. This determines the value of the feedback resistor R2 to be 10 V/100 µA = 100 kΩ.

Chapter 4

Using the diode capacitance, $C_D = 4$ pF, and the AD823 input capacitance, $C_{IN} = 1.8$ pF, the value of $C1 = C_D + C_{IN} = 5.8$ pF. Solving the above equations using $C1 = 5.8$ pF, $R2 = 100$ kΩ, and $f_u = 16$ MHz, we find that:

f_1 = 274 kHz
$C2$ = 0.76 pF
f_2 = 2.1 MHz.

In the final design (Figure 4.4.27), note that the 100 kΩ resistor is replaced with three 33.2 kΩ film resistors to minimize stray capacitance. The feedback capacitor, C2, is a variable 1.5 pF ceramic and is adjusted in the final circuit for best bandwidth/pulse response. The overall circuit bandwidth is approximately 2 MHz.

The full-scale output voltage of the preamp for 100 μA diode current is 10V, and the error (RTO) due to the photodiode dark current of 600 pA is 60 mV. The dark current error can be canceled using a second photodiode of the same type in the non-inverting input of the op amp as shown in Figure 4.4.27.

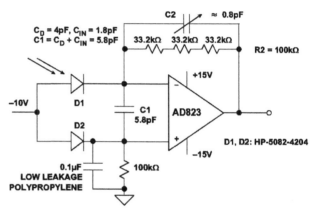

Figure 4.4.27: 2 MHz bandwidth photodiode preamp with dark current compensation.

High Speed Photodiode Preamp Noise Analysis

As in most noise analyses, only the key contributors need be identified. Because the noise sources combine in an RSS manner, any single noise source that is at least three or four times as large as any of the others will dominate.

In the case of the wideband photodiode preamp, the dominant sources of output noise are the input voltage noise of the op amp, V_N, and the resistor noise due to R2, $V_{N,R2}$ (see Figure 4.4.28). The input current noise of the FET-input op amp is negligible. The shot noise of the photodiode (caused by the reverse bias) is negligible because of the filtering effect of the shunt capacitance C1. The resistor noise is easily calculated by knowing that a 1 kΩ resistor generates about 4 nV/√Hz, therefore, a 100 kΩ resistor generates 40 nV/√Hz. The bandwidth for integration is the signal bandwidth, 2.1 MHz, yielding a total output rms noise of:

Sensor Signal Conditioning

$$V_{N,R2} \text{ RTO Noise} = 40\sqrt{1.57 \cdot 2.1 \cdot 10^6} = 73\mu Vrms \qquad \text{Eq. 4.4.8}$$

The factor of 1.57 converts the approximate single-pole bandwidth of 2.1 MHz into the *equivalent noise bandwidth*.

The output noise due to the input voltage noise is obtained by multiplying the noise gain by the voltage noise and integrating the entire function over frequency. This would be tedious if done rigorously, but a few reasonable approximations can be made which greatly simplify the math. Obviously, the low frequency 1/f noise can be neglected in the case of the wideband circuit. The primary source of output noise is due to the high-frequency noise-gain peaking which occurs between f_1 and f_u. If we simply assume that the output noise is constant over the entire range of frequencies and use the maximum value for AC noise gain [1 + (C1/C2)], then

$$V_N \text{ RTO Noise} \approx V_N\left(1+\frac{C1}{C2}\right)\sqrt{1.57 f_2} = 250\mu Vrms \qquad \text{Eq. 4.4.9}$$

The total rms noise referred to the output is then the RSS value of the two components:

$$\text{Total RTO Noise} = \sqrt{(73)^2 + (250)^2} = 260\mu Vrms \qquad \text{Eq. 4.4.10}$$

The total output dynamic range can be calculated by dividing the full scale output signal (10 V) by the total output rms noise, 260 µV rms, and converting to dB, yielding approximately 92 dB.

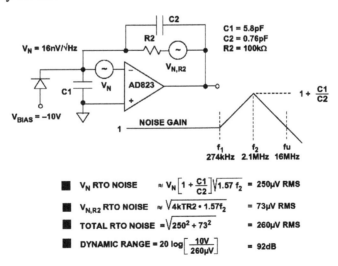

Figure 4.4.28: Equivalent circuit for output noise analysis

Chapter 4

High Impedance Charge Output Sensors

High impedance transducers such as piezoelectric sensors, hydrophones, and some accelerometers require an amplifier which converts a transfer of charge into a change of voltage. Because of the high DC output impedance of these devices, appropriate buffers are required. The basic circuit for an inverting charge sensitive amplifier is shown in Figure 4.4.29. There are basically two types of charge transducers: capacitive and charge-emitting. In a capacitive transducer, the voltage across the capacitor (V_C) is held constant. The change in capacitance, ΔC, produces a change in charge, $\Delta Q = \Delta C V_C$. This charge is transferred to the op amp output as a voltage, $\Delta V_{OUT} = -\Delta Q/C2 = -\Delta C V_C/C2$.

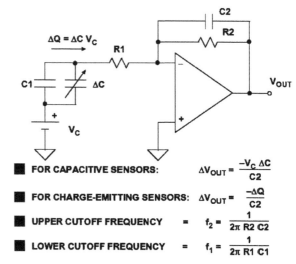

Figure 4.4.29: Charge amplifier for capacitive sensor.

Charge-emitting transducers produce an output charge, ΔQ, and their output capacitance remains constant. This charge would normally produce an open-circuit output voltage at the transducer output equal to $\Delta Q/C$. However, since the voltage across the transducer is held constant by the virtual ground of the op amp (R1 is usually small), the charge is transferred to capacitor C_2 producing an output voltage $\Delta V_{OUT} = -\Delta Q/C2$.

In an actual application, the charge amplifier only responds to AC inputs. The upper cutoff frequency is given by $f_2 = 1/2\pi R2C2$, and the lower by $f_1 = 1/2\pi R1C1$.

Low Noise Charge Amplifier Circuit Configurations

Figure 4.4.30 shows two ways to buffer and amplify the output of a charge output transducer. Both require using an amplifier which has a very high input impedance, such as the AD745. The AD745 provides both low voltage and low current noise. This combination makes this device particularly suitable in applications requiring very high charge sensitivity, such as capacitive accelerometers and hydrophones.

The first circuit (left) in Figure 4.4.30 uses the op amp in the inverting mode. Amplification depends on the principle of conservation of charge at the inverting input of

Sensor Signal Conditioning

the amplifier. The charge on capacitor C_S is transferred to capacitor C_F, thus yielding an output voltage of $\Delta Q/C_F$. The amplifier's input voltage noise will appear at the output amplified by the AC noise gain of the circuit, $1 + C_S/C_F$.

The second circuit (right) shown in Figure 4.4.30 is simply a high impedance follower with gain. Here the noise gain $(1 + R2/R1)$ is the same as the gain from the transducer to the output.

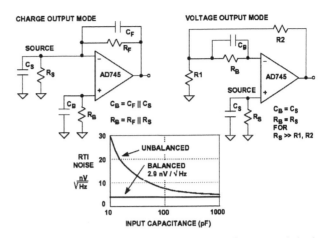

Figure 4.4.30: Balancing source impedances minimizes effects of bias currents and reduces input noise.

Resistor R_B, in both circuits, is required as a DC bias current return.

To maximize DC performance over temperature, the source resistances should be balanced on each input of the amplifier. This is represented by the resistor R_B shown in Figure 4.4.30. For best noise performance, the source capacitance should also be balanced with the capacitor C_B. In general, it is good practice to balance the source impedances (both resistive and reactive) as seen by the inputs of a precision low noise BiFET amplifiers such as the AD743/AD745. Balancing the resistive high impedance sensors components will optimize DC performance over temperature because balancing will mitigate the effects of any bias current errors. Balancing the input capacitance will minimize AC response errors due to the amplifier's nonlinear common mode input capacitance, and as shown in Figure 4.4.30, noise performance will be optimized. In any FET input amplifier, the current noise of the internal bias circuitry can be coupled to the inputs via the gate-to-source capacitances (20 pF for the AD743 and AD745) and appears as excess input voltage noise. This noise component is correlated at the inputs, so source impedance matching will tend to cancel out its effect. Figure 4.4.30 shows the required external components for both inverting and noninverting configurations. For values of C_B greater than 300 pF, there is a diminishing impact on noise, and C_B can then be simply a large mylar bypass capacitor of 0.01 µF or greater.

Chapter 4

A 40dB Gain Piezoelectric Transducer Amplifier Operates on Reduced Supply Voltages for Lower Bias Current

Figure 4.4.31 shows a piezoelectric transducer amplifier connected in the voltage-output mode. Reducing the power supplies to +5 V reduces the effects of bias current in two ways: first, by lowering the total power dissipation and, second, by reducing the basic gate-to-junction leakage current. The addition of a clip-on heat sink such as the Aavid #5801 will further limit the internal junction temperature rise.

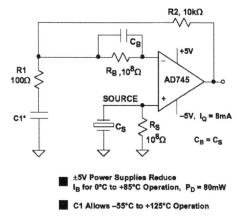

Without the AC coupling capacitor C1, the amplifier will operate over a range of 0°C to +85°C. If the optional AC coupling capacitor C1 is used, the circuit will operate over the entire –55°C to +125°C temperature range, but DC information is lost.

Figure 4.4.31: Gain of 100 piezoelectric sensor amplifier.

Hydrophones

Interfacing the outputs of highly capacitive transducers such as hydrophones, some accelerometers, and condenser microphones to the outside world presents many design challenges. Previously designers had to use costly hybrid amplifiers consisting of discrete low-noise JFETs in front of conventional op amps to achieve the low levels of voltage and current noise required by these applications. Now, using the AD743 and AD745, designers can achieve almost the same level of performance of the hybrid approach in a monolithic solution.

In sonar applications, a piezo-ceramic cylinder is commonly used as the active element in the hydrophone. A typical cylinder has a nominal capacitance of around 6,000 pF with a series resistance of 10 Ω. The output impedance is typically 10^8 Ω or 100 MΩ.

Since the hydrophone signals of interest are inherently AC with wide dynamic range, noise is the overriding concern among sonar system designers. The noise floor of the hydrophone and the hydrophone preamplifier together limit the sensitivity of the system and therefore the overall usefulness of the hydrophone. Typical hydrophone bandwidths are in the 1 kHz to 10 kHz range. The AD743 and AD745 op amps, with their low noise figures of 2.9 nV/Hz and high input impedance of 10^{10} Ω (or 10 GΩ) are ideal for use as hydrophone amplifiers.

Sensor Signal Conditioning

The AD743 and AD745 are companion amplifiers with different levels of internal compensation. The AD743 is internally compensated for unity gain stability. The AD745, stable for noise gains of five or greater, has a much higher bandwidth and slew rate. This makes the AD745 especially useful as a high-gain preamplifier where it provides both high gain and wide bandwidth. The AD743 and AD745 also operate with extremely low levels of distortion: less than 0.0003% and 0.0002% (at 1 kHz), respectively.

Op Amp Performance: JFET versus Bipolar

The AD743 and AD745 op amps are the first monolithic JFET devices to offer the low input voltage noise comparable to a bipolar op amp without the high input bias currents typically associated with bipolar op amps. Figure 4.4.32 shows input voltage noise versus input source resistance of the bias-current compensated OP27 and the JFET-input AD745 op amps. Note that the noise levels of the AD743 and the AD745 are identical. From this figure, it is clear that at high source impedances, the low current noise of the AD745 also provides lower overall noise than a high performance bipolar op amp. It is also important to note that, with the AD745, this noise reduction extends all the way down to low source impedances. At high source impedances, the lower DC current errors of the AD745 also reduce errors due to offset and drift as shown in Figure 4.4.32.

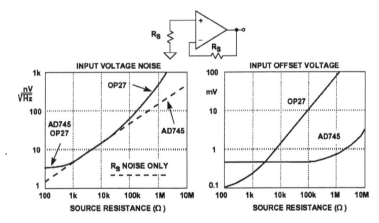

Figure 4.4.32: Effects of source resistance on noise and offset voltage for OP27 (bipolar) and AD745 (BiFET) op amps.

Chapter 4

A PH Probe Buffer Amplifier

A typical pH probe requires a buffer amplifier to isolate its 10^6 to 10^9 Ω source resistance from external circuitry. Such an amplifier is shown in Figure 4.4.33. The low input current of the AD795 allows the voltage error produced by the bias current and electrode resistance to be minimal. The use of guarding, shielding, high insulation resistance standoffs, and other such standard picoamp methods used to minimize leakage are all needed to maintain the accuracy of this circuit.

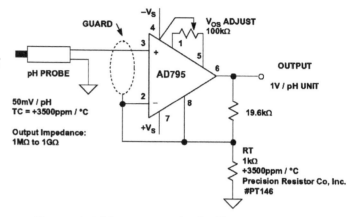

The slope of the pH probe transfer function, 50mV per pH unit at room temperature, has an approximate +3500 ppm/°C temperature coefficient. The buffer shown in Figure 4.4.33 provides a gain of 20 and yields an output voltage equal to 1 volt/pH unit. Temperature compensation is provided by resistor RT which is a special temperature compensation resistor, 1 kΩ, 1%, +3500 ppm/°C, #PT146 available from Precision Resistor Co., Inc. (Reference 18).

Figure 4.4.33: A pH probe buffer amplifier with a gain of 20 using the AD795 precision BiFET op amp.

CCD/CIS Image Processing

The *charge-coupled-device* (CCD) and *contact-image-sensor* (CIS) are widely used in consumer imaging systems such as scanners and digital cameras. A generic block diagram of an imaging system is shown in Figure 4.4.34. The imaging sensor (CCD, CMOS, or CIS) is exposed to the image or picture much like film is exposed in a camera. After exposure, the output of the sensor undergoes some analog signal processing and then is digitized by an ADC. The bulk of the actual image processing is performed using fast digital signal processors. At this point, the im-

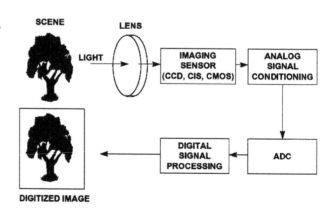

Figure 4.4.34: Generic imaging system for scanners or digital cameras.

Sensor Signal Conditioning

age can be manipulated in the digital domain to perform such functions as contrast or color enhancement/correction, etc.

The building blocks of a CCD are the individual light sensing elements called pixels (see Figure 4.4.35). A single pixel consists of a photo sensitive element, such as a photodiode or photocapacitor, which outputs a charge (electrons) proportional to the light (photons) that it is exposed to. The charge is accumulated during the exposure or integration time, and then the charge is transferred to the CCD shift register to be sent to the output of the device. The amount of accumulated charge will depend on the light level, the integration time, and the quantum efficiency of the photo sensitive element. A small amount of charge will accumulate even without light present; this is called dark signal or dark current and must be compensated for during the signal processing.

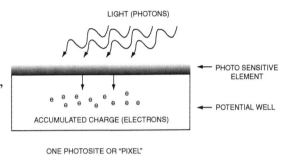

Figure 4.4.35: Light sensing element.

The pixels can be arranged in a linear or area configuration as shown in Figure 4.4.36. Clock signals transfer the charge from the pixels into the analog shift registers, and then more clocks are applied to shift the individual pixel charges to the output stage of the CCD.

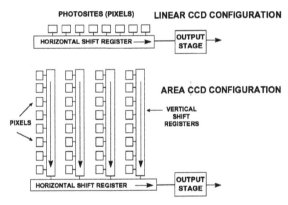

Figure 4.4.36: Linear and area CCD arrays.

Scanners generally use the linear configuration, while digital cameras use the area configuration. The analog shift register typically operates at frequencies between 1 and 10 MHz for linear sensors, and 5 to 25 MHz for area sensors.

A typical CCD output stage is shown in Figure 4.4.37 along with the associated voltage waveforms. The output stage of the CCD converts the charge of each pixel to a voltage via the sense capacitor, C_S. At the start of each pixel period, the voltage on C_S is reset to the reference level, V_{REF} causing a reset glitch to occur. The amount of light sensed by each pixel is measured by the difference between the reference and the video level, ΔV. CCD charges may be as low as 10 electrons, and a typical CCD output has a sensitivity of 0.6 μV/electron. Most CCDs have a saturation output voltage

131

Chapter 4

of about 500 mV to 1 V for area sensors and 2 V to 4 V for linear sensors. The DC level of the waveform is between 3 to 7 V.

Since CCDs are generally fabricated on CMOS processes, they have limited capability to perform on-chip signal conditioning. Therefore the CCD output is generally processed by external conditioning circuits. The nature of the CCD output requires that it be clamped before being digitized by the ADC. In addition, offset and gain functions are generally part of the analog signal processing.

CCD output voltages are small and quite often buried in noise. The largest source of noise is the thermal noise in the resistance of the FET reset switch. This noise may have a typical value of 100 to 300 electrons rms (approximately 60 to 180 mV rms). This noise, called "kT/C" noise, is illustrated in Figure 4.4.38. During the reset interval, the storage capacitor C_S is connected to V_{REF} via a CMOS switch. The on-resistance of the switch (R_{ON}) produces thermal noise given by the well known equation:

$$Thermal\ Noise = \sqrt{4kT \cdot BW \cdot R_{ON}} \qquad \text{Eq. 4.4.11}$$

The noise occurs over a finite bandwidth determined by the $R_{ON} C_S$ time constant. This bandwidth is then converted into equivalent noise bandwidth by multiplying the single-pole bandwidth by π/2 (1.57):

$$Noise\ BW = \frac{\pi}{2}\left[\frac{1}{2\pi R_{ON} C_S}\right] = \frac{1}{4 R_{ON} C_S} \qquad \text{Eq. 4.4.12}$$

Substituting into the formula for the thermal noise, note that the R_{ON} factor cancels, and the final expression for the thermal noise becomes:

$$Thermal\ Noise \sqrt{\frac{kT}{C}} \qquad \text{Eq. 4.4.13}$$

This is somewhat intuitive, because smaller values of R_{ON} decrease the thermal noise but increase the noise bandwidth, so only the capacitor value determines the noise.

Note that when the reset switch opens, the kT/C noise is stored on C_S and remains constant until the next reset interval. It therefore occurs as a *sample-to-sample* variation in the CCD output level and is common to both the reset level and the video level for a given pixel period.

A technique called *correlated double sampling* (CDS) is often used to reduce the effect of this noise. Figure 4.4.39 shows one circuit implementation of the CDS scheme, though many other implementations exist. The CCD output drives both SHAs. At the end of the reset interval, SHA1 holds the reset voltage level plus the kT/C noise.

Sensor Signal Conditioning

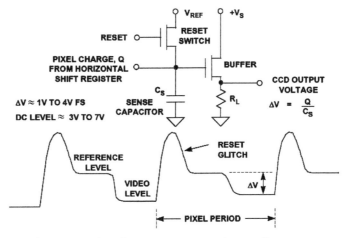

Figure 4.4.37: Output stage and waveforms.

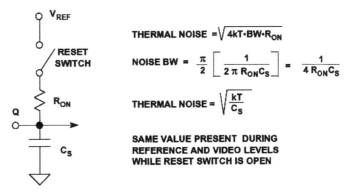

Figure 4.4.38: kT/C noise.

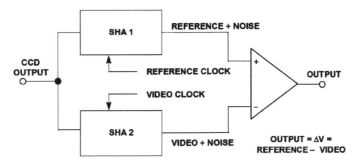

Figure 4.4.39: Correlated double sampling (CDS).

133

Chapter 4

At the end of the video interval, SHA2 holds the video level plus the kT/C noise. The SHA outputs are applied to a difference amplifier which subtracts one from the other. In this scheme, there is only a short interval during which both SHA outputs are stable, and their difference represents ΔV, so the difference amplifier must settle quickly. Note that the final output is simply the difference between the reference level and the video level, ΔV, and that the kT/C noise is removed.

Contact image sensors (CIS) are linear sensors often used in facsimile machines and low-end document scanners instead of CCDs. Although a CIS does not offer the same potential image quality as a CCD, it does offer lower cost and a more simplified optical path. The output of a CIS is similar to the CCD output except that it is referenced to or near ground (see Figure 4.4.40), eliminating the need for a clamping function. Furthermore, the CIS output does not contain correlated reset noise within each pixel period, eliminating the need for a CDS function. Typical CIS output voltages range from a few hundred mV to about 1 V full scale. Note that although a clamp and CDS is not required, the CIS waveform must be sampled by a sample-and-hold before digitization.

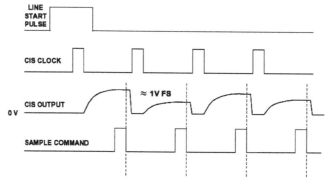

Figure 4.4.40: Contact image sensor (CIS) waveforms.

Analog Devices offers several *analog-front-end* (AFE) integrated solutions for the scanner, digital camera, and camcorder markets. They all comprise the signal processing steps described above. Advances in process technology and circuit topologies have made this level of integration possible in foundry CMOS without sacrificing performance. By combining successful ADC architectures with high performance CMOS analog circuitry, it is possible to design complete low cost CCD/CIS signal processing ICs.

The AD9816 integrates an analog-front-end (AFE) that integrates a 12-bit, 6 MSPS ADC with the analog circuitry needed for three-channel (RGB) image processing and sampling (see Figure 4.4.41). The AD9816 can be programmed through a serial interface, and includes offset and gain adjustments that gives users the flexibility to

Sensor Signal Conditioning

perform all the signal processing necessary for applications such as mid- to high-end desktop scanners, digital still cameras, medical x-rays, security cameras, and any instrumentation applications that must "read" images from CIS or CCD sensors.

The signal chain of the AD9816 consists of an input clamp, correlated double sampler (CDS), offset adjust DAC, programmable gain amplifier (PGA), and the 12-bit ADC core with serial interfacing to the external DSP. The CDS and clamp functions can be disabled for CIS applications.

The AD9814, takes the level of performance a step higher. For the most demanding applications, the AD9814 offers the same basic functionality as the AD9816 but with 14-bit performance. As with the AD9816, the signal path includes three input channels, each with input clamping, CDS, offset adjustment, and programmable gain. The three channels are multiplexed into a high performance 14-bit 6 MSPS ADC. High-end document and film scanners can benefit from the AD9814's combination of performance and integration.

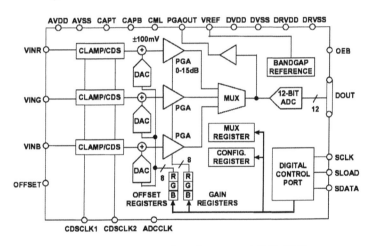

Figure 4.4.41: AD9816 Analog front end CCD/CIS processor.

Figure 4.4.42: AD9816 key specifications.

- Complete 12-Bit 6MSPS CCD/CIS Signal Processor
- 3-Channel or 1-Channel Operation
- On-Chip Correlated Double Sampling (CDS)
- 8-Bit Programmable Gain and 8-Bit Offset Adjustment
- Internal Voltage Reference
- Good Linearity: DNL = ±0.4LSB Typical, INL = ±1.5 LSB Typical
- Low Output Noise: 0.5 LSB RMS
- Coarse Offset Removal for CIS Applications
- 3-Wire Serial Interface
- Single +5V Supply, 420mW Power Dissipation
- 44-Lead MQFP Package

References

1. Ramon Pallas-Areny and John G. Webster, **Sensors and Signal Conditioning**, John Wiley, New York, 1991.
2. Dan Sheingold, Editor, **Transducer Interfacing Handbook**, Analog Devices, Inc., 1980.
3. Walt Kester, Editor, **1992 Amplifier Applications Guide**, Section 3, Analog Devices, Inc., 1992.
4. Walt Kester, Editor, **System Applications Guide**, Analog Devices, Inc., 1993.
5. Walt Kester, Editor, **Linear Design Seminar**, Analog Devices, 1994.
6. Walt Kester, Editor, **Practical Analog Design Techniques**, Analog Devices, 1994.
7. Walt Kester, Editor, **High Speed Design Techniques**, Analog Devices, 1996.
8. Thomas M. Fredrickson, **Intuitive Operational Amplifiers**, McGraw-Hill, 1988.
9. **Optoelectronics Data Book**, EG&G Vactec, St. Louis, MO, 1990.
10. Silicon Detector Corporation, Camarillo, CA, Part Number SD-020-12-001 Data Sheet.
11. **Photodiode 1991 Catalog**, Hamamatsu Photonics, Bridgewater, NJ.
12. *An Introduction to the Imaging CCD Array*, Technical Note 82W-4022, Tektronix, Inc., Beaverton, OR., 1987.
13. Lewis Smith and Dan Sheingold, *Noise and Operational Amplifier Circuits*, **Analog Dialogue 25th Anniversary Issue**, pp. 19-31, Analog Devices, 1991.
14. James L. Melsa and Donald G. Schultz, **Linear Control Systems**, pp. 196–220, McGraw-Hill, 1969.
15. Jerald G. Graeme, **Photodiode Amplifiers: Op Amp Solutions**, McGraw-Hill, 1995.
16. Erik Barnes, *High Integration Simplifies Signal Processing for CCDs,* **Electronic Design**, February 23, 1998, pp. 81–88.
17. Eric Barnes, *Integrated for CCD Signal Processing*, **Analog Dialogue** 32-1, Analog Devices, 1998.
18. Precision Resistor Co., Inc., 10601 75th St. N., Largo, FLA, 33777-1427, 727-541-5771, http://www.precisionresistor.com.

CHAPTER 5

Acceleration, Shock and Vibration Sensors

Craig Aszkler, Vibration Products Division Manager, PCB Piezotronics, Inc.

5.1 Introduction

Accelerometers are sensing transducers that provide an output proportional to acceleration, vibration[1] and shock. These sensors have found a wide variety of applications in both research and development arenas along with everyday use. In addition to the very technical test and measurement applications, such as modal analysis, NVH (noise vibration and harshness), and package testing, accelerometers are also used in everyday devices such as airbag sensors and automotive security alarms. Whenever a structure moves, it experiences acceleration. Measurement of this acceleration helps us gain a better understanding of the dynamic characteristics that govern the behavior of the object. Modeling the behavior of a structure provides a valuable technical tool that can then be used to modify response, to enhance ruggedness, improve durability or reduce the associated noise and vibration.

The most popular class of accelerometers is the piezoelectric accelerometer. This type of sensor is capable of measuring a wide range of dynamic events. However, many other classes of accelerometers exist that are used to measure constant or very low frequency acceleration such as automobile braking, elevator ride quality and even the gravitational pull of the earth. Such accelerometers rely on piezoresistive, capacitive and servo technologies.

5.2 Technology Fundamentals

Piezoelectric Accelerometer

Piezoelectric accelerometers are self-generating devices characterized by an extended region of flat frequency response range, a large linear amplitude range and excellent durability. These inherent properties are due to the use of a piezoelectric material as the sensing element for the sensor. Piezoelectric materials are characterized by their ability to output a proportional electrical signal to the stress applied to the material. The basic construction of a piezoelectric accelerometer is depicted in Figure 5.2.1.

---

[1] For information on machinery vibration monitoring, refer to Chapter 13.

Chapter 5

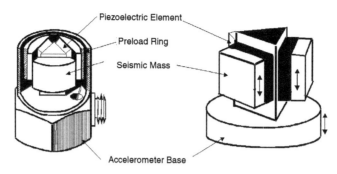

Figure 5.2.1: Basic piezoelectric accelerometer construction.

The active elements of the accelerometer are the piezoelectric elements. The elements act as a spring, which has a stiffness k, and connect the base of the accelerometer to the seismic masses. When an input is present at the base of the accelerometer, a force (F) is created on piezoelectric material proportional to the applied acceleration (a) and size of the seismic mass (m). (The sensor is governed by Newton's law of motion $F = ma$.) The force experienced by the piezoelectric crystal is proportional to the seismic mass times the input acceleration. The more mass or acceleration, the higher the applied force and the more electrical output from the crystal.

The frequency response of the sensor is determined by the resonant frequency of the sensor, which can generally be modeled as a simple single degree of freedom system. Using this system, the resonant frequency (ω) of the sensor can be estimated by: $\omega = \sqrt{k/m}$.

The typical frequency response of piezoelectric accelerometers is depicted in Figure 5.2.2.

Piezoelectric accelerometers can be broken down into two main categories that define their mode of operation. Internally amplified accelerometers or IEPE (internal electronic piezoelectric) contain built-in microelectronic signal conditioning. Charge mode accelerometers contain only the self-generating piezoelectric sensing element and have a high impedance charge output signal.

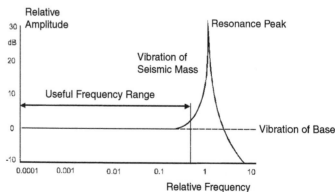

Figure 5.2.2 Typical frequency response of piezoelectric accelerometer.

Acceleration, Shock and Vibration Sensors

IEPE Accelerometers

IEPE sensors incorporate built-in, signal-conditioning electronics that function to convert the high-impedance charge signal generated by the piezoelectric sensing element into a usable low-impedance voltage signal that can be readily transmitted, over ordinary two-wire or coaxial cables, to any voltage readout or recording device. The low-impedance signal can be transmitted over long cable distances and used in dirty field or factory environments with little degradation. In addition to providing crucial impedance conversion, IEPE sensor circuitry can also include other signal conditioning features, such as gain, filtering and self-test features. The simplicity of use, high accuracy, broad frequency range, and low cost of IEPE accelerometer systems make them the recommended type for use in most vibration or shock applications. However, an exception to this assertion must be made for circumstances in which the temperature at the installation point exceeds the capability of the built-in circuitry. The routine upper temperature limit of IEPE accelerometers is 250°F (121°C); however, specialty units are available that operate to 350°F (175°C).

IEPE is a generic industry term for sensors with built-in electronics. Many accelerometer manufacturers use their own registered trademarks or trade name to signify sensors with built-in electronics. Examples of these names include: ICP® (PCB Piezotronics), Deltatron (Bruel & Kjaer), Piezotron (Kistler Instruments), and Isotron (Endevco), to name a few.

The electronics within IEPE accelerometers require excitation power from a constant-current, DC voltage source. This power source is sometimes built into vibration meters, FFT analyzers and vibration data collectors. A separate signal conditioner is required when none is built into the readout. In addition to providing the required excitation, power supplies may also incorporate additional signal conditioning, such as gain, filtering, buffering and overload indication. The typical system set-ups for IEPE accelerometers are shown in Figure 5.2.3.

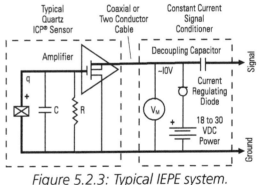

Figure 5.2.3: Typical IEPE system.

Chapter 5

Charge Mode Accelerometers

Charge mode sensors output a high-impedance, electrical charge signal that is generated directly by the piezoelectric sensing element. It should be noted that this signal is sensitive to corruption from environmental influences and cable-generated noise. Therefore it requires the use of a special low noise cable. To conduct accurate measurements, it is necessary to condition this signal to a low-impedance voltage before it can be input to a readout or recording device. A charge amplifier or in-line charge converter is generally used for this purpose. These devices utilize high-input-impedance, low-output-impedance charge amplifiers with capacitive feedback. Adjusting the value of the feedback capacitor alters the transfer function or gain of the charge amplifier.

Typically, charge mode accelerometers are used when high temperature survivability is required. If the measurement signal must be transmitted over long distances, it is recommended to use an in-line charge converter, placed near the accelerometer. This minimizes the chance of noise. In-line charge converters can be operated from the same constant-current excitation power source as IEPE accelerometers for a reduced system cost. In either case, the use of a special low noise cable is required between the accelerometer and the charge converter to minimize vibration induced triboelectric noise.

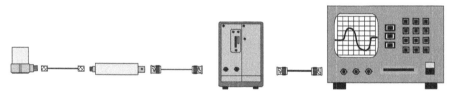

Figure 5.2.4: Typical in-line charge converter system.

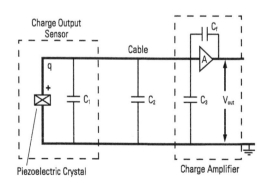

Figure 5.2.5: Laboratory charge amplifier system.

Sophisticated laboratory-style charge amplifiers usually include adjustments for normalizing the input signal and altering the feedback capacitor to provide the desired system sensitivity and full-scale amplitude range. Filtering also may be used to tailor the high and low frequency response. Some charge amplifiers provide dual-mode operation, which provides power for IEPE accelerometers and conditions charge mode sensors.

Because of the high-impedance nature of the output signal generated by charge mode accelerometers, several important precautionary measures must be followed. As noted above, always be attentive to motion induced (triboelectric) noise in the cable and mitigate by using specially treated cable. Also, always maintain high insulation resistance of the accelerometer, cabling, and connectors. To ensure high insulation resistance, all components must be kept dry and clean. This will help minimize potential problems associated with noise and/or signal drift.

Piezoelectric Sensing Materials

Two categories of piezoelectric materials that are predominantly used in the design of accelerometers are quartz and polycrystalline ceramics. Quartz is a natural crystal, while ceramics are man-made. Each material offers certain benefits. The material choice depends on the particular performance features desired of the accelerometer.

Quartz is widely known for its ability to perform accurate measurement tasks and contributes heavily in everyday applications for time and frequency measurements. Examples include everything from wristwatches and radios to computers and home appliances. Accelerometers benefit from several unique properties of quartz. Since quartz is naturally piezoelectric, it has no tendency to relax to an alternative state and is considered the most stable of all piezoelectric materials. This important feature provides quartz accelerometers with long-term stability and repeatability. Also, quartz does not exhibit the pyroelectric effect (output due to temperature change), which provides stability in thermally active environments. Because quartz has a low capacitance value, the voltage sensitivity is relatively high compared to most ceramic materials, making it ideal for use in voltage-amplified systems. Conversely, the charge sensitivity of quartz is low, limiting its usefulness in charge-amplified systems, where low noise is an inherent feature.

A variety of ceramic materials are used for accelerometers, depending on the requirements of the particular application. All ceramic materials are man-made and are forced to become piezoelectric by a polarization process. This process, known as "poling," exposes the material to a high-intensity electric field. This process aligns the electric dipoles, causing the material to become piezoelectric. If ceramic is exposed to

Chapter 5

temperatures exceeding its range, or large electric fields, the piezoelectric properties may be drastically altered. There are several classifications of ceramics. First, there are high-voltage-sensitivity ceramics that are used for accelerometers with built-in, voltage-amplified circuits. There are high-charge-sensitivity ceramics that are used for charge mode sensors with temperature ranges to 400°F (205°C). This same type of crystal is used in accelerometers that use built-in charge-amplified circuits to achieve high output signals and high resolution. Finally, there are high-temperature piezo-ceramics that are used for charge mode accelerometers with temperature ranges over 1000°F (537°C) for monitoring engine manifolds and superheated turbines.

Structures for Piezoelectric Accelerometers

A variety of mechanical configurations are available to perform the transduction principles of a piezoelectric accelerometer. These configurations are defined by the nature in which the inertial force of an accelerated mass acts upon the piezoelectric material. There are three primary configurations in use today: shear, flexural beam, and compression. The shear and flexural modes are the most common, while the compression mode is used less frequently, but is included here as an alternative configuration.

Shear Mode

Shear mode accelerometer designs bond, or "sandwich," the sensing material between a center post and seismic mass. A compression ring or stud applies a preload force required to create a rigid linear structure. Under acceleration, the mass causes a shear stress to be applied to the sensing material. This stress results in a proportional electrical output by the piezoelectric material. The output is then collected by the electrodes and transmitted by lightweight lead wires to either the built-in signal conditioning circuitry of ICP® sensors, or directly to the electrical connector for a charge

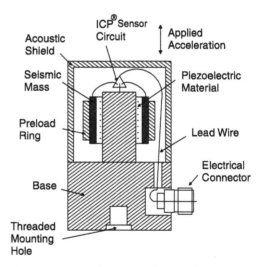

Figure 5.2.6: Shear mode accelerometer.

mode type. By isolating the sensing crystals from the base and housing, shear accelerometers excel in rejecting thermal transient and base bending effects. Also, the shear geometry lends itself to small size, which promotes high frequency response while minimizing mass loading effects on the test structure. With this combination of ideal characteristics, shear mode accelerometers offer optimum performance.

Acceleration, Shock and Vibration Sensors

Flexural Mode

Flexural mode designs utilize beam-shaped sensing crystals, which are supported to create strain on the crystal when accelerated. The crystal may be bonded to a carrier beam that increases the amount of strain when accelerated. The flexural mode enables low profile, lightweight designs to be manufactured at an economical price. Insensitivity to transverse motion is an inherent feature of this design. Generally, flexural beam designs are well suited for low frequency, low gravitational (g) acceleration applications such as those that may be encountered during structural testing.

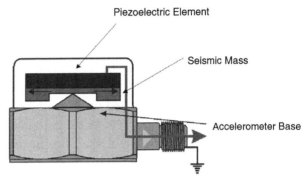

Figure 5.2.7: Flexural mode accelerometer.

Compression Mode

Compression mode accelerometers are simple structures which provide high rigidity. They represent the traditional or historical accelerometer design.

Upright compression designs sandwich the piezoelectric crystal between a seismic mass and rigid mounting base. A pre-load stud or screw secures the sensing element to the mounting base. When the sensor is accelerated, the seismic mass increases or decreases the amount of compression force acting upon the crystal, and a proportional electrical output results. The larger the seismic mass, the greater the stress and, hence, the greater the output.

This design is generally very rugged and can withstand high-g shock levels. However, due to the intimate contact of the sensing crystals with the external mounting base, upright compression designs tend to be more sensitive to base bending (strain). Additionally, expansion and contraction of the internal parts act along the sensitive axis making the accelerometer more susceptible to thermal transient effects. These effects can contribute to erroneous output signals when used on thin sheet-metal structures or at low frequencies in thermally unstable environments, such as outdoors or near fans and blowers.

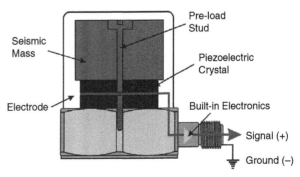

Figure 5.2.8: Compression mode accelerometer.

Chapter 5

Piezoresistive Accelerometers

Single-crystal silicon is also often used in manufacturing accelerometers. It is an anisotropic material whose atoms are organized in a lattice having several axes of symmetry. The orientation of any plane in the silicon is provided by its Miller indices. Piezoresistive transducers manufactured in the 1960s first used silicon strain gages fabricated from lightly doped ingots. These ingots were sliced to form small bars or patterns. The Miller indices allowed positioning of the orientation of the bar or pattern with respect to the crystal axes of the silicon. The bars or patterns were often bonded directly across a notch or slot in the accelerometer flexure. Figure 5.2.9 shows short, narrow, active elements mounted on a beam. The large pads are provided for thermal power dissipation and ease of electrical and mechanical connections. The relatively short web avoids column-type instabilities in compression when the beam bends in either direction. The gages are subsequently interconnected in a Wheatstone bridge configuration. This fact that the gages are configured in a bridge indicates that piezoresistive accelerometers have response down to DC (i.e., they respond to steady-state accelerations).

Since the late 1970s we have encountered a continual evolution of microsensors into the marketplace. A wide variety of technologies are involved in their fabrication. The sequence of events that occurs in this fabrication process are: the single crystal silicon is grown; the ingot is trimmed, sliced, polished, and cleaned; diffusion of a dopant into a surface region of the wafer is controlled by a deposited film; a photolithography process includes etching of the film at places defined in the developing process, followed by removal of the photoresist; and isotropic and anisotropic wet chemicals are used for shaping the mechanical microstructure. Both the resultant stress distribution in the microstructure and the dopant control the piezoresistive coefficients of the silicon.

Electrical interconnection of various controlled surfaces formed in the crystal, as well as bonding pads, are provided by thin film metalization. The wafer is then separated into individual dies. The dies are bonded by various techniques into the transducer housing, and wire bonding connects the metallized pads to metal terminals in the transducer housing. It is important to realize that piezoresistive accelerometers manufactured in this manner use silicon both as the flexural element and as the transduction element, since the strain gages are diffused directly into the flexure. Figures 5.2.10 and 5.2.11 show typical results of this fabrication process.

Acceleration, Shock and Vibration Sensors

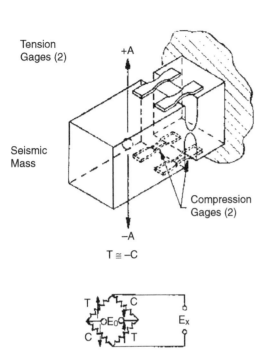

Figure 5.2.9: Bulk silicon resistors bonded to metal beam accelerometer flexure.

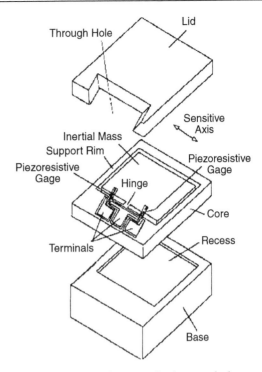

Figure 5.2.10: MEMS piezoresistive accelerometer flexure.

The advantages of an accelerometer constructed in this manner include a high stiffness, resulting in a high resonant frequency (ω) optimizing its frequency response. This high resonant frequency is obtained because the square root of the modulus-to-density ratio of silicon, an indicator of dynamic performance, is higher than that for steel. Other desirable byproducts are miniaturization, large signal amplitudes (semiconductor strain gages have a gage factor 25 to 50 times that of metal), good linearity, and improved stability. If properly temperature compensated, piezoresistive accelerometers can operate over a temperature range of –65 to +250°F. With current technology, other types of piezoresistive sensors (pressure) operate to temperatures as high as 1000°F.

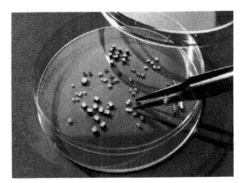

Figure 5.2.11: Multiple MEMS accelerometer flexure containing diffused and metallized piezoresistive gages in Wheatstone bridge configuration.

Chapter 5

Capacitive Accelerometers

Capacitive accelerometers are similar in operation to piezoresistive accelerometers, in that they measure a change across a bridge; however, instead of measuring a change in resistance, they measure a change in capacitance. The sensing element consists of two parallel plate capacitors acting in a differential mode. These capacitors operate in a bridge configuration and are dependent on a carrier demodulator circuit or its equivalent to produce an electrical output proportional to acceleration.

Several different types of capacitive elements exist. One type, which utilizes a metal sensing diaphragm and alumina capacitor plates, can be found in Figure 5.2.12. Two fixed plates sandwich the diaphragm, creating two capacitors, each with an individual fixed plate and each sharing the diaphragm as a movable plate.

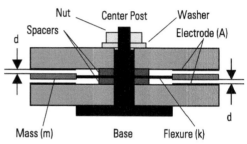

Figure 5.2.12: Capacitive sensor element construction.

When this element is placed in the Earth's gravitational field or is accelerated due to vibration on a test structure, the spring-mass experiences a force. This force is proportional to the mass of the spring-mass and is based on Newton's Second Law of Motion.

$F = ma$ where F = inertial force acting on spring-mass   Eq. 5.2.1
              m = distributed mass of spring-mass
              a = acceleration experienced by sensing element

Consequently, the spring-mass deflects linearly according to the Spring Equation.

$X = F/k$ where X = deflection of spring-mass   Eq. 5.2.2
              k = stiffness of spring-mass

The resulting deflection of the spring-mass causes the distance between the electrodes and the spring-mass to vary. These variations have a direct effect on each of the opposing capacitor gaps according to the following equation.

$C_2 = A_E [\varepsilon / (d + X)]$ and,
$C_2 = A_E [\varepsilon / (d - X)]$ where C = element capacitance   Eq. 5.2.3
                    A_E = surface area of electrode
                    ε = permittivity of air
                    d = distance between spring-mass and electrode

Acceleration, Shock and Vibration Sensors

A built-in electronic circuit is required for proper operation of a capacitive accelerometer. In the simplest sense, the built-in circuit serves two primary functions: (1) allow changes in capacitance to be useful for measuring both static and dynamic events, and (2) convert this change into a useful voltage signal compatible with readout instrumentation.

A representative circuit is shown in Figure 5.2.13 and Figure 5.2.14, which graphically depicts operation in the time domain, resulting from static measurand input.

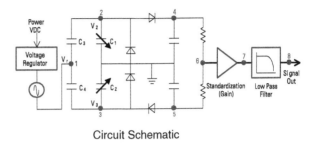

Circuit Schematic

The following explanation starts from the beginning of the circuit and continues through to the output, and describes the operation of the circuit.

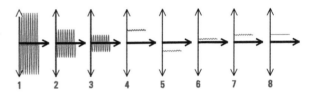

Response from Circuit due to applied +1g Static Acceleration
(x-axis = time and y-axis = voltage)

Figures 5.2.13 and 5.2.14: Operation of built-in circuit for capacitive accelerometer.

To begin, the supply voltage is routed through a voltage regulator, which provides a regulated dc voltage to the circuit. The device assures "clean" power for operating the internal circuitry and fixes the amplitude of a built-in oscillator, which typically operates at >1 MHz. By keeping the amplitude of the oscillator signal constant, the output sensitivity of the device becomes fixed and independent of the supply voltage. Next, the oscillator signal is directed into the capacitance-bridge as indicated by Point 1 in Figure 5.2.13. It then splits and passes through each arm of the bridge, which each act as divider networks. The divider networks cause the oscillator signal to vary in direct proportion to the change in capacitance in C_2 and C_4. (C_2 and C_4 electrically represent the mechanical sensing element.) The resulting amplitude-modulated signals appear at Points 2 and 3. Finally, to "demodulate" these signals, they are passed through individual rectification/peak-picking networks at Points 4 and 5, and then summed together at Point 6. The result is an electrical signal proportional to the physical input.

It would be sufficient to complete the circuit at this point; however, additional features are often added to enhance its performance. In this case, a "standardization" amplifier has been included. This is typically used to trim the sensitivity of the device so that it falls within a tighter tolerance. In this example, Point 7 shows how this amplifier can be used to gain the signal by a factor of two. Finally, there is a low pass filter, which is used to eliminate any high frequency ringing or residual affects of the carrier frequency.

If silicon can be chemically machined and processed as the transduction element in a piezoresistive accelerometer, it should similarly be able to be machined and processed into the transduction element for a capacitive accelerometer. In fact, MEMS technology is applicable to capacitive accelerometers. Figure 5.2.15 illustrates a MEMS variable-capacitance element and its integration into an accelerometer. As with the previously described metal diaphragm accelerometer, the detection of acceleration requires both a pair of capacitive elements and a flexure. The sensing elements experience a change in capacitance attributable to minute deflections resulting from the inertial acceleration force. The single-crystal nature of the silicon, the elimination of

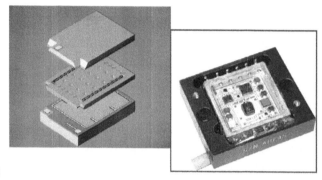

Figure 5.2.15: MEMS capacitor plates and completed accelerometer with top lid off.

mechanical joints, and the ability to chemically machine mechanical stops, result in a transducer with a high over-range capability. As with the previous metal diaphragm accelerometer, damping characteristics can be enhanced over a broad temperature range if a gas is employed for the damping medium as opposed to silicone oil. A series of grooves, coupled with a series of holes in the central mass, squeeze gas through the structure as the mass displaces. The thermal viscosity change of a gas is small relative to that of silicone oil. Capacitive MEMS accelerometers currently operate to hundreds of g's and frequencies to one kHz. The MEMS technology also results in accelerometer size reduction.

Most capacitive accelerometers contain built-in electronics that inject a signal into the element, complete the bridge and condition the signal. For most capacitive sensors it is necessary to use only a standard voltage supply or battery to supply appropriate power to the accelerometer.

Acceleration, Shock and Vibration Sensors

One of the major benefits of capacitive accelerometers is to measure low level (less than 2 g's), low frequency (down to dc) acceleration with the capability of withstanding high shock levels, typically 5,000 g's or greater. Some of the disadvantages of the capacitive accelerometer are a limited high frequency range, a relatively large phase shift and higher noise floor than a comparable piezoelectric device.

Servo or (Force Balance) Accelerometers

The accelerometers described to date have been all "open loop" accelerometers. The deflection of the seismic mass, proportional to acceleration, is measured directly using either piezoelectric, piezoresistive, or variable capacitance technology. Associated with this mass displacement is some small, but finite, error due to nonlinearities in the flexure. Servo accelerometers are "closed loop" devices. They keep internal deflection of the proof mass to an extreme minimum. The mass is maintained in a "balanced" mode virtually eliminating errors due to nonlinearities. The flexural system can be either linear or pendulous (C2 and C4 electrically represent opposite sides of the mechanical sensing element.) Electromagnetic forces, proportional to a feedback current, maintain the mass in a null position. As the mass attempts to move, a capacitive sensor typically detects its motion. A servo circuit derives an error signal from this capacitive sensor and sends a current through a coil, generating a torque proportional to acceleration, keeping the mass in a capture or null mode. Servo or "closed loop" accelerometers can cost up to ten times what "open loop accelerometers" cost. They are usually found in ranges of less than 50 g, and their accuracy is great enough to enable them to be used in guidance and navigation systems. For navigation, three axes of servo accelerometers are typically combined with three axes of rate gyros in a thermally-stabilized, mechanically-isolated package as an inertial measuring unit (IMU). This IMU enables determination of the 6-degrees of freedom necessary to navigate in space. Figure 5.2.16 illustrates the operating principal of a servo accelerometer. They measure frequencies to dc (0 Hertz) and are not usually sought after for their high frequency response.

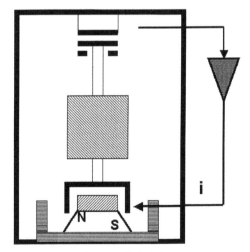

Figure 5.2.16:
Typical servo accelerometer construction.

5.3 Selecting and Specifying Accelerometers

Table 5.3.1 summarizes the advantages and disadvantages of different type of accelerometers along with some typical applications.

Table 5.3.1: Comparison of accelerometer types.

Accelerometer Type	Advantages	Limitations	Typical Applications
IEPE Piezoelectric Accelerometer	Wide Dynamic Range Wide Frequency Range Durable (High Shock Protection) Powered by Low Cost Constant Current Source Fixed Output Less Susceptible to EMI and RF Interference Can be Made Very Small Less Operator Attention, Training and Installation Expertise Required High Impedance Circuitry Sealed in Sensor Long Cable Driving without Noise Increase Operates into Many Data Acquisition Devices with Built-in Constant Current Input Operates across Slip Rings Lower System Cost per Channel	Limited Temperature Range Max Temperature of 175°C (350°F) Low Frequency Response is Fixed within the Sensor Built in amplifier is exposed to same test environment as the element of the sensor	Modal Analysis NVH Engine NVH Flight testing Body In White Testing Cryogenic Drop Testing Ground Vibration Testing HALT/HASS Seismic Testing Squeak and Rattle Helmet and Sport Equipment Testing Vibration Isolation and Control
Charge Piezoelectric Accelerometer	High operating temperatures to 700°C Wide dynamic Range Wide Frequency Range (Durable) High Shock Protection Flexible Output Simpler Design fewer parts Charge Converter electronics is usually at ambient condition, away from test environment	More Care/attention is required to install and maintain High impedance circuitry must be kept clean and dry Capacitive loading from long cable run results in noise floor increase Powered By Charge Amp which can be complicated and expensive Need to use Special Low Noise Cable	Jet Engine High Temperature Steam Pipes Turbo Machinery Steam Turbine Exhaust Brake

Table 5.3.1: Comparison of accelerometer types (continued).

Accelerometer Type	Advantages	Limitations	Typical Applications
Piezoresistive Accelerometer	DC Response Small Size	Lower Shock Protection Smaller Dynamic Range	Crash Testing Flight testing Shock testing
Capacitive Accelerometer	DC Response Better Resolution than PR Type Accelerometer	Frequency Range Average Resolution	Ride Quality Ride Simulation Bridge Testing Flutter Airbag Sensor Alarms
Servo Accelerometer	High Sensitivity Highest Accuracy for Low Level Low Frequency Measurements	Limited Frequency range, High Cost Fragile, Low Shock Protection.	Guidance Applications Requiring little or no DC Baseline Drift

Table 5.3.2 lists some of the *typical* characteristics of different sensors types.

Table 5.3.2: Typical accelerometer characteristics.

Accelerometer Type	Frequency Range	Sensitivity	Measurement Range	Dynamic Range	Size/weight
IEPE Piezoelectric Accelerometer	0.5 Hz to 50 000 Hz	.05 mV/g to 10 V/g	0.000001 g's to 100,000 g's	~120 dB	.2 Gram to 200 + grams
Charge Piezoelectric Accelerometer	0.5 Hz to 50 000 Hz	.01 pC/g to 100 pC/g	0.00001 g's to 100,000 g's	~110 dB	.14 grams to 200 + grams
Piezoresistive Accelerometer	0 to10000 Hz	0.0001 to 10 mV/g	0.001 to 100000 g's	~80 dB	1 to 100 grams
Capacitive Accelerometer	0 to 1000 Hz	10 mV/g to 1 V/g	0.00005 g's to 1000 g's	~90 dB	10 grams to 100 grams
Servo Accelerometer	0 to 100 Hz	1 to 10 V/g	<0.000001 g's to 10 g's	>120 dB	>50 grams

In order to select the most appropriate accelerometer for the application, you should look at a variety of factors. First you need to determine the type of sensor response required. The three basic functional categories of accelerometers are IEPE, Charge Mode and DC responding. The first two categories of accelerometers, the IEPE and Charge Mode type of accelerometers, work best for measuring frequencies starting at 0.5 Hz and above. The IEPE is a popular choice, due to its low cost, ease of use and low impedance characteristics, whereas the Charge Mode is useful for high temperature applications. There are advantages of each design.

When looking at uniform acceleration, as may be required for tilt measurement, or extremely low frequency measurements below 1 Hz, capacitive or piezoresistive accelerometers are a better choice. Both accelerometer types have been designed to

Chapter 5

achieve true 0 Hz (DC) responses. These sensors may contain built-in signal conditioning electronics and a voltage regulator, allowing them to be powered from a 5–30 VDC source. Some manufacturers offer an offset adjustment, which serves to null any DC voltage offset inherent to the sensor. Capacitive accelerometers are generally able to measure smaller acceleration levels.

The most basic criteria used to narrow the search, once the functionality category or response type of accelerometer has been decided, includes: sensitivity, amplitude, frequency range and temperature range. Sensitivity for shock and vibration accelerometers is usually specified in millivolts per g (mV/g) or picocoulombs per g (pC/g). This sensitivity specification is inversely proportional to the maximum amplitude that can be measured (g peak range.) Thus, more sensitive sensors will have lower maximum measurable peak amplitude ranges. The minimum and maximum frequency range that is going to be measured will also provide valuable information required for the selection process. Another important factor for accelerometer selection is the temperature range. Consideration should be given not only to the temperatures that the sensor will be exposed to, but also the temperature that the accelerometer will be stored at. High temperature special designs are available for applications that require that specification.

Every sensor has inherent characteristics, which cause noise. The broadband resolution is the minimal amount of amplitude required for a signal to be detected over the specified band. If you are looking at measuring extremely low amplitude, as in seismic applications, spectral noise at low frequency may be more relevant.

Physical characteristics can be very important in certain applications. Consideration should be given to the size and weight of the accelerometer. It is undesirable to place a large or heavy accelerometer on a small or lightweight structure. This is called "mass loading." Mass loading will affect the accuracy of the results and skew the data. The area that is available for the accelerometer installation may dictate the accelerometer selection. There are triaxial accelerometers, which can be utilized to simultaneously measure acceleration in three orthogonal directions. Older designs required three separate accelerometers to accomplish the same result, and thus add weight and require additional space.

Consideration should be given to the environment that the accelerometer will be exposed to. Hermetically sealed designs are available for applications that will be exposed to contaminants, moisture, or excessive humidity levels. Connector alternatives are available. Sensors can come with side connections or top connections to ease cable routing. Some models offer an integrated cable. Sensors with field-repairable cabling can prove to be very valuable in rough environments.

Accelerometer mounting may have an effect on the selection process. Most manufacturers offer a variety of mounting alternatives. Accelerometers can be stud mounted, adhesively mounted or magnetically mounted. Stud mounting provides the best stiffness and highest degree of accuracy, while adhesive mounts and magnetic mounting methods offer flexibility and quick removal options.

There are a wide variety of accelerometers to choose from. More than one will work for most applications. In order to select the most appropriate accelerometer, the best approach is to contact an accelerometer manufacturer and discuss the application. Manufacturers have trained application engineers who can assist you in selecting the sensor that will work best for your application.

5.4 Applicable Standards

In order to verify accelerometer performance, sensor manufacturers will test various characteristics of the sensor. This calibration procedure serves to help both the manufacturer and the end user. The end user will obtain a calibration certificate to confirm the accelerometer's exact performance characteristics. The manufacturer uses this calibration procedure for traceability, and to determine whether the product meets specifications and should be shipped or rejected. It can be viewed as a built-in quality control function. It provides a sense of security or confidence for both the manufacturer and the customer.

However, be aware that all calibrations are not equal. Some calibration reports may include terms such as "nominal" or "typical," or even lack traceability, or accredited stamps of approval. With the use of words like "nominal" or "typical," the manufacturer does not have to meet a specific tolerance on those specifications. This helps the manufacturer ship more products and reduce scrap, since fewer measured specifications means fewer rejections. While this provides additional profit for a manufacturer, it is not a benefit to the end customer. Customers have to look beyond the shiny paper and cute graphics, to make sure of the completeness of the actual measured data contained in each manufacturer's calibration certificate.

Due to the inconsistency of different manufacturer's calibration techniques and external calibration services, test engineers came up with standards to improve the quality of the product and certification that they receive. MIL-STD-45662 was created to define in detail the calibration system, process and components used in testing, along with the traceability of the product supplied to the government. The American National Standards Institute (ANSI) came up with its own version of specifications labeled ANSI/NCSL Z540-1-1994. This ANSI standard along with the International Organization for Standards (ISO) 10012-1, have been approved by the military as alternatives for the cancelled MIL standard. The ANSI Z540-1 and ISO 17025 require that uncertainty analyses for verifying the measurement process be documented and defined. Although the ANSI and ISO specifications are more common, the MIL specification, although cancelled in 1995, is still referenced on occasion.

Years ago, the National Bureau of Standards (NBS) recognized the inconsistencies in calibration reports and techniques, and developed a program for manufacturers to gain credibility and consistency. The National Institute of Standards and Technology (NIST) replaced the NBS in 1988, as the standard for accelerometer calibration approval. International compliance can be accredited by other sources. Physikalisch-Technische Bundesanstalt (PTB) in Germany, and United Kingdom Accreditation Service (AKAS) in Britain are popular organizations that supply this service. Manufacturers can send in their "Reference Sensors" that they utilize in back-to back testing to NIST or PTB in order to obtain certification from NIST or PTB (or both) to gain credibility and get the stamp of approval of these organizations.

The International Organization for Standardization initiated its own set of standards. The initial ISO standards concentrated on the documentation aspects of calibrating sensors. ISO 10012-1 addressed the MIL-45662 specifications, which added accuracy standards to the documentation. ISO17025 concentrated on traceability and accountability for the calibration work performed by the organization or laboratory, for a more complete set of standards. ISO standards that accelerometer customers should look for include:

    ISO 9001 – Quality systems for assurance in design, development and production
    ISO 10012-1 – Standards for measurement in management systems
    ISO 16063-21 – Methods for calibration of vibration and shock transducers
    ISO 17025 – General requirements for competence of testing and calibration laboratories
    RP-DTE011.1 – Shock and Vibration Transducer Selection, Institute of Environmental Sciences

Today, end users can purchase with confidence sensors that have traceable certifications that comply with the standards set by the NIST, PTB, ANSI, ISO and A2LA, provided that the data on the calibration is complete. The better manufacturers will reference most, if not all, of the above organizations.

5.5 Interfacing and Designs

One consideration in dealing with accelerometer mounting is the effect the mounting technique has on the accuracy of the usable frequency response. The accelerometer's operating frequency range is determined, in most cases, by securely stud mounting the test sensor directly to the reference standard accelerometer. The direct coupling, stud mounted to a very smooth surface, generally yields the highest mechanical resonant frequency and, therefore, the broadest usable frequency range. The addition of any mass to the accelerometer, such as an adhesive or magnetic mounting base, lowers the resonant frequency of the sensing system and may affect the accuracy and limits of the accelerometer's usable frequency range. Also, compliant materials, such as a rubber interface pad, can create a mechanical filtering effect by isolating and damping high-frequency transmissibility. A summary of the mounting techniques is provided below.

Stud Mounting

For permanent installations, where a very secure attachment of the accelerometer to the test structure is preferred, stud mounting is recommended. First, grind or machine on the test object a smooth, flat area at least the size of the sensor base, according to the manufacturer's specifications. For the best measurement results, especially at high frequencies, it is important to prepare a smooth and flat, machined surface where the accelerometer is to be attached. The mounting hole must also be drilled and tapped to the accelerometer manufacturer's specifications. Misalignment or incorrect threads can cause not only erroneous data, but can damage the accelerometer. The manufacturer's torque recommendation should always be used, measured with a calibrated torque wrench.

Adhesive Mounting

Occasionally, mounting by stud or screw is impractical. For such cases, adhesive mounting offers an alternative mounting method. The use of separate adhesive mounting bases is recommended to prevent the adhesive from damaging the accelerometer base or clogging the mounting threads. (Miniature accelerometers are provided with the integral stud removed to form a flat base.) Most adhesive mounting bases provide electrical isolation, which eliminates potential noise pick-up and ground loop problems. The type of adhesive recommended depends on the particular application. Wax offers a very convenient, easily removable approach for room temperature use. Two-part epoxies offer high stiffness, which maintains high-frequency response and a permanent mount.

Chapter 5

Magnetic Mounting

Magnetic mounting bases offer a very convenient, temporary attachment to magnetic surfaces. Magnets offering high pull strengths provide best high-frequency response. Wedged dual-rail magnetic bases are generally used for installations on curved surfaces, such as motor and compressor housings and pipes. However, dual-rail magnets usually significantly decrease the operational frequency range of an accelerometer. For best results, the magnetic base should be attached to a smooth, flat surface.

Probe Tips

Handheld vibration probes or probe tips on accelerometers are useful when other mounting techniques are impractical and for evaluating the relative vibration characteristics of a structure to determine the best location for installing the accelerometer. Probes are not recommended for general measurement applications due to a variety of inconsistencies associated with their use. Orientation and amount of hand pressure applied create variables, which affect the measurement accuracy. This method is generally used only for frequencies less than 1000 Hz.

Figure 5.5.1 summarizes the changes in the frequency response of a typical sensor using the various mounting methods previously discussed.

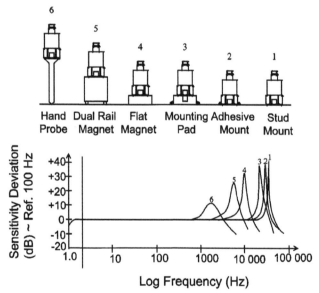

Figure 5.5.1: Relative frequency response of different accelerometer mounting techniques.

Acceleration, Shock and Vibration Sensors

Ground Isolation, Ground Noise, and Ground Loops

When installing accelerometers onto electrically conductive surfaces, a potential exists for ground noise pick-up. Noise from other electrical equipment and machines that are grounded to the structure, such as motors, pumps, and generators, can enter the ground path of the measurement signal through the base of a standard accelerometer. When the sensor is grounded at a different electrical potential than the signal conditioning and readout equipment, ground loops can occur. This phenomenon usually results in current flow at the line power frequency (and harmonics thereof), potential erroneous data, and signal drift. Under such conditions, it is advisable to electrically isolate or "float" the accelerometer from the test structure. This can be accomplished in several ways. Most accelerometers can be provided with an integral ground isolation base. Some standard models may already include this feature, while others offer it as an option. The use of insulating adhesive mounting bases, isolation mounting studs, isolation bases, and other insulating materials, such as paper beneath a magnetic base, are effective ground isolation techniques. Be aware that the additional ground-isolating hardware can reduce the upper frequency limits of the accelerometer.

Cables and Connections

Cables should be securely fastened to the mounting structure with a clamp, tape, or other adhesive to minimize cable whip and connector strain. Cable whip can introduce noise, especially in high-impedance signal paths. This phenomenon is known as the *triboelectric effect*. Also, cable strain near either electrical connector can lead to intermittent or broken connections and loss of data.

To protect against potential moisture and dirt contamination, use RTV sealant or heat-shrinkable tubing on cable connections. O-rings with heat shrink tubing have proven to be an effective seal for protecting electrical connections for short-term underwater use. RTV sealant is generally only used to protect the electrical connection against chemical splash or mist.

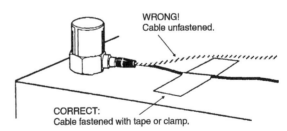

Figure 5.5.2: Cable strain relief of accelerometers.

Chapter 5

Latest and Future Developments

Manufacturers are continually trying to develop sensor packages that are smaller and lighter in weight than previous models. This minimizes mass loading effects and provides the user with the ability to test smaller and lighter components. Triaxial designs are becoming more popular and manufacturers have been designing improved versions of this product. A triaxial accelerometer can take the place of three single-axis accelerometers. The triaxial accelerometer can measure vibration in three orthogonal directions simultaneously, in one small lightweight convenient package, with a single cable assembly.

One of the emerging standards for accelerometers and other sensor types is the IEEE 1451 Smart Transducer Interface. This standard defines the hardware and communication protocol for interfacing a sensor onto a network. IEEE P1451.4 defines the architecture and protocol for compiling and addressing non-volatile memory that is imbedded within an analog measurement sensor. Accelerometers with this built-in digital memory chip are referred to as TEDS accelerometers. TEDS is an acronym for Transducer Electronic Data Sheet. TEDS allows the user to identify specific sensors within a large channel group, or determine output from different sensors at multiple locations very easily. TEDS can provide the user with technical information on the particular sensor. For instance, Model Number, Serial Number, Calibration Date and some technical specifications can be retrieved from each individual sensor that is TEDS compliant. Sensitivity specifications from calibration reports can be read through the TEDS and compensated for, so that the data acquisition system or readout can generate more accurate information.

With the need for high temperature models, manufacturers have been concentrating on developing new designs that will make accurate measurements in the most severe environments. End users have had requirements for sensors that can operate in colder and hotter temperatures than the standard accelerometers are specified for. As long as customers come up with new and unique applications, manufacturers will try to come up with products to satisfy their requirements.

The most exciting development that is likely to occur over the next decade is placing an analog-to-digital converter (ADC) directly inside the accelerometer. This will enable the accelerometer to provide digital output. Accelerometers enhanced with this type of output will be able to have features such as 24-bit ADC, wireless transmission, built-in signal processing, and the ability to be accessed over the world wide web.

References and Resources

1. C. M. Harris (ed), *Shock and Vibration Handbook, 4th edition*, McGraw-Hill, New York, NY 10020, 1996.
2. K.G. McConnell, *Vibration Testing Theory and Practice*, John Wiley & Sons Inc, New York, NY 10158, 1995.
3. Institute of Environmental Sciences and Technology, RP-DTE011.1, *Shock and Vibration Transducer Selection*.

CHAPTER 6

Biosensors

Young H. Lee and Raj Mutharasan,
Department of Chemical Engineering, Drexel University

6.1 Overview: What Is a Biosensor?

Biosensor = bioreceptor + transducer. A biosensor consists of two components: a bioreceptor and a transducer. The bioreceptor is a biomolecule that recognizes the target analyte, and the transducer converts the recognition event into a measurable signal. The uniqueness of a biosensor is that the two components are integrated into one single sensor (Figure 6.1.1). This combination enables one to measure the target analyte without using reagents (Davis et al, 1995). For example, the glucose concentration in a blood sample can be measured directly by a biosensor made specifically for glucose measurement, by simply dipping the sensor in the sample. This is in contrast to the commonly performed assays, in which many sample preparation steps are necessary and each step may require a reagent to treat the sample. The simplicity and the speed of measurements that require no specialized laboratory skills are the main advantages of a biosensor.

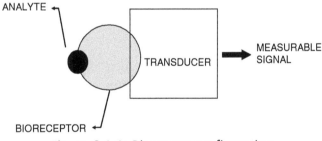

Figure 6.1.1: Biosensor configuration.

Enzyme is a Bioreceptor. When we eat food such as hamburgers and french fries, it is broken down into small molecules in our body via many reaction steps (these breakdown reactions are called *catabolism*). These small molecules are then used to make the building blocks of our body, such as proteins (these synthesis reactions are called *anabolism*). Each of these catabolism and anabolism reactions (the combination is called *metabolism*) are catalyzed by a specific enzyme. Therefore, **an enzyme is capable of recognizing a specific target molecule** (Figure 6.1.2). This biorecognition capability of the enzyme is used in biosensors. Other biorecognizing molecules (= bioreceptors) include antibodies, nucleic acids, and receptors.

Chapter 6

Immobilization of Bioreceptor. One major requirement for a biosensor is that the bioreceptor be immobilized in the vicinity of the transducer. The immobilization is done either by physical entrapment or chemical attachment. Chemical attachment often involves covalent bonding to transducer surface by suitable reagents. A comprehensive treatment of immobilization is available in Hermanson (1996). It is to be noted that only minute quantities of bioreceptor molecules are needed, and they are used repeatedly for measurements.

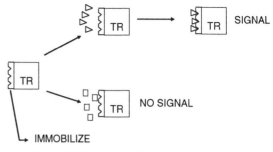

Figure 6.1.2: Specificity of biosensor (TR: transducer).

Transducer. A transducer should be capable of converting the biorecognition event into a measurable signal (Figure 6.1.3). Typically, this is done by measuring the change that occurs in the bioreceptor reaction. For example, the enzyme glucose oxidase is used as a bioreceptor in a glucose biosensor that catalyzes the following reaction:

$$\text{Glucose} + O_2 \xrightarrow{\text{Glucose Oxidase}} \text{Gluconic acid} + H_2O_2$$

To measure the glucose in aqueous solutions, three different transducers can be used:

1. An oxygen sensor that measures oxygen concentration, a result of glucose reaction
2. A pH sensor that measures the acid (gluconic acid), a reaction product of glucose
3. A peroxidase sensor that measures H_2O_2 concentration, a result of glucose reaction

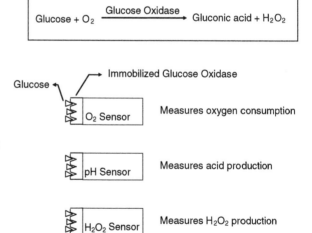

Figure 6.1.3: Three possible transducers for glucose measurement.

Note that an oxygen sensor is a transducer that converts oxygen concentration into electrical current. A pH sensor is a transducer that converts pH change into voltage change. Similarly, a peroxidase sensor is a transducer that converts peroxidase concentration into an electrical current. An excellent review of glucose sensing technologies was reported by Wilkins and Atansov (1996).

Biosensor Characteristics. Biosensors are characterized by eight parameters. These are: (1) *Sensitivity* is the response of the sensor to per unit change in analyte concentration. (2) *Selectivity* is the ability of the sensor to respond only to the target analyte. That is, lack of response to other interfering chemicals is the desired feature. (3) *Range* is the concentration range over which the sensitivity of the sensor is good. Sometimes this is called dynamic range or linearity. (4) *Response time* is the time required for the sensor to indicate 63% of its final response due to a step change in analyte concentration. (5) *Reproducibility* is the accuracy with which the sensor's output can be obtained. (6) *Detection limit* is the lowest concentration of the analyte to which there is a measurable response. (7) *Life time* is the time period over which the sensor can be used without significant deterioration in performance characteristics. (8) *Stability* characterizes the change in its baseline or sensitivity over a fixed period of time.

Considerations in Biosensor Development. Once a target analyte has been identified, the major tasks in developing a biosensor involve:

1. Selection of a suitable bioreceptor or a recognition molecule
2. Selection of a suitable immobilization method
3. Selection and design of a transducer that translates binding reaction into measurable signal
4. Design of biosensor considering measurement range, linearity, and minimization of interference, and enhancement of sensitivity
5. Packaging of the biosensor into a complete device

The first item above requires knowledge in biochemistry and biology, the second and third require knowledge in chemistry, electrochemistry and physics, and the fourth requires knowledge of kinetics and mass transfer. Once a biosensor has been designed, it must be packaged for convenient manufacturing and use. The current trend is miniaturization and mass production. Modern IC (integrated circuit) fabrication technology and micromachining technology are used increasingly in fabricating biosensors, as they reduce manufacturing costs. Therefore, an interdisciplinary research team, consisting of the various disciplines identified above, is essential for successful development of a biosensor.

Table 6.1.1: Considerations for biosensor development.

- Selection of a suitable biorecognition entity
- Selection of chemical immobilization method
- Selection and design of a suitable transducer
- Designing of biosensor for measurement range, linearity, and minimization of interference
- Packaging of biosensor into a complete unit

6.2 Applications of Biosensors

Health Care

Measurement of Metabolites. The initial impetus for advancing sensor technology came from the health care area, where it is now generally recognized that measurements of blood chemistry are essential and allow a better estimation of the metabolic state of a patient. In intensive care units, for example, patients frequently show rapid variations in biochemical composition and levels that require urgent remedial action. Also, in less severe patient handling, more successful treatment can be achieved by obtaining *instant* assays. At present, available *instant* analyses are not extensive. In practice, these assays are performed by analytical laboratories, where discrete samples are collected and shipped for analysis, frequently using the more traditional analytical techniques.

Market Potential. There is an increasing demand for inexpensive and reliable sensors for use in doctor's offices, emergency rooms, and operating rooms. Ultimately, patients themselves should be able to use biosensors in the monitoring of a clinical condition, such as diabetes. It is probably true that the major biosensor market may be found where an immediate assay is required. If the costs of laboratory instrument maintenance are included, then low-cost biosensor devices can be desirable in the whole spectrum of analytical applications from hospital to home.

Diabetes. The "classic" and most widely explored example of closed-loop drug control is found in the development of an artificial pancreas. Diabetic patients have a relative or absolute lack of insulin, a polypeptide hormone produced by the beta cells of the pancreas, which is essential for glucose uptake. Lack of insulin secretion causes various metabolic abnormalities, including higher than normal blood glucose levels. In patients who have lost insulin-secreting islets of Langerhan, insulin is supplied by subcutaneous injection. However, fine control is difficult to achieve and hyperglycaemia is often encountered. Further, even hypoglycaemia is sometimes induced, causing impaired consciousness and the serious long-term complications to tissue associated with this intermittent low glucose condition.

Insulin Therapy. Better methods for the treatment of insulin-dependent diabetes have been sought and infusion systems for continuous insulin delivery have been developed (Hall, 1991). However, regardless of the method of insulin therapy, its induction must be made in response to information on the current blood glucose levels in the patient. Three schemes are possible (Figure 6.2.1), the first two dependent on discrete manual glucose measurement and the third a "closed-loop" system, where insulin delivery is controlled by the output of a glucose sensor which is integrated with the insulin infuser. In the former case, glucose is estimated based on analysis of finger-prick blood samples with a colorimetric test strip or more recently with an amperometric pen-size biosensor device by the patients themselves. Clearly, these diagnostic kits must be easily portable, simple to use and require minimal skill and easy interpretation. However, even with the ability to monitor current glucose levels, intensive conventional insulin therapy requires multiple daily injections. This open-loop approach does not anticipate insulin dosage due to changes in diet and exercise. For example, it was shown that administration of glucose by subcutaneous injection, 60 minutes before a meal provides the best glucose/insulin management.

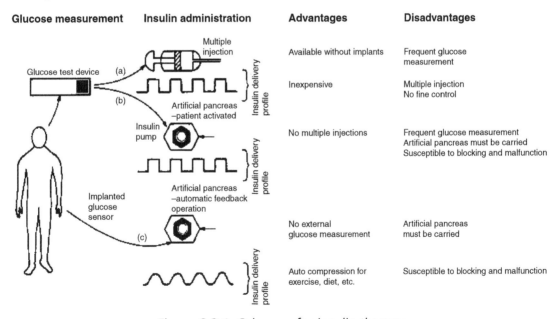

Figure 6.2.1: Schemes for insulin therapy.
From Hall (1991).

Chapter 6

Artificial Pancreas. The introduction of a closed-loop system, where integrated glucose measurements provide feedback control on a pre-programmed insulin administration based on habitual requirements, would therefore relieve the patient of frequent assay requirements and, perhaps more desirably, frequent injections. Ultimately, the closed-loop system becomes an artificial pancreas, where the glycemic control is achieved through an **implantable glucose sensor**. A proposed implantable sensor is given in Figure 6.2.2 (Turner et al., 1990). Clearly, the requirements for this sensor are very different from those for the discrete measurement kits. As summarized in Table 6.2.1, the prolonged lifetime and biocompatibility represent the major requirements.

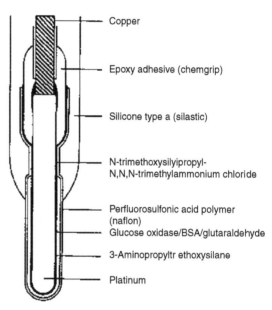

Figure 6.2.2: Cross-sectional view of an implantable glucose electrode in whole blood.
(From Turner et al., 1990.)

Table 6.2.1: Requirements for an implantable glucose sensor.

- Linear in physiological range 0–20 mM
- Specific for glucose; not affected by changes in blood chemistry
- Biocompatible
- Small—causes minimal tissue damage during insertion
- Response time < 1 min
- Prolonged lifetime ~ in years

Industrial Process Control

Bioreactor Control. Bioreactors are used to cultivate recombinant cells for production of therapeutic proteins such as insulin. The productivity of such systems depends on bioreactor conditions. Real-time monitoring of carbon sources, dissolved oxygen and carbon dioxide, and products of metabolism in fermentation processes could lead to optimization giving increased product yields at decreased processing and material cost (Scheper et al., 1996). While real-time monitoring with feedback control involving automated systems does exist, currently only a few common variables are measured on-line, (e.g., pH, temperature, CO_2, O_2) which are often only indirectly related to the culture activity under control. If cellular metabolic activity can be monitored in real-time using sensors, one can suitably alter the environmental variable to improve process productivity. The benefit of closed-loop control of a cellular state are many, such as improved product yield and quality. It is the lack of sensors that limits the use of closed loop online control of cell culture and fermentation processes.

Military and Homeland Security Applications

The requirement for rapid analysis is also present in military applications. Recent military engagement in the Middle East has caused, rather rapidly, the deployment of field-usable sensors for chemical and biological warfare agents. Many of the sensors are small portable analytical kits that may be termed "dipsticks." While they are reasonably robust, their performance in the field has been reported to be variable. Thus, there is a large need for robust sensors that require no maintenance. Both contact as well as remote sensing for warfare agents are currently under development. Distributed sensors and systems for monitoring hazards due to terrorist activity are currently being developed under the auspices of the Department of Homeland Security funding.

Environmental Monitoring

Environmental Protection Agency (EPA) Air and Water Monitoring. EPA's Environmental Monitoring and Assessment Program (EMAP) was established to provide a comprehensive report card on the condition of the nation's ecological resources and to detect trends in the condition of those resources. EPA routinely monitors both water and air in urban and rural areas. In recent times the Department of Homeland Security has initiated efforts to monitor the environment in urban areas, particularly large population centers, for potential bioterrorism agents. The primary measurement media is water or air, but the variety of target analytes is vast. At sites of potential pollution or terrorist activity, it would be desirable to install on-line real-time monitoring and alarm, targeted at specific analytes. Common environmental analytes are biological oxygen demand (BOD), atmospheric acidity, and river water pH, detergent,

Chapter 6

herbicides, and fertilizer concentrations in drainage and river (Leonard et al., 2003). The potential for biosensor technology for environmental monitoring is huge, and the potential impact is far-reaching. Although the principle of detection may be the same for a particular analyte, the actual technology platform used will be dependent on application. For example, a glucose sensor for on-line use in a fermenter has very different requirements from those used for monitoring glucose concentration in diabetic patients. Customization to meet application needs is often a very important part of technology development. For details on biosensors for environmental monitoring see Dennison and Turner (1995) and Wang et al., (1997).

6.3 Origin of Biosensors

Enzyme Electrode. The biosensor was first described by Clark and Lyons (1962), when the term *enzyme-electrode* was introduced. In this first enzyme electrode, an oxido-reductase enzyme, glucose oxidase, was held next to a platinum electrode in a membrane sandwich (Figure 6.3.1). The platinum anode polarized at +0.6 V responded to the peroxide produced by the enzyme reaction with substrate. The primary target substrate for this system was glucose:

$$\text{Glucose} + O_2 \xrightarrow{\text{Glucose Oxidase}} \text{Gluconic acid} + H_2O_2$$

and led to the development of the first glucose analyzer for the measurement of glucose in whole blood. This Yellow Springs Instrument (Model YSI 23) appeared on the market in 1974, and the same technique as employed here has been applied to many other oxygen-mediated oxido-reductase enzyme systems.

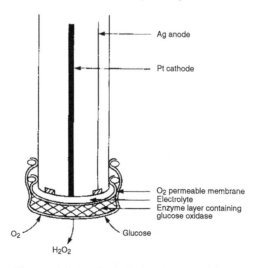

Figure 6.3.1: The Clark Enzyme Electrode.
From Hall (1991).

Use of Membrane for Selectivity. A key development in the YSI sensor was the employment of membrane technology in order to eliminate interference by other electro-active substances. Polarized at +0.6 V, the major interference to the peroxide measurement is ascorbic acid. Various combinations for a membrane-enzyme sandwich have been developed, all satisfying the following criteria:

- The membrane between electrode and enzyme layer should allow the passage of H_2O_2, but prevent the passage of ascorbate or other interferents
- The membrane between enzyme layer and sample should allow substrate/analyte to enter the enzyme layer

This was accomplished in the YSI, for example, with an enzyme layer sandwiched between a cellulose acetate membrane and a Nucleopore polycarbonate membrane. For further discussion on membranes and their properties see Davies et al., (1992).

6.4 Bioreceptor Molecules

Enzymes have been the most widely used bioreceptor molecules in biosensor applications, with antibodies and protein receptor molecules increasingly incorporated in biosensors. The specificity of a biosensor comes from the specificity of the bioreceptor molecule used. An enzyme is a good example. It has a three-dimensional structure that fits only a particular substrate (Figure 6.4.1a). An enzyme is a protein synthesized in the cell from amino acids according to the codings written in DNA. Enzymes act as catalysts for biochemical reactions occurring in the cell. To maintain high enzyme activity, the temperature and pH of the environment have to be maintained at proper levels.

Antibody. Antibodies represent one of the major classes of proteins. They constitute about 20% of the total plasma protein and are collectively called immunoglobulins (Ig). The simplest antibodies are described as Y-shaped molecules with two identical binding sites for antigen. An antigen can be any macromolecule that induces an immune response. The antibody has a basic structural unit consisting of four polypeptide chains: two light chains and two heavy chains (Figure 6.4.1b). The antibody binds reversibly with a specific antigen. Unlike the enzyme proteins, the antibodies do not act as catalysts. Their purpose is to bind foreign substances—antigens—so as to remove them from the circulatory system. Monoclonal antibodies belonging to the IgG class of immunoglobulins are usually used in sensor applications. In many instances, affinity-purified polyclonal antibodies are used as they show higher avidity to antigen. Polyclonal and monoclonal antibodies to an antigen can be developed in a few months. Currently, they cost $6,000 to $20,000 per gram when suitably manufactured in large quantities.

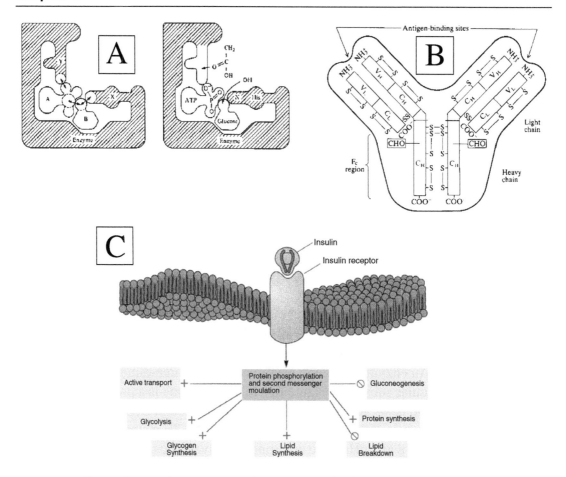

*Figure 6.4.1: Bioreceptor molecules used for biosensor applications:
(A) Enzyme; (B) Antibody; (C) Protein receptor.*
From Bailey and Ollis (1996) and Garrett and Grisham (1995).

Receptor Protein. Receptor proteins have specific affinity for biologically active compounds. These proteins are mostly bound to membrane (Figure 6.4.1c). There are hormone receptors, taste receptors, olfactory receptors for smelling, photoreceptors for eyes, and others. Receptor proteins activate opening and closing of membrane channels for transport of specific metabolites. They also play a key role in transducing intracellular messages for responsive action. Since the receptor proteins recognize specific biological entities, they are often used to measure target analytes. For example, death receptors on a cell surface transmit apoptosis signals initiated by specific ligands. They play an important role in apoptosis and can activate caspase cascade within seconds of ligand binding. Thus, when a death receptor is a sensing entity, one can potentially measure the presence of apoptotic-inducing chemicals in the environment.

Biosensors

Other Approaches. In principle, any biomolecules and molecular assemblies that have the capability of recognizing a target analyte can be used as a bioreceptor. In fact, membrane slices or whole cells have been used in biosensors. Figure 6.4.2 summarizes possible bioreceptors that can be utilized in a biosensor. Note that the bioreceptors require a suitable environment for maintaining their structural integrity and biorecognition activity. These requirements are described in Figure 6.4.2 along with the type of signal generated as a result of the biorecognition activity. The transducer in a biosensor should be responsive to this biochemical activity.

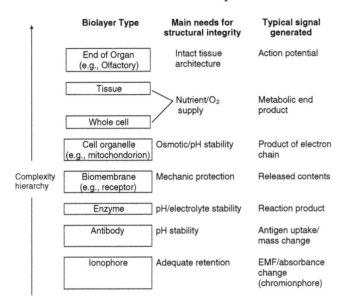

Figure 6.4.2: Possible bioreceptor molecules and molecular assemblies for biosensor applications; their requirements for structural integrity and signals generated.

6.5 Transduction Mechanisms in Biosensors

Conventional Transducers. The majority of biosensors in use today use three types of transducers for converting the action of the bioreceptor molecule into a measurable signal. These are: amperometry based on H_2O_2 or O_2 measurement; potentiometry based on pH or pIon measurement; and photometry utilizing optical fibers. Biorecognition reactions often generate chemical species that can be measured by electrochemical methods. In these, typically the reaction product is H_2O_2 (or the reactant is O_2) which can be measured by a pair of electrodes. When a suitable voltage is impressed on one of the electrodes against a reference electrode (typically Ag/AgCl or Calomel), the target species (H_2O_2 or O_2) is reduced at the electrode and this generates electrical current (hence the name *amperometry*). In potentiometry, a glass membrane or a polymeric membrane electrode is used for measuring the membrane potential (hence the name *potentiometry*) resulting from the difference in the concentrations of H^+ or other positive ions across the membrane. In photometry, the light from an indicator molecule is the measured signal. In this method, one of the reactants or products of the biorecognition reaction results in colorimetric, fluorescent or luminescent changes that are measured using photodetectors. Usually, an optical fiber is used for guiding the light signals from the source to the detector. Adaptation and exploitation of these three routes, (**amperometric, potentiometric and photometric**), where user acceptability is already established, has been an obvious approach to the development of reagentless biosensor devices.

Piezoelectric Transducers. The transducer of a biosensor is not restricted to the three described above. In principle, any variable that is affected by the biorecognition reaction can be used to generate the transduced signal. The **piezoelectric materials** and surface acoustic wave devices offer a surface that is sensitive to **changes in mass.** These transducers have been used where the biorecognition reaction causes a change in mass. For example, piezoelectric silicon crystals—called quartz crystal microbalance (QCM)—have been used to measure very small mass changes in the order of picograms. For example see Bunde et al., (1998). QCM with immobilized antibody to pathogens have been successfully used to measure the presence of pathogens in aqueous samples. Piezoelectrically driven cantilevers have also been used to measure adsorption of very minute quantities of biochemicals (Raiteri et al., 2001).

Conductimetric Transducers. Monitoring **solution conductance** was originally applied as a method of determining reaction rates. The technique involves the measurement of changes in conductance due to the migration of ions. Many enzyme-linked reactions result in a change in total ion concentration and this would imply that they are suitable for conductimetric biosensors.

Electrical Capacitance as Transducer. When the biorecognition reaction causes a **change in the dielectric measurement constant** of the medium in the vicinity of the bioreceptor, the capacitance measurement method can be used as a transducer. Antigen-antibody reaction is a good example. Suppose antibody molecules are immobilized between two metal electrodes of known area. When antigen is added and binds with the antibody, the dielectric constant of the medium between the two electrodes is expected to change significantly. This change translates into a change in capacitance.

Thermometric Transducer. All chemical reactions are accompanied by the absorption (endothermic) or evolution (exothermic) of heat. Measurements of ΔH, the enthalpy of reaction at different temperatures, allows one to calculate ΔS (entropy) and ΔG (Gibbs free energy) for a reaction and therefore collect basic thermodynamic data. The hydrolysis of ATP for example is exothermic:

$$ATP^{4-} + H_2O \sim ADP^{3-} + HPO^{4-} + H^+; \Delta H_{298} = -22.2 \text{ kJ (pH 7)}$$

or the immunoreaction between anti-HSA and its antigen HSA yields −30.5 kJ/mol. For this latter reaction, the total increase in temperature for 1 mmol of antibody is of the order of 10^{-5} K, but many enzyme-catalyzed reactions have greater ΔH, and produce more easily measurable changes in temperature.

Enzyme Thermistor. For a biosensor device, the biorecognition compound must be immobilized on a temperature-sensing element capable of detecting very small temperature changes. The major initiative in this area has come from the Mosbach group at the University of Lund. Initially, they immobilized glucose oxidase or penicillinase in a small column, so that temperature changes in the column effluent were monitored by thermistors to give an *enzyme thermistor* sensitive to glucose and penicillin, respectively. They have also applied the technique to other substrates and to immunoassay using an enzyme-labeled antigen.

FET as a Transducer. As advances are made in biosensors, a need has developed for miniaturization and mass production. Field effect transistors (FET) used extensively in the semiconductor industry in memory chips and logic chips respond to changes in electric field (in front of the "gate" of the FET). An FET is thus capable of detecting changes in ion concentration when the gate is exposed to a solution that contains ions. Therefore, pH and ion concentration can be measured with an FET. The advantage of this transducer is that it can be incorporated directly into the electronic signal processing circuitry. In fact, a pen-size FET-based pH sensor is being marketed commercially.

Table 6.5.1: Other transducers used in biosensors.

Category	Measured property	Examples
Piezoelectric	change in mass	microbalance based sensors
		Microcantilever based sensors
		SAW device based sensors
Conductive	conductivity change	Ion concentration
Capacitive	dielectric constant	antibody sensors
Thermometric	Temperature or heat flux	enzyme thermistor
		microcalorimeter

6.6 Application Range of Biosensors

Current Status. Since the development of Clark's glucose sensor for monitoring fermentation processes, many enzyme electrodes have been developed based on amperometry, potentiometry, and photometry. A sample of these biosensors is summarized in Tables 6.6.1, 6.6.2, and 6.6.3. The term "optode" (see Table 6.6.3) is used for sensors utilizing optical fiber for light signal transmission. Note that the bioreceptors used are all enzymes except the antibody sensor.

Table 6.6.1: Amperometric biosensors.

Substrate	Bioreceptor	Product detected	Range, g/L
Choline	choline oxidase	H_2O_2	0.5
Ethanol	alcohol oxidase	H_2O_2	3
Methanol	f. dehydrogenase	NADH	2.5
Glucose	glucose oxidase	H_2O_2, O_2	0-25
L-Glutamine	Glutaminase and glutamate oxidase	H_2O_2	10 mM
L-Glutamate		H_2O_2	10 mM
	Glutamate oxidase		
Hypoxanthine	Xanthine oxidase	H_2O_2	180
Lactate	Lactate oxidase	H_2O_2	40
Oligosaccharides	Glucoamylase, glucose oxidase	H_2O_2	2.5
Sucrose	Invertase, mutarotase, glucose oxidase	H_2O_2, O_2	25

Table 6.6.2: Potentiometric Biosensors.

Substrate	Bioreceptor	Product detected	Range, mM
Aspartamine	L-aspartase	NH_3	.5
Fats	lipase	fatty acids	0.05
Glucose	glucose oxidase	gluconic acid	2 g/L
Urea	urease	NH_4, CO_2	10
Nitrite	nitrite reductase	NH_4	1
Penicillin	penicillinase	H-+	70
Sulfate	sulfate oxidase	HS	
Antigen or antibody	partner of couple	Complex and various	100 ppm

Table 6.6.3: Enzyme sensors based on optodes

Substrate	Bioreceptor	Product detected	Range, mM
Ethanol	alcohol dehydrogenase	NADH	1
Glucose	glucose oxidase	O_2	20
Urease	urease	ammonia	3
Lactate	lactate monooxygenase	pyruvate	1
Penicillin	penicillinase	penicillinic acid	10

Products Detected. As is noted in Table 6.6.1–3, the analyte is indirectly measured by measuring the product of a reaction due to the enzyme. In Table 6.6.1, it is oxygen that is actually measured, and its concentration is directly proportional to the analyte concentration. For amperometry, the majority is H_2O_2 (with the exception of NADH), which is the common product for oxido-reductase enzymes. For potentiometric biosensors, the majority is acid which is detected by a pH sensor. In fermentor and cell culture reactors, products of metabolism CO_2 and NH_3 are indirectly detected by measuring the change in pH.

Biosensor Configurations. When bioreceptor molecules are combined with a suitable transducer, a biosensor is made. Figure 6.6.1 shows various biosensor configurations. Note that the bioreceptor molecules are immobilized in a suitable matrix to form a bioactive layer, which is then placed in the immediate vicinity of a transducer. The transducer's ion-selective electrode and FET belong to the potentiometric transducer category; the coated wire belongs to the amperometric sensor category; the surface plasmon detector and the surface acoustic wave detector belong to the piezo transducer category. The materials of construction for the transducers are also given in the figure.

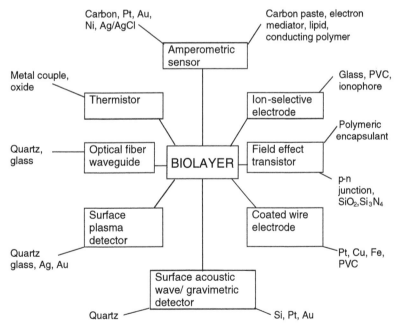

Figure 6.6.1: Various biosensor configurations.

Table 6.6.4: Biosensors based on FET (pH).

Substrate	Bioreceptor	Product detected	Range, mM
glucose	glucose oxidase	gluconic acid	0–20
urea	urease	CO_2, NH_3	0–6
penicillin	penicillinase	penicillic acid	0.2–20
triolein	lipase	fatty acids	0.6–3

Discriminative Membranes. Membranes are one of the most essential components of a biosensor. They are used for (1) barriers to non-analyte molecules, (2) protection of enzyme-immobilized membrane, thus preventing fouling; and (3) controlling the operating range of the biosensor. When a small molecule is the analyte, macromolecules such as proteins can be prevented from entering the active sensing area by using a small pore membrane. Note that proteins adsorb readily on most surfaces and thus foul sensing surfaces. The transport of charged molecules can be modulated by placing ion selective membranes. A combination of various discriminative membranes can be used for blocking the passage of different interfering molecules. It is to be noted that the use of a discriminative membrane increases time lag as it introduces diffusive transport resistance, and thus judicious choice of its thickness is essential for proper functioning. A summary is given in Table 6.6.5.

Table 6.6.5: Discriminative coatings for amperometric biosensors.

Transport mechanism	Seletive Membrane
Size exclusion	Cellulose acetate
	Ceramic membranes (Alumina, Zirconia)
	Polyaniline, Polypyrrole
	Polyphenol
Charge exclusion	Nafion
	Poly(vinylpyridine)
Polarity	Phospholipid
Gas permeability	silicone membrane

Sensitivity Requirements. The range and type of analytes fall into wide and varied values and thus cannot be considered using a single criterion. A particular application dictates the concentration range needed. It is often determined on the basis of expected target concentration range in samples. For example, a metabolite's concentration range is often in the uM range, whereas hormones are in nM range. Viruses and pathogens are found in 10–10,000 per mL. This vast range of concentrations is summarized in Figure 6.6.2. It is thus clear that often a varied approach is needed for a sensor designed for a metabolite compared with measuring tumor antigens.

Biosensors

Evolution of Biosensors. Biosensors can be classified into three generations according to the degree of integration of the separate components—i.e. the method of attachment of the biorecognition or bioreceptor molecule to the base transducer element. In the first generation, the bioreceptor is physically entrapped in the vicinity of the base sensor behind a discriminating membrane such as a dialysis membrane. In subsequent generations, immobilization is achieved via covalent bonds at a suitably modified transducer interface or by incorporation into a polymer matrix at the transduction surface. In the second generation, the individual components remain essentially distinct (e.g., control electronics—electrode—biomolecule), while in the third generation the bioreceptor molecule becomes an integral part of the base sensing element (Figure 6.6.3). While these definitions were probably intended for enzyme electrode systems, similar classifications appropriate to biosensors in general can be made. It is in the second and third generations of these families that the major development effort can now be seen.

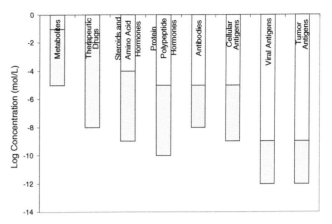

Figure 6.6.2: Detection ranges required for some clinically important analytes.
(Adapted from Hall [1991].)

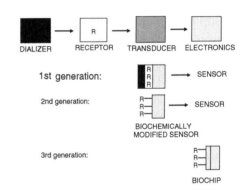

Figure 6.6.3: Three biosensor generations (R: Bioreceptor molecule).

6.7 Future Prospects

In recent years the emerging area of nanotechnology has produced very interesting materials, some of which provide opportunities for new sensing transduction technologies useful for biosensor development. In addition, use of self-assembly techniques and nano-electromechanical systems have produced new laboratory sensing methodologies. Some of these approaches, while not robust for common analytical instrumentation or field use, will emerge in future as practical sensors. The short discussion given below is to provide a snapshot of emerging methods.

Mass change sensors rely on changes in resonant frequency, as natural frequency depends on mass of oscillating mass. In this category, the quartz crystal microbalance (QCM) or thickness shear mode oscillator has been extensively used for detecting the presence of antigens by modifying the surface of QCM with an antibody specific to the target antigen. The same principle has been attempted in other forms of oscillating devices, such as silicon microcantilevers (Tang et al., 2004), piezoelectric-excited microcantilevers (Zhou et al., 2003), surface acoustic wave (SAW) sensors and others. In a SAW device, electrodes are on the same side of the crystal and interdigital transducers act as a transmitter and receiver to excite surface waves that travel across the crystal face. The changes to the wave caused by target antigen binding to the surface is confined to the crystal face, and is measured. SAW sensors are considerably more sensitive than QCM, but when aqueous phase is present on the surface, the signals are considerably attenuated. On the other hand, when no liquid solution is in contact, it provides very sensitive measurements for gas phase composition.

Raman spectroscopy is a useful tool for analysis because of its excellent chemical group identification capability; however, its limitation is low sensitivity. Recent observation that Raman scattering efficiency can be enhanced by many orders of magnitude when the analyte is adsorbed or near a gold or silver surface has made this technique a very powerful sensing methodology. This modified technique, known as surface-enhanced Raman scattering (SERS), has been shown to be suitable in a laboratory setting to observe DNA hybridization. Thus, single strands of DNA fragments can be labeled to SERS probes. The resulting SERG probes may be used to identify genes or detect bacterial and viral components. A further improvement can be achieved by using peptide nucleic acid (PNA), which was originally developed as a gene-targeting drug. PNA has demonstrated remarkable hybridization properties towards complementary oligonucleotides. Consequently, biosensors based on replacement of the DNA recognition layer with a PNA one, offer significantly improved distinction between closely related DNA sequences, as well as several other attractive advantages.

There has also been considerable interest in biophotonic sensors and DNA sensors (Junhui et al., 1997). For example, resonance enhancement due to gold nano-particles bound to recognition molecules has been shown to be effective in biophotonics (Homola et al., 1999). The advantage of photonics is the ability to measure without contacting the sample. Use of fiber optics and its variants have provided a rich source of transducing elements. In these cases, the surface of the fiber (glass) is derivatized with an amine group and then covalently linked to a bioreceptor via carboxylic group. When an analyte binds, the light transmission characteristics are altered and that is measured. Due to availability of inexpensive monochromic light sources (light emitting diodes) and inexpensive light sensing devices (photo diodes), biophotonic devices offer a relatively inexpensive sensing platform.

In this chapter an overview of biosensors was presented, various elements of biosensors were described, and a brief review of bioreceptors and transduction mechanisms were provided.

Acknowledgments

The author, Raj Mutharasan, wishes to thank the support of the Nanotechnology Institute of Ben Franklin Partnership.

References

Bailey, J. E. and Ollis, D. F., Biochemical Engineering Fundamentals, 2nd edition, McGraw Hill, New York (1986).

Bunde, R.L., Jarvi, E. J., and Rosentreter, J. J., Piezoelectric quartz crystal biosensors, Talanta, Volume 46, Issue 6, August 1998, pp. 1223–1236.

Clark, L.C. and C. Lyons Ann. N.Y.Academy of Sciences, 102, 29–45 (1962).

Davis, J., D. H. Vaughan and M. F. Cardosi, Elements of biosensor construction, Enzyme and Microbial Technology, Volume 17, Issue 12, December 1995, pp. 1030–1035.

Davies, M. L. , C. J. Hamilton, S. M. Murphy and B. J. Tighe, Polymer membranes in clinical sensor applications : I. An overview of membrane function, Biomaterials, Volume 13, Issue 14, 1992, pp. 971–978.

Dennison, M. J., and A. P. F. Turner, Biosensors for environmental monitoring, Biotechnology Advances, Volume 13, Issue 1, 1995, pp. 1–12.

Garrett, R. H. and C. M. Grisham, Biochemistry, Saunders College Publishing, Philadelphia, (1995).

Hall, E., Biosensors, Prentice Hall, Englewood Cliffs, New Jersey, (1991).

Hermanson G. T., 1996. Bioconjugate Techniques. Elsevier Science, San Diego, California, pp. 438–439.

Homola, J., S. S. Yee and G. Gauglitz, Surface plasmon resonance sensors: review, Sensors and Actuators B: Chemical, Volume 54, Issues 1–2, 25 January 1999, pp. 3–15.

Junhui, Z., C. Hong and Y. Ruifu, DNA based biosensors, Biotechnology Advances, Volume 15, Issue 1, 1997, pp. 43–58.

Leonard, P., Hearty, S., Brennan, J., Dunne, L., Quinn, J., Chakraborty, R., and O'Kennedy, R., Advances in biosensors for detection of pathogens in food and water, Enzyme and Microbial Technology, Volume 32, Issue 1, 2 January 2003, pp. 3–13.

Raiteri, R., Grattarola, M., Butt, H-J., and Skládal, P., Micromechanical cantilever-based biosensors, Sensors and Actuators B: Chemical, Volume 79, Issues 2–3, 15 October 2001, pp. 115–126.

Scheper, T. H., J. M. Hilmer, F. Lammers, C. Müller and M. Reinecke, Biosensors in bioprocess monitoring, Journal of Chromatography A, Volume 725, Issue 1, 19 February 1996, pp. 3–12.

Yanjun Tang, Ji Fang, Xiaodong Yan and Hai-Feng Ji, Fabrication and characterization of SiO2 microcantilever for microsensor application, Sensors and Actuators B: Chemical, Volume 97, Issue 1, 1 January 2004, pp. 109–113.

Wilkins, E., and P. Atanasov, Glucose monitoring: state of the art and future possibilities, Medical Engineering & Physics, Volume 18, Issue 4, June 1996, pp. 273–288.

Wang, J., Amperometric biosensors for clinical and therapeutic drug monitoring: a review, Journal of Pharmaceutical and Biomedical Analysis, Volume 19, Issues 1–2, February 1999, pp. 47–53.

Wang, J., G. Rivas, X. Cai, E. Palecek, P. Nielsen, H. Shiraishi, N. Dontha, D. Luo, C. Parrado, M. Chicharro et al., DNA electrochemical biosensors for environmental monitoring. A review, Analytica Chimica Acta, Volume 347, Issues 1–2, 30 July 1997, pp. 1–8.

Zhou, J., P. Li, S. Zhang, Y. Huang, P. Yang, M. Bao and G. Ruan, Self-excited piezoelectric microcantilever for gas detection, Microelectronic Engineering, Volume 69, Issue 1, August 2003, pp. 37–46.

CHAPTER 7

Chemical Sensors

Dr. Thomas Kenny, Department of Mechanical Engineering, Stanford University

Chemical sensors are used to detect the presence of specific chemical compounds or elements, and their concentrations. This chapter covers some basic concepts for sensing of chemical quantities and some important applications.

7.1 Technology Fundamentals

The Nose

The human nasal sensing apparatus contains a remarkably flexible and sensitive detection capability. Humans are capable of detecting and distinguishing thousands of different smells with almost instantaneous recognition. Odor detection is made very complicated because of the lack of uniqueness in the chemical basis of most smells. There is no "garlic molecule" that is distinct from the "enchilada molecule," yet a person can easily distinguish these smells.

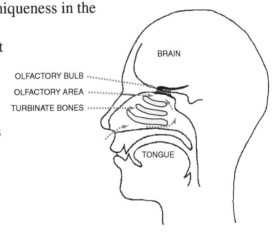

Figure 7.1.1: Human nose.

All biological odor detection systems are based on a fairly small number of distinguishable sensors. The smell recognition system is based on pattern-matching of the response of the different chemical sensors in the nose to various odors. Garlic and enchiladas produce a slightly different collective response in the entire set of sensors in your nose, and your brain has stored an extensive collection of these patterns which are used for comparisons. Psychologists have found that these odor patterns can be among the strongest of memories, and smells are often used to aid in the reconstruction of memories.

Chapter 7

Chemical sensing system designers need to draw lessons from these biological systems. An important lesson is that a multifunctional system will probably need to use a small set of distinct sensors and a pattern-matching algorithm to identify odors accurately.

Detectors of Particular Molecules

If a chemical sensing application requires detection of a particular molecule, several techniques are available. These techniques are based on the unique properties of particular molecules.

One set of properties is associated with the vibrational and rotational modes of molecules. The exact energies of these modes are generally unique to a particular molecule, and may be used for identification purposes. Most vibrations and rotations are "optically active," meaning that they may be excited by absorption of a photon, or may relax by emission of a photon. These photon absorptions are generally most likely to occur in the infrared, so infrared spectroscopy is a generally useful way to identify molecules.

For example, CO is a very simple molecule (visualize two balls and a spring), capable of oscillating at a single frequency (visualize them bouncing together and away) and rotating about two axes, both perpendicular to the line connecting the atoms. In quantum mechanics, a vibration is represented as a single frequency—the molecule may be in the ground state, or in any of a number of excited states, each of which is separated by the energy of the mode: $hw/(2\pi)$. Quantum mechanics includes "Selection Rules" which strongly favor relaxation a single step of $hw/(2\pi)$ at a time. This feature shows up in the infrared spectrum as a single absorption.

In the spectrum of the absorption of CO, shown in Figure 7.1.2, we see a pair of absorption peaks. This peak splitting is due to the fact that carbon exists in isotopes which have atomic mass of 12 or 13. The additional mass of the C13 reduces the vibrational energy simply because it lowers the resonant frequency $\left(w = \sqrt{k/m}\right)$.

In addition to vibrations, molecules also can have rotational energy. In the spectrum of NH_3 (ammonia), the rotational spectrum contributes a series of closely spaced lines centered about the vibrational peaks. These excitations involve the absorption of a photon and the change of both the vibrational and rotational energy.

Chemical Sensors

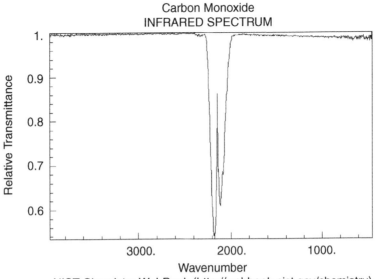

Figure 7.1.2: CO absorption spectrum.

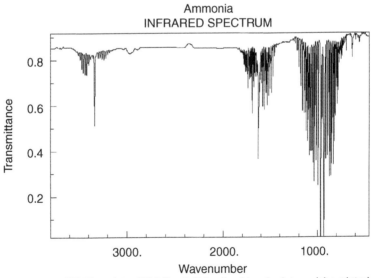

Figure 7.1.3: NH3 spectrum.

Chapter 7

Each molecule has a distinct infrared spectrum. Infrared spectrometers may be used to measure the absorption spectrum of a gas sample, and to look for features which indicate the presence of particular gas constituents. The simplest possible infrared instrument is shown in the next figure. In this instrument, light from a source passes through a sample-filled chamber, then through a chopper wheel, a filter selector, and is focused onto a detector. The filter selector consists of a pair of filters with very narrow infrared transmission bands. One of the filter's transmission bands is centered on the molecular absorption, and the other is centered just off the peak of the absorption band of interest. By alternating between these two filters, a difference signal may be detected only when the molecule of interest is present. This approach can be optimized only for a single molecule at a time, but is very reliable for such single-molecule applications. An example of such an application would be a CO detector for an automotive exhaust inspection/monitoring system.

Clearly, more complicated optical systems can be constructed. In addition to the use of a filter pair, it is possible to obtain complete spectra (as shown in the figures above) through use of an infrared spectrometer. Spectrometers are available in two basic designs, either of which tend to be very large and very expensive. One such instrument uses a rotating grating to disperse the light into separate wavelengths, and scan the spectrum across a detector.

Besides the vibrations and rotations, molecules may also be recognized by their mass (or mass spectra). Therefore, mass spectrometers are also used to detect and distinguish molecules. A diagram of a magnetic mass spectrometer is shown in Figure 7.1.4.

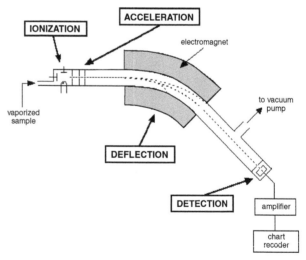

Figure 7.1.4: Mass spectrometer.

In this device, gas molecules are ionized by bombarding them with electrons from a heated filament. Some of the molecules in the gas become positively charged because the electron bombardment is generally more effective at stripping electrons than adding electrons (except for electronegative species like Cl.) Once the molecules are charged, they are accelerated to a constant kinetic energy by an electric field, and they enter the main body of the spectrometer. Here, there is a magnetic field, which exerts a Lorentz force on each molecule, tending to deflect the trajectory of the molecules. The amount of deflection depends on the velocity and on the mass, and therefore acts to separate the molecules by their mass. A series of detectors acts to record the molecule concentration vs. deflection angle, and the output can be displayed as a mass spectrum. One important drawback to such an instrument is that the molecules must travel their entire path without scattering from other molecules. In air at atmospheric conditions, the average distance between collisions is 1 micron. Since typical mass spectrometers need 10–100 cm of trajectory, the pressure must be 6–9 orders of magnitude lower than atmospheric pressure. Therefore, mass spectrometers generally require vacuum pumps.

In addition to the magnetic mass spectrometers, there are spectrometers based on oscillating electric fields (quadruple analyzers), and on the mass-velocity relationship for particles with the same kinetic energy (time of flight). These spectrometers all have similar requirements on pressure.

Finally, a series of chromatography instruments are available. In these instruments, the varying molecular diffusivities are used to detect specific molecules. A typical gas chromatograph is schematically shown in Figure 7.1.5.

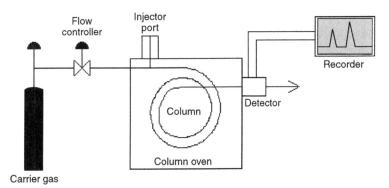

Figure 7.1.5: Gas chromatograph.

Here, a sample is added to a pressurized carrier gas, and forced to diffuse through a "column," which is essentially a very long narrow tube. The components of the sample diffuse at different rates through the column, and the detector at the end records a signal-vs.-time trace which contains peaks that may be identified as belonging to a specific sample. Chromatography has been in use for a long time, and is generally carried out in table-top instruments costing $10,000 to $100,000. Miniaturization of these instruments is currently the subject of much research in industry and academia.

Chapter 7

Microfabrication techniques can be used to manufacture the column and the detector area. An example is shown from research in the Kovacs group at Stanford, in the form of a capillary electrophoresis instrument shown in Figure 7.1.6.

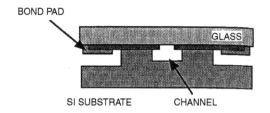

In electrophoresis, an electric voltage bias is applied to the column, and the carrier fluid is conductive. Therefore, there is a steady flow of carrier fluid ions through the column, which sweeps the sample along with it. Again, the diffusivity differences for different sample components causes the sample to be spread out into a spectrum, and the resulting output trace may be analyzed to identify the components and quantify their concentrations.

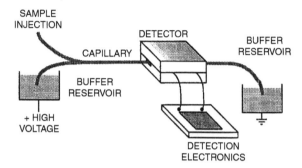

Figure 7.1.6: Electrophoresis instrument and its data.

Electrochemical Detection Techniques

The human body also does a great many chemical measurements on body fluids. The basic principles behind such measurements are discussed in the following paragraphs.

If a semi-permeable membrane separates two solutions, it may be possible for one component of the solutions to diffuse through from one side to the other. This notion of "semi-permeable membranes" may sound fanciful, but there are many biological examples of cell membranes that pass only a few nutrients.

In H2O solutions, many atoms and molecules exist in a charged state (Na generally has a +1 charge, for example), so the diffusion of these molecules also represents a diffusion of charge.

Now, when the two solutions are introduced, it is possible for a difference to exist in the concentration of the mobile ion on the two sides of the membrane. In general, one side is a reference sample, and the other side is a sample to be tested. Ions from each side begin immediately to diffuse through the membrane.

Chemical Sensors

Figure 7.1.7 shows a glass membrane separating a sample solution (on the outside) from a reference solution (inside), and electrodes to measure the potential difference.

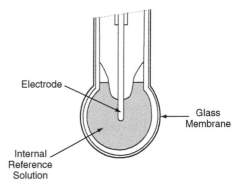

Figure 7.1.7: pH electrode.

If the concentrations are different on the two sides of the membrane, the amount of diffusion will be different, leading to a net diffusion. Since there is also a charge associated with these ions, there is a current. Very quickly, the motion of charge across the membrane causes the formation of an electric field, which opposes the flow of ions. An equilibrium is established when the electric field is large enough to overcome the concentration difference and the diffusion rates become balanced. At this point the potential difference across the membrane is absolutely related to the concentration ratio. The relationship is given in introductory chemistry textbooks as the Nernst equation:

$$V = \frac{RT}{nF} \ln\left(\frac{\text{concentration on one side}}{\text{concentration on other side}}\right)$$

In the normal units for chemistry, we have:

$$V = \frac{(8.31 \text{J}/\text{mol} \cdot \text{K})(300 K)}{(1 mol)(9.65 \times 10^4 \text{ valence}/\text{mol})} \ln\left(\frac{\text{concentration on one side}}{\text{concentration on other side}}\right)$$

$$= 2.6 \times 10^{-2} V \cdot \ln\left(\frac{\text{concentration on one side}}{\text{concentration on other side}}\right)$$

So, if there is a concentration ratio of 2, there will be a potential difference of about 20 mV. Such a potential difference is very easily measured.

In most actual situations, the thickness of the membrane is not too thin, so the potential difference builds up before there has been much actual transfer of concentration. This is important, because otherwise the required diffusion would alter the concentrations.

7.2 Applications

Automotive

An important automotive application for such a sensor is the automotive exhaust oxygen sensor. In an automobile, fuel and air are mixed and introduced into the cylinder for combustion. The mixture can contain too much fuel (rich) or too much air (lean). In older cars, this mixture was adjusted manually. Optimal performance was obtained by a slightly rich mixture, because this maximizes the compression ratio.

However, the EPA began to monitor automotive emissions, and determined that a rich mixture also leaves a large amount of undesirable hydrocarbons in the exhaust, which foul the air and have undesirable consequences for the ozone. In the 70s and 80s, it became required to have a control system that maintained the fuel-air ratio at the precise value needed for optimal combustion.

This legislative mandate was made possible by a very fortunate coincidence. Accurate measurement of the fuel and air input to the mixing chamber would be very expensive. However, after combustion, measurements can be carried out in the exhaust. During combustion of a rich mixture, very nearly all of the oxygen is consumed. During combustion of a lean mixture, the oxygen concentration is nearly the same as in the atmosphere (1–10%). So, it is possible to determine the state of the mixture very accurately by doing a simple measurement of the oxygen in the exhaust.

In a good automotive system, the concentration ratio for rich and lean may be different by 10–20 orders of magnitude. An electrochemical sensor that compared the oxygen concentration in the exhaust to that in the surrounding air would produce voltages near 1 V or near 0 V for those two conditions—easily distinguished by an engine controller.

O2 can diffuse in ceramic, so oxygen sensors can be made by producing a ceramic "nipple," whose inside surface is coated with a metal electrode. The potential between the inside surface and the surface exposed to the exhaust is measured and used to control the fuel/air mixture. Figure 7.2.1 shows the arrangement of electrodes.

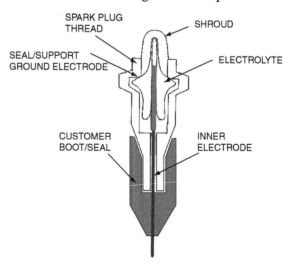

Figure 7.2.1: O2 sensor.

However, this basic technology has some undesirable characteristics. The sensor operates differently cold, and so cars are not optimized until the emissions system warms up. At present, most emissions from an automobile take place in the first few minutes after starting. In coming years, you should expect to see the EPA mandate heaters for more rapid warming of these sensors.

Other Miscellaneous Chemical Sensing Techniques and Applications

The measurement of impurities in water is becoming very important. Before using more expensive detection techniques, it is often easiest to simply measure the conductivity directly. Water's conductivity varies by several orders of magnitude as it varies from ultrapure to ordinary tap water. As for any other resistance measurement, a simple resistance bridge is often sufficient, as shown in Figure 7.2.2a.

If it is important to avoid direct electrical contact with the fluid because of chemical effects, the noncontact approach shown in Figure 7.2.2b will work. In this case, a measurement like the metal detector measurements is carried out. The conductivity of the fluid decreases the mutual inductance between a pair of coils wound around the fluid conduit. The disadvantage of this approach is that it is hard to detect the fluid conductivity if the pipes are themselves conductive.

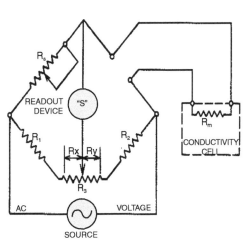

Figure 7.2.2a:
Conductivity measuring circuits.

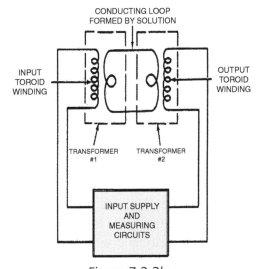

Figure 7.2.2b:
Noncontact conductivity measurement.

Chapter 7

CHEMFETs

Another method for chemical detection in solutions or in gas relies on the charge transfer that can occur during a chemical reaction on a surface. In these devices, the surface of interest is a metal electrode which is actually the exposed gate of a field effect transistor (FET). Since the conduction from source to drain in a FET is modified by charge on the gate, this device can be a remarkably sensitive detector of certain absorbed species. These sensors are called CHEMFETs.

Of course, to build a selective detector, it is necessary to select a metal electrode that allows only one chemical reaction to occur. Simple metals are not very selective, so simple CHEMFETs suffer from a lack of selectivity—meaning that they respond to many different chemical species. One way to improve the selectivity of such a sensor is to follow a biological example, and to coat the electrode with molecules that are indeed very selective. Antibodies are molecules that tend to react only with a particular (virus) molecule, and are more chemically selective than any simple metal electrode.

Finally, there are a number of medical applications that rely on detection of oxygen in the bloodstream. Unfortunately, the bloodstream is a difficult place to work because white blood cells interpret the presence of almost any foreign matter as an invading organism, and tend to form scabs on all surfaces of such objects.

Blood does exhibit a detectable change in color upon the absorption of oxygen, and blood oxygen may be crudely measured by looking at blood color. For example, a sensor that measures blood reflectivity at 700 nm and at 800 nm ought to be able to measure the blood oxygen content very accurately. The measurement at 800 nm is used to cancel out effects of scab overcoating.

One possible implementation is a fiber-optic system that transmits light of two colors (700 and 800 nm), and senses the reflected light intensity as a measure of blood oxygen. Such a system is often used during surgical procedures but is not typically used for long-term implants.

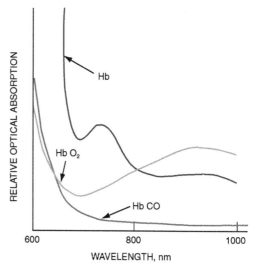

Figure 7.2.3: Optical absorption spectra.

Chemical Sensors

One device uses an LED emitter and a pair of detectors, each mounted looking out the side of a 1-mm thick catheter. The emitter and detector are separated by a few millimeters, so this instrument samples to a depth of a few millimeters, and is not badly affected by an overcoating of "scab."

This same technique can be applied to the measurement of skin color.

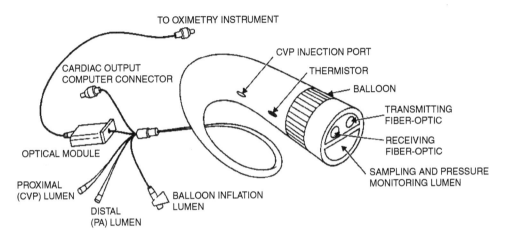

Figure 7.2.4: Opticath® catheter.
(Courtesy of Hospira, Inc.)

Summary

There are many different applications for chemical sensors, and many techniques which can be applied for any given application. In general, chemical sensing devices do not compare favorably to biologically developed detectors. Devices generally suffer from a lack of sensitivity, selectivity, and speed. For some applications, the signals of interest are large and easy to detect. For others, it is very tough going. Research can be expected to grow in detection of toxins in groundwater, vehicle emissions, biotoxins in public settings, and a large variety of chemicals in manufacturing process control.

CHAPTER 8

Capacitive and Inductive Displacement Sensors

Mark Kretschmar and Scott Welsby, Lion Precision

8.1 Introduction

Noncontact sensors and measurement devices—those that monitor a target without physical contact—provide several advantages over contacting devices, including the ability to provide higher dynamic response to moving targets, higher measurement resolution, and the ability to measure small fragile parts. Noncontact sensors are also virtually free of hysteresis, the error that occurs with contacting devices at the point where the target changes direction. With these noncontacting sensors there is no risk of damaging a fragile part because of contact with the measurement probe, and parts can be measured in highly dynamic processes and environments as they are manufactured.

Noncontact sensors are based on various technologies including electric field, electromagnetic field, and light/laser. Two complementary sensor technologies will be discussed in detail in this chapter: capacitive—electric field based, and inductive (eddy current)—electromagnetic field based.

A capacitive or inductive sensor consists of a *probe*, which is the actual physical device that generates the sensing field, and a *driver*, the electronics that drive the probe and generate the resulting output voltage proportional to the measurement. In some sensors, the driver is physically integrated in the probe itself.

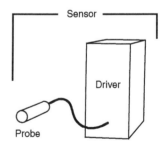

Figure 8.1.1: Noncontact sensor system.

Capacitive and inductive noncontact sensors have many similar characteristics as well as some characteristics unique to each technology. In the following pages we will discuss those things which are common to each of the technologies, compare those things which are different, and look at applications for each and at the unique solutions that are possible when using them together. We will start with capacitive sensors.

Chapter 8

8.2 Capacitive Sensors

Capacitive sensors are noncontact devices used for precision measurement of a conductive target's position or a nonconductive material's thickness or density. When used with conductive targets they are not affected by changes in the target material; all conductors look the same to a capacitive sensor. Capacitive sensors sense the surface of the conductive target, so the thickness of the material is not an issue; even thin plating is a good target. Capacitive sensors are widely applied in the semiconductor, disk drive and precision manufacturing industries where accuracies and high frequency response are important factors. When sensing nonconductors they are popular in packaging and other industries to detect labels, monitor coating thickness, and sense paint, paper, and film thicknesses.

Capacitive displacement sensors are known for nanometer resolutions, frequency responses of 20 kHz and higher, and temperature stability. They typically have measurement ranges of 10 μm to 10 mm although in some applications much smaller or larger ranges can be achieved.

Capacitive sensors are sensitive to the material in the gap between the sensor and the target. For this reason, capacitive sensors will not function in a dirty environment of spraying fluids, dust, or metal chips. Generally the gap material is air. Capacitive technology also works well in a vacuum, but the sensors must be properly designed for the peculiarities of a vacuum environment to prevent the probes from compromising the vacuum. Under some circumstances they can be used while immersed in a fluid but this is not common. When used with a conductive target, capacitive sensors are usually factory calibrated. Using capacitive sensors with nonconductive materials requires experimentation to determine the sensor's sensitivity to the material and the technology's suitability for the measurement.

Capacitive Technology Fundamentals

Capacitance is an electrical property that exists between any two conductors that are separated by a nonconductor. The simplest model of this is two metal plates with an air gap between them. When using capacitive sensors, the sensor is one of the metal plates and the target is the other. Capacitive sensors measure changes in the capacitance between the sensor and the target by creating an alternating electric field between the sensor and the target and monitoring changes in the electric field.

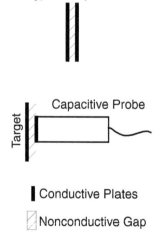

Figure 8.2.1: A capacitor is formed by the target and the capacitive probe's sensing surface.

Capacitive and Inductive Displacement Sensors

Capacitance is affected by three things: the sizes of the probe and target surfaces, the distance between them, and the material that is in the gap. In the great majority of applications, the sizes of the sensor and target do not change. When used with conductive targets, the gap material does not change. The only remaining variable is the distance between the sensor and target, so the capacitance is an indicator of the gap size, or the position of the target. Capacitive sensors are calibrated to produce a certain output change to correspond to a certain change in the distance between sensor and target. This is called the *sensitivity*.

Target Considerations

The electric field generated by a capacitive sensor typically covers an area on the target approximately 30 percent larger than the sensor area. Therefore, best results are obtained when the target is at least 30 percent larger than the sensing area of the probe. Sensors can be specially calibrated to smaller targets when the application demands it.

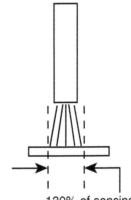

~130% of sensing surface diameter

Figure 8.2.2: "Spot size" on the target is about 30 percent larger than the probe's sensing surface area.

When used to measure nonconductive materials, the gap between the sensor and a conductive target is held constant and the material to be measured is passed through the

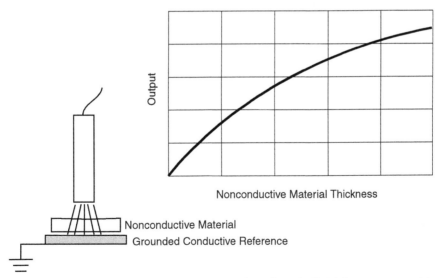

Figure 8.2.3: When measuring nonconductors, the electric field from a capacitive sensor passes through the nonconductive material on its way to a conductive target.

Chapter 8

gap. This way the gap is unchanging and the only remaining capacitance variable is the gap material. The output of the sensor will change with changes in the material's thickness, density, or composition. Holding two of these variables constant enables measurement of the third; for example, when a strip of plastic has a constant composition and density, changes in the capacitance can only indicate a change in thickness.

8.3 Inductive Sensors

Inductive sensors, also known as eddy current sensors, are noncontact devices used for precision measurement of a conductive target's position. Unlike capacitive sensors, inductive sensors are not affected by material in the probe/target gap so they are well adapted to hostile environments where oil, coolants, or other liquids may appear in the gap. Inductive sensors are sensitive to the type of target material. Copper, steel, aluminum and others react differently to the sensor, so for optimum performance the sensor must be calibrated to the correct target material.

Inductive sensors are known for nanometer resolutions, frequency responses of 80 kHz and higher, and immunity to contaminants in the measurement area. They typically have measurement ranges of 0.5mm to 15mm although in some applications much smaller and larger ranges can be achieved. Inductive sensors' tolerance of contaminants make them excellent choices for hostile environments or even for operating while immersed in liquid.

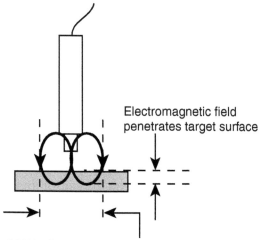

Figure 8.3.1: Inductive sensors use electromagnetic fields.

An inductive sensor's magnetic field creates electrical currents within the target material and therefore the targets have a minimum thickness requirement. Details are provided in the next section.

Inductive Technology Fundamentals

While capacitive sensors use an electric field for sensing the surface of the target, inductive sensors use an *electromagnetic* field that penetrates into the target. By passing an alternating current through a coil in the end of the probe, inductive sensors generate an alternating electromagnetic field around the end of the probe. When this alternating field contacts the target, small electrical currents are *induced* in the target

Capacitive and Inductive Displacement Sensors

material (eddy currents). These electrical currents, then, generate their own electromagnetic fields. These small fields react with the probe's field in such a way that the driver electronics can measure them. The closer the probe is to the target, the more the eddy currents react with the probes field and the greater the driver's output.

Inductive sensors are affected by three things: the sizes of the probe coil and target, the distance between them, and the target material. For displacement measurements the sensor is calibrated for the target material and the probe size remains constant, leaving the target/probe gap as the only variable. Because of its sensitivity to material changes, eddy current technology is also used to detect flaws, cracks, weld seams, and holes in conductive materials.

Target Considerations

Inductive sensors are sensitive to different conductive target materials. Sensors must be calibrated to the specific material with which they will be used. Some materials behave similarly and others differ significantly. There are two basic types of target materials: ferrous (magnetic) and nonferrous (not magnetic). Some inductive sensors will work with both materials, while others will only work with one type or the other. Some ferrous materials include iron, and most steels. Nonferrous materials include aluminum, copper, brass, zinc and others.

Inductive sensors are frequently used to monitor rotating targets such as crankshafts and driveshafts. However, measurements of rotating ferrous targets generate small errors because of tiny variations within the target material. This is called *electrical runout* or *magnetic runout*. These errors are quite small, on the order of 0.001 mm, which is negligible in the measurement of larger motions such as driveshafts. But inductive sensors are not well suited to high resolution measurement of rotating ferrous targets where they are expected to measure changes of 0.0001 mm.

Ideally, the target's measured surface must offer an area three times larger than the probe's diameter. This is because the electromagnetic field from an inductive sensor's probe is approximately three times the probe's diameter. Sensors can be specially calibrated to smaller targets when the application demands it.

Another target consideration is the thickness of the target material. Because electromagnetic fields penetrate the target, there is a minimum thickness requirement for the target. The minimum thickness is dependent on the electrical and magnetic properties of the material and on the frequency at which the probe is driven. As the frequency goes up, the minimum thickness goes down. This table lists some minimum thicknesses for common materials with a typical 1 MHz drive frequency.

Chapter 8

Copper	0.2 mm
Aluminum	0.25 mm
304 Stainless Steel	0.4 mm
Brass	1.6 mm
1040 Steel	0.008 mm
416 Stainless	0.08 mm
Iron	0.6 mm

8.4 Capacitive and Inductive Sensor Types

Capacitive and inductive sensors are available in three basic types: proximity switches, analog output, and linear output.

Proximity switches simply provide an on or off output to indicate whether or not the target is present in front of the probe. The distance from the probe to the target required to activate the proximity switch may be adjustable or may be fixed. Proximity switches do not provide any indication of the target's actual position, only whether or not its position is within the set proximity. Proximity sensors often have the driver electronics integrated in the probe body. They are inexpensive and readily available but they are not suited to precision positioning applications that require continuous readings of the target position.

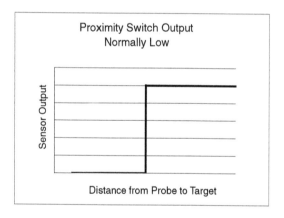

Figure 8.4.1: Proximity type sensors only provide off or on outputs which are triggered by the target position.

Analog output sensors provide a continuous analog output voltage that changes proportionately to the changes in the probe/target gap. Common output ranges are 0 to 10 VDC, ±10 VDC, 0–20 mA, or 4–20 mA. With analog sensors, the relationship of the output to the changing gap is not linear. While the output is not linear, it is repeatable, allowing for the accurate detection of a repeated position of the target. Analog output sensors frequently have gain and offset adjustments for adjusting the sensor to each application. Adjustable setpoint outputs are often provided on this type of sensor. These allow the user to set target position points at which digital outputs are activated.

Capacitive and Inductive Displacement Sensors

These sensors are useful where repeatability of position is more critical than knowing the position's exact dimension, and where the same sensor will be used in a variety of applications requiring recalibration in each. This type of sensor is used for coarse control of position or other production related servo controls where simple closer/farther information is sufficient as opposed to accurate, absolute position information.

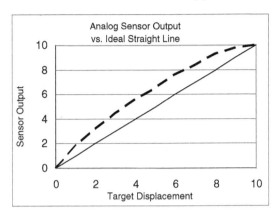

Figure 8.4.2: Analog sensor outputs are proportional to the target position but not in a linear fashion.

Linear output sensors produce a proportional output voltage whose relationship to the changing gap is linear. Common output ranges are 0 to 10 VDC, ±10 VDC, 0–20 mA, and 4–20mA. Linear output sensors are usually precisely calibrated at the factory to traceable calibration standards and therefore rarely have readily accessible calibration adjustments for the user.

These sensors are used when precise dimensional or position measurements are required throughout the range of the sensor. In critical dimensional measurement situations, these precision sensors are needed. They are used in positioning of photolithography stages in the production of semiconductor wafers, the disk drive industry, precision engineering applications, and anywhere that precise, continuous position information is required.

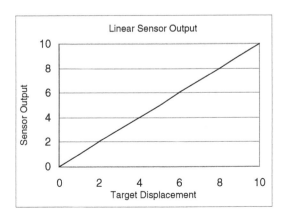

Figure 8.4.3: Linear output sensor provides accurate, linear representation of the target position.

8.5 Selecting and Specifying Capacitive and Inductive Sensors

Selecting the proper sensor starts by determining which of the three sensor types discussed previously is appropriate to the application. Proximity switches can be used to simply detect the presence of a target. Analog output sensors can be used for simple control of a process. Linear output sensors can be used for precision dimensional measurement of position, vibration, and motion.

Physical Configuration

Sensors are available in a large variety of shapes and sizes. The probes are usually cylindrical and are available with varied mounting schemes including threaded bodies for thru-hole or tapped hole mounting and smooth bodies for clamp mounting. Cylindrical probes range in size from 3mm to over 50mm. Probes are also available in other shapes as well, such as rectangular or flat, coin-like disks. These probes provide a lower profile design for sensing in areas where the length of a cylindrical probe may be prohibitive.

The physical size of the probe is directly related to the measurement range and offset of the sensor. Larger probes have a larger range and offset. Capacitive and inductive sensors are readily available with measurement ranges from 10 μm to 15 mm. Some manufacturers are willing to create custom probe designs for individual applications. These may be larger or smaller versions of standard probes or they may be incorporated into PCB designs or flex circuits for inclusion in equipment designs.

Drivers are available in various packages including: DIN rail mount, plug-in cards, bench top boxes, and small modules that are inline with the cable to the probe. Some drivers are integrated into the probe body itself, especially with the proximity switch type sensor.

Terminology

To make the best selection of sensor from the wide variety available you must first understand the terms used in the specifications. Unfortunately, not all manufacturers use precisely the same definitions but they are at least similar. These are some terms and definitions that you will encounter.

Output (or Output Range)

Output describes the type and range of the sensor's output which will be used to determine the measured dimension. Typical outputs are: 0–10 VDC, ±10 VDC, 4–20 mA, and 0–20 mA. The output indicates the total change of the output as the target moves through the total range. (Proximity switch sensors only have on/off switched outputs).

Capacitive and Inductive Displacement Sensors

Range (or Measurement Range)

Range is simply the operating range of the sensor. Sometimes the sensor's range is plainly stated as a range such as 2 mm–3 mm. This indicates that the sensor can measure the position of the target when its distance from the face of the probe is between 2 mm and 3 mm. However, range is sometimes given as a single dimension such as 1mm. This means that the total range over which the sensor can measure the target is 1mm but it gives no information as to where this 1mm range is located in terms of absolute distance from the probe face. In this case another specification is given called *offset* or *standoff*.

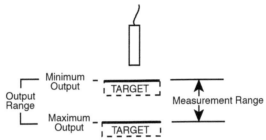

Figure 8.5.1: How the "Output" and "Range" of a sensor relate to target position.

Offset or Standoff

Offset or *Standoff* indicates where the active range is located relative to the probe's face. The range example above of 2 mm–3 mm may be listed as having a range of 1mm and an offset of 2 mm. This is typical for sensors with a single polarity output such as 0–10 VDC.

Some manufacturers may take a bipolar approach and define this same sensor as having a 2.5 mm standoff with a ±0.5 mm range. This is commonly used for sensors that have a bipolar output such as ±10 VDC.

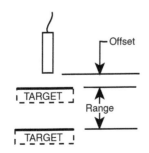

Figure 8.5.2: Some ranges are defined with an "offset" value.

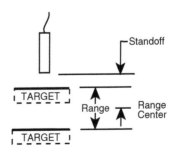

Figure 8.5.3: Some ranges are defined with a "standoff" value.

Chapter 8

Sensitivity

Sensitivity indicates how much the driver output changes as a result of a change in the gap between the target and the probe. If the sensitivity were 0.1 mm/1 V then for every 0.1 mm of change in the gap, the output voltage will change 1 V. When the output voltage is plotted against the gap size, the slope of the line is the sensitivity.

Linearity

This specification applies only to linear output type sensors, although it may be given occasionally for analog output sensors. This is a measure of how *straight* the line is when the target position is plotted against the driver's output. It describes how far the actual output varies from a perfect straight line drawn through the points, typically using a least squares fit calculation. It is usually given as a percent of full scale.

Linearity is important for precise measurements throughout the active range of the sensor. Linearity is only a measure of the straightness of the sensor's output. It is a major contributor to the accuracy of the sensor but it is not equivalent to accuracy. A sensor may be very linear, but be very inaccurate due to gross sensitivity errors, but a nonlinear sensor's accuracy will always be limited by the nonlinearity.

Bandwidth (Frequency Response)

When measuring a vibrating target the output is frequency dependent. As the frequency of the vibration increases, at some frequency the output begins to decrease due to frequency limitations within the driver electronics. Bandwidth usually specifies the frequency at which the output falls to −3 dB—approximately 70 percent; for example, 1 mm of vibration at the bandwidth frequency would appear as 0.7 mm at the output.

Resolution

Resolution is defined as the smallest reliable measurement that a system can make. The resolution of a measurement system must be smaller than the smallest measurement the sensor will be required to make. The primary determining factor of resolution is electrical noise. Electrical noise appears in the output causing small instantaneous errors in the output. Even when the probe/target gap is perfectly constant, the output of the driver has some small but measurable amount of noise that would seem to indicate that the gap is changing. This noise is inherent in electronic components and can be minimized, but not eliminated.

To measure resolution the noise voltage from the driver is viewed on an oscilloscope and measured. That measurement is listed as the resolution of the sensor. But there are two ways to calculate the measurement of the noise. The first is peak-to-peak (p-p). This simply measures the difference from the highest point to the lowest point. The other is RMS (root mean square) which is a mathematical calculation similar to but not the same as averaging. The RMS measurement of resolution is considerably lower than the p-p resolution. Both can be legitimate measurements, but be sure when comparing that all the sensors you are considering are using the same method.

Resolution is bandwidth dependent. Lower bandwidth means less electrical noise and therefore smaller resolution. Be careful when comparing resolutions that you know the bandwidth at which the measurement was taken. Many manufacturers list resolutions at several bandwidths, while others do not specify the bandwidth over which the resolution specification applies, resulting in an ambiguous specification.

Thermal Errors

All things electronic have the possibility of changing with temperature. In addition to electronic drift, physical changes in probes due to expansion and contraction can create output change that is related to temperature. Many of today's sensors are well designed to minimize and/or compensate for thermal errors but they are always present to some extent. Specifications may include thermal information listed as *temperature coefficient* or *thermal drift*. These specifications normally indicate the amount of change in the output per degree of temperature change.

Accuracy

Accuracy is the final result of the accumulated error sources that exist in any measurement system. Some error sources are part of the sensor itself, such as linearity, sensitivity errors, and thermal drift. Other sources are part of the measurement system as a whole, including fixturing, ambient temperature changes inducing differential thermal expansion within the system, mechanical alignment, electrical noise sources, and the accuracy of the equipment interpreting the sensor output. Because accuracy includes many factors outside of the sensor itself, it is rare for sensor specifications to include accuracy.

8.6 Comparing Capacitive and Inductive Sensors

Capacitive and inductive sensors each have unique characteristics. Below is a comparison of typical parameters for standard sensors. This is intended to provide a general idea of how the technologies differ. This data is by no means exhaustive; sensors are available with parameters exceeding those listed in Table 8.6.1.

Table 8.6.1: Comparing capacitive and inductive sensors.

Parameter	Capacitive	Inductive
Typical Range	.01 mm–10 mm	0.1 mm–15 mm
Resolution	2 nm	2 nm
Required Sensing Area	130% of probe diameter	300% probe diameter
Typical Probe Size	800% of Range	300% of range
Rotating Targets	Unaffected	Small errors on ferrous targets
Target Material	Conductive targets	Conductive targets only
	Not affected by material differences	Affected by conductive material differences
	Also measures nonconductors (i.e., plastics)	Does not measure nonconductors
Gap Material	Senses changes in nonconductive gap material	Ignores nonconductive gap materials
Cost	$$	$

8.7 Applications

A myriad of applications exist for these sensor technologies. Some of these applications can use either technology equally well, while others require one technology over the other.

Typical Sensor Operation

Noncontact sensors generally indicate a change from a known state. They are not frequently used for absolute measurements. The probes are mounted to make the measurement and the output is adjusted to some reference, usually zero, while the sensor is measuring the current state of the part. Measurement then proceeds with changes in the sensor output indicating changes from this initial condition.

Linear or Analog

Whether an analog sensor or linear sensor is necessary will depend on the required accuracies and specifications of the application. Generally, where the application is intended to produce a specific dimensional measurement of a particular feature or parameter, linear is the best option. When the application can operate with a simple "more or less" type of measurement, an analog type sensor is sufficient.

Interpreting the Output

Converting the output of the sensor into dimensional units is accomplished with this simple formula:

$$\text{Dimension} = \text{Output} \times \text{Sensitivity}$$

When using linear sensors the calculation is straightforward. A linear sensor's sensitivity is listed in calibration certificates or is otherwise listed on the sensor.

Capacitive and Inductive Displacement Sensors

Multiplying the change in the sensor output by the sensitivity yields the dimensional change; for example:

$$(2 \text{ V}) \times (1 \text{ mm}/1 \text{ V}) = 2 \text{ mm}$$

When using nonlinear sensors the sensitivity is not consistent throughout the range. The average sensitivity can be determined experimentally by measuring two part masters of known dimension, recording the sensor output for each part master and calculating the sensitivity:

$$\text{Sensitivity} = (\text{Change in Dimension})/(\text{Change in Output})$$

Because of the nonlinearity there will be measurement errors for parts that have a different dimension than the masters.

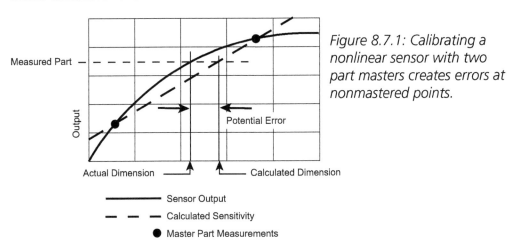

Figure 8.7.1: Calibrating a nonlinear sensor with two part masters creates errors at nonmastered points.

For more precise measurements with nonlinear sensors, an array of master parts is measured and recorded. The sensitivity is calculated for each sequential pair of parts The sensitivity will be different for each pair. These test measurements can be used to calculate accurate measurements with a computer program. The program can simply determine which sensitivity applies to the measured part, or, for maximum accuracy, a polynomial can be constructed based on the test measurements.

Dimension mm	Output volts	Sensitivity mm/volt
0.00	0.00	
		0.008
0.01	1.25	
		0.009
0.02	2.40	
		0.011
0.03	3.30	
		0.014
0.04	4.00	

Figure 8.7.2: Using multiple masters to calculate different sensitivities throughout the range of a nonlinear sensor.

Chapter 8

Multiple Channel Systems

Multiple channel measurements of the same physical target usually require that the sensors' drive oscillators be synchronized in frequency and phase. Sensor systems for multiple channel measurements need to be specified as such when ordered from the manufacturer so they are configured and calibrated to work together. Some less demanding applications may work with two independent, off-the-shelf sensors, but precision measurements will require synchronized sensors.

Applications for Capacitive or Inductive Sensors

These are typical applications in which either technology is effective assuming the measurement is taken in a relatively clean environment. Change the environment to include liquid, coolant, lubricant or other foreign material in the measurement gap and these all become applications for inductive sensors only.

All of the applications in this section assume a conductive target.

Relative Position (Displacement)

Figure 8.7.3: Measuring target position; the most fundamental application of noncontact sensors.

This is the most typical application for displacement sensors. This measurement is used in servo systems, part inspection, photolithography stages, and a host of other applications ranging from nanometers to millimeters.

The probe is mounted to monitor the position of the target.

Changes in the output of the sensor indicate changes in position of the target. When using linear sensors, the output change of the sensor is multiplied by the sensitivity of the sensor to produce a dimensional value. Some sensing systems are available with integral displays that convert the sensor output and display the dimensional value.

Position Window

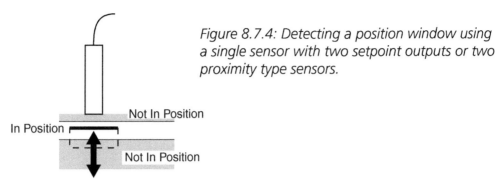

Figure 8.7.4: Detecting a position window using a single sensor with two setpoint outputs or two proximity type sensors.

This is a very specific application in which the position must be above a certain point but below another. This type of application is usually performed with an analog type sensor with two or more setpoints.

The probe is placed in the process to sense the position of the target.

A calibration routine follows in which each setpoint is adjusted to activate as a test or master target reaches the setpoint positions.

When setpoint 1 is active and setpoint 2 is inactive, the part is in the desired window. If the setpoint outputs are in any other condition the part is no longer in the acceptable window and the system takes appropriate action.

This same application can be accomplished by using two proximity type sensors. The outputs of the proximity sensors would be evaluated in the same way as the two setpoint outputs of the analog sensor. Cost, performance, and probe space determine the best solution. If there is room for two probes and the application doesn't require the higher performance and adjustability of analog sensors, proximity switches may be a more economical solution. Analog sensors are the choice if flexibility and adjustability drive the application.

Deflection, Deformation, Distortion

Applied essentially as a multichannel displacement measurement, this application specifically measures the intentional or unintentional distortion of an object.

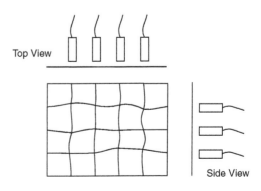

Figure 8.7.5: Measuring deformation of a surface with an array of sensors.

Several probes are mounted to measure the positions of different areas of the part. As distortion occurs, the position change for each channel can be collected. This data can be interpreted directly by the user, or can be fed to computer software to analyze and report on the distortion and predict the end effect of the distortion on the process being monitored.

Thermal Expansion

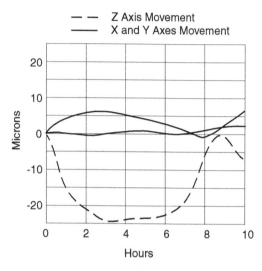

Figure 8.7.6: Capacitive and inductive sensors monitor thermal expansion in precision machine tools; here indicating 25 μm growth in the Z axis.

Thermal expansion and contraction can have profound effects on precision processes. For example, high-performance machine tools (mills, lathes, etc.) suffer from thermal expansion of the spindle and the machine as a whole. This is from machine generated heat and, to a lesser extent, ambient temperature changes.

There are a few approaches to solving thermal expansion problems: (1) measure the expansion throughout the temperature range and compensate during production based on the current temperature, (2) in the case of machine generated heat, measure the expansion over time and determine the amount of time required for the process to thermally stabilize and include an appropriate warm-up time, (3) if possible, use a sensor to monitor the expansion during production and compensate in real time.

Thickness

There are two ways to measure thickness with sensors. The first is a basic single channel position measurement of the top surface of the target while it rests on a reference surface. Surface finish, contaminants, and other factors affecting the way the target rests on the reference surface will create errors in the thickness measurement.

Capacitive and Inductive Displacement Sensors

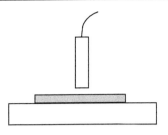

Figure 8.7.7a: A simple one channel thickness measurement depends on the quality of the contact between the target and the surface plate.

More precise thickness measurements which do not depend on a reference surface are accomplished with a two channel measurement. One channel measures the position of the top surface while the other measures the position of the bottom surface. The sum of the two indicate any changes in thickness while canceling any errors caused by the target moving up or down between the sensors.

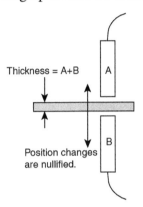

Figure 8.7.7b: A two-channel, differential thickness measurement nullifies errors created by changes in the target position.

As mentioned earlier in this chapter, inductive sensors require a minimum target thickness. Thin targets such as foils may not be thick enough for inductive sensors. This is dependent on the specific material and inductive sensor.

Assembly Inspection

Many assemblies include conductive and nonconductive parts. For critical assemblies 100 percent inspection is required to assure that the assembly includes the metallic component, which may or may not be visible.

Mount a probe such that the part to be inspected passes the probe such that the metallic part passes within the sensor's range.

Adjust the setpoint or a proximity switch to activate only when the metal is present.

Proximity type sensors can work in this application (especially inductive) but may not provide the level of adjustability required.

Chapter 8

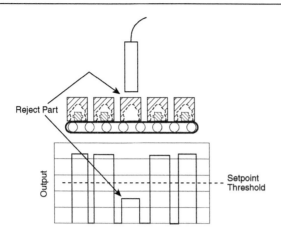

Figure 8.7.8: Incomplete assemblies are detected on a production line by a sensor with a switched output.

Vibration

Vibration is just a position measurement over time, but the output of the sensor is not only analyzed for position data but also for time-based information about the motion of the target. The sensor output is either viewed on an oscilloscope or collected by a computer data acquisition system for mathematical analysis.

Some manufacturers have basic signal processing modules designed to work with their sensors. These modules include peak capture functionality which can provide a TIR (total indicator reading) output. TIR is the difference between the maximum and minimum excursions of the output. A TIR output will indicate the total magnitude of the vibration of the target, but it provides no information on the time-based nature of the vibration.

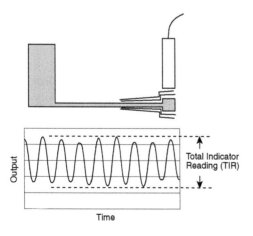

Figure 8.7.9: Vibration can be monitored over time for detailed analysis of simple TIR measurements.

Capacitive and Inductive Displacement Sensors

When measuring vibration, bandwidth (frequency response) is a critical consideration in selecting a sensor. If the measurement is only concerned with total magnitude of the vibration, bandwidth must be only slightly higher than the highest expected vibration frequency. If the measurement is to analyze detailed time-based behavior of the vibration, a bandwidth from three to ten times higher than the highest vibration frequency is required.

Applications for Capacitive Sensors

Precision Spindle Error Motion (Runout)

Precision spindles in disk drives and high-performance machine tools are achieving error motions of less than 100 nanometers. The error motion in disk drive spindles is directly related to the amount of information that can be packed on a disk. The error motion in a machine tool is directly related to hole roundness, feature location, and the quality of surface finish. The only way to properly measure the error motion of these spindles is when they are at full speed. This is a perfect application for capacitive sensing technology because of inductive sensors' rotating target errors (electrical runout) when measuring ferrous targets.

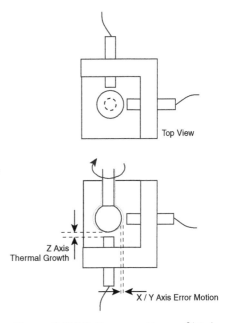

In the machine tool application, a precision master is mounted in the tool holder of the spindle to function as the target. In the disk drive application the disk spindle itself is the target.

Figure 8.7.10: Error motions of high-precision spindles are measured by capacitive probes in all three axes.

Two probes each are mounted in the X and Y plane (90° apart).

While the spindle rotates at operational speeds the outputs of the two channels are viewed on an oscilloscope or analyzed by computer software. When using an oscilloscope, the X channel is used to drive the horizontal axis and the Y channel is used to drive the vertical axis. This creates a *lissajous* pattern. The size and shape of the pattern gives indication of the amount and nature of the error motion of the spindle.

Chapter 8

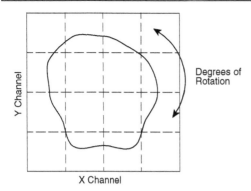

Figure 8.7.11: Driving the horizontal and vertical axes of an oscilloscope with the X and Y outputs indicates the motions of the rotating target.

For detailed analysis, computer software is required to calculate measurements for synchronous error motion (runout) and asynchronous error motion (nonrepeating runout NRR).

When precision rotating measurements of ferrous materials is required in a dirty environment, inductive sensors can be used by placing a nonferrous sleeve around the target or by characterizing the electrical runout of the target and using computer software to remove it from the result.

Rotor to Stator Gap (Embedded Sensors)

Stator to rotor gap can be a critical dimension in precision rotating devices such as air bearings and magnetic bearings. Capacitive probes can be designed to be embedded in the wall of the stators and monitor the position of the rotor. The sensor outputs are used as inputs to the servo system controlling the magnetic field or air pressure.

The specially designed probes are embedded by the bearing manufacturer and are in place for the final boring process of the stator. After final machining, the stator face with embedded sensors is as smooth as a standard stator. When the bearing is installed in the system, cables from the probes are connected to the driver electronics and the driver outputs connected to the system's servo inputs. When used in machine tools, these sensors provide extremely sensitive force measurements during the machining process.

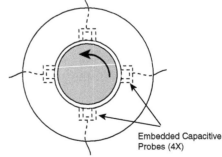

Figure 8.7.12: Embedded capacitive probes monitor rotor motion in air bearings and magnetic bearings.

Capacitive and Inductive Displacement Sensors

Nonconductive Material Thickness

Because capacitive sensors are sensitive to gap material they are effective at measuring nonconductive targets. Assuming that the composition and density of a nonconductive material are constant, changes in thickness can be measured by capacitive sensors.

The electric field from the sensing area of a capacitive probe must eventually return to ground. Nonconductors by definition cannot provide a ground. Nonconductive measurement is usually performed while the target material is between the probe and a grounded reference as shown in Figure 8.7.13. The gap between the probe and the grounded reference must be kept constant. Any change in that gap will appear as changes in the thickness of the target material.

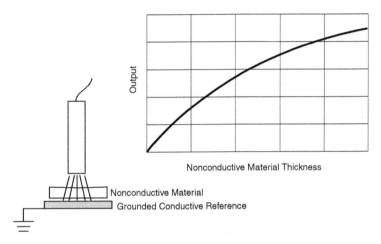

Figure 8.7.13: A nonconductive target can be measured against a conductive surface but the conductive surface must remain stationary.

In some less critical applications, measurements of nonconductive targets can be made without the grounded reference target. In this instance, the electric field from the probe wraps back to the grounded outside shell of the probe or the fixturing that is holding it.

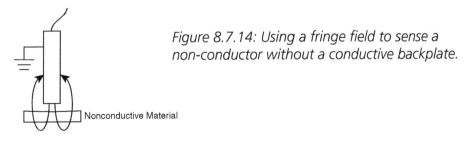

Figure 8.7.14: Using a fringe field to sense a non-conductor without a conductive backplate.

Chapter 8

As the nonconductive target nears the probe it interacts with the electric field causing a change in the output. But in this case the output will change in response to either changing thickness *or* changing position of the material. For measuring thickness, the gap between the probe and either the front or back surface must be held constant. Or, if the thickness is constant, changes in the output will indicate changes in position.

When measuring nonconductors, tests must be performed to determine the sensor's sensitivity to changes in the material thickness. See *Interpreting the Output* above. Manufacturers can determine and/or calibrate sensitivity when provided with a material sample, but the calibration will depend on consistency of test material with the sample material.

Double Feed Detection (Paper)

A frequent use of capacitive sensors is sensing double feeds of paper. This may be done in a top-of-the-line copier, a high-volume mailing machine, or a paper currency counter.

The application is straightforward. As paper is fed through the system, it passes between a capacitive probe and a grounded reference target. During setup a single thickness of paper is placed under the probe and the sensor is adjusted to some known output, usually zero. Then a double-thickness is placed under the probe and the sensor gain is adjusted for a known output, maybe 1 V or 5 V. A sensor used in this application will usually include an adjustable setpoint output. The setpoint is normally adjusted to activate at about half of the double sheet output. For example, if the double sheet output was 1 V, the setpoint would be adjusted to activate at 0.5 V. In this way, any output over 0.5 V would trigger a fault in the system.

As with all nonconductive applications, it is important to control the gap between the probe and the reference target behind the nonconductive target.

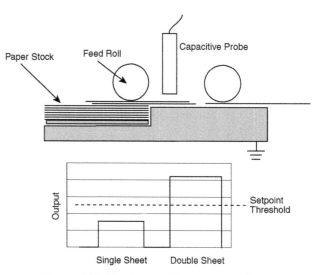

Figure 8.7.15: Capacitive sensors detect increases in the amount of material in the gap such as paper thickness.

Capacitive and Inductive Displacement Sensors

Label Sensing

The current state of the art in label sensing employs capacitive technology. During the process of placing labels on containers such as bottles, the location of the leading edge of the label is critical to locating the label in the center of the bottle. Optical sensors were used to perform this function. Then in the 1990s, clear labels on clear backing became common and optical sensors were no longer functional. Because capacitive sensing only senses changes in density or thickness, and is not affected by color, it is an ideal solution.

In label sensors, the detection of the label edge is accomplished using a differential sensor. The sensor actually has two sensing areas that are driven by the same circuit. The sensor only activates its output when there is a difference between the two sensors. (This particular configuration and application is covered under a U.S. Patent.)

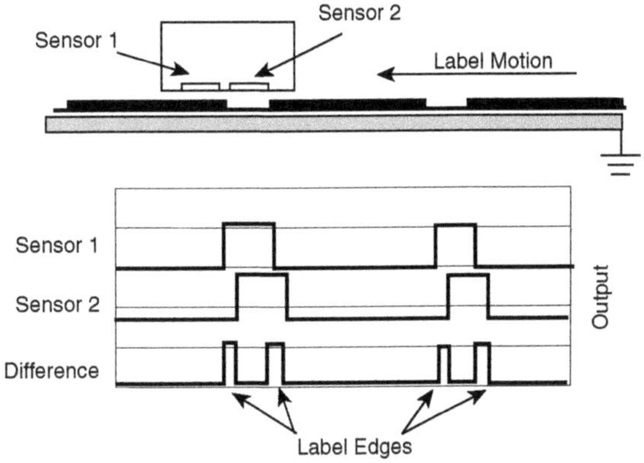

The advantage to the differential configuration is that the sensor is much more immune to changes in the gap between the sensor and the grounded reference. Any changes in the gap size are common to both sensors so there is no difference between them and the change does not affect the output.

Figure 8.7.16: Differential capacitive measurement senses the gap between labels on a web.

Glue or Paper Additive Sensing

The presence and/or amount of glue or other material deposited on a nonconductive material has been solved in many industries using capacitive technology. If the thickness/density of the underlying material is constant, changes in the amount of glue or other material is easily detected with capacitive sensors.

The probe is mounted in a position such that the applied material passes through its sensing area. Tests are performed to determine the sensor's sensitivity to the material and gain adjustments are made to an appropriate level. The output of the sensor is then monitored by the control system and either warns an operator if the material is no longer present, or in more sophisticated systems, the output is used to control the flow of the applied material.

Chapter 8

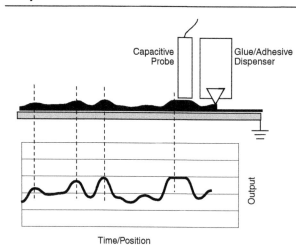

Figure 8.7.17: Capacitive sensors' sensitivity to varying amounts of nonconductive material make them ideal glue, adhesive, or additive sensors.

Applications for Inductive Sensors Only

These applications require a sensor that can operate in a dirty, hostile environment that would render a capacitive sensor ineffective or that exploit the unique nature of an electromagnetic field as opposed to an electric field.

Driveshaft Runout/Motion

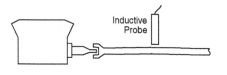

Figure 8.7.18: Inductive sensors are ideal for measuring rotating motion when the environment is dirty and accuracy less than a micron is not required.

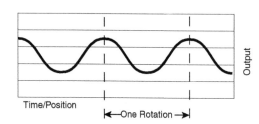

While inductive sensors have small errors when measuring rotating, ferrous targets, these errors are usually less than a micron. There are many rotating target applications that take place in a dirty environment for which inductive sensors are well suited that do not require that level of resolution.

Capacitive and Inductive Displacement Sensors

These measurements may be performed as part of a testing process or permanently installed in the final product. Examples are driveshafts or crankshafts running in an oil bath, or marine propeller driveshafts which may have water spraying or splashing in the sensing area, or power generator shafts which are monitored during operation to indicate bearing wear. The output of the sensor may be monitored by an operator or a computerized control system.

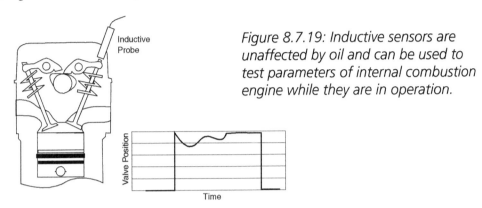

Figure 8.7.19: Inductive sensors are unaffected by oil and can be used to test parameters of internal combustion engine while they are in operation.

Valve Stroke, Piston Dynamics

Tests on internal combustion engines while they are running are ideal for inductive sensors. Their immunity to oil allows them to be installed directly in an engine for test during operation. One such test measures the position of the valve stem at its highest point. The probe is installed in the engine and positioned to sense the position of the end of the valve stem at its highest point. Measurements are taken while the engine is running and indicate the position of the valve stem as well as the repeatability of the valve stem position on consecutive excursions.

Calender Roller Gap

Rolling of sheets, coils, bars, and shapes is a process used to fabricate metals to proper thickness, shape, and texture. This rolling process also controls thickness, perforation, and texture on paper products, rubber, plastics, drywall, engineered wood, and other materials.

The calender rolling process squeezes the material to shape and thickness through the bite area between two parallel calender rollers, which can be up to three feet in diameter and ten or more feet long. Depending upon material characteristics and process requirements, the material might be pre-heated or process working fluids could be present such as steam and solvents.

Chapter 8

The process requires precise control of roller gap to control quality and prevent roller crash damage and premature wear.

Inductive position sensors can measure the roller gap accurately in the hostile process environment while the material is running. This enables precision machine setup, real-time roller gap control, and SPC data on the setup and running gap for all material produced.

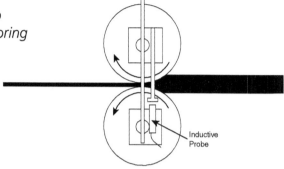

Figure 8.7.20: Inductive probes help control material thickness by monitoring the gap between calender rollers.

Thread Detection

Inductive sensors are a perfect technology for detecting the presence/absence or the quality of threads in tapped holes. A traditional probe design in which the electromagnetic field radiates from the end of the probe can function as a thread sensor, but sensitivity is increased with specially designed thread detection probes. These probes have internal coils turned 90° so that the field emanates from the side of the probe's cylindrical body. This gives deeper penetration of the field into the tapped surface. In its simplest application, these are used in automated production lines to check presence/absence of thread to check for a broken tap; especially when 0ppm defects are the requirement. A more sophisticated application uses the probes to monitor thread quality and depth to indicate the need to replace a worn tap. Thread detection probes are also monitored during insertion to detect a broken tap still present in the hole. The automation system stops the insertion if a broken tap is detected, thereby preventing damage to the probe.

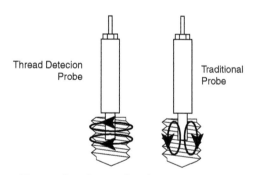

Figure 8.7.21: Inductive sensors can detect the presence or quality of threads in a tapped hole.

Capacitive and Inductive Displacement Sensors

Stamping/Molding Die Protection

When parts are formed through a stamping or molding process, offline measurement of finished product may not be desirable, since a significant quantity of product could be manufactured before a defect would be discovered.

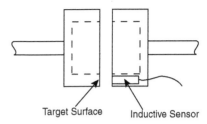

Using inductive sensors during the forming process, real-time measurements can be made on part dimensions, die-half position/movement, mold closure, and mold core movement. Inductive sensors are the only choice because a stamping process will produce significant shock, often in an oily environment; the molding process can impart high temperatures (200°C) and significant pressures (20,000 psi) on the sensor.

Figure 8.7.22: Inductive sensors monitor injection molds for proper closure.

In addition, these systems can be used to measure critical machine setup dimensions such as die closure position and the gap between the mold and core to ensure proper setup.

Application Using Capacitive and Inductive Sensors Together

This is a unique application that exploits the differences of the two technologies

Film/Paint Thickness of Conductive Part

For years measuring the thickness of paint or film against a moving metal substrate in a noncontact fashion has presented a difficult challenge. By using the unique aspects of capacitive and inductive sensors together, this can now be accomplished.

Capacitive sensors can measure the nonconductive paint or film, but they are also affected by any changes in the gap distance to the moving substrate. When the substrate is moving, it is not possible to discern if changes in the output are due to changes in the thickness of the paint or film or movement of the substrate.

Inductive sensors can measure movement of the substrate but they do not sense any changes in the paint or film thickness. (See Figure 8.7.23.)

If both technologies are used and their outputs combined differentially, any movement of substrate is cancelled through the differential operation, leaving only the changes due to thickness of the film or paint.

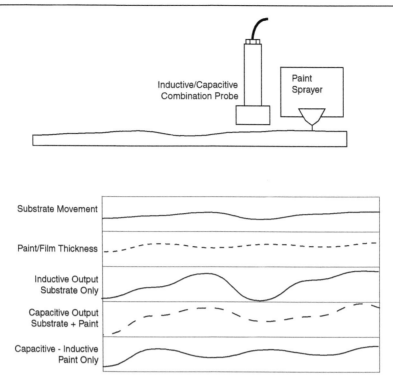

Figure 8.7.23: Combining inductive and capacitive technologies in one application to measure film or paint thickness on a metal substrate

Thickness Change = Capacitive Output − Inductive Output

which is essentially:

Thickness Change = (Substrate Movement + Thickness Change) − (Substrate Movement)

For greatest accuracy, the application should be performed with a specially designed probe that contains both capacitive and inductive sensing elements that are constructed to be coaxial. Coaxial construction assures that both sensors are reading precisely the same area of the target. If independent sensors are used, differences in the displacement of the substrate at the two different locations will appear as changes in the thickness of the film or paint. Don't underestimate this potential error source. When using sensors with sub-micron sensitivities, small differences can have profound effects.

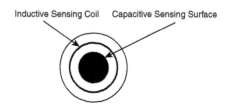

Figure 8.7.24: End view of a combination probe with capacitive and inductive elements.

Capacitive and Inductive Displacement Sensors

Considerations for Maximum Effectiveness

Temperature effects are usually the largest source of error in a precision measurement. Capacitive and inductive sensors exhibit some finite amount of drift associated with temperature. But with current design strategies these effects are minimized to very small amounts in comparison with temperature effects on fixturing and the physical properties of the target.

During design of the measurement system, care should be taken to minimize the *structural loop*—the total distance from the probe mount to the point of the measurement. The larger this loop, the more susceptible it is to temperature variations. There is also the potential for vibration. If not carefully designed, the probe mount system can function as a large tuning fork and vibrate at a resonant frequency. Stiffness of the mount is a key to minimizing these error sources.

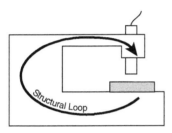

Figure 8.7.25: Larger structural loops are sensitive to vibration and thermal expansion. Keep them as small as possible.

Obviously, the best way to deal with temperature-related errors is to perform the measurement in a temperature stable environment.

8.8 Latest Developments

The physics of capacitive and inductive sensing has long been understood, not leaving much room for radical improvement in the physical design of probes. The relationship between the area of the sensor and its measurement range will remain essentially unchanged. Electronics, however, continues to advance at a tremendous pace. One advancement is in reducing the electrical noise inherent in electronic amplifiers. This has resulted in increasing resolutions, which can now be measured in nanometers for high-end systems.

Another advancement has occurred in miniaturization. As circuitry gets smaller, more and more sensors will be produced with the electronics contained in the actual probe body. That style of sensor can't offer as much adjustability, so external electronics with adjustment knobs and switches are still the order of the day for applications requiring user setup and adjustment. But with miniaturization of computer technology, probes with embedded electronics will become available in which the adjustments are made through a simple serial interface to internal, electronic adjustments instead of external, physical adjustments. Obviously, these sensors would need to be connected to an external computer or controller that can transmit the setup commands to the sensor.

8.9 Conclusion

Capacitive and inductive sensors have been solving measurement problems for several decades. With each new decade and advancement of technologies such as automation, the demand and opportunities for these sensors continues to increase. Every year capacitive and/or inductive sensing solutions are discovered for problems that have nagged engineers for years. Increasing awareness of these sensing technologies will lead to better solutions sooner and provide an abundant saving of resources.

Resources

1. **Capacitive Sensing Theory Website hosted by Lion Precision**
   www.lionprecision.com/theory

2. **Lion Precision Capacitance Displacement Measurement for 2003**
   Includes eleven page tutorial on capacitive sensing theory. Available from:
   Lion Precision
   563 Shoreview Park Road
   St. Paul, MN 55126
   USA
   651-484-6544
   www.lionprecision.com

CHAPTER 9

Electromagnetism in Sensing

Dr. Thomas Kenny, Department of Mechanical Engineering, Stanford University

9.1 Introduction

This chapter discusses the basic principles behind the use of electromagnetism in sensing. Since many established sensor types rely on electromagnetism, it covers a broad set of designs and reviews several product types.

9.2 Electromagnetism and Inductance

Before we get too far into electromagnetism, we begin with a review of the properties of an inductor as used in electronic circuits. An inductor is a passive circuit element that resists changes in current. The equation governing its behavior is:

$$V = -L\frac{\partial I}{\partial t}$$

where L is the inductance in units of henrys. As we can see from this equation, there is a voltage across the inductor whenever the current changes. The minus sign indicates that the voltage (V) opposes the change in current (I)—which is to say that whenever an external circuit tries to cause more current to flow, it must provide a voltage in order to overcome the voltage that arises in the inductor.

If we assume (as for all of the differential equations in this chapter) that the current and voltage are both oscillating quantities:

$$V = V_o e^{i\omega t}$$
$$I = I_o e^{i\omega t}$$

then we have

$$V_o = -i\omega L I_o$$

If we recognize Ohm's Law here ($V = IR$), then the effective resistance of the inductor is "R" = $-(i \omega L)$. As was the case for a capacitor ("R" = $1/(i \omega C)$), the i implies that there is a shift in phase between voltage and current, and the ω (Greek omega) implies that the effective resistance increases with frequency.

Chapter 9

Using this analogy, we can easily see how to build a large array of inductance-measuring circuits, which are exactly like resistance-measuring circuits (dividers and bridges). In fact, the most common approach to inductance measurement is a bridge.

How big is a typical inductance? The inductance of a coil is given by:

$$L = \frac{\mu_o N^2 A}{\text{length}}$$

Assume we have a coil with a 1-cm diameter and a 1-cm length, with 1000 loops of wire. Then,

$$L = \frac{\left(4\pi \times 10^{-7} \, H/m\right)\left(1 \times 10^3\right)^2 \left(\pi \left(5 \times 10^{-3} m\right)^2\right)}{\left(1 \times 10^{-2} m\right)} = 9.9 \times 10^{-3} \, H$$

Now, if we want to measure the effective resistance of this device, we will need to apply an oscillating voltage at some frequency. Normally, if we used a frequency of 1 kHz, the effective resistance of this device would only be about 60 ohms. Clearly, this is a small resistance, and measuring it to a high degree of accuracy will be difficult. We may go to a higher frequency to improve this situation.

What is really preferred is an inductor with a higher inductance. This is achieved in a way that is very much like what is done in capacitors. In an inductor, we can fill the coil with a material that has a higher magnetic permittivity. For example, iron has a *relative permittivity* of about 300, and permalloy has a relative permittivity of as much as 5000. So, we generally see coils filled with metallic materials such as iron.

To understand the situation with coils, we must review some electromagnetic theory. We will look at a few "laws" of electromagnetism and some example discussions of their effects.

Oersted discovered that whenever electrical charges move (current), a magnetic field is created. For the simple case of a straight wire, there is a magnetic field around the wire given by:

$$B = \frac{\mu_o I}{2\pi R}$$

Since the magnetic field is a vector, we need also to be concerned about the direction. For this case, the field lines form a loop around the wire, and the direction in the loop is determined by the *righthand rule*.

For a current of 1 ampere, the magnetic field 1 cm away from the wire is

$$B = \frac{(4\pi \times 10^{-7}\,H/m)(1A)}{(2\pi)(1\times 10^{-2}\,m)} = 2.0 \times 10^{-5}\, Tesla$$

The earth's magnetic field is generally about 5×10^{-5} Tesla, so this field is smaller than the earth's field. As we can see, the fields around wires are not generally large enough to be noticeable.

Whenever charges move in magnetic fields, there are forces on those charges. The expression for this force (known as the Lorentz force) is:

$$F = qV \times B$$

Since V, B, and F are vectors, we need to recognize this as a vector cross-product, and use a different *righthand rule* to determine the direction of the resulting force.

In this case, the force is always perpendicular to the velocity, and causes a freely moving electron to be deflected into a circular orbit. Since the force and velocity are perpendicular, the speed of the electron is unchanged.

Finally, Faraday's law of induction states that whenever the magnetic field that passes through a loop of wire is changed, there will be a voltage induced in the coil. We define the flux, ϕ (phi), as the product of the area of the loop and the component of the magnetic field perpendicular to the surface of the loop.

The induced voltage is therefore:

$$V = -\frac{\partial \phi}{\partial t}$$

We can put all of this together to see what is going behind the concept of inductance. If we have a simple coil, and try to suddenly cause a current to flow through it, the initial current flow causes a magnetic field to begin to form in the coil. Since this magnetic field is increasing, there is a change in the flux through the coil, and an opposing voltage appears. Eventually, the current rises to its limiting value, the magnetic field is stable, and the opposing voltage dies away.

Throughout all of these equations, there are many minus signs to keep track of. Rather than become accurate bookkeepers, it is easier to become familiar with Lenz's Law, which simply states that the combined effect of all of these interactions is to resist the modification of the magnetic field. In the case of an inductor, the electromagnetic interactions resist current changes as a way to resist any change of the magnetic field in the coil. We shall see this effect in all inductance-based sensors.

9.3 Sensor Applications

A good example of the Faraday law of induction is illustrated in Figure 9.3.1. Here a loop of wire is positioned between the pole faces of a permanent magnet. The magnetic field is confined to the region between the pole faces, and is essentially zero elsewhere. Therefore, the magnetic flux through the loops in this situation is simply the area of the loop that is within the magnetic field multiplied by the value of the field.

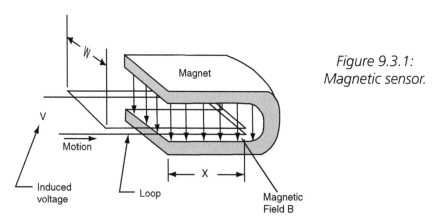

Figure 9.3.1: Magnetic sensor.

A voltage V is induced in the loop whenever it moves laterally. In this case, we assume it is confined to motion left and right in the figure, and that the flux at any moment is given by:

$$\phi = Bwx$$

where w is the width of the loop, x is the length of the loop within the coil, and B is the field in the magnet. Therefore, the voltage is:

$$V = -\frac{\partial \phi}{\partial t} = -Bw\frac{\partial x}{\partial t}$$

Electromagnetism in Sensing

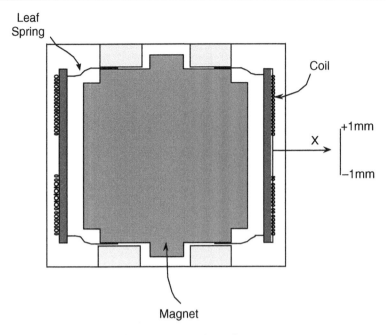

Figure 9.3.2: Geophone.

Since dx/dt is really the velocity of the coil, we see that this configuration is useful as a motion detector, and useless as a position detector. This approach is the basis of many so-called "moving coil" detectors, in which a voltage is generated whenever an external signal causes a coil to move relative to a permanent magnet. A good example of a commercial product based on such a device is the Geophone, as made by GeoSpace Corp (see Figure 9.3.2). In this device, a set of coils measures a differential voltage whenever a spring-supported magnet moves. This device is generally constructed with a fairly low frequency resonance—about 1 Hz. It is commonly used for detection of seismic signals or other low-frequency ground vibrations. It is also commonly used in the oil exploration business with buried explosive charges to map underground resource deposits.

Another very common approach to the use of electromagnetics in sensors is the inductive proximity sensor shown in Figure 9.3.3. Here, a pair of coils are wired in a bridge circuit and biased with an ac signal. If a conducting object is positioned near the end of the device, it is closer to the sense coil than the reference coil. The presence of a conductor has an important and complicated effect in this situation.

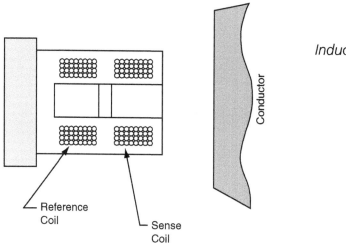

Figure 9.3.3:
Inductive proximity sensor.

A sheet of metal may be described as an assortment of essentially free electrons. If an electric field is applied to the metal, the electrons are free to move, and do so with very little resistance—hence the low resistance of a sheet of metal. In the presence of a magnetic field, the electrons feel a Lorentz force, which causes their trajectories to be curled into circles. These circular trajectories create a new magnetic field, which extends out of the sheet of metal back through the coils in the device. Since the sense coil is closer to the metal sheet, the flux through the coil due to the sheet is larger in the sense coil. These additional magnetic fields cause an additional opposing voltage in the coil, and the effect is to increase the inductance of the coils. Since the sense coil is closer to the sheet, its inductance is increased more.

Electromagnetism in Sensing

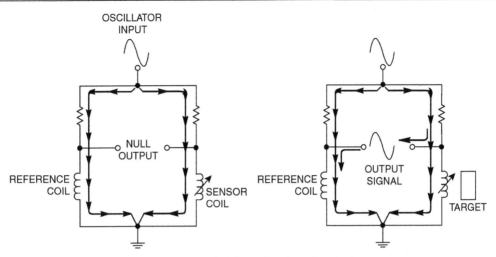

Figure 9.3.4: A bridge circuit using inductors.

A bridge circuit is used to measure these changes in inductance, as shown in Figure 9.3.4. Here, we can see that the effect of the sense coil changing more than the reference coil is that there will be an AC voltage appearing at the bridge outputs. The amplitude of this difference is proportional to the inductance difference, which is related to the distance to the metal sheet in a very complicated way. The actual relation between distance and inductance change is too complicated to derive in general, since it relies on the geometry of the situation, so this approach is not generally used for accurate position sensing.

However, there are a number of important applications that benefit from a signal that indicates the presence of a metal object. For example, many automotive interchanges these days rely on this sort of inductive proximity sensing to detect the presence of a car in the left-turn lanes or waiting in other lanes. Late at night, these systems control many intersections on streets. You will find that the speed with which left-turn signals are activated depends on the location of your car. If you look closely, you will find that a circular patch of the road surface has been cut open. These cuts allowed the burial of a pair of loops that sense the presence of the metal in a car. In order to activate the sensor, a car must be positioned so that there is a fair amount of metal in the region above the coils that are filled with a magnetic field when the coils are energized.

A number of commercial inductive proximity sensors are also available based on this general technique, some miniaturized for robotic applications.

An important but different sort of phenomena can be present if the metal object used to modify the inductance is a ferromagnetic material. Ferromagnetism is an effect in which the magnetic moments of the bound electrons and the nuclei also interact with external fields. In a ferromagnet, the orientation of the magnetic moments in the material can become aligned with external fields, causing an effective field amplification, which can be very large. In addition, this internal alignment can persist after the external field is removed, or even if its direction is changed.

Because the field amplification may be as much as 1000 times, the presence of ferromagnetic materials greatly increases the sensitivity of inductance coil bridges to their presence. Therefore, such sensors are very much more sensitive to ferromagnetic objects, and one often sees sensors which are tuned to detect a small piece of ferromagnet attached to a moving metal part.

A very common example of the use of a ferromagnetic element is as shown in Figure 9.3.5. In this system, the amount of magnetic field from one coil that is directed towards part of a second coil is dependent on the position of a ferromagnetic element. In the system shown in the drawing, the two halves of the pick-up coil (wired to V_{out}) are wound in opposite directions. If the ferromagnet were not present, the flux in each half of the pick-up coil would be equal and opposite, and V_{out} would be zero. When the ferromagnet is positioned in the middle, there is also a complete cancellation of the flux. However, whenever the ferromagnet is displaced, the flux balance is changed, and the net effect is that there will be a voltage across the pickup coil whose amplitude is proportional to the displacement of the ferromagnet from its center position.

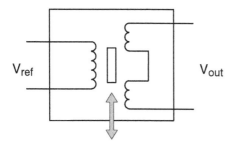

Figure 9.3.5: Inductive sensor circuit diagram.

Electromagnetism in Sensing

This sort of inductive position sensor is very commonly used in a class of devices called linear variable displacement transducers (LVDT). These transducers can have very good accuracy (much better than 1% of the total range of motion, which is commonly called "stroke"), and are often used for precision position measurement applications, such as flap and rudder position measurements on aircraft.

One method of analysis for these systems is based on a magnetic circuit analogy, in which the inductance is seen as the sum of a series of "reluctances," which are individually defined as

$$L = \sum_i \frac{1}{R_i} \text{ where } R_i = \frac{\text{length}}{\mu A_i}$$

where the length is of a segment of the magnetic circuit, the μ is the permittivity of that segment, and A_i is the cross-sectional area of that segment.

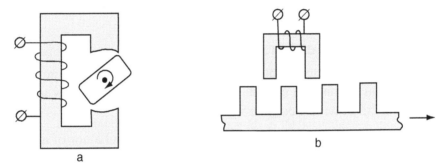

Figure 9.3.6: A magnetic circuit example #1.

An example of a device that relies on a magnetic circuit is shown in Figure 9.3.6. A coil is wrapped about one part of a ferromagnetic structure. The magnetic field created when there is a current in the loop is almost entirely confined within the magnetic material. A moving ferromagnet is placed within the extended legs of the structure. Since the magnetic field will mostly pass through the moving element, the length of the total magnetic circuit is dependent on the position of the moving element. Since the reluctance of the circuit is the sum of the reluctance of the elements, the total inductance that is measured at the coil is also dependent on the position of the moving element.

Chapter 9

Alternatively, these systems may be used in situations where the reluctance is a continuous function of the position of a moving element, as in Figure 9.3.7. The reluctance is inversely proportional to the permittivity, and the inductance is inversely proportional to the sum of reluctances, so the total inductance is maximized when the magnetic circuit has a minimum of air gap. In these systems, the inductance of the total element is increased by a large factor when the moving element is positioned so as to minimize the air gap.

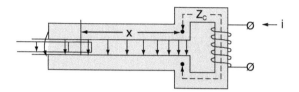

Figure 9.3.7: A magnetic circuit example #2.

9.4 Magnetic Field Sensors

There is another class of instruments used for detection of static magnetic fields. Magnetometers can be made in several ways, and here we will review a couple of specific devices that are of widest commercial use.

The flux gate magnetometer relies on measurement of the behavior of a ferromagnet-filled inductor. In the absence of external magnetic fields, current passed through the coil causes the formation of a magnetic field, which acts to polarize the ferromagnetic material. In general, the memory of the ferromagnetic material causes a hysteresis in the relation between the applied field and the polarization of the ferromagnet.

We can see this by thinking about the starting situation, where the ferromagnet is unpolarized and there is no current. If a current is applied, it polarizes the ferromagnet. As the current is increased, the polarization increases until saturation. Now, the current may be reduced to zero, and the resulting situation will include some residual polarization of the ferromagnet. If a current in the opposite direction is applied, the polarization is reversed, eventually saturated, and retains some residual reverse polarization when the current is again turned off. A graph of magnetization (B) versus applied magnetic field (H) is shown in Figure 9.4.1. The applied magnetic field is proportional to the current through the electromagnetic field and we see that the magnetization of the core responds to the applied field with hysteresis.

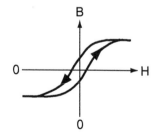

Figure 9.4.1: Hysteresis loop.

In the absence of an external magnetic field, the hysteresis loop is perfectly symmetric, and the graph of voltage versus current would only include the odd harmonics of the drive frequency.

If there is an external magnetic field, the hysteresis loop is shifted away from the origin. This is because there is a residual applied magnetic field when the current is off, due to the external magnetic field. One result of this is that the symmetry of the hysteresis loop is spoiled. If the I-V graph is analyzed, there will be a component at the second harmonic of the drive frequency, and the amplitude of this harmonic will be proportional to the component of the external magnetic field vector along the coil axis. Therefore, this device may be used to sense external magnetic fields.

In fact, when properly constructed and wired, this sort of sensor can be very sensitive to small changes in external magnetic field. This class of magnetometers is generally used for space science missions, and for all precision terrestrial applications.

A miniature flux-gate magnetometer is available from Applied Physics Systems, featuring resolution of less than 10^{-10} T/sqrt(Hz), very good linearity, and very small size. It is a fairly expensive instrument.

This style of magnetometer is also used for a number of prospecting applications. In several important applications, buried objects produce magnetic fields, and instruments that can measure local magnetic field gradients are very useful. Instruments available from Schonstedt feature a pair of flux gate magnetometers operated in a differential mode. If magnetic objects are positioned near the pair, the earth's magnetic field is distorted and the difference between the two magnetometers does not cancel perfectly. In this mode, the gradiometer can sense the presence of a magnetic object. This kind of instrument is commonly used by road-repair crews to locate buried cables prior to digging.

Another class of magnetometer is called the Hall effect sensor. In the Hall effect sensor, the transport of electrons through an electrical device is affected by the presence of an external magnetic field. As shown in Figure 9.4.2, current flowing from the top to the bottom of a device is deflected to the right, causing a charge build-up, and a measurable voltage. This sort of sensing approach offers ease of fabrication as a substantial advantage, but does not offer the performance of the flux-gate devices discussed above. In general, a Hall effect sensor can measure down to about 5% of the earth's magnetic field.

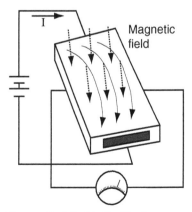

Figure 9.4.2: Hall effect sensor.

Chapter 9

Instead of measuring the build-up of a Hall voltage, it is also possible to measure the increased resistance of the device due to the deflected electrons. In this case, the Hall-based sensor is called a magnetoresistor. Recent years have seen much research on materials for magnetoresistors at Honeywell and elsewhere. A very important advantage of a Hall magnetoresistor is that the resistive film may be easily patterned into geometries that are easily connected into resistance bridges, as shown in Figure 9.4.3.

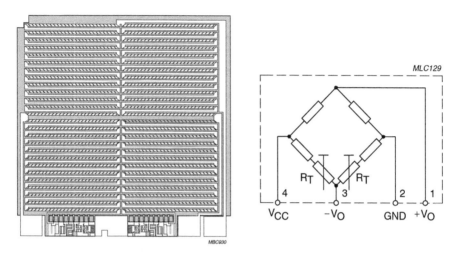

Figure 9.4.3: Magnetoresistors: Resistive film patterned into Wheatstone bridge on chip (offset trimmed to zero by R_T). (Courtesy Philips Semiconductors.)

A newer magnetoresistor material, which offers a larger magnetoresistance effect (called the Giant Magnetoresistance Effect or GMR for short), is being used for a number of applications. Several companies are marketing devices based on these new materials.

One very important application for miniature magnetic sensors is as the data-read head for disk drives. Clearly, improved sensitivity is important because it can enable increased storage density in disk drives. As a result, GMR read heads have become common on the newer generations of hard disk drives.

9.5 Summary

We have reviewed the basic principles of induction, and examined several examples of devices that use this principle to measure the position or presence of objects. We have also looked at flux-gate and magnetoresistance magnetometers, and looked at some products based on both. In general, a wide variety of magnetic sensor-based instruments are available. The emergence of thin-film sensors is important for the disk-drive industry, and other applications of thin-film magnetometers can be expected.

CHAPTER 10

Flow and Level Sensors

William Hennessy, BMT Scientific Marine Services, Inc.

Flow sensors are used in many monitoring and control applications, to measure both air and liquid flows. There are many ways of defining flow (mass flow, volume flow, laminar flow, turbulent flow). Usually the amount of a substance flowing (mass flow) is the most important, and if the fluid's density is constant, a volume flow measurement is a useful substitute that is generally easier to perform. There are numerous reliable technologies and sensor types used for this purpose. Some technologies have been applied to both air and liquid flow measurements, as their principles of operation hold true in either application. Other technologies lend themselves to being airflow or liquid flow specific. In this chapter, we will discuss several of the most commonly used techniques for measuring both airflow and liquid flow. Complementary to flow measurement is level measurement. Used together, flow and level sensors answer the basic question of "how much" in laboratories and industries worldwide. Both measurement processes also share the distinction of being fairly complicated.

10.1 Methods for Measuring Flow

Flow rate is typically obtained by first measuring the velocity of a fluid in a pipe, duct, or other structure and then multiplying by the known cross-sectional area at the point of measurement. Methods for measuring airflow include thermal anemometers, differential pressure measurement systems, and vortex shedding sensors. Methods used for measuring liquid flow include differential pressure measurement systems, vortex shedding sensors, positive displacement flow sensors, turbine based flow sensors, magnetic flow sensors, and ultrasonic flow sensors.

Thermal Anemometers

Thermal (or "hot wire") anemometers use the principle that the amount of heat removed from a heated temperature sensor by a flowing fluid can be related to that fluid's velocity. These sensors typically use a second, unheated temperature sensor to compensate for variations in the air temperature. Hot wire sensors are available as single point instruments for test purposes, or in multi-point arrays for fixed installation. These sensors are better at low airflow measurements than differential pressure types, and are commonly applied to air velocities from 50 to 12,000 feet per minute.

Differential Pressure Measurement

Differential pressure measurement sensor technologies can be used for both airflow and liquid flow measurements. A variety of application-specific sensors used for both airflow and pressure measurements are on the market, as well as differential pressure sensors used for liquid measurements. Differential pressure flowmeters are the most common type of unit in use, particularly for liquid flow measurement.

The operation of differential pressure flowmeters is based on the concept that the pressure drop across the meter is proportional to the square of the flow rate; the flow rate is found by measuring the pressure differential and taking the square root.

Differential pressure flow devices, like most flowmeters, have a primary and secondary element. The primary element causes a change in the kinetic energy, to create the differential pressure in the pipe. The unit must be correctly matched to the pipe size, flow conditions, and the properties of the liquid being measured. In addition, the measurement accuracy of the element must be good over a reasonable range. The secondary element measures the differential pressure and outputs the signal that is converted to the actual flow value.

For airflow measurements, common differential pressure flow devices include Pitot tubes (see Figure 10.1.1) and numerous other types of velocity pressure-sensing tubes, grids, and arrays. All of these sensing elements are combined with a low differential pressure transmitter to produce a signal proportional to the square root of the fluid velocity.

A Pitot tube consists of two tubes that measure pressure at different locations within a pipe. One tube measures static pressure, usually at the pipe wall, and the other measures impact pressure (static pressure plus velocity head). The faster the flow rate, the larger the impact pressure. Pitot tubes use

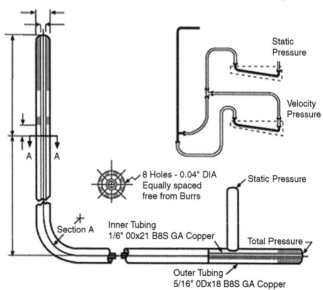

Figure 10.1.1: The Pitot tube.

the difference between impact and static pressure to calculate flow rate. Pitot tubes are low-cost devices, but they have the drawback that they only measure flow at a single point and need to be installed at the point of maximum flow. Changes in velocity profile can cause major errors. They are also prone to clogging. Averaging Pitot tubes have several ports for measuring flow at multiple locations; this makes it possible to take changing velocity profiles into account.

Figure 10.1.2 shows an example of a velocity pressure measurement with a U tube manometer.

Some differential pressure-based flow measurement systems include transmitters with the ability to extract the square root of the measured pressure electronically and provide a signal that is linear with respect to velocity. Others provide a signal proportional to measured pressure and depend on the control system to calculate the square root. Once the velocity is obtained, flow can then be found by multiplying by the cross-sectional area of the duct. Velocity range is limited by the range and resolution of the pressure transmitter used. Most differential pressure units are limited to a minimum velocity in the range of 400 to 600 feet per minute. Maximum velocity is limited only by the sensor's durability.

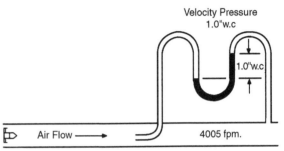

Figure 10.1.2: Velocity pressure measurement using U tube manometer.

To measure water flow, differential pressure flow devices typically either measure velocity pressure (insertion tube type), or measure the drop in pressure across a known restriction. Orifice plates, flow nozzles, Venturis, and Pitot tubes are types of restrictions commonly used.

Insertion tube flow sensors are typically made of a tube with multiple openings across the width of the flow stream, to give an average of the velocity differential across the tube and an internal baffle between upstream and downstream openings to obtain a differential pressure. Insertion tube meters have a low permanent pressure loss and can be satisfactory for many common applications.

A concentric orifice plate is the simplest and least costly of the differential pressure devices. The orifice plate constricts the flow of a fluid and produces a differential pressure across the plate (see Figure 10.1.3). This results in a high pressure upstream and a low pressure downstream that is proportional to the square of the flow velocity. An orifice plate usually produces a greater overall pressure loss than other flow elements. One advantage of this device is that cost does not increase significantly with pipe size.

Chapter 10

Venturi tubes are the largest and most expensive differential pressure device. They work by gradually narrowing the diameter of the pipe, and measuring the pressure drop that results (see Figure 10.1.4). An expanding section of the differential pressure device then returns the flow to close to its original pressure. As with the orifice plate, the differential pressure measurement is converted into a corresponding flow rate. Venturi tubes can typically be used only in those applications requiring a low pressure drop and a high accuracy reading. They are often used in large diameter pipes.

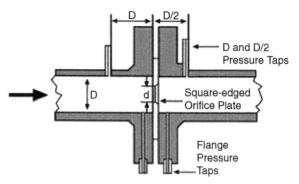

Fig. 10.1.3: The concentric orifice.

Flow nozzles are actually a variation on the Venturi tube, with the nozzle opening being an elliptical restriction in the flow, but having no outlet area for the pressure recovery (Figure 10.1.5). Pressure taps are located approximately 1/2 pipe diameter downstream and 1 pipe diameter upstream. The flow nozzle is a high-velocity flow meter used where turbulence is high (Reynolds numbers above 50,000), as in steam flow applications. The pressure drop of a flow nozzle is between that of a Venturi tube and the orifice plate (30 to 95 percent).

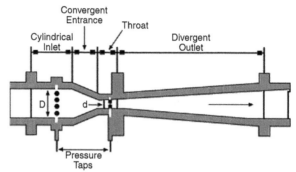

Figure 10.1.4: The Venturi tube.

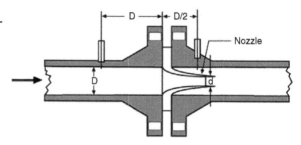

Figure 10.1.5: The flow nozzle.

Benefits of differential pressure instruments include their low cost, simplicity of operation and installation, and proven performance. It is a well-understood technology. Disadvantages of these devices can include permanent pressure loss; dirt buildup and clogging; large size and bulkiness of some devices; and unsuitability for use with certain types of fluids.

Vortex-Shedding Sensors

These flow sensors use the principle (Von Karman) that when a fluid flows around an obstruction in the flow stream (bluff object), eddies or vortices are shed alternately downstream of the object. The frequency of the vortex shedding is proportional to the velocity of the flowing fluid. Single sensors are used in small ducts, and arrays of sensors are applied to larger ducts, as with the other types of airflow measuring instruments. Vortex shedding airflow sensors are commonly used with air velocities in the range of 350 to 6000 feet per minute. These meters are equally suitable for flow rate or flow total measurements. Use with slurries or high viscosity fluids is not recommended.

Positive Displacement Flow Sensors

These flow measurement devices are used where high accuracy at high *turndown* (ratio of the full range of the device to the minimum measurable flow) is necessary and some permanent pressure loss will not cause excessive energy consumption. These units operate by separating liquids into measured segments, and then moving them on. Each segment is then counted by a connecting register. They are useful for viscous liquid flows or where a single mechanical meter is needed. Common types of positive displacement flow meters include lobed and gear type meters, nutating disk meters, rotary-vane meters, and oscillating piston meters. These meters are typically made of metals such as brass, bronze, and cast iron, but can also be constructed of engineered plastic, depending on the application.

Because of the close tolerances needed between moving parts of positive displacement flow meters, they can be prone to mechanical problems resulting from suspended solids in the flow stream. Positive displacement meters are available with flow indicators and totalizers that can be read manually. These devices are relatively costly.

Turbine-Based Flow Sensors

Turbine and propeller type meters use the principle that liquid flowing through the turbine or propeller will cause the rotor to spin at a speed directly related to flow rate. Electrical pulses can be counted and totaled. These devices are available in full bore, line-mounted versions and insertion types where only a part of the flow being measured passes over the rotating element. Turbine flow meters, when properly specified and installed, offer good accuracy, especially with low viscosity fluids. Insertion types are used for less critical applications; however, they are often easier to maintain and inspect because they can be removed without disturbing the main piping.

Chapter 10

Mass Flowmeters

Mass-related processes such as chemical reactions, heat transfer, etc. require more accurate flow measurements, and this has led to the development of mass flow meters. A number of designs are available, but the most common is the Coriolis meter, the operation of which is based on the phenomenon called the Coriolis force. Coriolis meters are true mass meters that measure the mass rate of flow directly, as opposed to measuring volume flow. Since mass does not change, the meter is linear without having to be adjusted for variations in liquid properties. Also, there is no need to compensate for changing temperatures and pressure conditions. This type of flowmeter is particularly useful for measuring liquids with a viscosity that varies with velocity at given temperatures and pressures.

Coriolis meters are available in various designs. One popular device consists of a U-shaped flow tube enclosed in a sensor housing connected to an electronics unit. The sensing unit can be installed directly into any process, and the electronics unit can be located up to 500 feet from the sensor. Inside the sensor housing, the U-shaped flow tube is vibrated at its natural frequency by a magnetic device located at the bend of the tube. This is similar to the vibration of a tuning fork, covering less than 0.1 in. and completing a full cycle about 80 times/sec. As the liquid flows through the tube, it is forced to take on the vertical movement of the tube. This causes the liquid to exert a force on the tube, causing the tube to twist. The amount of twist is directly proportional to the mass flow rate of the liquid flowing through the tube. Magnetic sensors located on each side of the flow tube measure the tube velocities, which change as the tube twists. The sensors feed this information to the electronics unit, where it is processed and converted to a voltage proportional to mass flow rate. This flowmeter has a wide range of applications, from adhesives and coatings to liquid nitrogen.

Electromagnetic Flow Sensors

Operation of these sensors is based upon Faraday's Law of electromagnetic induction, which says that a voltage will be induced when a conductor moves through a magnetic field.

The liquid is the conductor, and the magnetic field is created by energized coils outside the flow tube. The voltage produced is proportional to the flow rate. Electrodes mounted in the pipe wall sense the induced voltage, which is measured by the secondary element.

Electromagnetic flow meters are applied in measuring the flow rate of conducting liquids (including water) where a high quality, low maintenance system is needed. The cost of magnetic flow meters is high relative to other types of flowmeters. They do have many advantages, including: they can measure difficult and corrosive liquids and slurries, and they can measure reverse flow.

Ultrasonic Flow Sensors

Ultrasonic flow sensors can be divided into Doppler sensors and transit, or time-of-travel, sensors. Doppler sensors measure the frequency shifts caused by liquid flow. Two transducers are mounted in a case attached to one side of the pipe; a signal of a known frequency is delivered to the liquid to be measured. Bubbles, solids or any other discontinuity in the liquid cause the pulse to be reflected to the receiver element. Because the liquid that causes the reflection is moving, the frequency of the returned pulse shifts, and the frequency shift is proportional to the velocity of the liquid.

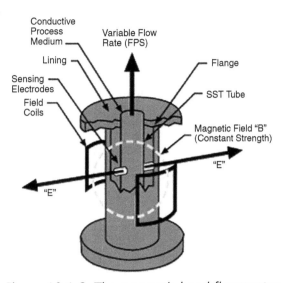

Figure 10.1.6: The magnetic head flow meter.

With transit, or time-of-travel, meters, transducers are mounted on each side of the pipe, such that the sound waves traveling between the devices are at a 45-degree angle to the direction of liquid flow. The speed of the signal moving between the transducers increases or decreases with the direction of transmission and the velocity of the liquid being measured. A time-differential relationship proportional to the flow can be obtained by transmitting the signal alternately in both directions. One limitation of this type of sensor is that the liquids that are being measured have to be relatively free of solids or entrained gases in order to minimize signal scattering and absorption.

Advantages of ultrasonic flow meters are that they are non-intrusive and their cost is moderate. Many models are designed to clamp onto existing pipe.

Laser Doppler Flow Measurement

The laser doppler anemometer (LDA) is a well-established technique that has been widely used for fluid dynamic measurements in liquids and gases for well over 30 years. The directional sensitivity and non-intrusiveness of LDA make it useful for applications with reversing flow, chemically reacting or high-temperature media, and rotating machinery, where physical sensors might be difficult or impossible to use. This technique does, however, require tracer particles in the flow.

Chapter 10

The main benefits LDA offers for flow measurement include the following:

- It is non-contacting
- No calibration is needed
- Measurement distance can range from centimeters to meters
- Velocity can range from zero to supersonic
- It can measure flow reversals
- It has high spatial and temporal resolution

The basic configuration of an LDA (Figure 10.1.7) consists of:

- a continuous wave laser
- transmitting optics, including a beam splitter and a focusing lens
- receiving optics, composed of a focusing lens, an interference filter, and a photodetector
- a signal conditioner and a signal processor.

In general, the LDA sends a monochromatic laser beam toward the target and collects the reflected radiation. According to the Doppler effect, the change in wavelength of the reflected radiation is a function of the targeted object's relative velocity. Therefore, the velocity of the object can be found by measuring the change in wavelength of the reflected laser light. This is done by forming an interference fringe pattern (pattern of light and dark stripes) by superimposing the original and reflected signals.

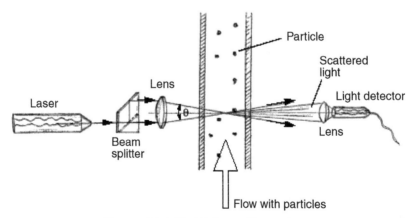

Figure 10.1.7: LDA configuration.

A Bragg cell is often used as the beam splitter. It is a glass crystal with a vibrating piezo crystal attached (Figure 10.1.8). The vibration generates acoustical waves which act like an optical grid.

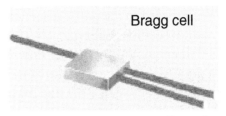

Figure 10.1.8. Bragg cell used as a beam splitter.

Two beams of equal intensity exit the Bragg cell, with frequencies *f0* and *fshift*, and these are focused into optical fibers that bring them to a probe. In the probe, the parallel exit beams from the fibers are focused by a lens to intersect in a region called the measurement volume, which is typically a few millimeters long and is where the measurement is made. The light intensity is modulated due to interference between the laser beams which produces parallel planes of high light intensity, called fringes. The fringe distance d_f is defined by the wavelength of the laser light and the angle between the beams:

$$d_f = \frac{\lambda}{2 \sin(\theta/2)}$$

Flow velocity information comes from light scattered by tiny "seeding" particles carried in the fluid as they move through the measurement volume. The scattered light contains a Doppler shift—the Doppler frequency *fD*—which is proportional to the velocity component perpendicular to the bisector of the two laser beams, which corresponds to the x axis of the measurement volume.

A receiver lens collects the scattered light and focuses it on a photo-detector. An interference filter mounted in front of the photo-detector passes only the required wavelength, removing noise from ambient light and from other wavelengths. The photo-detector converts the fluctuating light intensity into an electrical signal, the Doppler burst. The Doppler bursts are filtered and amplified in the signal processor, which determines *fD* for each particle, often by frequency analysis using the fast Fourier transform (FFT) algorithm.

The fringe spacing df provides information about the distance traveled by the particle. The Doppler frequency *fD* provides information about the time: $t = 1/fD$. Since velocity equals distance divided by time, the expression for velocity thus becomes:

$$V = d_f * fD$$

With regard to the particle seeding, it is often the case that liquids contain adequate natural seeding, but typically gases must be seeded. Ideally, the particles should be small enough to follow the flow, but large enough to scatter enough light to obtain a good signal-to-noise ratio at the photo-detector output. The size range of particles is usually between 1 μm and 10 μm. The particle material can be solid (powder) or liquid (droplets).

10.2 Selecting Flow Sensors

Improper device selection accounts for 90 percent of the problems encountered with flowmeters. The most important requirement in selecting a sensor is understanding exactly what the device is supposed to do. Here are important questions to ask during the selection process:

- Is the measurement for a process control application, where repeatability is the major concern, or for accounting or custody transfer, where high accuracy is important?

- Is local indication or a remote signal required? If a remote output is required, is it to be a proportional signal, or a contact closure to start or stop another device?

- Is the liquid to be measured clean, viscous, or a slurry?

- Is the liquid to be measured electrically conductive?

- What is the specific gravity or density of the liquid to be measured?

- What flow rates are involved in the application?

- What are the process's operating temperatures and pressures?

- What accuracy, range, linearity, repeatability, and piping requirements must be considered?

It is also important to know what a flowmeter *cannot* do before a final selection is made. Each type of sensor has advantages and disadvantages, and the degree of performance satisfaction is directly related to how well an instrument's capabilities and shortcomings are matched to the application's requirements. Most sensor suppliers are eager to help their customers select the appropriate flowmeter for a particular application. Many provide questionnaires, checklists, and specification sheets designed to obtain the critical information necessary to match the correct flowmeter to the job.

Technological advances must also be taken into consideration. One common mistake is to select a design that was popular for a given application years ago, assuming that it is still the best choice. Many changes and innovations may have occurred in recent years in flowmeter technology for that particular application, making the choice much broader.

10.3 Installation and Maintenance

Airflow sensors operate most efficiently in sections of pipes or ducts that have uniform, fully developed flow. In order for measurements to be reliable, all airflow sensing devices should be installed in accordance with the manufacturer's recommended straight runs of upstream and downstream duct. Some manufacturers offer flow-straightening elements that can be installed upstream of the sensing array to improve undesirable flow conditions. These should be considered when conditions do not permit installation with the required straight runs of duct upstream and downstream from the sensor.

All liquid flow sensors also operate best when measuring fully developed, uniform flow. Sensors should be installed in accordance with the manufacturer's recommended straight runs of upstream and downstream pipe in order to obtain the most reliable measurements.

Although most suppliers offer flowmeter installation service, a high percentage of users choose to install their own equipment. This can lead to installation errors, such as not allowing enough upstream and downstream straight-run piping, as described above. Every sensor design has a certain amount of tolerance for unstable velocity conditions in the pipe, but all devices require proper piping configurations to operate efficiently, so that a normal flow pattern for the device is provided. Without it, accuracy and performance are adversely affected. Flowmeters are also occasionally installed backwards (especially with orifice plates). Pressure-sensing lines may also be reversed.

With electrical components, intrinsic safety is an important consideration in hazardous areas. Stray magnetic fields are found in most industrial plants. Power lines, relays, solenoids, transformers, motors, and generators all contribute electromagnetic interference, and users must ensure that the flowmeter they have selected is immune to such interference. Most problems occur with the electronic components in secondary elements, which must be protected. Strict adherence to the manufacturer's recommended installation practices will usually prevent such problems.

Chapter 10

Calibration

An initial calibration is required with all flowmeters; usually the calibration is handled by the manufacturer. However, if qualified personnel are available, this can be handled by the sensor purchaser. The necessity of recalibration depends a great deal on how well the device suits the application. Some liquids that pass through flowmeters are abrasive, erosive, or corrosive, and in time, portions of the device will deteriorate enough to affect its performance. Some designs are more susceptible to damage than others. For example, wear of individual turbine blades can cause performance changes. If the application is critical, flowmeter accuracy should be checked often. In other cases, recalibration may not be necessary because the application is noncritical or because nothing in the environment affects the meter's performance. Some flowmeters require special equipment for calibration. Most manufacturers will provide this service in their own facility or in the user's facility, where they will bring the equipment for on-site calibration.

Maintenance

Flowmeters that have no moving parts usually need less attention than units with moving parts, but all flowmeters eventually require some type of maintenance.

With differential pressure flowmeters, primary elements require extensive valves, pipes, and fittings when they are connected to their secondary elements, so maintenance can be a recurring effort. Impulse lines can plug or corrode and have to be cleaned or replaced. Flowmeters with moving parts require periodic internal inspection, especially if the liquid being metered is dirty or viscous. Installing filters ahead of such units can help to minimize fouling and wear. Ultrasonic or electromagnetic flowmeters can develop problems with their secondary element's electronic components. Pressure sensors associated with secondary elements should be periodically removed and inspected.

Applications where coating can occur also represent potential problems for devices such as magnetic or ultrasonic units. If the coating is insulating, the operation of magnetic flowmeters will ultimately be impaired if the electrodes are insulated from the liquid. Periodic cleaning helps to prevent this. With ultrasonic flowmeters, refraction angles may change and the sonic energy absorbed by the coating will cause the meter to become inoperative.

10.4 Recent Advances in Flow Sensors

A recent study conducted by Flow Research and Ducker Worldwide showed that a major shift is taking place in the field of flow measurement, to "new technology" flowmeters. "New technology" flowmeters are defined in the study as magnetic, ultrasonic, Coriolis, vortex, and multivariable differential pressure meters. These have four features in common:

1. They were introduced in the last 50 years.

2. They make use of technological advances that solve some of the problems inherent in older types of flow measurement devices.

3. They are the predominant focus of new product development by manufacturers.

4. Their performance, including accuracy, is better than that of "traditional" flowmeter technology.

More features are typically provided with these types of flowmeters. This includes software capabilities, more application-specific packages, and extremely durable construction methods. Self-diagnostics is one such feature that users are looking for in flowmeters.

Recent innovations have taken place with differential pressure flowmeters, including the development of very high accuracy differential transmitters, the use of multivariable transmitters, and differential pressure transmitters that contain an integrated primary element. A very significant development has been the emergence of the integrated differential flowmeter. In the past, users purchased primary elements from one company and transmitters from another. Now transmitters are offered with an integrated primary element. The recent trend has been toward smarter and more accurate transmitters. However, it's important to understand that the transmitter is only one element in the system, and other variables contribute to flow accuracy.

With ultrasonic flowmeters, a new class of accurate and low-cost devices for small-diameter pipes (1/4 inch to 2 inches in diameter) is emerging, based on micro-electromechanical systems (MEMS) ultrasonic sensors and a mixed-signal, application-specific integrated circuit (ASIC) control chip. This technique offers an alternative to traditional mechanical meters or more expensive vortex or mass flowmeters for selected applications in smaller diameter pipes. The electronic output can be fed directly into process control and monitoring equipment. Also, the device can perform self-testing. It is very well-suited for measuring clean gas flow.

10.5 Level Sensors

Types of Liquid Level Sensors

As previously stated, level sensing is closely related to flow sensing. The most common application for level sensing is tank measurement and control operations. A number of level sensing technologies are currently available, including hydrostatic pressure, ultrasonic, RF capacitance, magnetorestrictive-based, and radar measurement systems.

Hydrostatic

Level measurement using differential hydrostatic pressure is based on the principle that the hydrostatic pressure difference between the top and bottom of a column of liquid is related to the density of the liquid and the height of the column. Pressure transmitters are available that are configured for level monitoring applications. Pressure instruments can also be remotely located. However, this requires field calibration of the transmitter to compensate for elevation difference between the sensor and the level being measured.

Bubbler type hydrostatic level instruments have been developed for use with atmospheric pressure underground tanks, sewage sumps and tanks, and other applications that cannot have a transmitter mounted below the level being sensed, or that are prone to plugging. Bubbler systems bleed a small amount of compressed air (or other gas) through a tube immersed in the liquid, with an outlet at or below the lowest monitored liquid level. The flow rate of the air is regulated so that the pressure loss of the air in the tube is negligible and the resulting pressure at any point in the tube is approximately equal to the hydrostatic head of the liquid in the tank.

The accuracy of hydrostatic level instruments is related to the accuracy of the pressure sensor used.

Smart differential-pressure (DP) transmitters can be adapted for level measurements and are being used extensively in industry, producing a renewed popularity of this sensor technology. Smart transmitters and 4–20 mA signals are used to communicate to/from remote distributed control systems (DCSs), programmable logic controllers (PLCs), and other control systems. Hydrostatic tank gauging (HTG) is an emerging application using this technique for accurately measuring liquid inventory and to monitor tank transfers. The level measurements can be digitally networked for remote computer access.

Ultrasonic

Ultrasonic level sensors emit sound waves, and the liquid surface reflects the sound waves back to the source. The transit time is proportional to the distance between the liquid surface and the transmitter. These sensors are ideal for noncontact level sensing of very viscous fluids such as heavy oil, latex, and slurries. Practically, there are limitations to this method, which include:

- foam on the surface can absorb sound
- speed of sound varies with temperature
- turbulence can cause inaccurate readings.

RF Capacitance

Capacitance level transmitters work on the principle that a capacitive circuit can be formed between a probe and a vessel wall. The capacitance changes with a change in fluid level, because all common liquids have a dielectric constant higher than that of air. The probe connects to an RF transmitter mounted externally on the tank. Transmission of the level measurement can be in various forms, and the receiving instrument could be a PC, a chart recorder, a distributed control system (DCS), a programmable logic controller (PLC), etc. These sensors are useful in sensing the levels of a variety of aqueous and organic liquids and slurries, and liquid chemicals such as quicklime. There are also dual probe capacitance level sensors that can be used to sense the interface between two liquids that have very different dielectric constants. These sensors are rugged, easy to use, contain no moving parts, and are simple to clean. They can be designed for very high temperature and and pressure applications.

There are variations of the basic RF systems called RF Impedance and RF Admittance. (Impedance is the total opposition to current flow in an AC or RF circuit; admittance is the reciprocal of impedance and measures how readily current flows in a circuit.) These techniques offer some improvements over the basic RF capacitance method, including better reliability and a larger range of applications.

Magnetostrictive

Magnetostrictive level transmitters (Figure 10.5.1) are based on the principle that an external magnetic field can be used to cause the reflection of an electromagnetic wave in a waveguide constructed of magnetostrictive material. The probe is composed of three concentric members. The outermost member is a protective outer pipe. Inside the outer pipe is a waveguide, which is a formed element constructed of a magnetostrictive material. A low-current interrogation pulse is generated in the transmitter electronics and transmitted down the waveguide, which creates an electromagnetic field along the length of the waveguide. When this magnetic field interacts with the permanent magnetic field of a magnet mounted inside the float, a torsional strain pulse, or waveguide twist, results. This waveguide twist is detected as a return pulse. The time between the initiation of the interrogation pulse and the detection of the return pulse is used to determine the level measurement with a high degree of accuracy and reliability.

In the past ten years, microwave or radar technology for level measurement has become more commonly used. This is a rapidly changing area. Formerly this technology was only used in high-accuracy applications, but the development of new techniques and mass production has led to radar becoming more affordable for more varied applications.

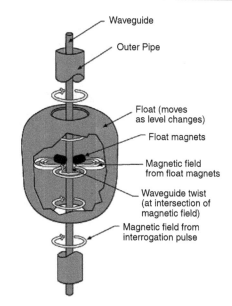

Figure 10.5.1: Magnetostrictive level sensors.

Microwave Radar Level Sensing

All of the types of radar level measurement use the basic principles of firing microwaves downward from a sensor located on top of the tank or other vessel; part of the energy is reflected back to the sensor from the surface of the substance being measured. The time of flight of the signal is used to measure the level.

Radar level gauging can be divided into two broad areas: through-air radar and guided wave radar (GWR), also called micropower impulse radar (MIR). Through-air radar can be further broken down into two categories: pulsed wave time of flight and frequency-modulated continuous wave (FMCW). Although both use microwave signals fired into the vapor space above the substance being measured, the return signal processing, manipulation and resultant distance calculation are different.

GWR

Unlike through-air radar systems, GWR is an invasive technology. It appears similar to RF Admittance sensor technology, but it does not have the same capabilities of coping with extremes of pressure, temperature or product coating. Pulses of electromagnetic energy are emitted from the base of the transmitter down the waveguide (a cable or rod). When the signal reaches a point where a change in dielectric constant occurs, usually at the surface of the substance, some of this signal is reflected back. The amount of signal reflection is therefore proportional to the difference in dielectric constant between the waveguide and the substance. In short, substances having a higher dielectric / conductivity provide stronger return signals.

The level measurement itself is a function of the time from when the electromagnetic signal is emitted to the time at which the resultant receive echo is received. This radar technology is often referred to as time domain reflectometry (TDR) radar. The propagation of the signal along a waveguide does eliminate false echoes and helps to minimize signal loss due to dust or vapors. Also, operation is possible in applications with changing vapor space humidity or fluctuating product dielectric constant. It should, however, be noted that, like all radar gauge types, low dielectric materials can cause problems. In addition, the guide can be damaged or corroded.

Systems are available with both single and dual waveguides, depending on the application. Dual waveguide systems tend to provide somewhat greater flexibility and are suitable for interface measurements, materials with a low dielectric, or where foam is present. The position of a liquid interface is measured by using that part of the initial electromagnetic pulse which is not reflected by the surface of the upper phase. This energy continues down the waveguide until it meets the liquid / liquid interface and a percentage of it is reflected back toward the gauge. The dielectric constant of the upper phase has to be known in order for an accurate measurement to be taken, since the gauge electronics must compensate for the resulting change in speed of the electromagnetic pulse through the upper phase, since it will be different from that of its speed through the air space.

Through-air Radar

Pulsed radar, or pulsed time-of-flight, is similar to the ultrasonic method of level measurement. A radar pulse is aimed at the surface of the substance being measured and the time for the pulse to return is used to find the level. This method uses lower power than FMCW and its performance can be affected by foam, obstructions in the vessel, and low-dielectric materials.

FMCW systems are non-invasive and continuously emit a swept frequency signal. In this type of radar technology, distance is inferred from the difference in frequency between the transmit and receive signals at any point in time. Although this technique provides an inferential measurement, it can be extremely accurate. The level of signal processing associated with FMCW radar gauges and the resulting power requirements means that two-wire gages of this type are suitable only for the simpler applications, with most applications requiring four-wire devices.

Selecting an appropriate antenna type and size is a crucial factor for achieving a focused beam and adequate return echo. Both cone and parabolic dish antennas are used for the through-air techniques. Cone antennas tend to keep the signals in a narrower, downward path while the dish antennas tend to produce a wider signal path. Factors such as foam in the tank, obstructions, and turbulence can affect antenna type and size choices.

Selecting a Level Sensing Technology

When choosing a method for any particular application, many factors beyond initial costs must be taken into consideration. The most important factors that sensor manufacturers need to know about a level measurement application are:

- name and characteristics of the material to be measured, whether solid, liquid, slurry, powder, or granular. The dielectric constant K is of particular importance, as well as viscosity, density, conductivity, and consistency (oily, watery, etc.).

- process information, such as pressure and temperature, degree of turbulence, and tank or vessel material.

- power requirements.

- main use of the vessel containing the material to be measured (storage, water separation, sump, etc.), and its size and shape, and location of any obstructions.

10.6 Applicable Standards

NIST Fluid Flow Group
Establishes, maintains, and disseminates standards for U.S. industry for fluid flowrate and fluid quantity measurements.

International Electrotechnical Commission
Establishes international standards for process control and flowmeters.

API Manual for Petroleum Measurement, Chapter 3.1A, "Manual Gauging of Petroleum and Petroleum Products." A commonly referenced U.S. standard for the manual level gauging of stationary tanks.

OIML R 85 (E)
A widely recognized international standard for custody transfer from the Organisation Internationale Metrologie Legale, specifying level measurement instrument performance.

CHAPTER 11

Force, Load and Weight Sensors

Ken Watkins, PCB Piezotronics, Inc.

11.1 Introduction

Measurement of a force, load, or a weight can be accomplished by a multitude of different sensors. While other technologies exist, the most commonly used sensors are generally based on either piezoelectric quartz crystal or strain gage sensing elements, which will be the focus of this chapter.

Before proceeding, it is important to recognize the difference between *force*, *load*, and *weight*, as these terms are often incorrectly used interchangeably with one another.

Force: The measurement of the interaction between bodies.
Load: The measurement of the force exerted on a body.
Weight: The measurement of gravitational forces acting on a body.

11.2 Quartz Sensors

Technology Fundamentals

Quartz force sensors are ideally designed and suited for the measurement of dynamic oscillating forces, impact, or high speed compression/tension forces. The basic design utilizes the piezoelectric principle, where applied mechanical stresses are converted into an electrostatic charge that accumulates on the surface of the crystal. See Figure 11.2.1.

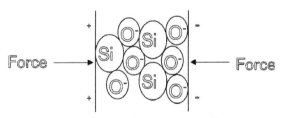

Figure 11.2.1: Piezoelectric effect.

The quartz crystals of a piezoelectric force sensor generate an electrostatic charge only when force is applied to or removed from them. In other words, if you apply a static force to a piezoelectric force sensor, the electrostatic charge output initially generated will eventually leak away and the sensor output ultimately will return to zero.

Chapter 11

The rate at which the charge leaks back to zero is exponential and based on the sensor's discharge time constant (DTC). DTC is defined as the time required for a sensor or measuring system to discharge its signal to 37% of the original value from a step change of measurand. This value (seconds) is generally known and is determined by multiplying the lowest insulation resistance path (ohms) by the total capacitance (farads) of the system prior to the amplifier circuit. (This is true of any piezoelectric sensor, whether the operation is force, pressure or vibration monitoring.)

The DTC of a system directly relates to the low frequency monitoring capabilities of a system. It is because of this characteristic that piezoelectric force sensors can only be used for "quasi-static" measurements and are not generally used for weighing applications.

To help clarify the DTC characteristic, assume a weight is placed on top of a piezoelectric force sensor. Initially, the piezoelectric sensing crystals will generate a charge (Q), which is immediately seen at the input to the built-in (or external) amplifier. However, after this initial step input, the charge signal decays according to the equation:

$$q = Qe^{-t/RC}$$

where:

q = instantaneous charge (Coulomb)
Q = initial quantity of charge (Coulomb)
R = resistance prior to amplifier (ohm)
C = total capacitance prior to amplifier (Farad)
e = base of natural log (2.718)
t = time elapsed after time zero (Second)

This equation is graphically represented in Figure 11.2.2.

The product of R and C represents the DTC (in seconds) of the sensor. It can be inferred that the longer the DTC, the more accurately the sensor will be able to track longer duration events. Generally, piezoelectric force sensors with built-in electronics can have time constants that vary from just a few seconds to >2000 seconds. Special time constants can be supplied by altering the resistor value, R, in the sensor's built-in circuitry. Charge mode

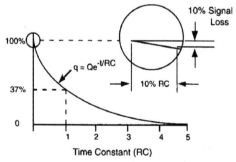

Figure 11.2.2: Decay due to discharge time constant (DTC).

Force, Load and Weight Sensors

sensors, which do not contain built-in circuitry. require an external amplifier ("charge amplifier") to set the DTC. Some charge amplifiers provide an adjustable DTC.

It should also be mentioned how the range of the force sensor is determined. For sensors that contain built-in circuitry, the range is set by the manufacturer and can be determined by examining the performance specifications for the sensor. For sensors which do not contain built-in circuitry, the range is set by the external amplifier. There are a number of trade-offs between each type. See Reference 2 for additional information.

Sensor Types

Charge Mode, High-Impedance, Piezoelectric Force Sensor

A charge mode piezoelectric force sensor, when stressed, generates an electrostatic charge from the crystals. For accurate analysis or recording purposes, this high impedance charge must be routed through low noise cable to an impedance converting amplifier, such as a laboratory charge amplifier or source follower. (Connection of the sensor directly to a readout device such as an oscilloscope may be possible for high frequency impact indication, but is not suitable for most quantitative force measurements.)

The primary function of the external amplifier is to convert the high impedance charge output to a usable low impedance voltage signal for analysis or recording purposes. Figure 11.2.3 shows a typical charge mode sensor system schematic including: sensor, low noise cable, and charge amplifier.

As previously mentioned, in a charge mode system, the sensors do not contain built-in amplifiers. Therefore, the sensor range and DTC are determined by the settings on an external charge amplifier.

A feedback resistor working together with a capacitor on the operational amplifier determines these characteristics. Laboratory-style charge amplifiers generally feature a variety of ranging options as well as short, medium and long time constant selections. (It is assumed that the electrical insulation resistance of the force sensor and cable connecting to the charge amplifier are larger than that of the feedback resistor in the charge amplifier; otherwise, performance degradation issues, such as signal drift, may arise. Therefore, to assure this, the force sensor connection point and cable must be kept clean and dry.)

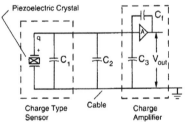

q = charge signal
C_1 = sensor capacitance
C_2 = cable capacitance
C_3 = amplifier input capacitance
C_f = amplifier feedback capacitor

Figure 11.2.3:
Charge mode sensor system.

Chapter 11

Voltage Mode, Low-Impedance, Piezoelectric Force Sensor

Voltage mode or low impedance force sensors share the same basic design used in charge mode force sensors, but incorporate a built-in microelectronic amplifier. This amplifier serves to convert the high impedance charge output from the quartz crystals into a low impedance voltage signal for analysis or recording. This type of sensor, powered by separate constant current source, operates over long ordinary coaxial or ribbon cables without signal degradation. Additionally, the low impedance voltage signal is not affected by triboelectric cable noise (noise caused by vibration or movement of the cable) or environmental contaminants. Figure 11.2.4 shows a complete voltage mode sensor system schematic.

The sensor range and DTC are fixed by the components in the voltage mode sensor's internal amplifier. (It is assumed that the DTC of the signal conditioner is greater than that of the force sensor.)

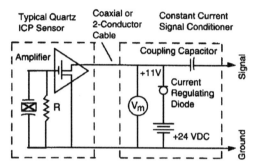

Figure 11.2.4: Voltage mode sensor system.

Piezoelectric Force Sensor Construction

The basic mechanical construction of general purpose quartz forces sensors consist of thin quartz discs that are "sandwiched" between upper and lower base plates. A relatively elastic, beryllium-copper stud (or sometimes a sleeve) holds the upper and lower plates together and preloads the crystals. Preloading of the crystals is required to assure that the upper and lower plates are in intimate contact with the quartz crystals, ensuring good linearity and the capability for tension as well as compression measurements. This "sensing element" configuration is then packaged into a rigid, stainless steel housing and welded to provide hermetic sealing of the internal components against contamination. Figure 11.2.5 depicts the typical construction of a general-purpose quartz force sensor.

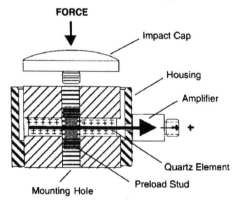

Figure 11.2.5: Cross section of a general purpose force sensor.

Selecting and Specifying

Selecting and specifying a quartz based force sensor is primarily application driven. Application considerations typically include the magnitude, frequency and the direction of the force or forces to be measured. Additional considerations may include size constraints, environmental conditions and mechanical integration requirements. Given the multitude of applications and application requirements, most manufacturers offer a range of models to fit typical applications. These include:

General Purpose – Internally preloaded for measurement of compression and/or tension forces. Typical applications include impact testing, punching and forming, drop testing, materials testing, fatigue testing, material fracture, machinery studies, modal analysis force input and biomechanics.

Figure 11.2.6:
Typical general-purpose force sensor.

Penetration – Penetration style sensors are specifically designed for compression and impact force measurements in materials testing applications such as helmet testing. Smooth, cylindrical housings and curved impact caps avoid cutting through specimens permitting yield, deformation, and break point measurements of polymers, composites, and other materials.

Figure 11.2.7:
Typical penetration force sensor.

Miniature – The miniature sensor configuration permits low-amplitude, dynamic compression, tension, and impact force measurements. Two configurations, one with a tapped mounting hole and impact cap and the other with tapped holes on both ends of the sensor, are typical. Link, integrated link, and free-standing installations are possible. Applications include matrix print-head studies, wire bonding, and high frequency, low level impulse testing.

Figure 11.2.8:
Typical miniature force sensor.

Impact – Impact style sensors are specifically designed for impact force measurements. The sensor is typically mounted in a freestanding manner with the impact cap directed toward the oncoming object with which it will collide. Applications include crash testing, wire crimping and metal forming, machinery studies, impact testing, drop testing, and laboratory shock test machines.

Figure 11.2.9:
Typical impact force sensor.

Rings – Ring sensor configurations measure dynamic compression. Tension measurements are also possible if the unit has been installed with proper pre-load. The through-hole mounting supports platform, integrated link, and support style installations using either a through bolt or the supplied stud. Tension range is dependent upon the amount of applied pre-load and strength of mounting stud used. Applications include tablet presses, stamping, punching and forming operations, balancing, machinery studies, and force-controlled vibration testing.

Figure 11.2.10:
Typical ring force sensor.

Links – Link style sensors measure dynamic compression and tension. They are constructed using a force ring that is under compressive pre-load between threaded mounting hardware. The threaded mounting on both ends of the sensor supports integrated link style installations. Applications include tablet presses, tensile testing, stamping, punching and forming operations, balancing, machinery studies, and force-controlled vibration testing.

Figure 11.2.11:
Typical link force sensor.

Multi-component – Multi-component sensors permit simultaneous measurement of dynamic force vector components in three orthogonal directions. Additionally, some sensors have been designed to measure moments as well. The through-hole mounting supports platform, integrated member, and support style installations using either a through bolt or the supplied stud. Applications include machine tool cutting forces, stamping, punching and forming operations, machinery studies, and force controlled vibration testing.

Chapter 11

Figure 11.2.12:
Typical multi-component force sensor.

Applicable Standards

The basic design of quartz-based force sensors is not governed by a specific standard. However, applicable standards do exist for calibration and certification. Most manufacturers comply or conform to standards such as ISO 10012-1 (former MIL-STD-45662A), ISO 9001 and ISO/IEC 17025.

Latest and Future Developments

The basic design of modern quartz force sensors has not changed significantly since the 1960's. New applications have driven the continued development of the mechanical packaging of this technology to fit specific applications. While improvements have been made in the performance and manufacturability, the basic design has not changed.

Major Manufacturers

PCB Piezotronics, Inc. – 3425 Walden Avenue, Depew, NY 14043

Kistler Instrument Corporation – 75 John Glenn Drive, Amherst, NY 14228

Dytran Instruments, Inc. – 21592 Marilla St., Chatsworth, CA 91311

Endevco Corporation – 30700 Rancho Viejo Rd., San Juan Capistrano, CA 92675

11.3 Strain Gage Sensors

Technology Fundamentals[1]

Sensors based on foil strain gage technology are ideally designed for the precise measurement of a static weight or a quasi-dynamic load or force. The design of strain gage-based sensors consists of specially designed structures that perform in a predictable and repeatable manner when a force, load or weight is applied. The applied input is translated into a voltage by the resistance change in the strain gages, which are intimately bonded to the transducer structure. The amount of change in resistance indicates the magnitude of deformation in the transducer structure and hence the load that is applied.

Figure 11.3.1: Typical Wheatstone bridge configuration.

[1] See Chapter 19 for more on strain gages.

The strain gages are connected in a four-arm Wheatstone bridge configuration, which acts as an adding and subtracting electrical network and allows for compensation of temperature effects as well as cancellation of signals caused by extraneous forces.

A regulated 5 to 20 volt DC or AC rms excitation is required and is applied between A and D of the bridge. When a force is applied to the transducer structure, the Wheatstone Bridge is unbalanced, causing an output voltage between B and C proportional to the applied load.

Most load cells follow a wiring code established by the Western Regional Strain Gage Committee as revised in May 1960. The code is as follows:

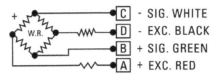

Figure 11.3.2: Standard electrical connection color code.

Most load cells comply with the Axis and Sense Definitions of NAS-938 (National Aerospace Standard-Machine Axis and Motion) nomenclature and recommendations of the Western Regional Strain Gage Committee. These axes are defined in terms of a "right handed" orthogonal coordinate system as shown below. A (+) sign indicates force in a direction which produces a (+) signal voltage and generally defines a tensile force. The primary axis of rotation or axis of radial symmetry of a load cell is the z-axis, as defined by the following figure:

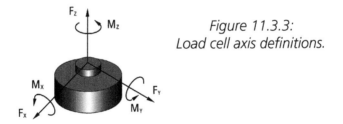

Figure 11.3.3: Load cell axis definitions.

Sensor Types

The most critical mechanical component in any strain gage-based sensor is the "spring element." In general terms, the spring element serves as the reaction mechanism to the applied force, load or weight. It also focuses it into a uniform, calculated strain path for precise measurement by the bonded strain gage. Three common structure designs used in the industry are bending beam, column and shear.

Chapter 11

Bending Beam

Sensor spring elements that employ the bending beam structure design are the most common. This is because the bending beam is typically a high-strain, low force structural member that offers two equal and opposite surfaces for strain gage placement. The bending beam design is typically used in lower capacity load cells.

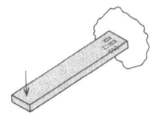

Figure 11.3.4:
Bending beam structure.

Column

The column type load cell is the earliest type of strain gage transducer. Although simple in its design, the column spring element requires a number of design and application considerations. The column should be long enough with respect to its cross section so that a uniform strain path will be applied to the strain gage. In application, the end user must beware of second-order effects, as the column load cell is susceptible to the effects of off-axis loading.

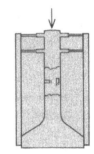

Figure 11.3.5:
Column Structure.

Shear-Web

The principle of a shear-web load cell typically takes the form of a cantilever beam that has been designed with a cross section larger than normal with respect to the rated load to be carried in order to minimize structure deflection. Under this condition, the surface strain along the top of the beam would be too low to produce an adequate electrical output from the strain gage. However, if the strain gages are placed on the sides of the beam at the neutral axis, where the bending stress is zero, the state of stress on the beam side is one of pure shear, acting in the vertical and horizontal direction.

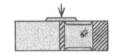

Figure 11.3.6:
Shear-web structure.

Force, Load and Weight Sensors

Classification

General Purpose
General-purpose load cells are designed for a multitude of applications in the automotive, aerospace, and industrial markets. The general-purpose load cell, as the name implies, is designed to be utilitarian in nature. Within the general-purpose load cell market there are several distinct categories: precision, universal, weigh scale, and special application. Universal load cells are the most common in industry.

Fatigue Rated
Fatigue rated load cells are specially designed and manufactured to withstand millions of cycles. Many manufacturers utilize premium fatigue resistant steel and special processing to insure mechanical and electrical integrity as well as accuracy. Fatigue rated load cells typically are guaranteed to last 100 million fully reversed cycles (full tension through zero to full compression). An added benefit of fatigue rated load cells is that they are extremely resistant to extraneous bending and side loading forces.

Special Application
Special application load cells are load cells that have been designed for a specific unique force measurement task. Special application load cells can be single axis or multiple axes. They include but are not limited to:

- Pedal Effort
- Seat Belt
- Steering Column
- Crash Barrier
- Hand Brake
- Road Simulator
- Tow Ball
- Femur
- Skid Trailer
- Bumper Impact
- Tire Test
- Gear Shift

Selecting and Specifying
Selecting and specifying a strain gage-based sensor is primarily application driven. Application considerations typically include environmental concerns, the magnitude of the force, load, or weight to be measured, accuracy, size constraints, mechanical integration, present extraneous forces or loads, and certification requirements. Given the multitude of application requirements, most manufacturers offer a range of models to fit typical applications. These include:

General Purpose – General-purpose load cells are suitable for a wide range of general force measurement applications, including weighing, dynamometer testing, and material testing machines. Most of these designs operate in both tension and compression, and offer excellent accuracy and value.

Figure 11.3.7:
Typical general-purpose load cell.

Fatigue Rated – Fatigue rated load cells are specifically designed for fatigue testing machine manufacturers and users, or in any application where high cyclic loads are present. All fatigue rated load cells are guaranteed against fatigue failure for over 100 million fully reversed cycles. These rugged load cells are manufactured using premium, fatigue-resistant, heat treated steels. Internal flexures are carefully designed to eliminate stress concentration areas. Close attention is paid to the proper selection and installation of internal strain gages and wiring to ensure maximum life. Fatigue rated load cells are available in a variety of configurations.

Figure 11.3.8:
Typical fatigue rated load cell.

Hollow, Washer, Button – Hollow compression load cells, load washers and load buttons are popular choices when the application requires a compact sensor. These sensors are available in a wide range of sizes and capacities. Hollow compression load cells are useful for applications requiring a load to pass directly through the load cell itself. Load washers may be used to determine clamping forces with respect to applied torques on fasteners. Load buttons have an integral, spherical loading surface and are commonly used for compressive loading in production and testing applications.

Figure 11.3.9:
Typical hollow load cell.

Figure 11.3.10:
Typical washer load cell.

Figure 11.3.11:
Typical button load cell.

Multi-component – These are sensors that measure multiple forces and moments simultaneously with a single transducer. Multi-component applications include fatigue testing machines, tire uniformity testing and automotive testing requirements. Typically separate receptacles or cables are provided for each axis, or channel of measurement. A careful structural analysis is required to isolate the forces and moments, which result in cross-talk. A variety of configurations and combinations of force and moment measurements are available.

Chapter 11

Figure 11.24:
Typical multi-component load cell.

Applicable Standards

The basic design of strain gage-based load cells is not governed by a specific standard. However, individual applicable standards do exist for calibration and certification of strain gage-based sensors that are used for specific measurement of a force, load, or weight.

Strain gage sensors used for measurement of a weight are governed by more specific applicable standards than sensors used for measurement of a load or a force, as strain gage sensors used to measure a weight are typically used for trade or sale of goods. These standards are typically Factory Mutual, OMIL, and Underwriters Laboratory. Most manufacturers comply or conform to standards such as ISO 10012-1 (former MIL–STD-45662A), ISO 9001 and ISO/IEC 17025.

Latest and Future Developments

The basic design of strain gage-based force, load, or weight sensors has not changed significantly since the original development in the 1950s. Several manufacturers have recently added on-board electronics for signal conditioning at the sensor itself. This improves the overall quality of the sensor and reduces the system cost. Typical conditioned outputs that are available include voltage and current output options.

TEDS, which stands for "Transducer Electronics Data Sheet," is based on IEEE P1451, which is an emerging standard defining the architecture and protocol for compiling and addressing nonvolatile memory that is embedded within an analog measurement sensor. TEDS allows the user to retrieve specific information such as manufacturer name, sensor type, model number, serial number, and calibration data as well as write information such as channel ID, location, position, direction, tag number, etc. Some manufacturers are offering sensors with the TEDS feature with the overall intent of simplifying and reducing the customer's set up time and reducing errors.

Future developments within load cells and force sensors will most likely include wireless technology for transmission of digitized data over short distances. Development of this technology is underway. However, cost will be a major hurdle to overcome.

Other future developments may include digital temperature compensation.

Major Manufacturers

PCB Piezotronics, Inc. – 3425 Walden Avenue, Depew, NY 14043
Lebow Products – 1728 Maplelawn Road, Troy, MI 48084
Interface, Inc. – 7401 East Butherus Drive, Scottsdale, AZ 85260
Sensotec – 2080 Arlingate Lane, Columbus, OH 43228
Tedea Huntleigh – 677 Arrow Grand Circle, Covina, CA 91722
HBM, Inc. – 19 Bartlett Street, Marlborough, MA 01752

References and Resources

[1] PCB Piezotronics, Inc. Force/Torque Division Product Catalog FTQ-200C.
[2] PCB Piezotronics, Inc., ICP® Guide.
[3] Wheatstone, Charles: An account of several new Instruments and Processes for determining the Constants of a Voltaic Circuit. Philosophical Transaction of the Royal Society of London, 1843.
[4] Jerome Johnston and Keith Coffey: Getting the Most out of Strain Gauge Load Cells. Sensors Magazine, May 2000.

CHAPTER 12

Humidity Sensors

John Fontes, Senior Applications Engineer,
Honeywell Sensing and Control

12.1 Humidity

Humidity is defined as the water vapor content in air or other gases. Humidity is usually measured in terms of absolute humidity (the ratio of the mass of water vapor to the volume of air or gas), dew point (the temperature and pressure at which a gas begins to condense into a liquid), and relative humidity, or RH (the ratio of the moisture content of air compared to the saturated moisture level at the same temperature or pressure).

Thermal conductivity humidity sensors, also known as absolute humidity sensors, are capable of measuring absolute humidity using a system that employs two thermistors in a bridge connection, even at high temperatures or in polluted environments.

Since the early 1960s, chilled mirrors have been used to measure dew point, but the development of thin film capacitive sensors now allows measurement of dew points at temperatures as low as –40°F at far less cost and with greater accuracy.

Relative humidity was once determined by measuring the change in moisture absorption in silk, human hair, and later, nylon and synthetics. Mechanical methods for measuring RH were introduced in the 1940s. Recently, polymer-based resistive and capacitive sensors have been developed.

12.2 Sensor Types and Technologies

Recent developments in semiconductor technology have made possible humidity sensors that are highly accurate, durable, and cost effective. The most common humidity sensors are capacitive, resistive, and thermal conductivity. The following sections discuss how each sensor type is constructed and used to measure humidity.

Chapter 12

Capacitive RH Sensors

Capacitive RH sensors are used widely in industrial, commercial, and weather telemetry applications. They dominate both atmospheric and process measurements and are the only types of full-range RH measuring devices capable of operating accurately down to 0% RH. Because of their low temperature effect, they are often used over wide temperature ranges without active temperature compensation.

In a capacitive RH sensor, change in dielectric constant is almost directly proportional to relative humidity in the environment. Typical change in capacitance is 0.2–0.5 pF for 1% RH change. Bulk capacitance is between 100 and 500 pF at 50% RH at 25°C.

These sensors have low temperature coefficient and can function at high temperatures up to 200°C. They are able to fully recover from condensation and resist chemical vapors. Response time ranges from 30 to 60 seconds for a 63% RH step change.

Thermoset polymer-based capacitive RH sensors directly detect changes in relative saturation as a change in sensor capacitance with fast response, high linearity, low hysteresis, and excellent long-term stability. Relative saturation is the same as ambient relative humidity when the sensor is at ambient temperature. Because this is almost always the case, sensor capacitance change is then a measure of RH change.

These sensors use an industrially proven thermoset polymer, three layer capacitance construction, platinum electrodes and except for high temperature versions, on-chip silicon integrated voltage output signal conditioning. (See Figure 12.2.1.)

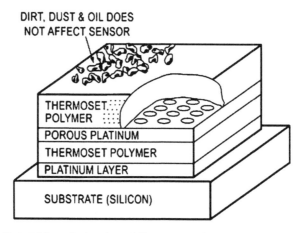

Figure 12.2.1: This relative humidity sensor has three-layer capacitance construction and consists of thermoset polymer, platinum electrodes, and a silicon chip with integrated voltage output signal conditioning.

In operation, water vapor in the active capacitor's dielectric layer equilibrates with the surrounding gas. The porous platinum layer shields the dielectric response from external influences while the protective polymer overlay provides mechanical protection for the platinum layer from contaminants such as dirt, dust and oils. A heavy contaminant layer of dirt will slow down the sensor's response time because it will take longer for water vapor to equilibrate in the sensor.

Thermoset polymer-based capacitive sensors, as opposed to thermoplastic-based capacitive sensors, allow higher operating temperatures and provide better resistivity against chemical liquids and vapors such as isopropyl, benzene, toluene, formaldehydes, oils, common cleaning agents, and ammonia vapor in concentrations common to chicken coops and pig barns. In addition, thermoset polymer RH sensors provide the longest operating life in ethylene oxide-based (ETO) sterilization processes.

Thermoset thin film polymer capacitive sensors have been shown to have an almost ideal response to RH, as opposed to absolute moisture (i.e., water vapor pressure). This response is due to the driving force free energy for absorption, G:

$$G = R\,T\,\mathrm{Ln}(P/P_0)$$

where
G = driving force
R = gas constant
P = partial water vapor pressure
P_0 = saturation water vapor pressure

P/P_0 is the same as ambient RH when the sensor is at ambient temperature. The relative saturation level driving sensor response is 100% at the sensor temperature T.

Research has also demonstrated that the RH sensor calibration in air applies to relative saturation measurement in oil to within 0.3% (a result which can be extended to other chemically compatible liquids).

Resistive Humidity Sensors

Resistive humidity sensors measure the impedance change, which usually has an inverse exponential relationship to humidity. (Figure 12.2.2.) Typically, the impedance change of a medium such as a conductive polymer, salt, or treated substrate is measured.

The first mass-produced humidity sensor was the Dunmore type. Produced in 1940, it is still widely used in precision air conditioning controls and for monitoring transmission lines, antennas, and waveguides used in telecommunications.

Chapter 12

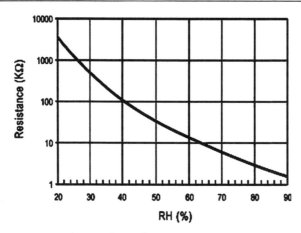

Figure 12.2.2: Relationship of impedance change to humidity.
Source: Roveti, Denes. "Choosing a Humidity Sensor: A Review of Three Technologies."
Sensors Online. July 2001. http://www.sensorsmag.com/articles/0701/54/main.shtml

The latest resistive humidity sensors use ceramic coating to provide protection in environments where condensation occurs. These sensors are constructed with noble metal electrodes deposited by a photoresist process, and a substrate surface coated with a conductive polymer/ceramic binder mixture. The sensor is protected in a plastic housing.

Interchangeability is better than 3% RH over the 15%–95% RH range, while precision is confirmed to ±2% RH. The recovery time for resistive sensors from full condensation to 30% is a few minutes. Voltage output is directly proportional to the ambient relative humidity when a signal conditioner is used. For most resistive sensors, response time is from 10 to 30 seconds for a 63% step change, while impedance range varies from 1 kΩ to 100 MΩ.

Thermal Conductivity Humidity Sensors

Thermal conductivity humidity sensors (also known as absolute humidity sensors) measure absolute humidity by calculating the difference between the thermal conductivity of dry air and air containing water vapor.

These sensors are constructed using two negative temperature coefficient (NTC) thermistor elements in a DC bridge circuit. One of the elements is sealed in dry nitrogen, while the other is exposed to the environment. (See Figure 12.2.3.) The difference in the resistance between the two thermistors is directly proportional to absolute humidity.

Humidity Sensors

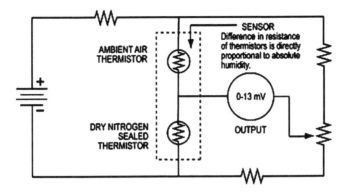

Figure 12.2.3: Thermal conductivity (or absolute) humidity sensors.
Source: Roveti, Denes. "Choosing a Humidity Sensor: A Review of Three Technologies."
Sensors Online. July 2001. http://www.sensorsmag.com/articles/0701/54/main.shtml

12.3 Selecting and Specifying Humidity Sensors

The following sections address what differentiates each sensor from another, including temperature, accuracy, and interchangeability. The advantages and disadvantages of each sensor type are also identified.

Selecting Humidity Sensors

Important considerations when selecting a humidity sensor include:

- Accuracy
- Interchangeability
- Repeatability
- Stability
- Condensation recovery
- Contaminant resistance
- Size and packaging
- Cost effectiveness
- Cost to replace sensor
- Calibration
- Complexity and reliability of signal conditioning and data acquisition circuitry

In general, environmental conditions for the given application will dictate the choice of sensor.

Chapter 12

Selecting Capacitive RH Sensors

Applications for capacitive RH sensors are wide ranging, including

- Automotive onboard devices such as windshield defoggers
- Computer printers
- Medical devices such as ventilators and incubators
- Appliances such as microwave ovens, refrigerators, and clothes dryers
- HVAC
- Recorders and data loggers
- Leak detection
- Weather stations
- Industrial and food processing equipment
- Environmental test chambers

Taking advantage of cutting-edge principles in semiconductor design, many capacitive sensors have minimal long-term drift and hysteresis. Incorporating a complementary metal oxide semiconductor (CMOS) timer pulses the sensor producing near-linear voltage output. (See Figure 12.3.1.)

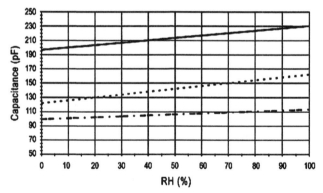

Figure 12.3.1: Voltage output with a CMOS timer.
Source: Roveti, Denes. "Choosing a Humidity Sensor: A Review of Three Technologies."
Sensors Online. July 2001. http://www.sensorsmag.com/articles/0701/54/main.shtml

Typical uncertainty is ±2% RH from 5% to 95% RH using two-point calibration. The capacitive effect of the connecting cable relative to the small capacitance changes of the sensor limits the distance the sensing element can be located from the signal conditioning circuitry to a practical range of less than 10 feet.

Laser trimming reduces variance ±2%, improving direct field interchangeability. Computer recalibration programs are also capable of compensating for sensor capacitance from 100 to 500 pF.

Capacitive RH sensors are not linear below a few percent RH, which is why many sensors incorporate a dew-point measurement system that employs microprocessor-based circuitry to store calibration data. This development has reduced the cost of hygrometers and transmitters in HVAC and weather telemetry applications.

Advantages

- Near-linear voltage output
- Wide RH range and condensation tolerance
- Interchangeable, if laser trimmed
- Stable over long-term use

Disadvantages

- Distance from sensing element to signal conditioning circuitry limited

Selecting Resistive Humidity Sensors

Resistive sensors are small, low-cost humidity sensors that provide long-term stability and are highly interchangeable. They are suitable for many industrial, commercial, and residential applications, especially control and display products.

Resistive sensors respond nonlinearly to humidity changes, but they may be linearized by analog or digital methods. Typical variable resistance ranges from a few kilohms to 100 MV. Nominal excitation frequency is from 30 Hz to 10 kHz.

RH sensors are highly interchangeable (within ±2% RH). Electronic signal conditioning circuitry can be calibrated at a fixed RH point, eliminating the need for humidity calibration standards. Accuracy can be tested in an RH calibration chamber or by a computer-based system referenced to a standardized environment. Resistive sensors have a nominal operating temperature of –40°C to 100°C.

Life expectancy is less than five years in residential and commercial applications, but exposure to contaminants may cause premature failure. Resistive sensors also tend to shift values during exposure to condensation when water-soluble coatings are used.

Advantages

- No calibration standards, so highly interchangeable and field replaceable
- Long-term stability
- Usable from remote locations
- Small size
- Low cost

Disadvantages

- Exposure to chemical vapors and contaminants may cause premature failure
- Values may shift when water-soluble coatings are used

Selecting Thermal Conductivity Humidity Sensors

Thermal conductivity humidity sensors are commonly used in appliances, including clothes dryers and microwave ovens. They are used in many industrial applications including wood-drying kilns, drying machinery, pharmaceutical production, cooking, and food dehydration.

Constructed with glass, semiconductor material, high-temperature plastics, and aluminum, thermal conductivity sensors are very durable and resistant to chemical vapors. They provide better resolution than capacitive and resistive sensors in temperatures greater than 200°F. Typical accuracy is +3 g/m^3, which converts to approximately ±5% RH at 40°C and ±0.5% RH at 100°C.

Advantages

- Very durable
- Work well in corrosive and high-temperature environments up to 575°F
- Better resolution than capacitive and resistive sensors

Disadvantages

- Responds to any gas with thermal properties different than dry nitrogen, which may affect measurement.

12.4 Applicable Standards

Standards Bodies

American National Standards Institute (ANSI): http://www.ansi.org
A private, non-profit organization responsible for administering the U.S. voluntary standardization and conformity assessment system.

American Society of Testing and Materials (ASTM): http://www.astm.org
One of the largest voluntary standards development organizations in the world. Develops and publishes voluntary consensus standards for materials, products, systems, and services.

Canadian Standards Association (CSA): http://www.csa.ca
Not-for-profit membership-based association serving business, industry, government, and consumers in Canada and around the world. Develops standards for enhancing public safety and health, advancing the quality of life, helping to preserve the environment, and facilitating trade.

Instrumentation, Systems, and Automation Society (ISA): http://www.isa.org
Helps advance the theory, design, manufacture, and use of sensors, instruments, computers, and systems for measurement and control in a variety of applications.

International Electrotechnical Commission (IEC): http://www.iec.ch
Prepares and publishes international standards for all electrical, electronic, and related technologies.

International Organization for Standardization (ISO): http://www.iso.ch/iso/en/ISOOnline.openerpage
A network of national standards institutes from 146 countries working in partnership with international organizations, governments, industry, business and consumer representatives.

Japanese Standards Association (JSA): http://www.jsa.or.jp/default_english.asp
Objective is "to educate the public regarding the standardization and unification of industrial standards, and thereby to contribute to the improvement of technology and the enhancement of production efficiency."

National Institute of Standards and Technology (NIST): http://www.nist.gov
Founded in 1901, NIST is a non-regulatory federal agency within the U.S. Commerce Department's Technology Administration. Its mission is to develop and promote measurement, standards, and technology to enhance productivity, facilitate trade, and improve the quality of life.

Industry Organizations

American Society for Quality (ASQ): http://www.asq.org/
> Purpose is to improve workplace and communities by advancing learning, quality improvement, and knowledge exchange. Advises the U.S. Congress, government agencies, state legislatures, and other groups and individuals on quality-related topics.

International Measurement Confederation (IMEKO): http://www.mit.tut.fi/imeko/
> Non-governmental federation of 36 member organizations. Promotes international interchange of scientific and technical information in the field of measurement and instrumentation and the international cooperation among scientists and engineers from research and industry.

National Conference of Standards Laboratories International (NCSL International):
http://www.ncsli.org/
> A professional association for individuals involved in all aspects of measurement science.

Underwriter's Laboratories (UL): http://www.ul.com
> An independent, not-for-profit product-safety testing and certification organization.

12.5 Interfacing and Design Information

Temperature and Humidity Effects

The output of all absorption-based humidity sensors (capacitive, bulk resistive, conductive film, etc.) is affected by both temperature and %RH. Because of this, temperature compensation is used in applications that call for either higher accuracy or wider operating temperature ranges.

When calibrating a humidity sensor to compensate for temperature, it is best to make the temperature measurement as close as possible to the humidity sensor's active area—i.e., within the same moisture microenvironment. This is especially true when combining RH and temperature as a method for measuring dew point. Industrial grade humidity and dew point instruments incorporate a 1000-ohm platinum RTD on the back of the ceramic sensor substrate for unmatched temperature compensation measurement integrity. No on-chip signal conditioning is provided in these high temperature sensors.

Humidity Sensors

Voltage Output

A humidity sensor with a relative humidity integrated circuit (RHIC) has a linear voltage output that is a function of V_{supply}, %RH and temperature. The output is "ratiometric," so as the supply voltage rises, the output voltage rises in the same proportion. A surface plot of the sensor behavior for temperatures between 0°C and 85°C is shown in Figure 12.5.1.

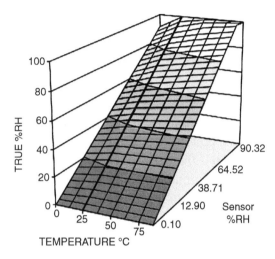

Figure 12.5.1: Surface plot of sensor behavior.

This surface plot is well approximated by a combination of two equations:

1. A "Best Fit Line at 25°C," or a similar, sensor-specific equation at 25°C. The sensor independent "typical" Best Fit Line at 25°C (bold line in graph) is:

    $V_{out} J V_{supply} (0.0062 (\%RH) + 0.16)$

    A sensor-specific equation can be obtained from an RH sensor printout. The printout equation assumes V_{supply} J 5VDC and is included or available as an option on the sensors.

2. A sensor-independent equation, which corrects the %RH reading (from the Best Fit Line Equation) for temperature, T:

    True RH = (%RH)/(1.0546 − .00216 T); T = °C
    Or True RH = (%RH)/(1.093 − .0012 T); T = °F

The previous equations match the typical surface plot (Best Fit Line at 25°C) or the actual surface plot (sensor-specific equation at 25°C) to within the following tolerances:

$\pm 1\%$ for $T>20°C$
$\pm 2\%$ for $10°C<T<20°C$
$\pm 5\%$ for $T<10°C$

Dewpoint instruments may account directly for a sensor-specific version of the surface plot via a look-up table.

Note: Convert the observed output voltage to %RH values via the first equation before applying the second equation.

Condensation and Wetting

Condensation occurs whenever the surface temperature of the sensor's active area drops below the ambient dew point of the surrounding gas. Condensation forms on the sensor (or any surface) even if the surface temperature only momentarily drops below the ambient dew point. Small temperature fluctuations near the sensor can unknowingly cause condensation to form when operating at humidity levels above 95%.

While quick to condense, water is slow to evaporate in high humidity conditions (i.e., when the surface temperature of the sensor is only slightly above the ambient dew point.) Because of this, a sensor's recovery period from either condensation or wetting is much longer than its normal time response. During recovery, the sensor outputs a constant 100% RH signal regardless of the ambient RH.

When an application calls for continuous monitoring of RH at humidity levels of 90% and above, take steps to avoid intermittent condensation. Some strategies are:

1. Maintain a good air mixing to minimize local temperature fluctuations.

2. Some sensors use a sintered stainless steel filter to protect the sensor from splashing. A hydrophobic coating further suppresses condensation and wetting in rapidly saturating and de-saturating or splash-prone environments.

3. Heat the RH sensor so that the active area is hotter than the local dew point. This can be done through an external heater or by self heating of the CMOS RH chip by operating it at a higher supply voltage.

Note: Heating an RH sensor above ambient temperature changes its calibration and makes it sensitive to thermal disturbances such as airflow.

Integrated Signal Conditioning

All RH sensors quickly recover from condensation or wetting with no shift in calibration. However, after 24-hour or longer exposures to either high (>95%) RH or continuous condensation, an upward shift of 2% to 3% RH may occur. This shift is repeatable and can be reversed by placing the sensor in a low (10%) RH environment for a 10-hour period.

Silicon integrated humidity sensors incorporate signal conditioning circuitry on-chip with the sensing capacitor. These RHIC humidity sensors are laser trimmed so that at V_{supply} J 5 V, the output voltage typically spans 0.8 V to 3.9 V for the 0% RH to 100% RH range at 25°C. (Sensor-specific calibration data printouts and Best Fit Lines at 25°C are either included or available as an option on these sensors.)

RHIC-based sensors are factory calibrated, micro-power devices with either individual calibration and/or good unit-to-unit interchangeability. These features help OEM manufacturers avoid in-house humidity calibration costs and extend battery life in portable instruments. Improved accuracy can be obtained by tuning system electronics to account for an individual sensor's Best Fit Line at 25°C.

Interchangeability defines the range of voltages for any population of sensors at an RH point. An interchangeability of ±5% @ 0%RH is compared to the baseline output for the RHIC chip, which is 0.8 V to 3.9 V (0 to 100% RH) with an excitation voltage of 5 VDC.

If you take the baseline slope, 0.031 V/%RH times ± 5%RH you get ± 0.155 V. This means that the output voltage for this device is 0.8 V ± 0.155 V or a range of 0.645 V to 0.955 V. When exposed to an RH of 0%, the output of the entire population of sensors will fall within this range.

Interchangeability increases with increasing RH since the RHIC die is actively trimmed only at 0% RH. Trimming at other RH values is impractical.

Interchangeability lets you lower design cost by avoiding calibrating your system to each individual sensor. The RHIC sensor keeps its interchangeability and accuracy advantages at higher humidity.

Accuracy is based on the specific calibration curve for any individual sensor and equals ±2% RH. For example, if a specific sensor has an output voltage of .850 V at 0%RH (5 VDC supply assumed), then this sensor should always output this voltage ±0.062 V*** or a range of 0.788 V to 0.912 V. Accuracy equals interchangeability ±2% when you don't calibrate your system to each sensor. If you calibrate to each RHIC sensor, then total accuracy can be ±1–2% RH.

Chapter 12

References and Resources

Christian, Stephan. "New generation of humidity sensors." Sensor Review 22 (2002):300-2.

Honeywell web site, humidity sensor information: http://content.honeywell.com/sensing/prodinfo/humiditymoisture/#technical

Measurements Science Conference (MSC): http://www.msc-conf.com/

Quelch, D. "Humidity Sensors for Industrial Applications." International Conference on Sensors and Transducers, Vol. 1. Tavistock, UK: Trident Exhibitions, 2001.

Rittersma, ZM. "Recent Achievements in Miniaturized Humidity Sensors: A Review of Transduction Techniques." Sensors and Actuators 96 (2002):196-210.

Roveti, D.K. "Choosing a Humidity Sensor: A Review of Three Technologies." Sensors 18 (2001):54-8.

CHAPTER 13

Machinery Vibration Monitoring Sensors

Timothy J. Geiger, Division Manager of IMI Sensors, PCB Piezotronics, Inc.

13.1 Introduction

The ears and hands are very subjective when sensing vibrations. The days of judging a machine's health by sound and touch (or listening to a screwdriver placed against the machine) have quickly transitioned to a more scientific approach, allowing data trending and early prediction of machinery failure.

In order to make objective, informed decisions, most measurement engineers prefer to have consistent, trendable data that can be regularly referred to. Machinery vibration sensors, along with a readout instrument, can provide this objective information, allowing for a more precise assessment of machinery health.

Due to the piezoelectric accelerometer's wide frequency response, it is an excellent sensor to replace human subjectivity in most machinery health monitoring. In addition, permanent mounting of these sensors provides continuous monitoring or shut-down functions for critical machinery. This allows plant and production management to be predictive in their maintenance strategy. The result is cost-effective, scheduled down time to repair equipment as opposed to a reactive approach toward maintenance, which often involves expensive loss of production and repairs. In the future, it is expected that a greater percentage of machines will continue to be monitored for earlier prediction of failures. This will allow maintenance personnel to focus on the smaller percentage of machines actually having problems as opposed to manually monitoring and wasting expensive labor on healthy machinery.

Some key applications and industries in which machinery vibration sensors are utilized are as follows:

- Aluminum Plants
- Automotive Manufacturing
- Balancing
- Bearing Analysis & Diagnostics
- Bearing Vibration Monitoring

Chapter 13

- Bridges and Civil Structures
- Coal Processing
- Cold Forming Operations
- Concrete Processing Plants
- Condition Based Monitoring
- Compressors
- Cooling Towers
- Crushing Operations
- Diagnostics of Machinery
- Engines
- Floor Vibration Monitoring
- Food, Dairy & Beverage
- Foundations
- Gearbox Monitoring
- Geological Exploration
- Heavy Equipment & Machinery
- Helicopters
- Hull Vibration Monitoring
- HVAC Equipment
- Impact Measurements
- Impulse Response
- Machine Tools
- Machinery Condition Monitoring
- Machinery Frames
- Machinery Mount Monitoring
- Machinery Vibration Monitoring
- Manufacturing
- Mining
- Modal Analysis
- Motor Vibration
- Off-Road Equipment
- Paper Machinery Monitoring
- Petrochemical
- Pharmaceutical
- Power Generation
- Predictive Maintenance
- Printing
- Pulp & Paper
- Pumps

- Quality Control
- Seismic Monitoring
- Shipboard Machinery
- Shock Measurements
- Shredding Operations
- Site Vibration Surveys
- Slurry Pulsation Monitoring
- Spindle Vibration & Imbalance
- Squeak & Rattle Detection
- Steel & Metals
- Structure-Borne Noise
- Structural Testing
- Submersible Pumps
- Transportation Equipment
- Turbines
- Turbomachinery
- Underwater Pumps
- Vibrating Feeders
- Vibration Control
- Vibration Isolation
- Water Treatment Plants
- Wastewater Treatment Plants

The typical faults that machinery vibration sensors are able to detect, along with their approximate percent of occurrence, are as follows:

Imbalance	40%
Misalignment	30%
Resonance	20%
Belts and Pulleys	30%
Bearings	10%
Motor Vibration	8%
Pump Cavitation	5%

Accelerometers are the most common machinery vibration sensor used today. Applications for accelerometers in the industrial sector are primarily focused on extending service life of machinery by predicting failures and allowing maintenance to be conducted in a planned manner. By doing so, operators can make more intelligent decisions about spare parts purchases and keep critical equipment up-and-running longer and faster to increase product output and profitability. Accelerometers are ideal for small to medium sized machines with rolling element bearings, where the casing and bearings move with the rotor. In this case, an accelerometer (sometimes referred to as a seismic transducer) on the bearing housing will also be a good indicator of shaft motion. They are convenient to use and easily attach directly to the outside of the bearing housing. (These can still be used on journal bearing machines on the outside casings, but the motion will be smaller on the outside casing because of the mass of the casing and the reduced transmissibility through the fluid film.)

Technically speaking, accelerometers are transducers that are designed to produce an electrical output signal that is proportional to applied acceleration. Several sensing technologies are utilized to construct accelerometers, including resistive (foil or silicon strain gages), capacitive and piezoelectric. Resistive and capacitive types possess the ability to measure constant acceleration, such as that of earth's gravity, and are generally only used for measuring very low frequency vibration, such as that encountered with massive, slow-speed rollers. On the other hand, piezoelectric types possess an extremely wide frequency range and are suited to accurately measure most other types of vibration. They are ideal for machinery vibration monitoring and are recommended for nearly every industrial vibration application on rotating machinery.

Piezoelectric accelerometers are generally durable, protected from contamination, impervious to extraneous noise influences, and are easy to implement. Accelerometers that are designed with these features are classified as Industrial Accelerometers and are normally suitable for use in rigorous industrial, field or submersible environments.

13.2 Technology Fundamentals

Piezoelectric accelerometers rely on the piezoelectric effect. This basically means that the internal sensing material will provide an electrical charge when squeezed or strained.

There are several methods used to stress the piezoelectric material. (All of these methods utilize a seismic mass attached to the sensing material and rely on Newton's Second Law: Force = Mass × Acceleration. Refer to Chapter 5 for additional information.) These methods are categorized in the following terms that define the geometry of the sensing element—compression, flexural, and shear. Each has advantages and disadvantages; however, for best overall performance in industrial machinery vibration applications, leading manufacturers utilize, almost exclusively, the shear structured geometry for its piezoelectric industrial accelerometers. (See Figure 13.2.1.)

Machinery Vibration Monitoring Sensors

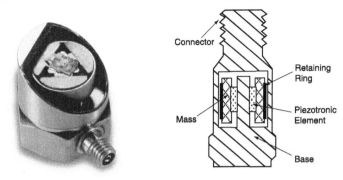

Figure 13.2.1: Photo and cross section diagram of shear accelerometer.

The shear sensing element design exhibits compact construction, good stiffness, good durability and high resonant frequency. This design is less susceptible to extraneous inputs from base strain and thermal transients. Older compression geometries have been superseded by the shear type for industrial applications, primarily due to the thermal transient induced base strain error susceptibility of compression designs, particularly at low frequencies. This low frequency error, or drift, causes what is commonly termed "ski slope" error, which becomes evident when acceleration signals are integrated into velocity signals. Regrettably, some manufacturers still rely on this aged design in their product line. Volume production of shear style accelerometers has also reduced their cost to a level acceptable to the industrial market, further justifying their use.

Regardless of the sensing element design, as the piezoelectric material is stressed, charged ions accumulate at their surfaces. Electrodes that are in intimate contact with the crystals collect these ions, which result in an electrical signal. The signal is then transmitted via a small wire to a signal conditioning circuit that is positioned within the accelerometer's housing. The signal conditioning circuit serves to convert the difficult-to-use, high-impedance charge signal from the crystal into a low-impedance voltage signal that can be transmitted outside of the accelerometer housing, over a relatively long distance, without degradation. This low-impedance voltage signal can then be interrogated by data collection equipment, displayed on an oscilloscope, stored on a recording instrument, or analyzed with a data acquisition system or FFT analyzer. This type of accelerometer is generically termed an IEPE accelerometer for Integrated Electronic PiezoElectric, or also commonly referred to as an ICP® (Integrated Circuit Piezoelectric) accelerometer, which is a registered trademark of PCB Group, Inc.

Additional circuitry can be added to enhance or tune the output signal for specific purposes. Such enhancements include filtering, rms conversion, 4–20 mA transmitters, integrators, and TEDS (Transducer Electronic Data Sheet—an on-board, addressable memory with stored, self-identifying information).

To properly operate, sensors with built-in electronics require a constant current power source. Normally, this constant current power supply is built into the industrial readout equipment (data collector or on-line monitor). If it is not included, sensor manufacturers offer a variety of different power supplies.

While the fundamental technology associated with accelerometers is very similar, the specific design of the sensor is critical. Industrial accelerometers have to endure severe environments and tough operating conditions. In order to achieve these requirements, there are several design and construction characteristics to be aware of. These design criteria include:

Welded, stainless steel housing – This proven, corrosive-resistant material stands up well against dirt, oil, moisture, and harsh chemicals. Stainless steel is also non-magnetic, which minimizes errors induced when used in the vicinity of electric motors and other sources of electromagnetic interference. For durability, all mating housing parts are precision laser welded. No epoxies are used which can eventually fatigue or cause leaks.

Internal Faraday shield – Accelerometers that utilize an internal electrical shield to guard against radio frequency interference (RFI), electromagnetic interference (EMI), and other extraneous noise influences. The result is an electrically case-isolated sensing element that is insulated from ground, which also insures that there won't be any ground loops in the measurement system that can cause signal drift, noise, and other hard-to-trace electrical problems.

Durable, military style electrical connectors or sealed, integral cables – It is highly recommended to use electrical connectors that offer a true hermetic seal with glass-to-metal fusing of connector pins and shells. All connectors should be laser welded to the sensor housing. Sensors with integral cables incorporate hermetic feed-through pins and high-pressure, injection-molded sealing of the cable to a metallic shell that is laser welded to the sensor housing. Integral polyurethane jacketed cables offer 750 psi submersibility along with excellent pull strength and strain relief characteristics.

Rugged Cables – Interconnect cables should utilize shielded, twisted conductor pairs and an outer jacket material that will survive exposure to most industrial media, or excessive temperatures present in the environment in which it will be used. Stainless steel, armored cables are recommended for installations where machined debris or chips may damage cable jackets, or where cables have the potential of being stepped on or pinched. Electrical connectors for cables are offered in a variety of styles and configurations to suit the application. Take care to note the temperature rating of the connector and material of construction. Environmentally sealed cable connectors provide a boot to protect the integrity of the connection in outdoor installations or during washdown episodes.

13.3 Accelerometer Types

Low-Cost, Industrial ICP Accelerometers for Permanent Installation

Low-cost, industrial ICP® accelerometers are recommended for permanent installation onto machinery to satisfy vibration trending requirements in predictive maintenance and condition monitoring applications. In addition to sound design and high volume manufacturing techniques, lower cost is achieved by relaxing the tolerance on sensitivity from unit to unit and by calibrating at only one reference frequency point, typically 100 Hz. Measurement accuracy is compromised only if the sensor's nominal sensitivity is used. If the provided single-point sensitivity is used, accuracy is very good.

Since low-cost sensors carry a wider sensitivity tolerance, the actual measurement obtained using the nominal sensitivity value may not be as quantitatively accurate as could be achieved if one uses the supplied reference sensitivity value. This disparity, however, may be irrelevant since when trending, the user is primarily interested in recognizing changes in the overall measured vibration amplitude, or frequency signature of the machinery. When comparing against previously acquired data obtained with the same sensor in the same location, the excellent repeatability of these piezoelectric vibration sensors becomes the vital attribute for successful trending requirements. The user benefits by being able to employ a lower cost sensor, which in turn makes monitoring additional measurement points a more attractive undertaking.

Low-Frequency Industrial ICP Accelerometers

Low-frequency accelerometers are generally designed for use on large slow-rotating equipment. Applications include:

- Vibration measurements on slow rotating machinery
- Paper machine rolls
- Large structures and machine foundations
- Large fans and air handling equipment
- Cooling towers
- Buildings, bridges, foundations, and floors

Low-acceleration levels go hand-in-hand with low-frequency vibration measurements. For this reason, these accelerometers combine extended low frequency response with high output sensitivity in order to obtain the desired amplitude resolution characteristics and strong output signal levels necessary for conducting low-frequency vibration measurements and analysis.

The most sensitive low-frequency accelerometers are known as seismic accelerometers. These models are larger in size to accommodate their larger seismic, internal masses, which are necessary to generate a stronger output signal. These sensors have a limited amplitude range, which renders them unsuitable for many general purpose industrial vibration measurement applications. However, when measuring the vibration of slow, rotating machinery, buildings, bridges and large structures, these low-frequency, low-noise accelerometers will provide the characteristics required for successful results.

All low-frequency industrial accelerometers benefit from the same advantages offered by general-purpose industrial accelerometers including: rugged, laser-welded, stainless steel housing with the ability to endure dirty, wet, or harsh environments; hermetically sealed military connector or sealed integral cable; and a low-noise, low-impedance, voltage output signal with long-distance signal transmission capability.

High-Frequency Industrial ICP® Accelerometers

High-frequency accelerometers are generally used for high-speed rotating machinery. Applications include:

- Vibration measurements on high-speed rotating machinery
- Gear mesh studies and diagnostics
- Bearing monitoring
- Small mechanisms
- High-speed spindles

Successful vibration measurements begin with sensors that have adequate capabilities for the requirement. If the sensor's frequency response characteristics are inadequate, the user risks corrupted or insufficient data to achieve a proper analysis and diagnosis.

For vibration monitoring, testing, and frequency analysis of high-speed rotating machinery, spindles, and gear mesh, it is imperative to utilize a sensor with a sufficient high frequency range to accurately capture the vibration signals within the bandwidth of interest. Miniature sized units are also suitable for vibration measurements on small mechanisms where sensor size and weight become important factors.

4–20 mA Vibration Sensing Transmitters

These sensors are accelerometers which have been configured to have a loop-powered, 4–20 mA output. Generally, vibration sensing transmitters provide output signals that are representative of the overall vibration. (Typically, an internal integrator is included so that the output is in velocity, such as in/sec rms or mm/s pk.) This vibration signal may be interfaced with many types of commercially available current-loop monitoring equipment, such as recorders, alarms, programmable logic controllers (PLCs) and digital control systems (DCSs).

The vibration-sensing transmitters capitalize on the use of existing process control equipment and human machine interface (HMI) software for monitoring machinery vibration and alarming of excessive vibration levels. This allows the operator to monitor machinery using the existing process control system as opposed to having to purchase specific vibration monitory software. This practice offers the ability to continuously monitor machinery and provide an early warning detection of impending failure. With this approach, existing process control technicians can monitor the vibration levels while skilled vibration specialists are called upon only in the event that the vibration signal warrants more detailed signal analysis. It is envisioned that someday machinery vibration measurements will be monitored in process facilities via 4–20 mA transmitters as widely as pressure, temperature, flow and load are currently done so today.

A choice of velocity or acceleration measurement signals are offered with a variety of amplitude and frequency ranges to suit the particular application. Most models also feature an optional analog output signal connection (raw vibration option) for conducting frequency analysis and machinery diagnostics.

DC Response, Industrial Capacitive Accelerometers – Applications

DC response accelerometers are generally only used on very large, slow-rotating equipment or on extremely large civil structures. Applications include:

- Paper machine rolls
- Large structures and machine foundations
- Cooling towers
- Buildings, bridges, foundations, and floors

Unlike piezoelectric accelerometers that have a low frequency limit, industrial capacitive accelerometers are capable of measuring to true DC, or 0 Hz.

Capacitive accelerometers utilize two opposed plate capacitors that share a common flexible plate in the middle. As the flexible plate responds to acceleration, differential output signals from the two capacitors are created. These signals are conditioned to reject common mode noise and combined to provide a standardized output sensitivity. Built-in voltage regulation and optional low-power versions permit operation from a wide variety of DC power sources. The result is a sensor that is easy to implement and delivers a high integrity, low noise output signal.

DC response, industrial capacitive accelerometers offer many of the same advantages as general purpose industrial accelerometers including: rugged, laser-welded, stainless steel housing with the ability to endure dirty, wet, or harsh environments; hermetically sealed military connector or sealed integral cable; and a low noise, low impedance, voltage output signal with long distance signal transmission capability.

13.4 Selecting Industrial Accelerometers

There will usually be several accelerometer models that meet the required measurement parameters, so the question naturally arises, Which should be used? By answering the following questions as accurately as possible, the user will be able to determine a set of key specifications required for the accelerometer.

1. **Measurement Range / Sensitivity** – Determine the maximum peak vibration amplitude that will be measured and select a sensor with an appropriate measurement range. For a typical accelerometer, the maximum measurement range is equal to ±5 volts divided by the sensitivity. For example, if the sensitivity is 100 mV/g then the measurement range is (5 V / 0.1 V/g) = ±50 g. Allow some overhead in case the vibration is a little higher than expected. The vibration severity chart shown in Figure 13.4.1 may help when determine the expected measurement range.

Machinery Vibration Monitoring Sensors

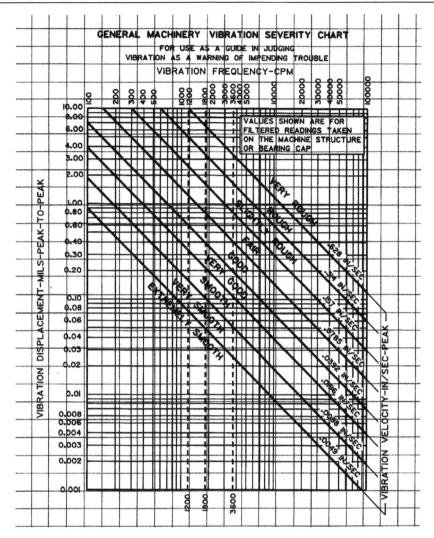

Figure 13.4.1: Vibration severity chart.

2. **Frequency Range** – Determine the lowest and highest frequencies to be analyzed. If you are not sure what the upper frequency range should be, use the following table showing "Recommended Frequency Spans" as a guideline.

Table 13.4.1: Frequency range guidelines [Ref. 4].

Recommended Frequency Spans (Upper Frequency)

Shaft Vibration	10 × RPM
Gearbox	3 × GMF
Rolling Element Bearings	10 × BPFI
Pumps	3 × VP
Motors / Generators	3 × (2 × LF)
Fans	3 × BP
Sleeve Bearings	10 × RPM

RPM – Revolutions Per Minute
GMF – Gear Mesh Frequency
BPFI – Ball Pass Frequency Inner race
VP – Vane Pass frequency
LF – Line Frequency (60 Hz in USA)
BP – Blade Pass frequency

Select an accelerometer that has a frequency range that encompasses both the low and high frequencies of interest. In some rare cases, it may not be possible to measure the entire range of interest with a single accelerometer.

High Frequency Caution – Many machines, such as pumps, compressors, and some spindles, generate high frequencies beyond the measurement range of interest. Even though these vibrations are out of the range of interest, the accelerometer is still excited by them. Since high frequencies are usually accompanied by high accelerations, they will often drive higher sensitivity accelerometers (100 and 500 mV/g models) into saturation causing erroneous readings. If a significant high frequency vibration is suspected or if saturation occurs, a lower sensitivity (typically 10 or 50 mV/g) accelerometer should be used. For some applications, higher sensitivity accelerometers are available with built-in low pass filters. These sensors filter out the unwanted high frequency signals and thus provide better amplitude resolution at the frequencies of interest.

To determine if you have a condition that will overdrive (saturate) the accelerometer, look at the raw vibration signal in the time domain on a data collector, spectrum analyzer, or oscilloscope. Set the analyzer for a range greater than the maximum rated output of the accelerometer. If the amplitude exceeds the maximum rated measurement range of the accelerometer (typically 5 volts or 50 g for a 100 mV/g unit), then a lower sensitivity sensor should be selected. If the higher sensitivity sensor is used, clipping of the signal and saturation of the electronics is likely to occur. This will result in false harmonics, "ski slope" as well as many other serious measurement errors.

3. **Broadband Resolution (Noise)** – Determine the amplitude resolution that is required. This will be the smaller of either the lowest vibration level or the smallest change in amplitude that must be measured. Select a sensor that has a broadband resolution value equal to or less than this value. For example, if measuring a spindle with 0.001 g minimum amplitude, choose an accelerometer with a resolution better than 1 mg. If the known vibration levels are in velocity (in/s) or displacement (mils), convert the amplitudes to acceleration (g) at the primary frequencies. **Note:** The lower the resolution value, the better the resolution is. Generally, ceramic sensing elements have better resolutions (less noise) than quartz based sensors.

4. **Temperature Range** – Determine the highest and lowest temperatures that the sensor will be subjected to and verify that they are within the specified range for the sensor.

   Temperature Transients – In environments where the accelerometer will be subjected to significant temperature transients, quartz sensors may achieve better performance than ceramic. Ceramic sensing elements are subject to the pyroelectric effect, which can cause erroneous outputs with changes in temperature. These outputs typically occur as drift (very low frequency) and usually cause significant "ski slope" when the signal is integrated and displayed in the velocity spectrum.

5. **Size** – In many cases, the style of the sensor used can be restricted by the amount of space that is available on a machine to mount the sensor. There are typically two parameters that govern which sensors will fit, the footprint and the clearance. The footprint is the area covered by the base of the sensor. The clearance is the height above the surface required to fit the sensor and cable. As an example, a top exit sensor will require more clearance than a side exit.

Chapter 13

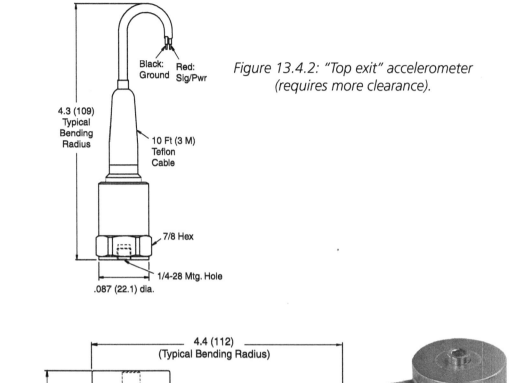

Figure 13.4.2: "Top exit" accelerometer (requires more clearance).

Figure 13.4.3: "Side exit" accelerometer (requires less clearance).

Orientation – Cable orientation is another consideration. Ring-style, side exit models can be oriented 360°, however, in some very tight spaces, even these may be difficult to install. For example, there may not be enough height clearance to fit a wrench to tighten the unit. In that case, a swivel mount style accelerometer may be required.

*Figure 13.4.4:
"Swivel mount" accelerometer.*

6. **Duty (Accuracy, Sensitivity Tolerance, and Safety)** – The duty refers to the type of use that a sensor will see. Most typical predictive maintenance applications are either in a walk-around application, as with a portable data collector, or permanently mounted to a particular machine. In permanent mount applications, the sensor may terminate at a junction box where measurements are taken with a portable data collector or tied to an on-line monitoring system. 4–20 mA output sensors would usually be tied to existing plant systems such as a PLC.

   Sensitivity Tolerance (Absolute Accuracy) — Sensitivity tolerance is the maximum deviation that the actual sensitivity of an accelerometer can vary from its published nominal sensitivity and still be within specification. Most manufacturers offer accelerometers with ± 5%, ± 10%, ± 15%, and ± 20% tolerances on sensitivity. Thus, a nominal 100 mV/g sensor with a ± 5% tolerance could have an actual sensitivity between 95 and 105 mV/g. A ± 20% tolerance unit could vary between 80 and 120 mV/g. If the nominal sensitivity is used to convert the actual data to engineering units (e.g., entered into the data collection device), then a looser tolerance sensor will be less accurate, in general, than a tighter tolerance model. However, if the actual calibration value that is supplied with the sensor is used, then both readings will be equally accurate. In applications were absolute accuracy is important (e.g., in acceptance testing) then either higher tolerance sensors or the actual calibration factors should be used.

   Lower tolerance sensors are typically provided with a single point calibration rather than full calibration. This, coupled with the looser tolerance, helps keep costs down and allows them to be offered at a much more economical price. Normally, these sensors are selected for permanent mount applications where larger numbers of accelerometers are needed.

Calibration Interval – Due to the inherent stability of quartz, accelerometers with quartz sensing elements have a longer recommended calibration interval than do ceramic sensors. The recommended time between calibrations is 1 year for ceramic sensors and 5 years for quartz. As a practical matter, however, it may not be possible to send ceramic sensors in for yearly recalibration. As long as the sensor is permanently mounted and not going through severe thermal transients on a regular basis, its sensitivity should remain fairly stable. However, if it is seeing repeated shocks (as with magnetic mounting in a walk around system) or severe thermal transients, it is highly recommended that the sensor be recalibrated yearly. One advantage of quartz sensors is its long-term stability even in high shock and thermally transient environments. A typical calibration certificate is shown in Figure 13.4.5.

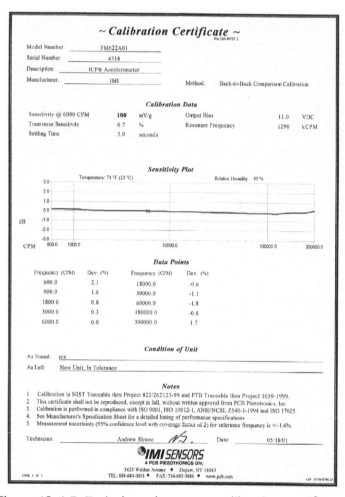

Figure 13.4.5: Typical accelerometer calibration certificate.

Accessibility, Safety, and Production Considerations – Monitoring locations on machines are often inaccessible due to shrouds, safety requirements, space constraints, or other physical obstacles. Additionally, they may be in hazardous areas or have limited access due to pressing production schedules. In cases like these, permanent mount accelerometers should be selected. This provides a fast, easy, and safe way to collect vibration data.

7. **Cabling** – It is recommended, in most cases, that connector style accelerometers be used rather than ones with integral cable. Cables are very susceptible to damage and are usually the source of most sensor problems. Therefore, it is much easier and more cost effective to replace a cable rather then the entire accelerometer/cable assembly. It is important to recognize that cables are vulnerable to damage and should be installed out of harms way. Having spare cables on hand is recommended as they can help troubleshoot system performance and keep a measurement system up and running in the event of a cable failure.

    Integral cable models are recommended in submersible applications where sealing is of prime importance. Armored cable is recommended in applications where sharp objects could cut the cable, such as metal chips in machining operations.

Figure 13.4.6: Armor cabling.

8. **Intrinsically Safe / Explosion Proof** – Many sensor models are approved for use in hazardous areas when used with a properly installed intrinsic safety (I.S.) barrier or enclosure. Approval authorities include Canadian Standards Association, CENELEC, Factory Mutual, and Mine Safety Administration.

Chapter 13

9. **Mounting Considerations** – There are several methods for securing vibration sensors to machinery. These include stud mounting, adhesive mounting, and magnetic mounting. Additionally, the use of a probe tip may be useful for very inaccessible measurement points, for locations where physical attachment of a sensor is just not practical, or for determining installation locations where vibration is most prevalent.

Each mounting method will affect the frequency response attainable by the vibration sensor since the mounted resonant frequency of the sensor/hardware assembly will be dependent upon its mass and stiffness. The following diagram depicts how resonant frequency is affected by each mounting technique.

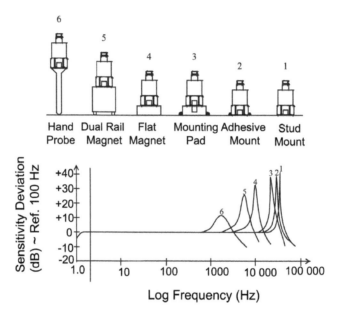

Figure 13.4.7: Effect of mounting technique on frequency response.

Stud mounting is the best technique to use to achieve the maximum frequency range. All sensor specifications and calibration information supplied with the sensor are based upon stud mounting during qualification tests. For best results, a smooth, flat surface should be prepared on the machinery surface as well as a perpendicular tapped hole of proper dimension. Spot face tools are available to simplify surface preparation. A thin layer of silicone grease or other lubricant should be applied to the surface and the sensor should be installed with the recommended mounting torque. Be sure to follow all installation recommendations including torque specifications supplied with each specific sensor model.

If drilling and tapping mounting holes into machinery structures is not practical, adhesive mounting is the next best technique. Sensors may be secured directly to the machine with adhesive or to a mounting pad with suitable tapped hole.

Mounting pads can be adhesively bonded or welded to machinery surfaces at specific vibration sensor installation points. The pads ensure that periodic measurements are always taken from the exact same location, lending to more accurate and repeatable measurement data. Pads with tapped holes are for use with stud mounted sensors, whereas the untapped pads are intended for use with magnetically mounted sensors.

For permanent installations, the pads facilitate mounting of sensors without actually machining the surface onto which they are to be installed. Also, the untapped pads may be utilized to achieve magnetic attraction on non-ferrous surfaces.

Magnetic mounting offers the most convenient method of temporary sensor installation for route-based measurements and data collection. Magnetic mounting bases consist of rare-earth magnet elements to achieve high attraction forces to the test structure. This aids in high-frequency transmissibility and assures attraction for large heavy sensors and conditions of high vibration.

Rail mount styles are utilized for curved surfaces, such as motor housings and pipes. Knurled housings aid in gripping for removal. Hex-shaped magnetic bases are designed for smaller high-frequency sensors. All magnetic mounting bases should be manufactured from resilient, stainless steel.

Note: Exercise caution when installing magnetically mounted sensors by engaging the edge of the magnet with the structure and carefully rolling the sensor/magnet assembly to an upright position. Never allow the magnet to impact against the structure as this may damage the sensor by creating shock acceleration levels beyond survivable limits.

13.5 Applicable Standards

ANSI S2.41, 1985 (R 1990), "Mechanical Vibrations of Machines with Operating Speeds from 10 to 200 Rev/s – Measurment and Evaluation of Vibration Severity in Situ" American National Standards Institute, NY.

ANSI Standards can be obtained from the Acoustical Society of America, Standards and Publications Fulfillment Center, P.O. Box 1020, Sewickley, PA 15143-9998.

API 670, 1986, *Vibration, Axial Position, and Bearing Temperature Monitoring System*, 2nd Ed., American Petroleum Institute, Washington, D.C.

API 678, 1981, *Accelerometer-Based Vibration Monitoring System*, API, Washington, D.C.

European Standard – EN 13980: Potentially Explosive Atmospheres, Application of Quality Systems.

ISO 7919, 1986, "Mechanical Vibrations of Non-Recipricating Machines – Measurements on Rotating Shafts and Evaluation," International Standards Organization, Geneva, Switzerland.

ISO 2372, 1974, "Mechanical Vibrations of Machines with Operating Speeds from 10 to 200 RPM – Basis for specifying Evaluation Standards," International Standards Organization, Geneva Switzerland.

ISO Standards can be obtained from the Director of Publications, American National Standards Institute, NY, NY 10005-3993.

Open Standards for Operations and Maintenance. Mimosa – 4259 Niagara Ave., San Diego, California 92107.

13.6 Latest and Future Developments

TEDS (Transducer Electronic Data Sheet) allows the operator to retrieve stored data from within the sensor. Data includes sensor numbers (for warranty and traceability) calibration info, date of manufacture, name of manufacturer and installation date, and location on the machine in which plant.

Wireless Sensors – There has been a strong interest in implementing wireless technology in industrial applications. Until price, the smaller overall capacity of a wireless system versus a cable system, and powering of the sensor can be addressed, wireless technologies will only be used on very specialized niche applications.

13.7 Sensor Manufacturers

Endevco
GE Bently Nevada
IMI Sensors, a Division of PCB Piezotronics, Inc.
Kaman Instrumentation
Kistler Instruments
Metrix
Monitran
PCB Piezotronics, Inc.
Vibrometer
Wilcoxon Research

13.8 References and Resources

1. Baxter, Nelson L. Machinery Vibration Analysis III, Vibration Institute, Willowbrook, IL (1995).

2. Crawford, A. R. and Crawford, S., *The Simplified Handbook of Vibration Analysis,* Volume One, Computational Systems, Inc. (1992).

3. Eisenmann, Sr. R.C. and Eisenmann, Jr. R.C., Machinery Malfunction Diagnosis and Correction, Prentice Hall PRT (1998).

4. Eshleman, Ronald L., Basic Machinery Vibrations: An Introduction to Machine Testing, Analysis, and Monitoring, VIPress, Incorporated, 1999.

5. Harris, Cyril M., *Shock and Vibration Handbook,* McGraw-Hill, Inc. ISBN 0-07-026801-0.

6. Maedel, P.H., Jr., "Vibration Standards and Test Codes," Shock and Vibration Handbook, 4th Edition, C.M. Harris, ed, McGraw-Hill, NY (1996).

7. Taylor, James I., *The Vibration Handbook*, Vibration Consultants, Inc., Tampa, FL (1994).

8. Wowk, Victor – Machinery Vibration, Measurements and Analysis. McGraw-Hill, New York, 1990.

CMVA/ACVM (Canadian Machinery Vibration Association)
Suite 877, 105-150 Crowfoot Crescent NW
Calgary, AB T3G 3T2
Ph: (403) 208-9618   Fax: (403) 208-9619   Web: www.cmva.com

MFPT (Machinery Failure Prevention Technology)
1877 Rosser Lane
Winchester, VA 22601
Ph: (540) 678-8678   Fax: (540) 678-8799   Web: www.mfpt.org

Vibration Institute
6262 S. Kingery Highway
Suite 212
Willowbrook, IL 60527
Ph: (630) 654-2254   Fax: (630) 654-2271   Web: www.vibeinst.org

Chapter 13

Trade Magazines:

Intech	www.isa.org
Maintenance Technology	www.mt-online.com
Reliability Magazine	www.reliability-magazine.com
Sensors	www.sensorsmag.com
Sound & Vibration	www.sandv.com
Turbomachinery International	www.turbomachinerymag.com
Vibrations	www.vibeinst.org

Companies:

GE Bently website	www.bently.com
Kaman Instrumentation website	www.kamaninstrumentation.com
PCB Piezotronics, Inc. website	www.pcb.com
IMI Sensors website	www.imi-sensors.com

CHAPTER 14

Optical and Radiation Sensors

Dr. Thomas Kenny, Department of Mechanical Engineering, Stanford University

This chapter offers an overview of the basic types of sensors used to detect optical and near-infrared radiation.

14.1 Photosensors

Detection of light is a basic need for everything from devices to plants and animals. In the case of animals, light detection systems are very highly specialized, and often operate very near to thermodynamic limits to detection. Device researchers have worked on techniques for light detection for many years, and have developed devices that offer excellent performance as well.

Clearly, the military has been a major sponsor of light detection device research. Devices for light detection are of fundamental importance throughout military technology, and the maturity and widespread availability of inexpensive photosensors is a direct result of this DOD research investment over many years.

Light is a quantum-mechanical phenomena. It comes in discrete particles called photons. Photons have a wavelength λ, a velocity $c = 3 \times 10^8 \, m/s$, a frequency $\omega = \frac{2\pi c}{\lambda}$, energy $E = \frac{hc}{\lambda}$ where $h = 6.67 \times 10^{-34} \, Js$, and even a momentum $P = \frac{h}{\lambda}$. Among all of this, it is important to remember the relationship between energy and wavelength. In all cases, the energy of the photon determines how we detect it.

Light detectors may be broken into two basic categories. The so-called *quantum detectors* all convert incoming radiation directly into an electron in a semiconductor device, and process the resulting current with electronic circuitry. The *thermal detectors* simply absorb the energy and operate by measuring the change in temperature with a thermometer.

Chapter 14

Quantum Detectors

We will start by examining the quantum detectors, since they offer the best performance for detection of optical radiation.

In all of the quantum detectors, the photon is absorbed and an electron is liberated in the structure with the energy of the photon. This process is very complicated, and we will not examine it in detail. It is important to recognize that semiconductors feature the basic property that electrons are allowed to exist only at certain energy levels. If the device being used to detect the radiation does not allow electrons with the energy of the incident photon, the photon will not be absorbed, and there will be no signal.

On the other hand, if the photon carries an amount of energy which is "allowed" for an electron in the semiconductor, it can be absorbed. Once it is absorbed, the electron moves freely within the device, subject to electric fields (due to applied voltages) and other effects. Many such devices have a complicated "band structure" in which the allowed energies in the structure change with location in the device. One example of such a "band structure" is that offered by a p-n diode. In a diode, the p-n junction produces a step in the allowed energy levels, resulting in a direction in which currents flow easily and the opposite direction in which current flow is greatly reduced.

A photodiode is simply a diode, biased against its easy flow direction ("reverse-biased") so that the current is very low. If a photon is absorbed and an electron is freed, it may pass over the energy barrier if it possesses enough energy. In this respect, the photodiode only produces a current if the absorbed photon has more energy than that needed to traverse the p-n junction. Because of this effect, the p-n photodiode is said to have a cutoff wavelength—photons with wavelength less than the cutoff produce current and are detected, while photons with wavelength greater than the cutoff do not produce current and are not detected.

Photodiodes may be biased and operated in two basic modes: photovoltaic and photoconductive. In the photovoltaic mode, the diode is attached to a virtual ground preamplifier as shown in Figure 14.1.1, and the arrival of photons causes the generation of a voltage which is amplified by the op-amp. The primary feature of this approach is that there is no dc-bias across the diode, and so there is no basic leakage current across the diode aside from thermally generated currents. This configuration does

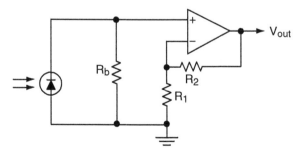

Figure 14.1.1: Connection of a photodiode in a photovoltaic mode.

suffer from slower response because the charge generated must charge the capacitance of the diode, causing an R-C delay.

In the photoconductive mode, the diode is biased, and the current flowing across the diode is converted to a voltage (by a resistor), and amplified. A photoconductive circuit is shown in Figure 14.1.2. The primary advantage of this approach is that the applied bias decreases the effective capacitance of the diode (by widening the depletion region), and allows for faster response. Unfortunately, the dc bias also causes some leakage current, so detection of very small signals is compromised.

In addition to making optical detectors from diodes, it is also possible to construct them from transistors. In this case, the "photocurrent" is deposited in the base of a bipolar junction transistor. When subjected to a collector-emitter bias (for npn), the current generated by the photons flows from the base to the emitter, and a larger current is caused to flow from the collector to the emitter. For an average transistor, the collector-emitter current is between 10 and 100× larger than the photocurrent, so the phototransistor is fundamentally more sensitive than the diode.

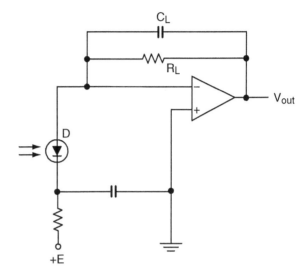

Figure 14.1.2: Photoconductive operating mode.

Photodiodes and phototransistors are very widely available. Most semiconductor device manufacturers also offer photodiodes and transistors, so there are nearly 100 suppliers. More than 10 manufacturers specialize in photosensors. As a result, optimized photodiodes and transistors are available at very low cost.

These devices are also available in packages designed for particular applications. For example, it is common to use a light-emitting diode and a detector mounted in a pair so that passing objects can interrupt the optical beam between them. *Opto-interruptors* consisting of such emitter-detector pairs are available in a wide variety of configurations. *Proximity detectors* situated side by side sense the presence of a reflecting surface by causing reflected light to strike the detector.

Other applications of optical detector-emitter pairs include measurement of the rotation rate of electric motors. In this case, a disk is mounted on the shaft of the motor with a large number of slits cut through it. The detector emitter pair is mounted so that the slits cause an oscillation in the signal—and the rotary position can be determined by counting the peaks in the signal. This is called an optical encoder, or an incremental encoder, and it is widely used in electric motors, as shown in Figure 14.1.3.

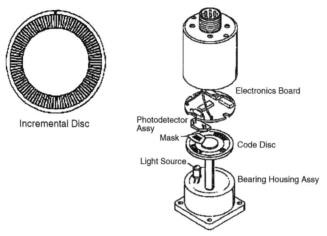

Figure 14.1.3: Incremental encoders.
(Courtesy BEI Technologies, Inc.)

Most phototransistors and photodiodes have their peak sensitivity in the near infrared (see Figure 14.1.4). The peak sensitivity occurs near the cutoff wavelength (near 1 µm) and extends to shorter wavelengths. The location of this peak sensitivity is due to the energy of the "bandgap" in silicon, and is not easily adjusted.

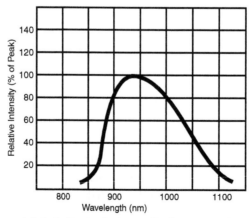

Figure 14.1.4: Typical photodiode spectral response.

Table 14.1.1: Bandgaps of some semiconductors.

Material	Bandgap (eV)
ZnS	3.6
CdS	2.41
CdSe	1.8
CdTe	1.5
Si	1.12
Ge	0.67
PbS	0.37
InAs	0.35
Te	0.33
PbTe	0.3
PbSe	0.27
InSb	0.18

Photosensors can be made from other electronic materials with different bandgaps, as shown in Table 14.1.1. None of these materials are as widely available as silicon, and costs for detectors made from InSb can be substantially higher.

There is another important consideration to keep in mind when selecting photosensors. In addition to the photocarriers in the device, thermally generated carriers can be produced. The distribution of energies generated by thermal processes is dependent on the thermodynamics of the device, and on the temperature. Because of this relationship, increasing the temperature causes an increase in the number of thermally generated carriers. Conversely, reducing the bandgap of a room-temperature device will also cause an increase in the number of thermally generated carriers. Silicon detectors work well at room temperature, but heating to more than 100°C starts to cause substantial increases in "dark current." Detectors made from materials other than silicon may offer increased cutoff wavelength, but may also require cooling below room temperature.

In general, there is a nearly linear relationship between the maximum operating temperature and the cutoff energy for the detector. By selecting a material with a cutoff energy one-fifth that of silicon (such as InSb), it is necessary to cool the device to about one-fifth of the maximum operating temperature of silicon (cooling to 77K is optimal for InSb). This tradeoff between cutoff and operating temperature imposes severe cost issues for operation of devices at fairly long wavelengths.

If cooling is affordable, a large selection of materials and devices with "engineered band-gaps" is available. The tremendous interest in devices with cutoff wavelengths near 10–20 μm is a direct result of the DOD interest in infrared detectors for night

Chapter 14

vision. It turns out that the peak of the infrared spectrum for objects at room temperature is in this region, and so the maximum contrast in thermal detection is available by producing devices with sensitivity in this region.

There is a simple relationship between the temperature of an infrared source and the peak wavelength of the blackbody spectrum.

$$\lambda_m = \frac{2898}{T}$$

where the wavelength is in microns, and the temperature is in Kelvin. So, for room temperature, the maximum wavelength is near 10 microns.

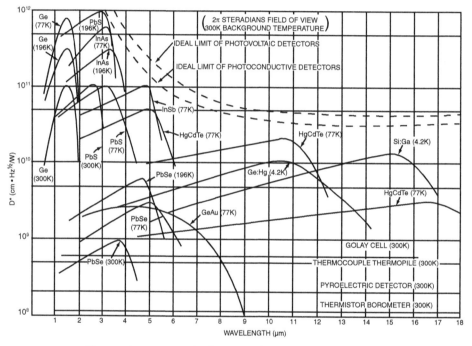

Figure 14.1.5: Typical spectral response of IR detectors.
(Courtesy of Electro Optical Industries, Inc.)

Of the materials most studied, the clear winner is Mercury Cadmium Telluride (MCT). It may be formulated to have cutoff between 10 and 20 microns, and offers excellent properties for infrared detection. In particular, it offers low dark current, high absorptivity, and low carrier scattering. Unfortunately, it is difficult and expensive to manufacture. As should be expected for anything containing mercury, its fabrication process is an environmental nightmare, and the basic material is not compatible with electronics. As a result, it is "bump-bonded" onto silicon substrates for readout and signal processing. In addition, it must be operated at or below 77K, which imposes operational complications. A commercial imaging system based on MCT detector arrays generally costs near $100,000 at the present time.

Most military applications (ballistic missiles, aircraft imaging systems, satellite systems) can afford this set of costs and complications, but commercial and civilian applications are generally cost-constrained. Therefore, recent research activities have focused on other materials which might be less expensive to make and operate.

InSb does not offer sensitivity in the 10–20 μm region, but is more easily made than MCT, is electronics compatible, and can be operated near 100K. Research to extend the operation to higher temperatures is underway throughout academia and industry.

Overall, the relationship between cutoff and operating temperature is fairly strict. MCT, which has been the focus of billions of dollars of materials research effort, has only been slightly extended to higher temperatures. There is not tremendous hope that InSb or other materials will benefit from a large change in operating requirements.

The other type of optical detector, the thermal detector, does offer some hope for this problem.

Thermal Detectors

Thermal detectors operate by absorbing the infrared radiation and measuring the temperature rise of the detector with a thermometer. Generally, the performance of thermal detectors is limited by the availability of sensitive and small heat capacity thermometers.

An important advantage of thermal infrared detectors is due to the absence of any relationship between the wavelength of the absorbed radiation and the response of the detector. Any energy absorbed causes a response in the detector. Therefore, it is possible to use a thermal infrared detector at room temperature to detect radiation from room temperature blackbodies.

However, it is important to note that if the conditions allow use of a quantum detector, such a detector will outperform a thermal detector by several orders of magnitude. Thermal detectors come into their own in situations that don't allow quantum detectors.

Since the thermometer is mounted within the infrared detector structure, it is connected to a temperature reference by a finite thermal conductance. This finite conductance imposes dynamic constraints on the system behavior, and we may analyze the situation as follows:

Chapter 14

Assume we have a thermometer that is a thermistor with a temperature coefficient given by:

$$\alpha = \frac{1}{R}\frac{\Delta R}{\Delta T}$$

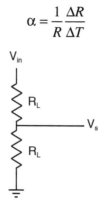

Figure 14.1.6: Voltage divider.

This thermistor is mounted in an electrical circuit with a load resistor with a resistance of R_L, and is biased by a dc voltage of V_{in}. The electrical circuit is shown in Figure 14.1.6.

As in all voltage dividers, we have:

$$V_{out} = V_{in}\frac{R_t}{R_L + R_t}$$

$$V_{out} \approx V_{in}\frac{R_t}{R_L} \text{ for } R_L \gg R_t$$

The sensitivity of this system is given by:

$$\text{Sensitivity} = \frac{\Delta V_{out}}{\Delta T} = V_{in}\frac{\alpha R_t}{R_L}$$

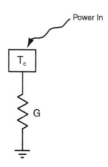

Figure 14.1.7: Thermal circuit.

Optical and Radiation Sensors

However, we must consider the thermal characteristics of this system as well. In this case, we model the thermometer as a finite heat capacity attached to an object by a finite thermal conductance. Infrared power is deposited into the thermometer, causing the temperature of the thermometer to oscillate. This thermal situation can be modeled as a thermal circuit as shown in Figure 14.1.7.

By energy balance, the energy gained is equal to the change in energy of the thermometer:

$$P_{in} - G(T_c - T_O) = C\frac{\Delta T}{\Delta t}$$

Now, we assume that the power and the thermometer temperature oscillate:

$$P_{in} = P_{in1} + P_{in2}e^{i\omega t}$$
$$T_c = T_{c1} + T_{c2}e^{i\omega t}$$

We insert these expressions into the energy balance equation, and we have:

$$P_{in1} + P_{in2}e^{i\omega t} - G(T_{c1} + T_{c2}e^{i\omega t} - T_O) = i\omega C T_{c2} e^{i\omega t}$$

We can take the constant and oscillating parts to be independent, and we have:

$$P_{in1} - G(T_{c1} - T_O) = 0$$
$$P_{in2} - G(T_{c2}) = i\omega C T_{c2}$$

These reduce to:

$$T_{c1} = T_O + \frac{P_{in1}}{G}$$
$$T_{c2} = \frac{P_{in2}}{i\omega C + G}$$

So the sensitivity of this device to changes in infrared absorbed power is:

$$\text{Sensitivity} = \frac{\Delta V}{\Delta P} = \frac{\Delta V}{\Delta T}\frac{\Delta T}{\Delta P} = V_{in}\frac{\alpha R_t}{R_L}\frac{1}{i\omega C + G}$$

To improve the sensitivity, it is important to choose a thermometer with a large temperature coefficient and a small heat capacity. We can see from this expression that the response of the detector will have a simple 1-pole response, which is to say that it is frequency-independent below the cutoff frequency and decreases as $1/f$ above the cutoff. Its response is exactly the same as that of an electrical low-pass filter.

Chapter 14

There are several different infrared detectors based on this detection concept. In fact, almost every well-established thermometer has also been optimized as an infrared detector.

Phototransistor Example

The phototransistor is a device that operates by converting incoming photons to electrons in the base of a bipolar transistor. As for any such transistor, the base current causes a larger collector-emitter current to flow, which is detected by a circuit. The easiest way to detect a current is to use a resistor to convert it to a voltage, as shown in Figure 14.1.8.

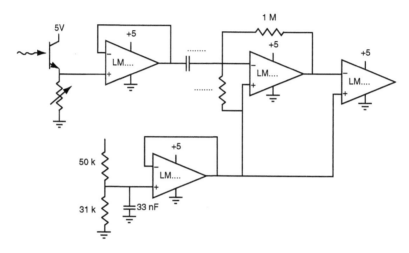

Figure 14.1.8: Phototransistor circuit.

In this case, an oscillator circuit is powering a light-emitting diode, causing a light oscillation at 1 kHz. The phototransistor is pointed at the LED, and detects the oscillation in the incident light. The circuit converts the current to a voltage with a pull-down resistor, buffers the signal, high-pass filters the signal, and then converts it to a square wave with a comparator. This circuit is one of many possible such circuits, and may be considered typical.

We can see that the signal at the beginning of the circuit reflects the oscillation as well as the background illumination (dc and 60-Hz components). Some of the filter design is intended to reduce the sensitivity to these "noise" components while preserving the sensitivity to the signal at 1 kHz.

Optical and Radiation Sensors

The variable resistor at the front of the circuit is an important degree of freedom. The current flowing through the transistor cannot exceed the saturation current:

$$I_{sat} = \frac{V_{bias} - 0.6V}{R_{pd}}$$

where V_{bias} is the total collector-emitter voltage, and R_{pd} is the value of the pull-down resistor. If the background is very bright, the current flowing in the device may already be very close to the saturation value, and any additional signal illumination will not produce much additional signal.

Depending on the amount of background illumination, it is possible to saturate the detector, thereby reducing the sensitivity to signals. We use a variable resistor here to allow adjustment so that the detector is biased at a point of good performance.

It is possible to obtain such detectors in side-by-side emitter-detector pairs, which cause a signal only if a reflective object is nearby. Depending on the biasing and the background illumination, it is possible to detect objects at a range of more than 1 cm.

14.2 Thermal Infrared Detectors

In recent years, the Department of Defense (DOD) has invested a great deal of research and development funds into detection techniques that allow long-wave detection from uncooled platforms. An additional focus of this work has been techniques that are compatible with the formation of dense arrays. One interesting device that has emerged due to this investment has been the Uncooled Detector arrays made by Honeywell.

These detectors are based on the simplest thermal design—a resistance thermometer. What is novel about this device is that it combines the best microfabrication technology with good thermometer technology and electronics integration.

A drawing of the microbolometer is shown in Figure 14.2.1. The basic idea is to use silicon microfabrication techniques (such as those in many accelerometers) to make an isolated thermal structure with very little heat capacity. As we saw in the thermometer discussion, the thermal infrared detector is improved by minimizing the heat capacity.

Chapter 14

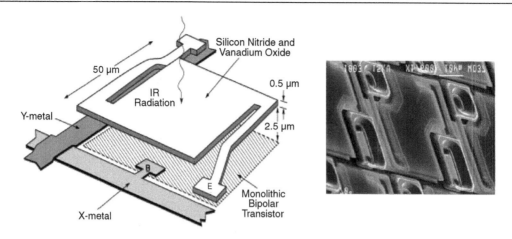

Figure 14.2.1: Microbolometer.
(Courtesy Infrared Solutions, www.infraredsolutions.com)

In the final device, a flake of silicon nitride with dimensions of 50 μm × 50 μm × 0.5 μm is floated above a silicon substrate. This flake is supported by a pair of legs, and is coated with a resistive material with a good thermal coefficient of resistance. Underneath the flake is a transistor that is used to connect the current-measuring circuit to the device using a conventional row-column addressing technique. The device current is passed out to a processing circuit on the perimeter of the device by the x and y metal leads.

In this device, much research went into developing a technique for depositing the nitride on top of a transistor, for releasing the devices with very high yield, and for obtaining a sensitive thermometer in the form of a deposited metal film. This resistor is made from vanadium oxide, which offers a TCR of about 1% near room temperature. The resistance change is a result of a structural phase transition in vanadium oxide above room temperature, so this device must be held near room temperature to allow operation with good sensitivity.

Having developed this technology, Honeywell has gone on to make dense arrays (200 × 200), and to continue optimizing the performance of the devices. In the last couple of years, a complete camera system has been demonstrated. This base technology has been offered for licensing, and is presently being commercialized by several manufacturers of infrared imaging systems.

This device does not out-perform the MCT imager, but it does enable operation at room temperature, and might be available at low cost with further development.

Another very important technology for low-cost uncooled infrared detectors has emerged in recent years in the form of pyroelectric plastic material. As with piezoelectricity, pyroelectricity is a phenomenon in which a change in temperature causes thermal expansion, which causes the appearance of charge (through the piezoelectric effect) (see Figure 14.2.2).

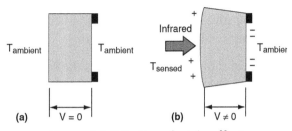

Figure 14.2.2: Pyroelectric effect.

Infrared detectors have been available for many years based on other specialized piezoelectric materials. The best of them is Deuterated Tri-Glycine Sulfide (DTGS). This very expensive material offers the best pyroelectric coefficients, and is commonly used for IR detection in laboratory spectrometers, and in early motion detection systems.

A variety of other pyroelectric materials are also available—it is generally true that any material which is piezoelectric is also pyroelectric. There are many applications requiring good performance (lab spectroscopy, for example), and these applications generally justify use of the best material available.

On the other hand, there are other applications in which the best detector performance is not required. In these applications, PVDF film has become the best choice available, primarily due to the extremely low cost of the device material. A good example of a low-performance application is an infrared motion detector. Nowadays, it is common to offer backyard lighting systems or door opening systems that detect the presence of a moving object with elevated infrared emission. If you wave your hand about, the infrared scene that can be detected features a variation in the infrared signal to some pixel of an imaging system. What is needed is an array of detector elements and some sort of focused optics. Without the focused optics, moving your hand about does not produce a change in the total illumination—and would not produce a variable signal. Remember that the pyroelectric detectors do not detect heat, only *changes* in heat.

It has become common to package a PVDF detector array in a low-cost optical package that uses a Teflon lens to focus the light. Teflon lens material is also inexpensive, and is transmissive enough in the IR that it does an acceptable job.

Chapter 14

Typical Teflon lenses used in motion detection systems are made with a surface texture that includes several circular bumps. These bumps act as focusing lenses, and will bring light from a particular part of the scene to the detector. As a warm object moves through the scene, radiation is occasionally focused on the detector, causing a transient in signal that is detected.

A typical motion detector allows the setting of a threshold, which is simply an electrical threshold in the detection circuit, and an output voltage which indicates the threshold has been crossed recently. Usually, you can also set the duration of illumination after a detection event.

Many such products are now available on the market at very low prices. There has been a recent substantial improvement in the availability of low-quality, inexpensive IR sensors, and a family of devices adequate for imaging systems are emerging.

CHAPTER 15

Position and Motion Sensors

15.1 Contact and Non-contact Position Sensors

Adolfo Cano Muñoz, Product Manager, Honeywell Sensing and Control

Introduction

Position sensors play an increasing role in our daily lives. They are abundant in our homes, in our cars, and in our work places. As sensing technology improves, positioning devices continue to get smaller, better and more inexpensive, opening the way for more applications than ever before.

As their name implies, position sensors provide position feedback. They are able to perform precise motion control, encoding and counting functions by determining the presence or absence of a target or by detecting its motion, speed, direction or distance. Position sensors detect a target object, a person, a substance or the disturbance of a magnetic or an electrical field and convert that physical parameter to an electrical output to indicate the target's position.

There are many ways to sense the position of a target. Some of them, such as limit switches and potentiometers, involve physical contact with the object being sensed. These are called contact position sensors. Contact position sensors often prove to be the simplest, lowest cost solution in applications where contact with the target is acceptable.

Sensor manufacturers have employed a much wider variety of approaches and technologies to develop non-contact position sensors, which have no physical contact with target objects and don't "wear out" from repeated contact. This chapter does not cover every possible technology that can be applied to sense position. The aim here is to provide a good understanding of the most commonly used technologies and the reasons one technology might be preferred over another in specific applications.

Chapter 15

Types of Position Sensors

Every commonly applied position sensing technology has its own characteristic benefits and limitations. Obviously, some of these technologies provide a better fit than others in different applications. The goal is to find the most cost-effective solution for the performance parameters that are important in your specific application and environment. The types of position sensors covered here include:

- Contact devices
  - Limit switches
  - Resistive position transducers
- Non-contact devices
  - Magnetic sensors, including Hall effect and magneto-resistive sensors
  - Ultrasonic sensors
  - Proximity sensors
  - Photoelectric sensors

Limit Switches

Limit switches are electromechanical contact devices. Easy to understand and apply, they are the cost-effective switches of choice for detecting objects that can be touched. These rugged, dependable switches are offered in a variety of sizes with different seals, enclosures, actuators, circuitries and electrical ratings.

Limit switches contain a set of contacts. When a target object comes into contact with a limit switch's actuator on the conveyor in Figure 15.1.1, the switch operates.

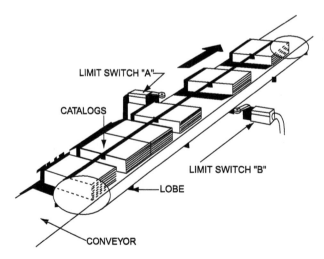

Figure 15.1.1: Limit switches on a conveyor.

Various limit switches provide years of reliable operation even in the most demanding environmental conditions. They are appropriate for:

- Material handling
- Breweries
- Packaging machinery
- Wood products
- Special machinery
- Garbage compactors/trucks
- Valves
- Foundry equipment

Limit switches are available in explosion-proof versions to contain and cool escaping hot gases that otherwise could cause an explosion outside the switch in:

- Petroleum plants
- Chemical plants
- Waste treatment facilities
- Power generating stations
- Hazardous material handling
- Grain storage/handling
- Deep sea oil well platforms

Selecting and Specifying Limit Switches

Limit switches can be ordered with a variety of actuators, including plungers, rotary levers, and "wobble levers," which are flexible spring-like levers that operate by any movement except direct pulling.

Figure 15.1.2 shows how characteristics are measured for rotary actuation, and Figure 15.1.3 applies to limit switches actuated by in-line plungers.

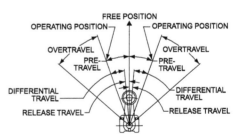

Figure 15.1.2: Operating characteristics for rotary-actuated limit switches.

The characteristics of rotary-actuated limit switches are shown in degrees of angular rotation. The operating characteristic dimensions on enclosed switches for harsh environments are often shown in linear dimensions with the adjustable lever in one extreme position.

Linear dimensions for in-line actuation are from the top of the plunger to a reference line, usually the center of the mounting holes. In the case of flange- or bottom-mounted switches, the reference line is the bottom of the switch.

Chapter 15

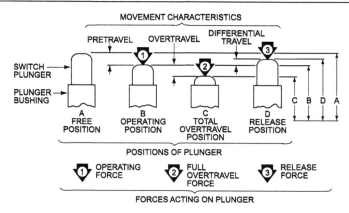

Figure 15.1.3: Operating characteristics for limit switches with in-line plunger actuators.

To select the right limit switch for your application, consider:

- Actuator type
- Circuitry
- Ampere rating
- Supply voltage
- Housing material
- Termination type

Applicable Standards for Limit Switches

IEC (International Electrotechnical Commission), *www.iec.ch/*, especially JIC 60947-1 and IEC 60947-5-1, which explain the general rules relating to low-voltage switch and control gear for industrial use; IEC 529 rates the level of protection provided by enclosures, using an IP (International Protection) rating system.

CENELEC (The European Committee for Electrotechnical Standardization), *www.cenelec.org*, especially EN 50041 and EN 50047, which define characteristics and dimensions for limit switches.

NEMA (National Electrical Manufacturer's Association), *www.nema.org*. NEMA rates the protection level of enclosures as does IEC 529, but includes tests for environmental conditions, such as rust, oil, etc. that are not included in IEC 529.

Position and Motion Sensors

Interfacing and Design Information for Limit Switches

Here are some considerations for incorporating limit switches in your designs.

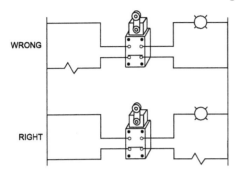

Figure 15.1.4: Opposite polarities should not be connected to the contacts of one limit switch unless the switch is specifically designed for this.

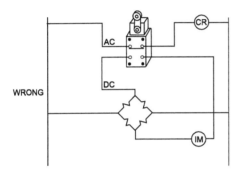

Figure 15.1.5: Power from different sources should not be connected to the contacts of one switch unless it is specifically designed for this.

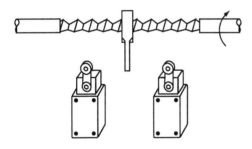

Figure 15.1.6: Where relatively slow motion operates a limit switch, you'll generally want to apply a snap-action switch.

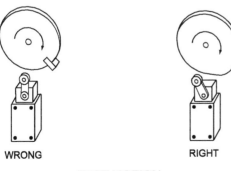

Figure 15.1.7: Where relatively fast motion operates the switch, cams should be arranged so that the switch does not receive a severe impact.

Chapter 15

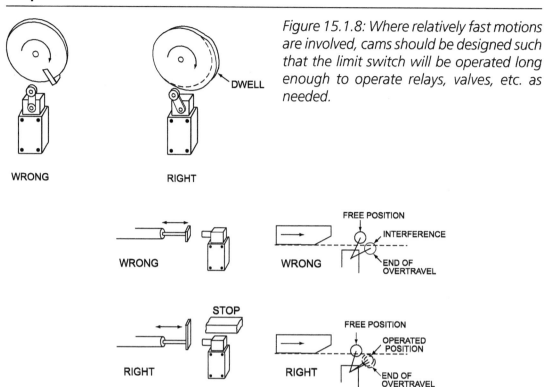

Figure 15.1.8: Where relatively fast motions are involved, cams should be designed such that the limit switch will be operated long enough to operate relays, valves, etc. as needed.

Figure 15.1.9: Operating mechanisms for limit switches should be designed so that under any operating or emergency conditions, the switch is not operated beyond its overtravel limit position. A limit switch should not be used as a mechanical stop.

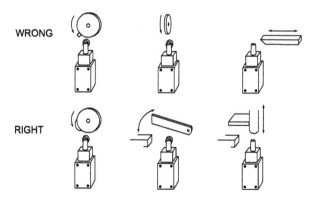

Figure 15.1.10: For limit switches with pushrod actuators, the actuating force should be applied as nearly as possible in line with the pushrod axis. The same holds true for other actuators. For instance, the actuating force of a lever actuator should be applied as nearly perpendicular to the lever as practical and perpendicular to the shaft axis about which the lever rotates.

Position and Motion Sensors

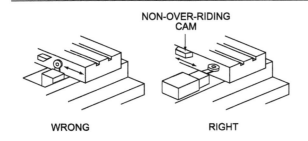

Figure 15.1.11: Mount the switch rigidly in a readily accessible location. Cover plates should face the maintenance access point.

Resistive Position Sensors

Resistive position sensors, also called potentiometers or simply position transducers, were originally developed for military applications. They were widely used as panel-mounted adjustment knobs on radios and televisions in the years before pushbuttons and remote controls. Today, potentiometers are most commonly found in industrial applications that range from forklift throttles to machine slide sensing.

Potentiometers are passive devices, meaning they require no power supply or additional circuitry to perform their basic linear or rotary position sensing function. They are typically operated in one of two basic modes: rheostat and voltage divider (true potentiometric operation). As resistance varies with motion, rheostat applications make use of the varying resistance between a fixed terminal and the sliding contact wiper. In voltage divider applications, a reference voltage signal is applied across the resistive element track so that the voltage "picked up" by the movable contact wiper can be used to determine the wiper's position.

Advantages of potentiometers include low cost, simple operational and application theory, inherent absolute measurement even through power-off cycles, and robust EMI emission/susceptibility performance. Disadvantages include eventual wear-out due to the sliding contact wiper and sensing angles that are limited to less than 360 degrees, although vernier drives can be applied to give them "multi-turn" sensing capability.

Selecting and Specifying Resistive Position Sensors

For many general-purpose position sensing applications, a low-cost potentiometer is often more than adequate. In some cases (certain electric motor controllers, for example), a changing resistance is expected as the control signal, and thus a potentiometer is really the only choice. In most other applications, a fixed reference voltage is applied, and a voltage reading at the wiper terminal senses the position. Rotary potentiometers can offer linearity errors on the order of 1% maximum and rotational lifetimes of a million or more cycles. For more critical applications, higher-precision potentiometers can offer linearity errors a hundred or more times better than this, and rotational lifetime ratings in the tens or even hundreds of million cycles.

Chapter 15

In looking for the best cost versus performance trade-off, consider the system as well as the transducer. Usually you will take one of two paths to select a transducer:

- Find a transducer that works with the power supplies and amplifiers or controllers in your system.
- Select the position transducer and then match your system components to it.

What is the length to be measured?

For potentiometers this is called the theoretical electrical travel. For most applications this is straightforward; however, there are times where you may want to measure only a portion of the total travel of your system. For example, it may be advantageous to have the highest possible resolution at one end of the system's travel. Consider a 10-inch total travel, but by monitoring the last inch, you can increase your resolution tenfold.

What accuracy can I get?

Accuracy can have several meanings. Do you mean how small a motion you can pick up (resolution)? How much error there is at any point along the electrical travel compared to a reference line (independent linearity)? Is the output the same at a given point from one cycle to another (repeatability)? These are not the same, but any one is often called "accuracy."

It is important to differentiate and prioritize these performance parameters. Many conductive plastic position transducers have infinite resolution. The repeatability with most transducers is excellent and will rarely be more than the signal-to-noise ratio that will likely exist in your system. Look for independent linearity, the error versus the reference line, guaranteed to be less than 0.1%.

How rugged does the transducer need to be?

The unit's ability to hold up under high levels of shock, vibration, moisture, dirt, oil or temperature extremes is sometimes more important than accuracy. What good is a high-accuracy device if it won't last long in the application? Conductive plastic withstands many harsh chemicals and will work immersed in hydraulic oil.

What excitation should I use?

For potentiometers, the excitation is determined by the input of your signal conditioner or the controller you are using. You will need to decide whether to use regulated or unregulated voltage, depending on your conditioner or controller.

Position and Motion Sensors

What mounting factors should I consider?

These vary from application to application. Proper alignment of the shaft of the potentiometer is important for maximum life from the unit. Mounting should allow for minor adjustment to minimize any misalignment. Rod ends or shaft couplings are an effective way to compensate for misalignment. Side load on a potentiometer will wear out the bearings long before the wiper or element fail, so careful mounting is to your advantage. If there is heavy hose down or spray from oil or water, you should use a water-resistant or waterproof potentiometer.

Does the transducer need to be compensated for temperature effects?

Potentiometers, while they have measurable temperature coefficients, will not usually require temperature compensation because they are most often used as voltage dividers.

Applicable Standards for Resistive Position Sensors

The Variable Electronics Component Institute (*http://www.veci-vrci.com*) has developed a number of testing and performance standards governing potentiometers. While their standards are not binding on manufacturers, they help ensure practical, meaningful and consistent terminology and methodologies. Many potentiometers also conform to military standards such as MIL-STD-202F.

Interfacing and Design Information for Resistive Position Sensors

For the most part, using a resistive position sensor is quite straightforward. First, establish what sort of electrical signal is needed from the mechanism being sensed. If a resistance change is needed, connect one end of the potentiometer and the wiper terminal into the circuit. Resistance will then vary with motion in this mode (rheostat mode); contact noise (an unpredictably varying resistance) will appear superimposed on the expected smooth change of resistance. Contact noise results from the variation in mechanical contact between the wiper and the resistive element's surface. It manifests as a series resistance between the element surface (contact point) and the wiper terminal. It can vary from a fraction of a percent up to 5% or more of the total element resistance.

Voltage divider mode operation is by far more prevalent in position sensing applications, and in this mode, the effects of contact resistance variation are diminished or eliminated. Its output is a voltage ratio determined from the wiper voltage divided by the applied voltage. The output voltage is taken from the wiper terminal. Thus, if little or no current is flowing through this path (as is the case when the voltage feeds a high-impedance measurement circuit), the series contact resistance variation creates no change in the voltage between the contact point and the wiper terminal.

When an excitation voltage is applied across the resistive element, the wiper moves from the zero voltage end of the element toward the maximum output end. The voltage between the wiper and the resistive element varies linearly with position. (See Figure 15.1.12.)

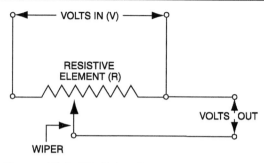

Figure 15.1.12: Potentiometer component.

The maximum output voltage of the sensor never exceeds the applied voltage. For example, if you apply 10 volts to the resistive element, the maximum output voltage is 10 volts. The wiper will vary from 0 to 10 volts, with 5 volts at the center position. If you apply 5 volts, an output of 2.5 volts indicates the center position.

Magnetic Position Sensors

Magnetic properties can be used to determine position by detecting the presence, strength, or direction of magnetic fields not only from the Earth but also from magnets, fields generated from electric currents, and even brain wave activity. Magnetic sensors can measure these properties without physical contact and have become the eyes of many industrial and navigation control systems.

The magnetic field is a vector quantity that has both magnitude and direction. Magnetic sensors measure this quantity in various ways. Some magnetometers measure total magnitude but not direction of the field (scalar sensors). Others measure the magnitude of the component of magnetization along their sensitive axis (uni-directional sensors). This measurement may also include direction (bi-directional sensors). Vector magnetic sensors have two or three bi-directional sensors. Some magnetic sensors have a built-in threshold and produce an output only when that threshold is passed.

A Hall effect device derives its output from magnetic field *strength,* while a magneto-resistive device measures the *angle direction* of a magnetic field, so its output is based on the electrical resistance of the field. Some of the advantages of measuring field direction vs. field strength include:

- Insensitivity to the temperature coefficient of the magnet,
- Less sensitivity to shock and vibration, and
- The ability to withstand large variations in the gap between the sensor and the magnet.

Hall effect position sensors are very affordable and accurate. They contain a Hall element constructed from a thin sheet of conductive material with output connections perpendicular to the direction of current flow. When subjected to a magnetic field, a Hall sensor responds with an output voltage proportional to the strength of the field. The voltage output is very small and requires additional electronics to achieve useful voltage levels. These signal-conditioning electronics are combined with a Hall element on an integrated circuit (IC) to form a basic Hall effect sensor.

Magneto-resistive (MR) sensors are usually made of a nickel-iron (Permalloy) thin film deposited on a silicon wafer and patterned as a resistive strip. The properties of the MR thin film cause it to change resistance by 2 to 3% in the presence of a magnetic field. For a typical MR sensor, the bandwidth is in the 1 to 5 MHz range. Reaction is very fast and not limited by coils or oscillating frequencies.

MR sensors measure both linear and angular position and displacement in the Earth's magnetic field (below 1 gauss). They are an excellent solution for locating objects in motion. By affixing a magnet or sensor element to an angular or linear moving object with its complementary sensor or magnet stationary, the relative direction of the resulting magnetic field can be quantified electronically.

The demand for Hall effect magnetic sensors remains high because the Hall effect is the only magnetic effect that can be implemented in a standard complementary metal oxide semiconductor (CMOS) technology, which makes them very affordable. Some of the advantages of Hall effect sensing devices include their long life and high speed. They operate with stationary input (zero speed), and have a broad temperature range (–40 to +150 degrees C) for industrial and automotive applications.

While Hall effect is highly linear with no saturation effects out to extremely high fields, magneto-resistance is roughly 200 times more sensitive than the Hall effect in silicon, so it can be used for longer sensor-to-magnet distances.

A unique aspect of using magnetic sensors is that measuring magnetic fields is usually not the primary intent. Another parameter, such as wheel speed or the position of a part, is sought. Magnetic sensors don't measure these parameters directly but extract them from changes in magnetic fields.

The enacting input has to create or modify a magnetic field. Once the sensor detects the field or change to a field, the output signal requires some signal processing to translate the sensor output into the desired parameter value. This makes magnetic sensing a little more difficult to apply in most applications, but it also allows for reliable and accurate sensing of parameters that are difficult to sense otherwise.

Chapter 15

Selecting and Specifying Magneto-resistive Position Sensors

MR sensors can detect the presence of a vehicle from a distance of about 50 feet and are often used in traffic and toll way applications. Other common applications include automotive wheel speed, crankshaft sensing and compass navigation. An area of growth for MR sensors is high-density read heads for tape and disk drives. MR sensors are beneficial in applications for linear position or linear displacement, LVDT (linear variable differential transformer) replacements, proximity detection, valve positioning, shaft travel, automotive steering, robotics, brake and throttle position systems. By utilizing an array of magnetic sensors or magnets, the capability of extended range angular or linear position measurements can be achieved.

MR sensors come in a variety of shapes and forms and can sense DC static fields as well as the strength and direction of the field. The long span absolute linear position (Honeywell patented) sensing solution illustrated in Figure 15.1.13 utilizes an array of MR sensors, a magnet and signal conditioning electronics. The sensors are used to determine the position of a magnet that is attached to a moving object. In addition to the mechanical benefits of no moving parts to wear out and no dropped signals from worn tracks, this solid state solution provides high accuracy with low power and works in rugged environments, such as high temperature.

This array device is designed to be sensitive to the direction of a magnetic field when it operates in saturation mode. The saturation mode is when external magnetic fields are above a certain field strength level (called saturation field). The magnetic moments in the device are aligned in the same direction of the field. Therefore, the output of the device only reflects the direction of the external magnetic field and not its strength. The incentive of operating in saturation mode includes:

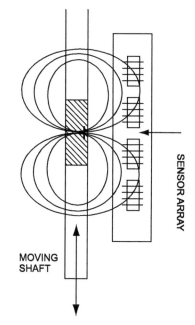

Figure 15.1.13: System design with an array of four magneto-resistive sensors.

- Immunity to temperature coefficient of magnet,
- Insensitivity to the gap between the magnet and the sensor array, and
- Insensitivity to magnetic field strength when the field is greater than saturation level.

The saturation field of this type of sensor is 30–50 gauss. Magnetic fields in this magnitude can be easily provided by any permanent magnet, including low cost AlNiCo or ceramic magnets, unlike Hall effect devices that require a stronger magnet.

Magneto-resistive sensors bring a unique feature to this solution by measuring the angle direction of a field from a magnet versus the strength of a magnetic field. A permanent magnet provides the magnetic field, which has the function of keeping the sensors in saturation mode, minimizing effects of stray magnetic fields and providing a linear operating range for selected sensor pairs.

Unlike other incremental sensors, this technology is absolute reading; no reference point is required. Position is accurately known at any time as well as at power-on. When each sensor in the array is connected to the supply voltage, it converts any ambient, or applied magnetic field to a voltage output.

MR sensors provide highly predictable outputs. Their high sensitivity, small size, noise immunity, and reliability are advantages over mechanical or other electrical alternatives.

Highly adaptable and easy to assemble, these sensors solve a variety of problems in custom applications. One of the key benefits of MR sensors is that they can be bulk manufactured on silicon wafers and mounted in commercial integrated circuit packages. This allows magnetic sensors to be auto-assembled with other circuit and systems components.

Selecting and Specifying Hall Effect Position Sensors

Hall sensors find a wide variety of position applications, especially as proximity sensors. They are used extensively in automobiles to sense everything from the positions of pistons and the angles of throttles, to power window and door interlock positions. They show up in office machines where they sense things like paper placement and in cameras to detect shutter position. On factory floors they are used to control everything from motor speeds to drilling machines.

Both digital and analog Hall effect position sensors are available. Digital output sensors are in one of two states: ON or OFF. Analog sensors provide a continuous voltage output that increases with a strong magnetic field and decreases with a weak magnetic field. Both are operated by the magnetic field from a permanent magnet or an electromagnet with actuation mode, depending on the type of magnets used.

There are two types of digital sensors: bipolar and unipolar. Bipolar sensors require positive gauss (south pole) to operate and negative gauss (north pole) to release. Unipolar sensors require a single magnetic pole (south) to operate. Release is obtained by moving the south pole away from the sensor. Analog sensors operate by proximity to either magnetic pole.

Ratiometric linear Hall effect sensors are small and versatile. The ratiometric output voltage is set by the supply voltage and varies in proportion to the strength of the magnetic field. An integrated Hall effect chip provides increased temperature stability and sensitivity. Laser trimmed thin film resistors on the chip provide high accuracy and temperature compensation to reduce null and gain shift over temperature.

Ratiometric linear sensors respond to either positive or negative gauss and can be used to monitor either or both magnetic poles. Their quad Hall sensing elements make them stable and predictable by minimizing the effects of mechanical or thermal stress on the output. The positive temperature coefficient of the sensitivity (+0.02% per °C typical) helps compensate for the negative temperature coefficients of low cost magnets, providing a robust design over a wide temperature range. Rail-to-rail operation (over full voltage range) provides a more usable signal for higher accuracy.

Figures 15.1.14 and 15.1.15 illustrate two concepts for developing a proximity sensor that can be used for accurate positioning.

In the first example, event signals are generated by the sensors, which represent distances measured from a reference surface. These signals define the acceptable dimensional limits between which the item under test must generate electrical pulses. In a known application, each of the sensors has accumulated at least 8 million operate/release cycles per month and is still operating, without replacement or maintenance.

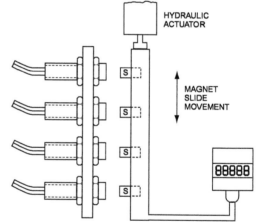

The second example achieves linear positioning accuracy of .002 inch. Sensing various lens locations for photo processing equipment is an ideal application for this concept. It also could be used to sense the precise location of a moving table for a 35 mm slide mounter.

Figure 15.1.14: Four digital output, unipolar sensors are threaded into an aluminum housing and actuated individually by four magnetic actuators.

Vector Hall sensors have two sensing axes, so they can detect absolute position over 360 degrees of rotation while other designs are limited to 180 degrees. Mostly considered for rotary applications, such as motors, the sensors can also be applied to one- and two-axis linear applications.

Position and Motion Sensors

Vector Hall sensors have four Hall elements on a single die (two for the X axis and two for the Y axis) to provide a differential output voltage for each axis. The signal pair is "gain matched" so that the mathematical ratio of the two signals cancels out any variation in gain or field strength due to temperature or a mechanical shift, such as shock or vibration. This not only reduces offset but also allows for relatively wide tolerance of axial and radial misalignment of the magnet and sensor, making the installation process less critical. The time required for the microprocessor to process this ratio determines

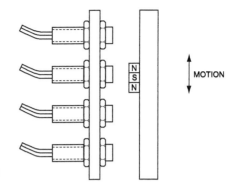

Figure 15.1.15: Four digital output, bipolar sensors are actuated by one magnet mounted on a rod.

the maximum rate of position updates and does limit the usable rotational speed of the sensor somewhat.

Some manufacturers offer automatic indexing to set a mechanical "zero" or "index" position whenever a particular signal is applied to the circuit. For example, this can be accomplished by shorting the output to I-V while bringing up the power. An indexing feature eliminates significant rigging and alignment at installation.

Interfacing and Design Information for Magnetic Position Sensors

Begin your design by determining sensing specifications. For sensing systems actuated by a magnet, specifications include:

- The minimum and maximum gap between the magnet and the position sensor,
- The limits of magnet travel,
- Special requirements for the magnet such as high coercive force due to adverse magnetic fields in the system,
- Mechanical linkages (if required),
- Sensor output type (NPN or PNP),
- Operating temperature range,
- Storage temperature range, and
- Various input/output specifications from the system specification.

The next step is to choose the magnetic mode, magnet, sensor, and functional interface. These four items are interdependent. The required magnet strength is dependent on the gap and the limits of magnet travel (magnetic mode). The sensor is dependent on the strength of the magnetic field and therefore, on the magnetic mode and the magnet chosen. The functional interface is dependent on the sensor output type and electrical characteristics

Chapter 15

Figures 15.1.16–20 illustrate a few of the ways a magnetic system can be presented to a linear output sensor for position measurement as listed in the comparison chart. The method of actuation is determined based upon cost, performance, accuracy and other requirements for a given application.

A simple method of position sensing is shown in Figure 15.1.16. One pole of a magnet is moved directly to or away from the sensor. This is a unipolar head-on position sensor. When the magnet is farthest away from the sensor, the magnetic field at the sensing face is near zero gauss. In this condition, the sensor's nominal output voltage will be 6 volts with a 12-volt supply. As the south pole of the magnet approaches the sensor, the magnetic field at the sensing surface becomes more and more positive. The output voltage will increase linearly with the magnetic field until a +400 gauss level or nominal output of 9 volts is reached. The output as a function of distance is nonlinear, but over a small range may be considered linear.

Figure 15.1.16:
Unipolar head-on position sensor.

Bipolar head-on sensing is shown in Figure 15.1.17. When the magnets are moved to the extreme left, the sensor is subjected to a strong negative magnetic field by magnet #2, forcing the output of the sensor to a nominal 3.0 volts. As magnet #1 moves toward the sensor, the magnetic field becomes less negative until the fields of magnet #1 and magnet #2 cancel each other (at the midpoint between the two magnets). The sensor output will be a nominal 6.0 volts. As magnet #1 continues toward the sensor, the field will become more and more positive until the sensor output reaches 9.0 volts. This approach offers high accuracy and good resolution as the full span of the sensor is utilized. The output from this sensor is linear over a range centered on the null point.

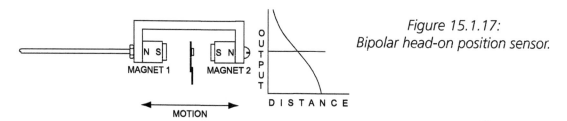

Figure 15.1.17:
Bipolar head-on position sensor.

Biased head-on sensing, a modified form of bipolar sensing, is shown in Figure 15.1.18. When the moveable magnet is fully retracted, the sensor is subjected to a negative magnetic field by the fixed bias magnet. As the moveable magnet approaches the sensor, the fields of the two magnets combine. When the moveable magnet is close enough to the sensor, it will "see" a strong positive field. This approach features mechanical simplicity.

Figure 15.1.18: Biased head-on position sensor.

Slide-by actuation is shown in Figure 15.1.19. A tightly controlled gap is maintained between the magnet and the sensor. As the magnet moves back and forth at that fixed gap, the field seen by the sensor becomes negative as it approaches the north pole, and positive as it approaches the south pole. This type of position sensor features mechanical simplicity, and when used with a long enough magnet, can detect position over a long magnet travel. The output characteristic of a bipolar slide-by configuration is the most linear of all systems illustrated, especially when used with a pole piece at each pole face. However, tight control must be maintained over both vertical position and gap to take advantage of this system's characteristics.

Figure 15.1.19: Slide-by position sensor.

Some descriptive terms:

Motion type refers to the manner in which the system magnet may move. These types include:

- Continuous motion (motion with no changes in direction),
- Reciprocating motion (motion with direction reversal), and
- Rotational motion (circular motion that is either continuous or reciprocating).

Mechanical complexity refers to the level of difficulty in mounting the magnet(s) and generating the required motion.

Chapter 15

Symmetry refers to whether or not the magnetic curve can be approached from either direction without affecting operate distance.

Digital refers to the type of sensor, either unipolar or bipolar, recommended for use with the particular mode.

Linear refers to whether or not a portion of the gauss versus distance curve (angle relationship) can be accurately approximated by a straight line.

Precision refers to the sensitivity of a particular magnetic system to changes in the position of the magnet.

A definite relationship exists between the shape of a magnetic curve and the precision that can be achieved. Assume the sloping lines in Figure 15.1.20 are portions of two different magnet curves; 01 and 02 represent the range of actuation levels (unit to unit) for digital output Hall effect sensors. It is evident from this illustration that the curve with the steep slope (B) will give the smaller change in operate distance for a given range of actuation levels. Thus, the steeper the slope of a magnetic curve, the greater the accuracy that can be achieved.

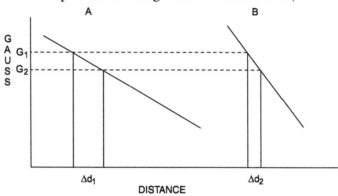

Figure 15.1.20: Effect of slope

Figure 15.1.21 shows the output of a digital Hall effect sensor. The sensor in this particular example is NPN (current sinking, so the current flows from the load into the sensor, open collector) in the actuated (ON) state. Current sinking devices contain NPN integrated circuit chips. Like a mechanical switch, the digital sensor allows current to flow when turned ON, and blocks current flow when turned OFF. Unlike an ideal switch, a solid-state sensor has a voltage drop when turned ON, and a small current (leakage) when turned OFF. The sensor will only switch low-level DC voltage (30 VDC max.) at currents of 20 mA or less. In some applications, an output interface may be current sinking output.

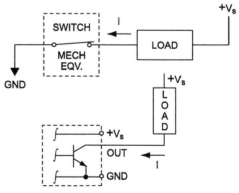

Figure 15.1.21: Typical digital Hall effect NPN output.

Figure 15.1.22 illustrates supply for an NPN (current sinking) sensor. In this circuit configuration, the load is generally connected between the supply voltage and the output terminal (collector) of the sensor. When the sensor is actuated, turned ON by a magnetic field, current flows through the load into the output transistor to ground.

The sensor's output voltage is measured between the output terminal (collector) and ground (–). When the sensor is not actuated, current will not flow through the output transistor (except for the small leakage current). The output voltage, in this condition, will be equal to V_{LS} (neglecting the leakage current). When the sensor is actuated, the output voltage will drop to ground potential if the saturation voltage transistor is neglected. In terms of the output voltage, an NPN sensor in the OFF condition is considered to be normally high.

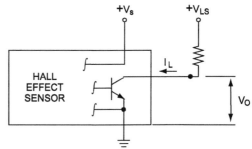

Figure 15.1.22: The supply voltage (Vs) need not be the same as the load supply (V_{LS}); however, a single supply is usually convenient.

There are several methods for linearizing output for analog magnet sensors. The output of the sensor as a function of magnetic field is linear, while the output as a function of distance may be quite nonlinear.

Sensor output can be converted to one that compensates for the non-linearities of magnetics as a function of distance. One method involves converting the analog output of the sensor to digital form. The digital data is fed to a microprocessor, which linearizes the output through a ROM look-up table or transfer function computation techniques (Figure 15.1.23).

A second method involves implementing an analog circuit that has the necessary transfer function to linearize the sensor's output (Figure 15.1.24).

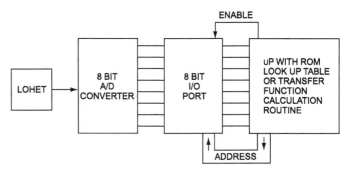

Figure 15.1.23: Microprocessor linearization.

Chapter 15

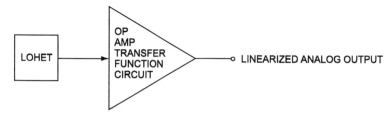

Figure 15.1.24: Analog linearization

A third method for linearizing the sensor output can be realized through magnetic design by altering the geometry and position of the magnets used. These types of magnetic assemblies are not normally designed using theoretical approaches. In most instances, it is easier to design magnetics empirically by measuring the magnetic curve of the particular assembly. By substituting a calibrated Hall element for the variety of magnetic systems available, the designer can develop systems that perform a wide variety of sensing functions.

Ultrasonic Position Sensors

Ultrasonic sensors provide precise no-touch presence/absence sensing and distance sensing or tracking. They are particularly useful where other sensing technologies have difficulty, such as with clear or shiny target objects, foggy or particle-laden air, and environments with splashing liquids. Ultrasonic solutions are often used where larger sensors or longer sensing distances are required.

Factory noise does not affect ultrasonic sensor operation because their operating frequency is well above the frequency of ambient sound. And because sound is used, air pressure, humidity, and smoke, dust, vapors and other airborne contaminants have little effect on accuracy. They provide an alternative to photoelectric or proximity sensors in applications where target characteristics or environmental factors interfere with performance, such as:

- Material handling,
- Packaging,
- Paper processing,
- Food and beverage,
- Chemical,
- Plastics industry, Rubber/tire processing, and
- Steel processing.

Ultrasonic sensors work by exciting an acoustic transducer with voltage pulses, causing the transducer to vibrate ultrasonically. These oscillations are directed at a target and, by measuring the time for the echo to return to the transducer, the target's distance is calculated. Ultrasonic sensors generally provide an accuracy of 1 mm at distances ranging from 100 mm to 6,000 mm (over 19 feet).

Ultrasonic sensors have no difficulty working with round, moving targets, such as film, paper, rubber or steel. (See Figure 15.1.25.)

Ultrasonic sensors often are used for fill-level control for both solids and liquids in the food and beverage, chemical and plastics industries. They can detect the presence of glass parts and provide a simple but highly effective warning system to prevent damage in warehouses when a forklift or other vehicle is in danger of a collision.

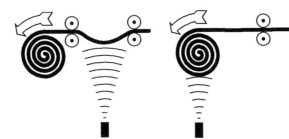

Figure 15.1.25: Two ultrasonic sensors provide roll diameter and tension control by providing outputs directly proportional to distance and roll diameter.

Selecting and Specifying Ultrasonic Sensors

Since ultrasonic sensors operate on an elapsed time measurement system, when the sensor is adjusted to sense a target at a given distance, a timing window is established. The sensor accepts or acknowledges *only* the echoes received within this window. Signals echoing from background material take longer and will not be acknowledged.

The maximum switching frequency is the rate at which the sensor is capable of turning on and off. It depends on several variables. The most significant are target size, target material and distance to the target. The smaller the target, the more difficult it is to detect. Thus, maximum frequency for a small target will be lower than for a large target. Materials that absorb high frequency sound (cotton, sponge, etc.) are more difficult to sense than steel, glass, or plastic. Thus, they also have a lower maximum switching frequency.

Target-to-sensor distance is very important in determining maximum switching frequency. The sensor sends an ultrasonic beam though the air. It takes a finite time for the signal to leave the sensor, travel to the target, strike the target, and return to the sensor as an echo. The farther a target is from the sensor, the longer it takes the sound to complete this cycle, and the lower the switching frequency.

Surface finish is also a consideration. If a smooth flat target is inclined more than +3 degrees to the normal of the beam axis, part of the signal is deflected away from the sensor and the sensing distance is decreased. However, for small targets located close to the sensor, the deviation from normal may be increased to +8 degrees. If the target is inclined more than approximately 12 degrees to the normal of the beam axis, the entire signal is deflected away from the sensor, and the sensor will not respond. A beam striking a target with a coarse surface (such as granular material) is diffused and reflected in all directions and some of the energy returns to the sensor as a weakened echo. (See Figure 15.1.26.)

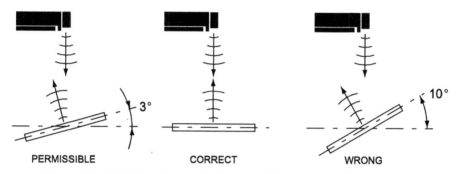

Figure 15.1.26: Maximum inclination for smooth, flat targets.

The velocity of sound in air is temperature dependent. An internal temperature sensor adapts the clock frequency of the elapsed time counter and carrier frequency to help compensate for variations in air temperature. However, larger temperature fluctuations within the beam path can cause dispersion and refraction of the ultrasonic signal, adversely affecting the measurement accuracy and stability (Figure 15.1.27). If a hot object must be detected, experiment by positioning the sensor and target on a vertical plane and aim at the lower (cooler) portion of the target. In this way, it may be possible to avoid the warm air currents and achieve satisfactory operation.

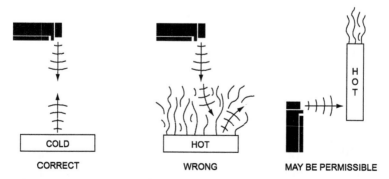

Figure 15.1.27: Sound dispersion due to warm air currents.

Position and Motion Sensors

Ultrasonic sensors have a "dead zone" in which they cannot accurately detect the target (Figure 15.1.28). This is the distance between the sensing face and the minimum sensing range. If the target is too close, the tone burst's leading edge can travel to the target and strike it before the trailing edge has left the transducer. Echo information returning to the sensor is ignored because the transducer is still transmitting and not yet receiving. The echo generated could also reflect off the face of the sensor and again travel out to the target. These multiple echoes can cause errors when the target is in the dead zone.

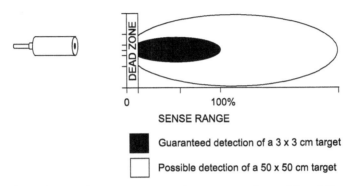

Figure 15.1.28: Ultrasonic sensors must be mounted outside of the dead zone.

Maximum sensing distance or range for each target and application is determined experimentally. Figures 15.1.29–31 show sensitivity characteristics and typical sensing distances.

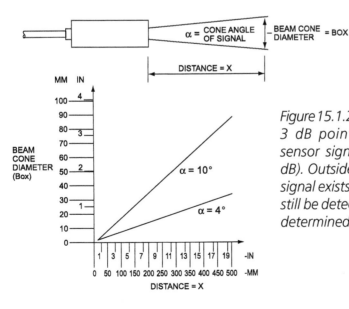

Figure 15.1.29: Beam cone angle values are the 3 dB points (i.e., points at which the sensor signal is attenuated by at least 3 dB). Outside this cone angle, the ultrasonic signal exists but is rather weak. Targets may still be detected. This can be experimentally determined.

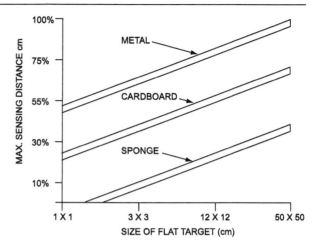

Figure 15.1.30: Sensing range with maximum sensitivity.

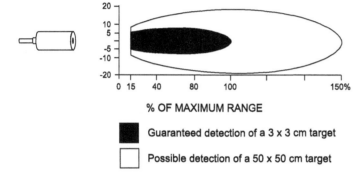

Figure 15.1.31: Target size versus beam spot.

The ultrasonic sensor emits a sound beam in a beam cone angle that eliminates side lobes. Target size versus beam spot size is important. Theoretically, the smallest detectable target is one half the wavelength of the ultrasonic signal. At 215 kHz, the signal wavelength is 0.063 in. Under ideal conditions, these sensors are capable of sensing targets as small as 0.032in. Targets usually are larger and are sensed at various distances. In order to estimate the area covered by the ultrasonic signal at a given distance, use the formula: Box $- 2 \times 1 \tan (K/2)$

Where:

Box = Beam Cone Diameter at distance

X = Distance, target to sensor

K = Beam Cone Angle

The beam cone angle can be halved using a beam cone concentrator.

Normal atmospheric air pressure changes have no substantial effect on measurement accuracy. Ultrasonic sensors are not intended for use in areas with high or low air pressure changes.

The effect of humidity on measurement is virtually negligible, amounting to only 0.07% for a change of relative humidity of 20%. Absorption of sound increases, however, with increasing humidity. Thus, the maximum measurement distance is slightly reduced.

Air turbulence, air currents and layers of different densities cause refraction of the sound wave. An echo may be produced, and the signal weakened or diverted to the extent that the echo is not received. Maximum sensing range, measurement accuracy and measurement stability can deteriorate under these conditions.

Protective measures for the sensor may include a silicone rubber coating when a device is used in an aggressive acid or alkaline atmosphere. To maintain operating efficiency, care must be taken to prevent solid or liquid deposits of these potentially destructive materials from forming on the sensor face.

Interfacing and Design Information for Ultrasonic Sensors

A feature to consider in your design includes an inhibit/sync signal setting. This disables the transmitter, preventing it from sending out any signals. This signal can be used to multiplex/synchronize two or more sensors that are mounted close to each other where acoustic interference is possible. Inhibit multiplexes the sensors so that only one transmits the ultrasonic signal at a given time. Also, the inhibit signal wires from all the sensors can be connected together, synchronizing the sensors to transmit at the same time.

In reflective applications, optional beam deflectors can be used to deflect the beam 90 degrees, helping to reduce the space required for mounting the sensors. Beam concentrators reduce the beam diameter approximately one half. This improves accuracy when measuring liquids, while extending the sensing range as much as 125%. Concentrators are also helpful when sensing small parts.

Proximity Sensors

Proximity sensors are low cost, solid-state devices that come in a variety of technologies, configurations and sensing ranges. They feature fast operation, choice of AC or DC, inherent long life, and compatibility with industrial controllers.

Inductive proximity sensors detect all metals, ferrous metals only, or non-ferrous metals only. Capacitive sensors detect all materials.

Chapter 15

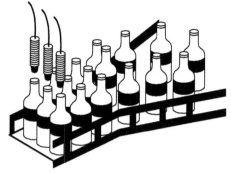

Figure 15.1.32:
Proximity sensors detect bottle caps.

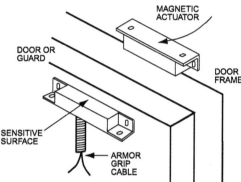

Figure 15.1.33:
A proximity sensor acts as a door guard.

Capacitive proximity sensors have an oscillating electric field, sensitive to all materials: dielectric materials, such as glass, rubber and oil; and conductive materials, such as metals, salty fluids and moist wood. Capacitance (C) is a function of the size of the electrodes (A), the distance between them (d), and the dielectric constant (D) of the material between the electrodes (air =1.0). See Figure 15.1.34.

Capacitance = $C = \dfrac{D \times A}{d}$

Figure 15.1.35 illustrates a simple capacitive sensor. The top electrode is the face of the sensor. A seal ring, the target, passes between it and the ground electrode (a metal conveyor belt). The sensor housing insulates the electrode from galvanic coupling to ground. The rubber seal ring has a dielectric constant (D) of 4.0. When it enters the electric field, the capacitance increases. The sensor detects the change in capacitance and provides an output signal.

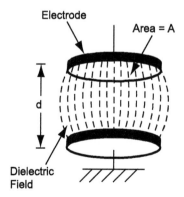

Figure 15.1.34: The oscillating electric field of a capacitive proximity sensor responds to the dialect constant of the target.

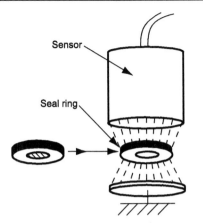

Figure 15.1.35: A capacitive proximity sensor detects a change in capacitance incurred by the presence of a seal ring target.

Figure 15.1.36 illustrates a metal target, or some other conductive material, entering the electric field. The resulting increase in capacitance is detected and converted to an output signal.

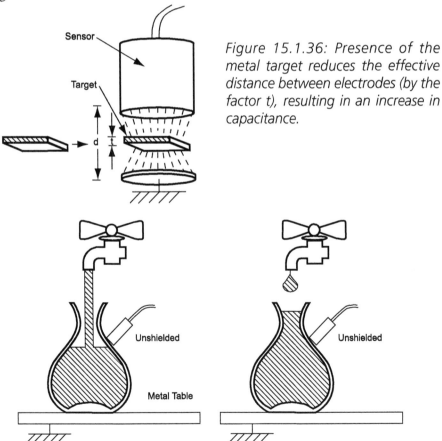

Figure 15.1.36: Presence of the metal target reduces the effective distance between electrodes (by the factor t), resulting in an increase in capacitance.

Figure 15.1.37: Once the ground electrode provided by the fluid is in place, the circuit closes and a signal results.

Chapter 15

The level of conductive fluid pouring into a glass bottle is below the sensor (Figure 15.1.37). With no change in capacitance, there is no output. Once the fluid has reached the level of the sensor, it provides the ground electrode. This happens even though the glass of the bottle separates the fluid and the metal table. The three materials form a capacitor. The alternating current provides a path to ground.

Figure 15.1.38 shows an unshielded sensor and a shielded sensor working together. The shielded sensor locates the glass bottle so it can be filled with liquid. The unshielded sensor indicates that the fill level is reached and it can be turned off. Note that shielded sensors may be flush mounted in any solid material.

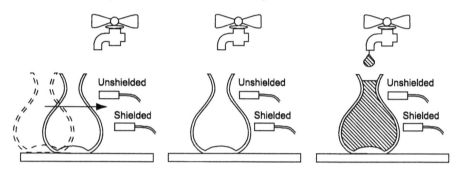

Figure 15.1.38: Neither sensor switches as the bottle approaches. Once the shielded sensor detects the entrance of the glass into its electric field, it switches. When the fluid reaches the level of the unshielded sensor, it switches.

Without some sort of compensation, any material entering the sensing field can cause an output signal. This includes water droplets on the sensor face, dirt or dust, and other contaminants. A compensation electrode in the sensor solves this problem by creating a compensation field. (See Figure 15.1.39.) When contaminants lie directly on the sensor face, both fields are affected, and the capacitance increases by the same ratio. The sensor does not see this as a change in capacitance, and an output is not produced.

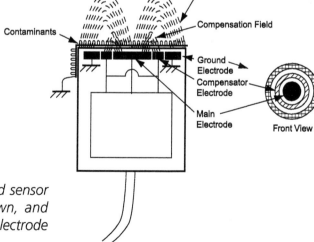

Figure 15.1.39: Shows a shielded sensor with two sensing fields—its own, and the compensation field that the electrode created.

The compensation field is very small and does not extend very far from the sensor. When a target enters the sensing field, the compensation field is unchanged. The disproportionate change in the sensing field (with respect to the compensation field) is detected and converted to an output.

Most inductive proximity switches consist of an oscillator, demodulator, level and switching amplifier as shown in Figure 15.1.40.

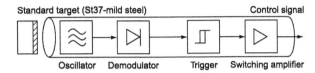

Figure 15.1.40: Inductive proximity sensor components.

If a metal object enters the electromagnetic field of the oscillator coil, eddy currents are induced in this coil, which change the amplitude of oscillation. The demodulator converts the change in the amplitude to a DC signal, causing the trigger stage to trip and the semiconductor output stage to switch.

Inductive proximity sensors operate contactors, electromagnetic clutches, valves, brakes and so forth, without additional interface components. The actuator may be a machine-tool carriage, a metal work piece, a pneumatic cylinder or a drill bit. In fact, nearly any piece of metal of almost any shape or material will actuate an inductive proximity sensor. A broad variety of housings, sensing ranges, basic functions, supply voltages, installation and wiring possibilities are available to enable optimum selection for any particular application.

Selecting and Specifying Proximity Sensors

When you choose an inductive proximity sensor for a particular application, there are several things to consider. The first is the usable sensing distance. The sensing distance published in most order guides is the nominal distance. Your target material, size, method of actuation, manufacturing tolerance, and temperature tolerance must be considered. Each one of these parameters affects the usable sensing distance.

The nominal sensing distance (Sn) is the nominal target-to-sensor distance at which the sensor will turn on. Actual sensing distance may be affected by several considerations, especially manufacturing and temperature tolerances.

Manufacturing tolerance may result in variations of as much as ±10% from unit to unit. The effective sensing distance (Sr) is the difference between the nominal sensing distance and the 10% manufacturing tolerance.

Chapter 15

$$Sr = Sn \pm (.10 \times Sn)$$

Temperature drift tolerances must be also calculated. Over a range of −25 to +70°C, a sensing distance drift of +10% can be expected. Over −25 to +85°C, the tolerance increases to +15%.

$$Su = Sr \pm (.10 \times Sr) \quad -25 \text{ to } +70°C$$

$$Su = Sr \pm (.15 \times Sr) \quad -25 \text{ to } +85°C$$

Usable sensing distance (Su) of any sensor can now be estimated. Su is the distance at which the sensor will always operate. If the target-to-sensor range is greater, the sensor may or may not operate reliably. (See Figure 15.1.41.)

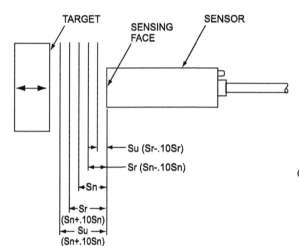

Figure 15.1.41: Nominal sensing distance (Sn) versus usable sensing distance (Su).

Once the usable sensing distance is determined, you need to figure in the actual application conditions. There are three factors to take into account:

- Target material,
- Target size, and
- Target presentation mode.

The nominal sensing distance given in inductive proximity sensor specifications is determined with a target made of mild steel (in accordance with EN 60947-5-2). Whenever a target is a different metal, a correction needs to be made to the usable sensing distance (Su). The formula is:

New Su = Old Su × M

M = material correction factor

The standard target is a square of steel, 1mm (.04 in.) thick, with sides equal to sensor diameter. To determine the sensing distances for materials other than standard, a corresponding correction factor is used. Some common materials and their correction factors are listed in Table 15.1.1.

Table 15.1.1: Correction factors for non-standard target materials.

Correction Factor:	
400 series stainless steel	1.15
Cast iron	1.10
Mild steel (Din 1623)	1.00
Aluminum foil (0, 0.5mm)	0.90
300 series stainless steel	0.70
Brass MS63F38	0.40
Aluminum ALMG3F23	0.35
Copper CCUF3O	0.30

Mild steel targets of "standard" sizes are used to establish published sensing distances. The standard size for each size and style of sensor usually is given in the manufacturer's order guides. If your desired target is the same size or larger than the standard target, no correction factor is necessary. However, a smaller target affects sensing distance. The surface area of the application target versus the surface area of the standard target provides the correction factor. (See Table 15.1.2.) The formula is:

New Su = Old Su × T

T = target correction factor

Table 15.1.2: Correction factors for non-standard target sizes

Surface Area Target	Percent of Standard: Sensing Distance	
	Shielded	Unshielded
25%	56%	50%
50%	83%	73%
75%	92%	90%
100%	100%	100%

When working with capacitive sensors, the dielectric constant of the target must be determined. All materials have a dielectric constant. This constant is what increases the capacitance level of the sensor to a set trigger point. The larger the dielectric constant, the easier a material will be to detect. Materials with high dielectric constants can be detected at greater distances than those with low constants. This allows materials with high dielectric constant to be sensed through the walls of containers made of a material with a lower constant. An example is the detection of salt (6) through a glass wall (3.7).

Each application should be tested. The list of dielectric constants in Table 15.1.3 is provided to help determine the feasibility of the application.

Table 15.1.3: Dielectric constants for different targets.

Material	Dielectric Constant
Acetone	19.5
Acrylic Resin	2.7–4.5
Air	1.000264
Ammonia	15–25
Aniline	6.9
Aqueous Solutions	50–80
Benzene	2.3
Carbon Dioxide	1.000~85
Carban Tetrachloride	2.2
Cement Powder	4
Cereal	3–5
Chlorine Liquid	2.0
Ebonite	2.7–2.9
Epoxy Resin	2.5-6
Ethanol	24
Ethylene Glycol	38.7
FiredAsh	1.5–1.7
Flour	2.5–3.0
Freon R22 & 502 (liquid)	6.11
Gasoline	2.2
Glass	3.7–10
Glycerin	47
Marble	8.5
Melamine Resin	4.7–10.2
Mica	5.7–6.7
Nitrobenzene	96
Nylon	4-5
Paper	1.6–2.6
Paraffin	1.9–2.5
Perspex	3.5
Petroleum	2.0–2.2
Phenol Resin	4–12
Polyacetal	3.6–3.7
Polyester Resin	2.8–8.1
Polypropylene	2.0–2.2
Polyvinyl Chloride Resin	2.8–3.1
Porcelain	5-7
Powdered Milk	3.5–4
Press board	2-5
Rubber	2.5–35
Salt	6
Sand	3–5
Shellac	2.5–4.7

Position and Motion Sensors

Shell Lime	1.2
Silicon Varnish	2.8–3.3
Soybean Oil	2.9–3.5
Styrene Resin	2.3-3.4
Sugar	3.0
Sulpher	3.4
Tetraflouroethylene Resin	2.0
Toluene	2.3
Turpentine	2.2
Urea Resin	5-8
Vaseline	2.2–2.9
Water	80
Wood Dry	2-6
Wood Wet	10–30

As shown in Figure 15.1.42, there are two target presentation modes. Published sensing distances usually are determined by using the head-on mode of actuation. The target can also approach the sensor in the slide-by mode. However, the slide-by method reduces actual sensor-to-target distance by 20%.

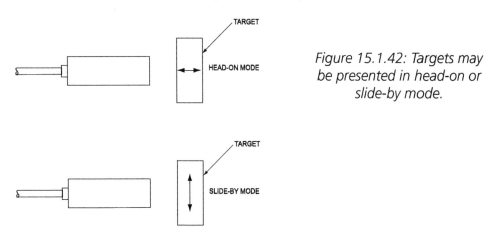

Figure 15.1.42: Targets may be presented in head-on or slide-by mode.

Inductive proximity switches are available with a choice of switching functions. Normally open circuitry causes output current to flow when a target is detected; normally closed circuitry produces zero output current when a target is detected. Changeover circuitry has two sensing outputs; one conducts when a target is detected while the other will not.

Applicable Standards for Proximity Sensors

CENELEC (The European Committee for Electrotechnical Standardization), *www.cenelec.org*.

Chapter 15

IEC (International Electrotechnical Commission), www.iec.ch/, especially IEC 60947-1 and IFC 60947-5-1, which explain the general rules relating to low-voltage switch and control gear for industrial use; IEC 529 rates the level of protection provided by enclosures, using an IP (International Protection) rating system.

Description of protective classes (EN 60529) common to proximity sensors:

- IP 65: Protection against ingress of dust and liquid
- IP 67: Protection against limited immersion in water and dust ingress under predetermined pressure and time conditions (1 meter of water for 30 minutes minimum)
- IP 68: Protection against the effects of continuous immersion in water

NEMA (National Electrical Manufacturer's Association), *www.nema.org*. NEMA rates the protection level of enclosures as does IEC 529, but includes tests for environmental conditions, such as rust, oil, etc. that are not included in IEC 529.

UL (Underwriters Laboratories), *www.ul.com*.

Interfacing and Design Information for Proximity Sensors

When applying capacitive sensors, it's important to note that while shielded capacitive sensors may be flush-mounted, unshielded sensors require isolation—a material-free zone around the sensing face. Materials immediately opposite both shielded and unshielded sensors must be removed to avoid false actuation. See Figure 15.1.43.

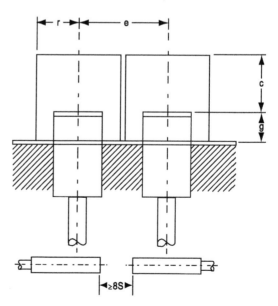

Figure 15.1.43: Unshielded proximity sensors require isolation.

Device-to-device isolation is used when two or more sensors are mounted near each other to prevent cross talk and interference between the devices. Mounting distance between shielded capacitive proximity sensors (center to center) should be at least the diameter of the sensing face. Distance between unshielded sensors will vary and be three to four times the nominal sensing distance.

When shielded or unshielded sensors are facing each other, distance between sensing faces should be at least eight times the sensing distance. To ensure that both shielded and unshielded proximity switches function properly, and to eliminate the possibility of false signals from nearby metal objects, plan for minimum distances as shown in Figure 15.1.44.

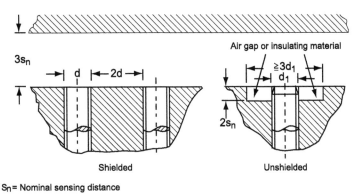

S_n = Nominal sensing distance

Figure 15.1.44: Minimum distances for proximity sensors.

For unshielded proximity switches mounted opposite to each other or side by side, the minimum allowable distances in Figure 15.1.45 apply:

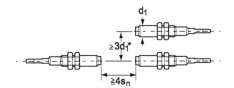

*Shielded devices may be mounted at distances $\geq 2d_1$

Figure 15.1.45: Minimum mounting distances for unshielded sensors

The switching hysteresis (Figure 15.1.46) represents the difference between the switch ON and switch OFF points for axial or radial approach to a target and the subsequent retreat. Usually it will be 3 to 15% of the real sensing distance (Sr).

Chapter 15

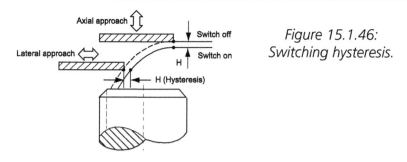

Figure 15.1.46: Switching hysteresis.

To measure the maximum switching frequency, two tests (performed in accordance with EN 60947-5-2) enable the maximum switching frequency

$$f = 1/(t1 + t2)$$

to be determined exactly from the duration of the "switch ON" period (t1) and the "switch OFF" period (t2). (See Figure 15.1.47.)

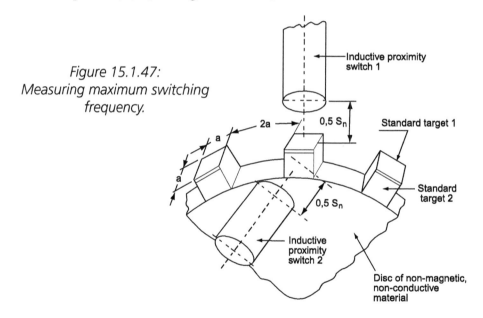

Figure 15.1.47: Measuring maximum switching frequency.

Most DC versions employ normally open, normally closed or changeover circuitry and are available with either NPN or PNP open collector outputs.

- Operating voltage (VB)
  A 5% residual ripple must not cause the operating voltage to fall below the minimum stated value. Correspondingly, a 10% ripple must not cause the operating voltage to exceed the maximum value quoted.

- Voltage drop (Vd)
  Maximum voltage drop at the proximity switch if the output drops to zero.

- Residual voltage (Vr)
  Voltage drop at the load if the sensing output is not conducting.

- Maximum load current (la)
  Under nominal conditions, the output of the proximity switch cannot be driven by a current greater than this value.

- Residual current (lr)
  If the output is not conducting, Ir is the maximum current flowing through the load.

- Current consumption without load (lo)
  Current consumption of the switch under nominal conditions without load.

- Standby delay (tv)
  Period between the application of the operating voltage and the sensor reaching the "ready" state. It is determined by the transient behavior of the oscillator.

- Series and parallel circuitry

If required, inductive proximity switches can be connected in series or in parallel. For series connection, the voltage drops Vd of two or more 3-wire switches (DC) or 2-wire switches (AC or DC) can be significant. Care should be taken that the output voltage is large enough to drive the load. With the NPN-version, the 3-wire switches must be connected to a common positive terminal. With the PNP-version, connect the switches to a common negative terminal. Series connection results in an AND function.

Parallel connection of 2-wire switches (AC) and 3-wire switches (DC) with open collector outputs is possible. The sum of the residual currents must be negligible enough to prevent the load (the holding current of a relay or magnetic switch) from being activated. For 3-wire switches with a collector resistor, it is recommended to decouple the sensing outputs with diodes. An OR function is obtained by connecting the switches in parallel.

Logic cards can be added to inductive proximity sensors. They receive the proximity sensor signal, amplify it and modify the output to respond in a particular way (as determined by time delay, pulse, or other logic). Besides operating output devices, the logic card output signal can be used as input to another card for customer logic. This is done most often with a modular control base.

One-shot (pulsed) logic gives a single fixed pulse in response to a change at the sensor. This is often used as a leading edge detector for moving parts, where the first indication of presence requires a single operation to take place, but where the continued presence will not cause recycling to occur.

Maintained (latching) logic might be used to detect parts for manual reject. The output is continuous until the operator resets. After resets, the output will not trigger if the original target is still in front of the sensor.

ON delay does not trigger immediately with a change at the sensor, but will trigger only if the input signal exceeds a preset time delay. For example, it can provide jam-up detection on a conveyor for parts feeding at specific intervals. A slow down or stoppage downstream will cause a slower rate of passage, recognized as overloading or jam-up, and will cause an output to give warning or shut down the equipment until the cause is eliminated. A similar type provides an output which stays ON even when the cause is corrected, until manually reset by the operator.

ON/OFF delay is used especially for jam-up detection on vibration feeders and conveyors. The ON delay detects a jam-up, and the OFF delay allows the needed time for the jam to clear the sensing area.

Zero-speed detection provides shutdown for universal jam-up detection where the product may end up in front of the sensor for too long an interval, depending on whether the jam is upstream or downstream. If the interval exceeds a preset time, the output turns OFF or shuts down the equipment.

Photoelectric Sensors

Photoelectric sensors respond to the presence of all types of objects, be it large or small, transparent or opaque, shiny or dull, static or in motion. They can sense targets from distances of a few millimeters up to 100 meters. Photoelectric sensors use an emitter unit to produce a beam of light that is detected by a receiver. When the beam is broken, a "presence is detected.

The emitter light source is a modulated, vibration-resistant LED. This beam, which may be infrared, visible red or green, is switched at high currents for short time intervals so as to generate a high-energy pulse to provide long scanning distances or penetration in severe environments. Pulsing also means low power consumption.

The receiver contains a phototransistor that produces a signal when light falls upon it. A phototransistor is used because it has the best spectral match to the LED, a fast response, and is temperature stable. By tuning the receiver circuitry to respond to a narrow band around the LED pulsing frequency, very high ambient light and noise rejection can be achieved. Tuning the receiver to respond only to a specific phase of the pulsed beam can further enhance this effect.

The availability of various fiber optic cables with sensing elements permits photoelectric sensors to be used in many applications where space is limited or where there is a hazardous environment. These sensors also are capable of sensing objects traveling at high speeds with the option of detection at up to 8 kHz if necessary.

Selecting and Specifying Photoelectric Sensors

There are different scanning techniques available for photoelectric controls.

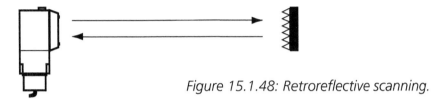

Figure 15.1.48: Retroreflective scanning.

Retroreflective scanning uses an emitter and receiver housed in the same unit with the beam reaching the receiver via a reflector (Figure 15.1.48). Advantages are single side mounting, easy alignment and the ability to mount a reflector in spaces too small for a receiver unit. Reflectors are either acrylic discs or panels, or reflective tape cut to a convenient size. The larger the reflector, the more light reaches the receiver, giving longer scanning distances.

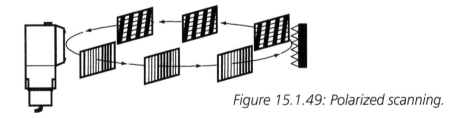

Figure 15.1.49: Polarized scanning.

Polarized scanning involves all the features of retroreflective scanning with the addition of a polarized lens (Figure 15.1.49). When the light wave hits the prismatic reflector, it is turned 90 degrees and, on return, allowed to pass through the receiving lens. This prevents false reflections when detecting shiny surfaces.

To reliably activate retroreflective and polarized scanning techniques, approximately 80 percent of the effective beam needs to be blocked. (See Figure 15.1.50.) The diameter of the effective beam is the same as the reflector on one end and the lens of the photoelectric.

Chapter 15

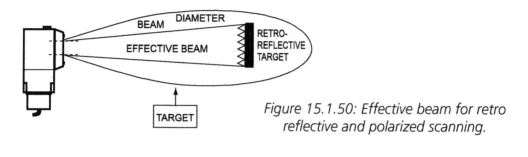

Figure 15.1.50: Effective beam for retro reflective and polarized scanning.

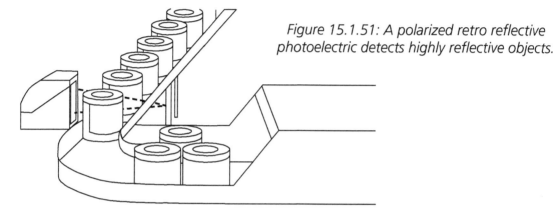

Figure 15.1.51: A polarized retro reflective photoelectric detects highly reflective objects.

Using polarized retroreflective photoelectrics, highly reflective objects (Figure 15.1.51) are detected for conveyor control. Polarized controls respond only to corner-cubed reflectors and ignore light reflected from the target, ensuring that the target always blocks the beam.

In automated assembly, the proper orientation of parts can be controlled by memorizing the reflectivity difference of the target sides. With the microprocessor-based photoelectric in Figure 15.1.52, this is achieved by simply pushing an auto-tuning button.

With a through-scan technique (Figure 15.1.53), the emitter and receiver are separate and positioned opposite one another, so that the light from the emitter shines directly on the receiver. This scanning mode gives maximum reliability (little chance of false reflections to the receiver), high penetration in contaminated environments, and long scanning distances. When installing adjacent through-scan systems, the emitter of one should be positioned next to the receiver of the next, to avoid one system detecting light from the other.

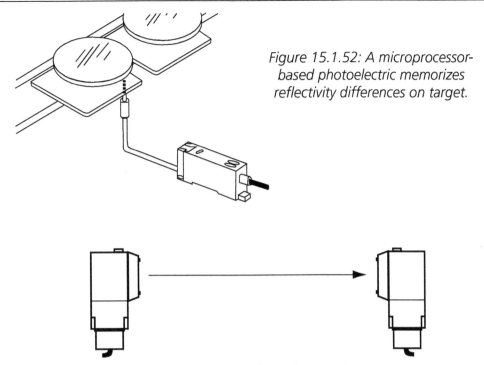

Figure 15.1.52: A microprocessor-based photoelectric memorizes reflectivity differences on target.

Figure 15.1.53: Through scanning.

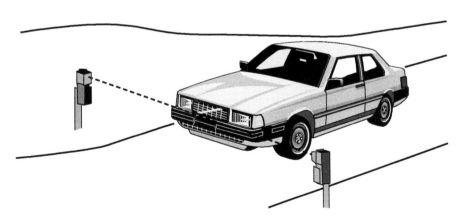

Figure 15.1.54: Long distance, harsh duty photoelectrics withstand outdoor environments to solve such applications as traffic control at toll ways and automatic security gates.

Chapter 15

To reliably activate through scanning, approximately 80 percent of the effective beam needs to be blocked. The diameter of the effective beam is the same size as the emitter and receiver lenses as shown in Figure 15.1.55.

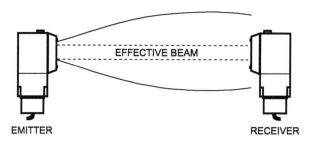

Figure 15.1.55: Effective beam for through scanning.

In diffuse scanning, the emitter and receiver share the same housing, and the emitted beam is reflected to the receiver directly from the target (Figure 15.1.56). This mode is used in cases where it is impractical to use a reflector, due to space considerations or when detection of a specific target is required. Because the reflected light is diffuse, a cleaner environment is necessary and scanning distances are shorter. The maximum scan distance of a diffuse-scan sensor is rated to a 10×10 cm white card. If the actual target is less reflective than a white card, the scan distance will be reduced. If the target is more reflective, the distance will be increased.

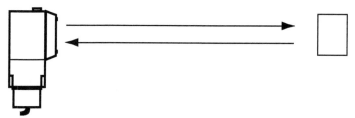

Figure 15.1.56: Diffuse scanning.

Diffuse with background suppression is a special variety of diffuse scan. Using dual receivers and adjustable optics, targets can be reliably detected while backgrounds directly behind the targets are ignored (Figure 15.1.58). They can be very useful when dark-colored objects are placed in front of highly reflective backgrounds (stainless steel, white conveyors, etc.).

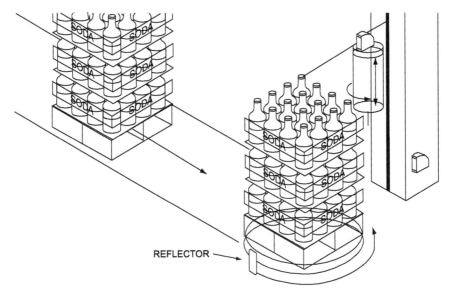

Figure 15.1.57: Polarized and diffuse photoelectrics with time delays are used to detect both the presence and the height of the target to control wrapping on this palletizing and wrapping machine.

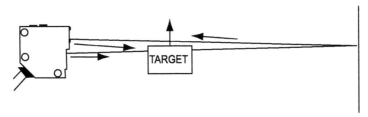

Figure 15.1.58: Diffuse scanning with background suppression.

A convergent beam is another special variety of diffuse scan. Special lenses converge the beams to a fixed focal point in front of the control (Figure 15.1.59). Convergent beams are useful for product positioning and ignoring background reflections. Convergent beams using visible red or green light produce a concentrated, small light spot on the target that can be used to detect color marks. Targets are detected within the "sensing window" of convergent beam controls. This window will increase with targets of higher reflectivity and decrease with targets of lower reflectivity.

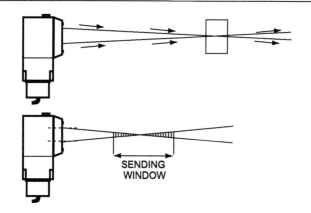

Figure 15.1.59: Convergent beam scanning

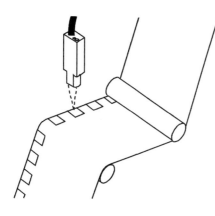

Figure 15.1.60: A visible red and green convergent beam photoelectric provides a small beam spot that enables accurate detection of color marks used in packaging.

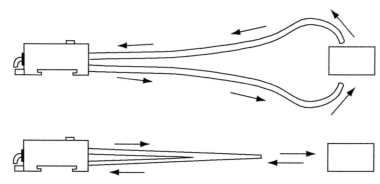

Figure 15.1.61: Fiber optic photoelectrics.

Fiber optic photoelectric sensors use either through scan or diffuse scan fiber optic cables (Figure 15.1.61). These cables allow sensing in very space-restricted areas and provide detection of very small targets. Cables are available with either plastic or glass fibers that the user can cut to length. Glass and stainless steel cables provide rugged protection and high-temperature capability. Many different cable end tips help solve many different applications.

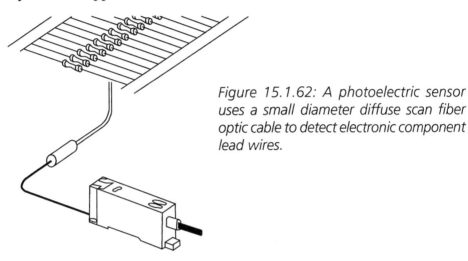

Figure 15.1.62: A photoelectric sensor uses a small diameter diffuse scan fiber optic cable to detect electronic component lead wires.

The specified scanning distance for a photoelectric sensor is the guaranteed minimum operating distance in a clean environment. For retroreflective units, this distance is that obtained using a reflector of 100 percent efficiency. For diffuse units, this distance is that obtained using white Kodak paper with specified dimensions, usually 10×10 cm. Use of other materials affects the diffuse scanning distance as follows:

- Kodak white paper, 100%
- Aluminum, 120–150%
- Brown Kraft paper, 60–70%

Response time is the time between optical change of the system and the output changing to ON or OFF.

Frequency of operation is measured in cycles per second (Hz) and is calculated by:

Frequency of Operation = 1

(Response time ON + Response time OFF)

Chapter 15

Interfacing and Design Information for Photoelectric Sensors

Photoelectric sensors have light and dark operation (LO/DO) modes. In LO, the output is ON when light is incident on the receiver and OFF when there is no light at the receiver; in DO, the output is ON when there is no light incident on the receiver and OFF when there is light at the receiver.

Today, many photoelectric sensors have self-diagnostic LED indicators and outputs. Most are equipped with LED indicators that provide early warning of malfunctions due to misalignment or contaminants on the lens surface, Generally, the LEDs indicate a stable light or unstable light condition (see Figure 15.1.63).

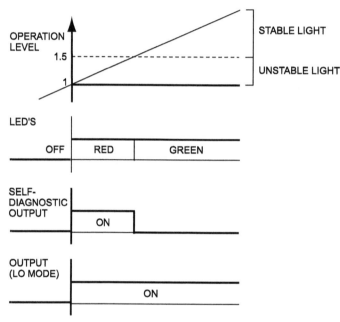

Figure 15.1.63: LEDs indicate stable and unstable light conditions.

Stable light: The Green LED illuminates to show that the photoelectric is receiving at least 1.5 times the minimum operating light level of the sensor (normal operation).

Unstable light: The Green LED changes to Red (or turns OFF) to show that the photoelectric is receiving an amount of light less than 50% extra but still greater than the minimum operating point. The sensor is still operating but marginally.

Position and Motion Sensors

Certain photoelectric sensors also are equipped with an additional wire that provides a remote self-diagnostic output. This output activates when the sensor is operating in the unstable light condition. This signal can be connected to a PLC or directly to an alarm circuit to inform an user at a remote location about an unstable sensor. Adjustment to the sensor (cleaning the lens, realignment, etc.) can then be made to prevent downtime.

Some newer photoelectrics have LED indicators that provide information on the both the "dark" conditions as well as the "light" conditions. On these sensors, the Green LED indicates whether the sensor is operating in a stable dark or unstable dark condition in addition to stable and unstable light (Figure 15.1.64).

Stable dark: The Green LED illuminates to show that the emitted light beam is fully blocked from the receiving element of the photoelectric (normal operation).

Unstable dark: The Green LED turns OFF to show that some light is still reaching the photoelectric receiver. It is not a level high enough to operate the sensor, but it is a marginal condition. If the marginal condition continues for one full cycle of operation, the Green LED will flicker and activate a remote self diagnostic output.

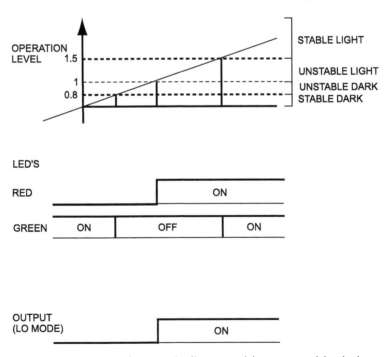

Figure 15.1.64: LEDs also can indicate stable or unstable darkness.

Chapter 15

Latest and Future Developments

Position sensors indicate the precise location of an object, a defined target or even a human being to control a surrounding process or improve its effectiveness. New electronic parts have improved the overall characteristics of sensors, and more functionality is being added at the sensor level. Diagnostic functions and easy-to-use calibration features are improving control systems and reducing installation time.

Communication modes are increasingly important in determining the right sensing technology for an application, and the ability of manufacturers to offer a combination of technologies is a major advantage. The focus is and will be on the application and how to best solve it. Sensing technology is the enabler and, therefore, the emphasis should not be on the technology itself but on the most effective way to meet the needs of the application.

The demand for communications, especially the ability to receive real-time data from remote locations to improve process control, continues to grow. Wireless technology is a "hot" topic as it significantly improves the flow of real-time data. It is quite possible that in the near future, sensors will not only be able to communicate to remote control areas, but also start communicating amongst themselves. Some local control loops will also be available to optimize processes and ensure that quality and safety standards are being met at all times.

References and Resources

"Hall Effect Sensing and Application," Honeywell, Inc.
http://content.honeywell.com/sensing/prodinfo/solidstate/technical/hallbook.pdf

"Applying Linear Output Hall Effect Transducers," Honeywell, Inc.
http://content.honeywell.com/sensing/prodinfo/solidstate/technical/c20084.pdf

"Current Sink and Current Source Interfacing for Solid State Sensors," Honeywell, Inc.
http://content.honeywell.com/sensing/prodinfo/solidstate/technical/c20078.pdf

"Interfacing Digital Hall Effect Sensors," Honeywell, Inc.
http://content.honeywell.com/sensing/prodinfo/solidstate/technical/c20083.pdf

"Interpreting Operating Characteristics for Solid State Sensors," Honeywell, Inc.
http://content.honeywell.com/sensing/prodinfo/solidstate/technical/c20082.pdf

"Gear Tooth Sensor Target Guidelines," Honeywell, Inc.
http://content.honeywell.com/sensing/prodinfo/solidstate/technical/005838_1.pdf

"Magnet Conversion Chart," Honeywell, Inc.
   http://content.honeywell.com/sensing/prodinfo/solidstate/technical/c20103.pdf

"Magnets," Honeywell, Inc.
   http://content.honeywell.com/sensing/prodinfo/solidstate/technical/c20099.pdf

"Method of Magnet Actuation," Honeywell, Inc.
   http://content.honeywell.com/sensing/prodinfo/solidstate/technical/c20102.pdf

"Solid State Sensors Glossary of Terms," Honeywell, Inc.
   http://content.honeywell.com/sensing/prodinfo/solidstate/technical/c20119.pdf

Chapter 15

15.2 String Potentiometer and String Encoder Engineering Guide

Tom Anderson, SpaceAge Control, Inc.

This section reviews the advantages and disadvantages of string potentiometers and string encoders, hereafter referred to as CPTs (cable position transducers). Other names often used to refer to these transducers are:

- cable actuated position sensor
- cable extension transducer
- cable position transducer
- cable sensor
- cable-actuated sensor
- CET
- CPT
- stringpot
- string potentiometer
- draw wire encoder
- draw wire transducer
- wire rope transducer
- wire sensor
- wire-actuated transducer
- yo yo pot
- yo yo potentiometer

These names all refer to devices that measure displacement via a flexible displacement cable that extracts from and retracts to a spring-loaded drum. This drum is attached to a rotary sensor (see Figure 15.2.1). By understanding the strengths and weaknesses of CPT technology, designers, engineers, and technicians can specify and design the best displacement measurement solution for their application.

Technology Review

CPTs were first developed in the mid-1960s in concert with the growth of the aerospace and aircraft industries. The first applications involved the monitoring of aircraft flight control mechanisms during flight testing.

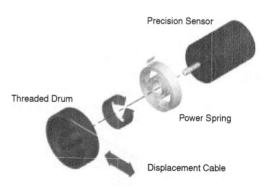

Figure 15.2.1: How CPTs work.

Position and Motion Sensors

While the technology is proven and mature, it is certainly not dated. A broad range of high-performance and cost-conscious applications use CPTs as the basis for key control and monitoring operations. Recent examples include:

- Delta IV missile thrust vectoring system
- Military fighter level sensor
- Diesel engine fuel index measurement
- International Space Station environmental control systems
- commercial and military aircraft flight data recorder input sensors
- excavator hydraulic cylinder control
- medical table actuation feedback system
- V-22 flight control surface monitoring
- Global Hawk UAV landing gear stroke measurement
- logistics sorting and positioning equipment
- earth borer positioner

Advantages of CPTs

CPTs have numerous advantages over other types of position sensors:

Multi-axis Capability. As Figure 15.2.2 below shows, CPTs can be used to track linear, rotary, 2-dimensional, and 3-dimensional displacements. This capability makes CPTs ideal in test engineering as well as in OEM applications where size and mounting restrictions eliminate other choices.

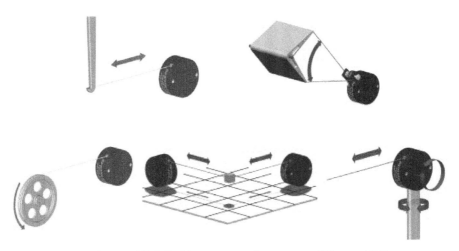

Figure 15.2.2: Linear, angular, rotary, 2D, and 3D displacements can be monitored with CPTs.

Chapter 15

Flexible Mounting. The flexible displacement cable inherent in CPT technology allows for flexible mounting. The cable can be attached to the application in a number of ways as shown in Figure 15.2.3. Other methods include magnets and eyebolts or other threaded fasteners.

Figure 15.2.3: A few displacement cable terminations.

The cable can also be routed around barriers using pulleys (see Figure 15.2.4) and flexible conduits.

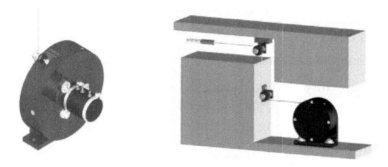

Figure 15.2.4: Pulleys and idlers allow displacement cable to be routed to the application.

Finally, innovative transducer mounting bases and cable exit options (see Figures 15.2.5 and 15.2.6) give additional mounting flexibility, eliminating the expense associated with special fixturing and adapters.

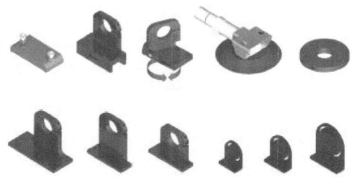

Figure 15.2.5: A few mounting base options.

Position and Motion Sensors

Figure 15.2.6: Cable exit choices provide ease of installation and application flexibility.

Fast Installation. The flexible mounting features combined with the broad tolerance for displacement cable misalignment provide for fast installation, often in less than 2 minutes (see Figure 15.2.7). This reduces installation costs and can be particularly valuable in test and research and development applications.

Figure 15.2.7: Installation is fast.

Small Size. CPT technology gives the user a small size relative to measurement range. The world's smallest CPT measures a 1.5 inches (38.1 mm) displacement with a size of only 0.75 inch square by 0.38 inch (19 mm × 19 mm × 10 mm) as shown in Figure 15.2.8. As the measurement range increases, the CPT's relative small size advantage becomes more obvious as shown in Figure 15.2.9.

Figure 15.2.8: World's smallest CPT: The Series 150.

Chapter 15

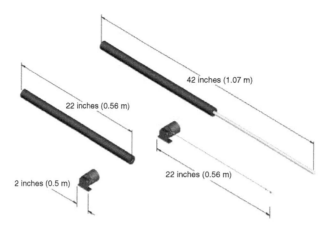

Figure 15.2.9: Size comparison of CPTs to rod-and-cylinder devices such as LVDTs and linear potentiometers.

Lightweight. CPTs measure displacement with a reliable, low mass stainless steel or high-strength fabric-based cable. This feature along with generally anodized aluminum components results in a product that has a low-weight-to-range ratio. This benefit can be important in space, missile, aircraft, racing, robotic, and biomedical applications. Low mass can also increase survivability in high shock and vibration environments encountered in industrial machinery and equipment applicatoins. Table 15.2.1 shows a comparison of weight to range for various displacement measurement sensors.

Table 15.2.1: CPTs have a weight-to-range advantage over other sensor types (mass shown in oz (g)).

Range (in (mm))	CPT	LVDT	Linear encoder	Magnetostrictive
2 (51)	1 (28)	13 (369)	20 (567)	15 (425)
10 (254)	2 (57)	41 (1162)	25 (709)	29 (822)
40 (1016)	8 (227)	67 (1900)	44 (1247)	51 (1446)

Rugged. Properly designed and manufactured CPTs have performed reliably for over 35 years in harsh industrial, aerospace, testing, and outdoor environments. The design of CPTs can be mechanically and electrically simple, resulting in high reliability, low maintenance, and years of service. CPT environmental testing has demonstrated the effective operation of CPTs in environments containing high shock, vibration, humidity, corrosion, moisture, and other parameters.

Variety of Electrical Outputs. What does your controller or data acquisition device require? 4–20 mA? 0 to 5 VDC? 0 to 10 VDC? ±5 VDC? ±10 VDC? Strain gage compatible? Quadrature? RS-232? LVDT or RVDT type signals? Synchro or resolver type signals? Because CPTs can incorporate a broad range of rotary sensors and related signal conditioning, the electrical output of your choice is typically available.

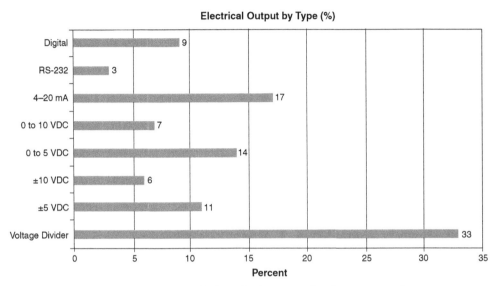

Figure 15.2.10: Typical electrical output by frequency of use.

Low Power, Simple Signal Conditioning. CPTs, particularly analog potentiometer types, generally have low power and simple signal conditioning requirements. 5 VDC power or less is suitable and no special signal conditioning is required for the majority of applications. The CPT signal input and output requirements can reduce total system cost, allow for fast setup, and allows for relatively inexperienced personnel to work with the devices.

Unobtrusive. The small size and low mass of CPTs give them a small profile. This provides for aesthetically pleasing designs. Additionally, this unobtrusiveness reduces the possibility for unwanted application interaction (accidental interference) with the transducer. Finally, the displacement cable can be designed to "break away" in applications where it is paramount that sensor failure does not interfere with the measured object. This can be quite important in flight control applications where flight safety can be affected if a failure of sensor's actuation mechanism makes it impossible for the aircraft to be properly controlled.

Broad Operating Temperature. Analog-output CPTs can operate in temperate ranges from –65°C to +125°C while digital-output CPTs can range from –40°C to +85°C or –20°C to +100°C. This broad operating range reduces operating and installation issues and makes the products well-suited for outdoor, aerospace, and industrial control applications. Broader operating ranges can be attained with the use of custom sensors.

Accuracy. Using non-backlash connections and threaded drums, analog-output CPTs can offer linearity-compensated accuracy exceeding ±0.025% of full scale. As an example, a 10-inch (254-mm) range analog-output CPT can give an accuracy of ±0.0025 inch (±0.0635 mm).

Broad Measurement Range. CPTs measurement ranges are broad: 1.5 inches to over 2000 inches (38 mm to over 50 m). For some long-range measurement challenges, the CPT is the only feasible solution.

Cost Effective. CPTs are generally cost-effective devices, especially when considering lifetime costs. Flexible mounting, fast installation, high reliability, and minimal signal conditioning requirements reduce costs relative to many alternatives. The CPT's cost effectiveness increases at longer ranges because the displacement cable is the primary cost adder to the design in going from shorter ranges to longer ranges.

Other Design Factors

There is no perfect sensing technology. CPTs are no exception. Even Superman has his Kryptonite. When considering a CPT for an application, keep the following items in mind:

Frequency Response. CPTs are used extensively in vehicle impact testing where accelerations exceed 50 g's. CPTs have been used in other applications where accelerations approach 100 g's. However, applications involving extreme acceleration can exceed the frequency response capabilities of CPTs. Examples where CPT frequency response can be an issue include applications that require cable accelerations in excess of 100 g's and environments using analog-output CPTs with very high frequency, low displacement motion that tends to cause dithering wear of the potentiometric element. During extreme high accelerations, it is possible the cable reel's inertia and the power spring's insufficient torque will cause the reel to rotate to a point where the displacement cable goes slack.

Lifetime. CPTs with optical encoder or conductive plastic sensing technologies can operate in excess of 100 million shaft revolutions. However, longer range analog-output CPTs require multi-turn potentiometers with lifetimes of less than 10 million shaft revolutions. While long-displacement, high-cycle applications are not common, you should do a thorough cost and reliability analysis before specifying a CPT for this type of use.

Cable Tension on Application. Non-contact sensing devices such as ultrasonic, Hall effect, or laser do not mechanically affect the application. The CPT's cable tension imparts a load on the application. While this load can be minimized to as little as 1 oz. (0.278 N), it cannot be eliminated. Hence, for applications sensitive to external loads, other technologies should be considered.

Cable Interference. The CPT's small, lightweight displacement cable is one of its key advantages. In some applications, it can be a disadvantage. For example, inadvertent damage can occur if technicians and operators do not know of the cable's presence. While signage and protective tubing or other devices can be installed, damage can still occur from unplanned events.

Accuracy. CPTs that use no-backlash mechanical connections and threaded drums deliver excellent accuracy. Nevertheless, this is sometimes insufficient for some applications. For those situations, you should consider using LVDTs, laser-based devices, or other higher-accuracy technologies.

Accuracy is a broadly-used yet not always fully defined word. When you are determining what "accuracy" you require for your application, ensure you really need the accuracy you are specifying. See if your application does not require raw accuracy but rather good linearity, resolution, repeatability, or hysteresis.

Catenary Curve Error Effects. A catenary curve describes the shape the displacement cable takes when subjected to a uniform force such as gravity. Because the mass of the cable per unit length is so small and the cable tension is relatively high, cable sag does not produce any significant error unless the cable length is exceptionally long (over 60 feet (18 meters)). The cable sag error is minor compared to other error sources (generally less than ± 0.0025%). Nevertheless, some applications, (typically involving high angular rotation or long measurements), produce sufficient forces to cause significant cable sag error. To calculate the catenary curve error effect for your application, visit http://spaceagecontrol.com/calccabl.htm.

Chapter 15

Conclusion

Cable position transducers can be flexible, robust, cost-effective displacement measurement devices for a broad range of applications. At the same time, these products have limitations that may result in other technologies being more appropriate for specific applications.

Reference Materials

These documents provide additional information on the selection and use of CPTs:

- Application Note for Aircraft/Aerospace (S004A)
  http://spaceagecontrol.com/s004a.pdf

- Application Note for Ground Vehicles/Transportation (S005A)
  http://spaceagecontrol.com/s005a.pdf

- Selecting Position Transducers
  http://spaceagecontrol.com/selpt.htm

- Sensor and Transducer Total Cost of Ownership (S054A)
  http://spaceagecontrol.com/s054a.htm

15.3 Linear and Rotary Position and Motion Sensors

Analog Devices Technical Staff
Walt Kester, Editor

Modern linear and digital integrated circuit technology is used throughout the field of position and motion sensing. Fully integrated solutions which combine linear and digital functions have resulted in cost effective solutions to problems which in the past have been solved using expensive electro-mechanical techniques. These systems are used in many applications including robotics, computer-aided manufacturing, factory automation, avionics, and automotive.

This chapter is an overview of linear and rotary position sensors and their associated conditioning circuits. An interesting application of mixed-signal IC integration is illustrated in the field of AC motor control. A discussion of micromachined accelerometers ends the chapter.

- Linear Position: Linear Variable Differential Transformers (LVDT)
- Hall Effect Sensors
- Proximity Detectors
- Linear Output (Magnetic Field Strength)
- Rotational Position:
- Rotary Variable Differential Transformers (RVDT)
- Optical Rotational Encoders
- Synchros and Resolvers
- Inductosyns (Linear and Rotational Position)
- Motor Control Applications
- Acceleration and Tilt: Accelerometer

Figure 15.3.1: Position and motion sensors.

Linear Variable Differential Transformers (LVDTS)

The linear variable differential transformer (LVDT) is an accurate and reliable method for measuring linear distance. LVDTs find uses in modern machine-tool, robotics, avionics, and computerized manufacturing. By the end of World War II, the LVDT had gained acceptance as a sensor element in the process control industry largely as a result of its use in aircraft, torpedo, and weapons systems. The publication of *The Linear Variable Differential Transformer* by Herman Schaevitz in 1946 (Proceedings of the SASE, Volume IV, No. 2) made the user community at large aware of the applications and features of the LVDT.

Excerpted from *Practical Design Techniques for Sensor Signal Conditioning*, Analog Devices, Inc., www.analog.com.

Chapter 15

The LVDT (see Figure 15.3.2) is a position-to-electrical sensor whose output is proportional to the position of a movable magnetic core. The core moves linearly inside a transformer consisting of a center primary coil and two outer secondary coils wound on a cylindrical form. The primary winding is excited with an AC voltage source (typically several kHz), inducing secondary voltages which vary with the position of the magnetic core within the assembly. The core is usually threaded in order to facilitate attachment to a nonferromagnetic rod which in turn in attached to the object whose movement or displacement is being measured.

The secondary windings are wound out of phase with each other, and when the core is centered the voltages in the two secondary windings oppose each other, and the net output voltage is zero. When the core is moved off center, the voltage in the secondary toward which the core is moved increases, while the opposite voltage decreases. The result is a differential voltage output which varies linearly with the core's position. Linearity is excellent over the design range of movement, typically 0.5% or better. The LVDT offers good accuracy, linearity, sensitivity, infinite resolution, as well as frictionless operation and ruggedness.

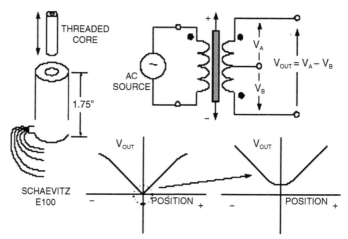

Figure 15.3.2: Linear variable differential transformer (LVDT).

A wide variety of measurement ranges are available in different LVDTs, typically from ±100 µm to ±25 cm. Typical excitation voltages range from 1 V to 24 V rms, with frequencies from 50 Hz to 20 kHz. Key specifications for the Schaevitz E100 LVDT are given in Figure 15.3.3.

Position and Motion Sensors

- **Nominal Linear Range:** ±0.1 inches (± 2.54 mm)
- **Input Voltage:** 3V rms
- **Operating Frequency:** 50 Hz to 10 kHz (2.5 kHz nominal)
- **Linearity:** 0.5% Fullscale
- **Sensitivity:** 2.4 mV Output / 0.001 inches / Volt Excitation
- **Primary Impedance:** 660Ω
- **Secondary Impedance:** 960Ω

Figure 15.3.3: Schaevitz E100 LVDT specifications.

Note that a true null does not occur when the core is in center position because of mismatches between the two secondary windings and leakage inductance. Also, simply measuring the output voltage V_{OUT} will not tell on which side of the null position the core resides.

A signal conditioning circuit which removes these difficulties is shown in Figure 15.3.4 where the absolute values of the two output voltages are subtracted. Using this technique, both positive and negative variations about the center position can be measured. While a diode/capacitor-type rectifier could be used as the absolute value circuit, the precision rectifier shown in Figure 15.3.5 is more accurate and linear. The input is applied to a V/I converter which in turn drives an analog multiplier. The sign of the differential input is detected by the comparator whose output switches the sign of the V/I output via the analog multiplier. The final output is a precision replica of the absolute value of the input. These circuits are well understood by IC designers and are easy to implement on modern bipolar processes.

The industry-standard AD598 LVDT signal conditioner shown in Figure 15.3.6 (simplified form) performs all required LVDT signal processing. The on-chip excitation frequency oscillator can be set from 20 Hz to 20 kHz with a single external capacitor. Two absolute value circuits followed by two filters are used to detect the amplitude of the A and B channel inputs. Analog circuits are then used to generate the ratiometric function [A–B]/[A+B]. Note that this function is independent of the amplitude of the primary winding excitation voltage, assuming the sum of the LVDT output voltage amplitudes remains constant over the operating range. This is usually the case for most LVDTs, but the user should always check with the manufacturer if it is not specified on the LVDT data sheet. Note also that this approach requires the use of a 5-wire LVDT.

Chapter 15

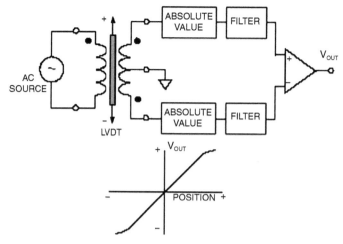

Figure 15.3.4: Improved LVDT output signal processing.

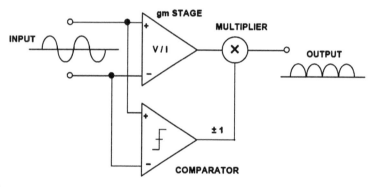

Figure 15.3.5: Precision absolute value circuit (full-wave rectifier).

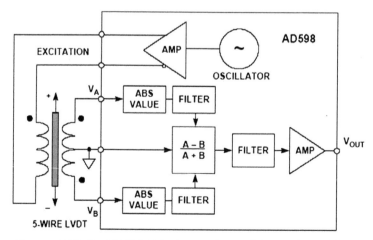

Figure 15.3.6: AD598 LVDT signal conditioner (simplified).

Position and Motion Sensors

A single external resistor sets the AD598 excitation voltage from approximately 1 V rms to 24 V rms. Drive capability is 30 mA rms. The AD598 can drive an LVDT at the end of 300 feet of cable, since the circuit is not affected by phase shifts or absolute signal magnitudes. The position output range of V_{OUT} is ±11 V for a 6 mA load and it can drive up to 1000 feet of cable. The V_A and V_B inputs can be as low as 100 mV rms.

The AD698 LVDT signal conditioner (see Figure 15.3.7) has similar specifications as the AD598 but processes the signals slightly differently. Note that the AD698 operates from a 4-wire LVDT and uses synchronous demodulation. The A and B signal processors each consist of an absolute value function and a filter. The A output is then divided by the B output to produce a final output which is ratiometric and independent of the excitation voltage amplitude. Note that the sum of the LVDT secondary voltages does not have to remain constant in the AD698.

The AD698 can also be used with a half-bridge (similar to an auto-transformer) LVDT as shown in Figure 15.3.8. In this arrangement, the entire secondary voltage is applied to the B processor, while the center-tap voltage is applied to the A processor. The half-bridge LVDT does not produce a null voltage, and the A/B ratio represents the range-of-travel of the core.

It should be noted that the LVDT concept can be implemented in rotary form, in which case the device is called a *rotary variable differential transformer* (RVDT). The shaft is equivalent to the core in an LVDT, and the transformer windings are wound on the stationary part of the assembly. However, the RVDT is linear over a relatively narrow range of rotation and is not capable of measuring a full 360° rotation. Al-

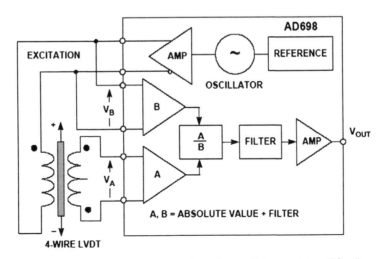

Figure 15.3.7: AD698 LVDT signal conditioner (simplified).

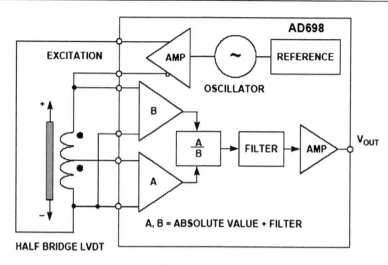

Figure 15.3.8: Half-bridge LVDT configuration.

though capable of continuous rotation, typical RVDTs are linear over a range of about ±40° about the null position (0°). Typical sensitivity is 2–3 mV per volt per degree of rotation, with input voltages in the range of 3 V rms at frequencies between 400 Hz and 20 kHz. The 0° position is marked on the shaft and the body.

Hall Effect Magnetic Sensors

If a current flows in a conductor (or semiconductor) and there is a magnetic field present which is perpendicular to the current flow, then the combination of current and magnetic field will generate a voltage perpendicular to both (see Figure 15.3.9). This phenomenon, called the *Hall Effect*, was discovered by E. H. Hall in 1879. The voltage, V_H, is known as the *Hall Voltage*.

V_H is a function of the current density, the magnetic field, and the charge density and carrier mobility of the conductor.

As discussed earlier in this chapter and elsewhere in this handbook, he Hall effect may be used to measure magnetic fields (and hence in contact-free current measurement), but its commonest application is in motion sensors where a fixed Hall sensor and a small magnet attached to a moving part can replace a cam and contacts with a great improvement in reliability. (Cams wear and contacts arc or become fouled, but magnets and Hall sensors are contact free and do neither.)

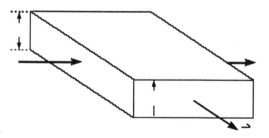

Figure 15.3.9: Hall effect sensors.

Position and Motion Sensors

Since V_H is proportional to magnetic field and not to rate of change of magnetic field like an inductive sensor, the Hall effect provides a more reliable low-speed sensor than an inductive pickup.

Although several materials can be used for Hall effect sensors, silicon has the advantage that signal conditioning circuits can be integrated on the same chip as the sensor. CMOS processes are common for this application. A simple rotational speed detector can be made with a Hall sensor, a gain stage, and a comparator as shown in Figure 15.3.10. The circuit is designed to detect rotation speed as in automotive applications. It responds to small changes in field, and the comparator has built-in hysteresis to prevent oscillation. Several companies manufacture such Hall switches, and their usage is widespread.

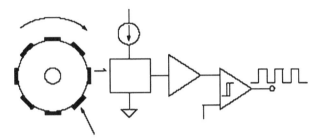

Figure 15.3.10: Hall effect sensor used as a rotation sensor.

There are many other applications, particularly in automotive throttle, pedal, suspension, and valve position sensing, where a linear representation of the magnetic field is desired. The AD22151 is a linear magnetic field sensor whose output voltage is proportional to a magnetic field applied perpendicularly to the package top surface (see Figure 15.3.11). The AD22151 combines integrated bulk Hall cell technology and conditioning circuitry to minimize temperature related drifts associated with silicon Hall cell characteristics.

The architecture maximizes the advantages of a monolithic implementation while allowing sufficient versatility to meet varied application requirements with a minimum number of external components. Principal features include dynamic offset drift cancellation using a chopper-type op amp and a built-in temperature sensor. Designed for single +5 V supply operation, low offset and gain drift allows operation over a –40°C to +150°C range. Temperature compensation (set externally with a resistor R1) can accommodate a number of magnetic materials commonly utilized in position sensors. Output voltage range and gain can be easily set with external resistors. Typical gain range is usually set from 2 mV/Gauss to 6 mV/Gauss. Output voltage can be adjusted from fully bipolar (reversible) field operation to fully unipolar field sensing. The voltage output achieves near rail-to-rail dynamic range (+0.5 V to +4.5 V), capable of supplying 1 mA into large capacitive loads. The output signal is ratiometric to the positive supply rail in all configurations.

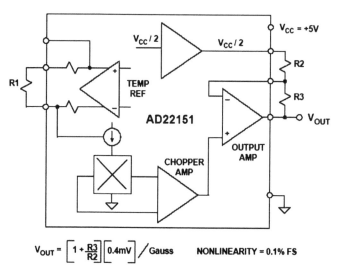

Figure 15.3.11: AD22151 Linear output magnetic field sensor.

Optical Encoders

Among the most popular position measuring sensors, optical encoders find use in relatively low reliability and low resolution applications. An *incremental* optical encoder (left-hand diagram in Figure 15.3.12) is a disc divided into sectors that are alternately transparent and opaque. A light source is positioned on one side of the disc, and a light sensor on the other side. As the disc rotates, the output from the detector switches alternately on and off, depending on whether the sector appearing between the light source and the detector is transparent or opaque. Thus, the encoder produces a stream of square wave pulses which, when counted, indicates the angular position of the shaft. Available encoder resolutions (the number of opaque and transparent sectors per disc) range from 100 to 65,000, with absolute accuracies approaching 30 arc-seconds (1/43,200 per rotation). Most incremental encoders feature a second light source and sensor at an angle to the main source and sensor, to indicate the direction of rotation. Many encoders also have a third light source and detector to sense a once-per-revolution marker. Without some form of revolution marker, absolute angles are difficult to determine. A potentially serious disadvantage is that incremental encoders require external counters to determine absolute angles within a given rotation. If the power is momentarily shut off, or if the encoder misses a pulse due to noise or a dirty disc, the resulting angular information will be in error.

Position and Motion Sensors

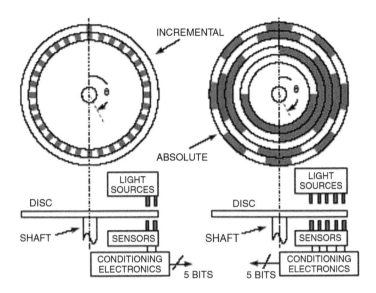

Figure 15.3.12: Incremental and absolute optical encoders.

The *absolute* optical encoder (right-hand diagram in Figure 15.3.12) overcomes these disadvantages but is more expensive. An absolute optical encoder's disc is divided up into N sectors (N = 5 for example shown), and each sector is further divided radially along its length into opaque and transparent sections, forming a unique N-bit digital word with a maximum count of $2^N - 1$. The digital word formed radially by each sector increments in value from one sector to the next, usually employing Gray code. Binary coding could be used, but can produce large errors if a single bit is incorrectly interpreted by the sensors. Gray code overcomes this defect: the maximum error produced by an error in any single bit of the Gray code is only 1 LSB after the Gray code is converted into binary code. A set of N light sensors responds to the N-bit digital word which corresponds to the disc's absolute angular position. Industrial optical encoders achieve up to 16-bit resolution, with absolute accuracies that approach the resolution (20 arc seconds). Both absolute and incremental optical encoders, however, may suffer damage in harsh industrial environments.

Resolvers and Synchros

Machine-tool and robotics manufacturers have increasingly turned to resolvers and synchros to provide accurate angular and rotational information. These devices excel in demanding factory applications requiring small size, long-term reliability, absolute position measurement, high accuracy, and low-noise operation.

A diagram of a typical synchro and resolver is shown in Figure 15.3.13. Both synchros and resolvers employ single-winding rotors that revolve inside fixed stators. In the case of a simple synchro, the stator has three windings oriented 120° apart and electrically connected in a Y-connection. Resolvers differ from synchros in that their stators have only two windings oriented at 90°.

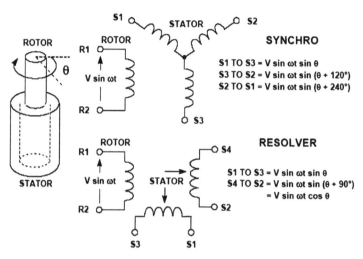

Figure 15.3.13: Synchros and resolvers.

Because synchros have three stator coils in a 120° orientation, they are more difficult than resolvers to manufacture and are therefore more costly. Today, synchros find decreasing use, except in certain military and avionic retrofit applications.

Modern resolvers, in contrast, are available in a brushless form that employ a transformer to couple the rotor signals from the stator to the rotor. The primary winding of this transformer resides on the stator, and the secondary on the rotor. Other resolvers use more traditional brushes or slip rings to couple the signal into the rotor winding. Brushless resolvers are more rugged than synchros because there are no brushes to break or dislodge, and the life of a brushless resolver is limited only by its bearings. Most resolvers are specified to work over 2 V to 40 V rms and at frequencies from 400 Hz to 10 kHz. Angular accuracies range from 5 arc-minutes to 0.5 arc-minutes. (There are 60 arc-minutes in one degree, and 60 arc-seconds in one arc-minute. Hence, one arc-minute is equal to 0.0167 degrees.)

In operation, synchros and resolvers resemble rotating transformers. The rotor winding is excited by an AC reference voltage, at frequencies up to a few kHz. The magnitude of the voltage induced in any stator winding is proportional to the sine of the angle, θ, between the rotor coil axis and the stator coil axis. In the case of a synchro, the voltage induced across any pair of stator terminals will be the vector sum of the voltages across the two connected coils.

For example, if the rotor of a synchro is excited with a reference voltage, Vsinωt, across its terminals R1 and R2, then the stator's terminal will see voltages in the form:

S1 to S3 = V sinωt sinθ

S3 to S2 = V sinωt sin (θ + 120°)   S2 to S1 = V sinωt sin (θ + 240°),

where θ is the shaft angle.

In the case of a resolver, with a rotor AC reference voltage of Vsinωt, the stator's terminal voltages will be:

S1 to S3 = V sinωt sin θ

S4 to S2 = V sinωt sin(θ + 90°) = V sinωt cosθ.

It should be noted that the 3-wire synchro output can be easily converted into the resolver-equivalent format using a Scott-T transformer. Therefore, the following signal processing example describes only the resolver configuration.

A typical resolver-to-digital converter (RDC) is shown functionally in Figure 15.3.14. The two outputs of the resolver are applied to cosine and sine multipliers. These multipliers incorporate sine and cosine lookup tables and function as multiplying digital-to-analog converters. Begin by assuming that the current state of the up/down counter is a digital number representing a trial angle, φ. The converter seeks to adjust the digital angle, φ, continuously to become equal to, and to track θ, the analog angle being measured. The resolver's stator output voltages are written as:

V_1 = V sinωt sinθ

V_2 = V sinωt cosθ

where θ is the angle of the resolver's rotor. The digital angle φ is applied to the cosine multiplier, and its cosine is multiplied by V_1 to produce the term:

V sinωt sinθ cosφ.

The digital angle φ is also applied to the sine multiplier and multiplied by V_2 to product the term:

V sinωt cosθ sinφ.

Chapter 15

These two signals are subtracted from each other by the error amplifier to yield an AC error signal of the form:

V sinωt [sinθ cosφ – cosθ sinφ]. Using a simple trigonometric identity, this reduces to:

V sinωt [sin (θ –φ)].

The detector synchronously demodulates this AC error signal, using the resolver's rotor voltage as a reference. This results in a DC error signal proportional to sin(θ – φ).

The DC error signal feeds an integrator, the output of which drives a voltage-controlled-oscillator (VCO). The VCO, in turn, causes the up/down counter to count in the proper direction to cause:

sin (θ – φ) → 0.

When this is achieved,

θ – φ → 0,

and therefore

φ = θ

to within one count. Hence, the counter's digital output, φ, represents the angle θ. The latches enable this data to be transferred externally without interrupting the loop's tracking.

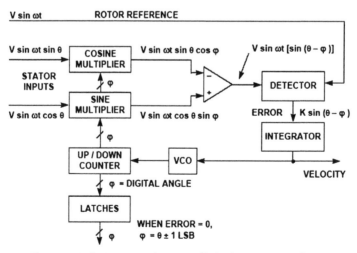

Figure 15.3.14: Resolver-to-digital converter (RTD).

This circuit is equivalent to a so-called type-2 servo loop, because it has, in effect, two integrators. One is the counter, which accumulates pulses; the other is the integrator at the output of the detector. In a type-2 servo loop with a constant rotational velocity input, the output digital word continuously follows, or tracks the input, without needing externally derived convert commands, and with no steady state phase lag between the digital output word and actual shaft angle. An error signal appears only during periods of acceleration or deceleration.

As an added bonus, the tracking RDC provides an analog DC output voltage directly proportional to the shaft's rotational velocity. This is a useful feature if velocity is to be measured or used as a stabilization term in a servo system, and it makes tachometers unnecessary.

Since the operation of an RDC depends only on the ratio between input signal amplitudes, attenuation in the lines connecting them to resolvers doesn't substantially affect performance. For similar reasons, these converters are not greatly susceptible to waveform distortion. In fact, they can operate with as much as 10% harmonic distortion on the input signals; some applications actually use square-wave references with little additional error.

Tracking ADCs are therefore ideally suited to RDCs. While other ADC architectures, such as successive approximation, could be used, the tracking converter is the most accurate and efficient for this application.

Because the tracking converter doubly integrates its error signal, the device offers a high degree of noise immunity (12 dB-per-octave rolloff). The net area under any given noise spike produces an error. However, typical inductively coupled noise spikes have equal positive and negative going waveforms. When integrated, this results in a zero net error signal. The resulting noise immunity, combined with the converter's insensitivity to voltage drops, lets the user locate the converter at a considerable distance from the resolver. Noise rejection is further enhanced by the detector's rejection of any signal not at the reference frequency, such as wideband noise.

The AD2S90 is one of a number of integrated RDCs offered by Analog Devices. Key specifications are shown in Figure 15.3.15. The general architecture is similar to that of Figure 15.3.14. The input signal level should be 2 V rms ± 10% in the frequency range from 3 kHz to 20 kHz.

Chapter 15

- 12-Bit Resolution (1 LSB = 0.08° = 5.3 arc min)
- Inputs: 2 V rms ±10%, 3 kHz to 20 kHz
- Angular Accuracy: 10.6 arc min ±1 LSB
- Maximum Tracking Rate: 375 revolutions per second
- Maximum VCO Clock Rate: 1.536 MHz
- Settling Time:
   - 1° Step: 7 ms
   - 179° Step: 20 ms
- Differential Inputs
- Serial Output Interface
- ±5 V Supplies, 50 mW Power Dissipation
- 20-Pin PLCC

Figure 15.3.15: Performance characteristics for AD2S90 resolver-to-digital converter.

Inductosyns

Synchros and resolvers inherently measure rotary position, but they can make linear position measurements when used with lead screws. An alternative, the Inductosyn™ (registered trademark of Farrand Controls, Inc.) measures linear position directly. In addition, Inductosyns are accurate and rugged, well-suited to severe industrial environments, and do not require ohmic contact.

The linear Inductosyn consists of two magnetically coupled parts; it resembles a multipole resolver in its operation (see Figure 15.3.16). One part, the scale, is fixed (e.g., with epoxy) to one axis, such as a machine tool bed. The other part, the slider, moves along the scale in conjunction with the device to be positioned (for example, the machine tool carrier).

The scale is constructed of a base material such as steel, stainless steel, aluminum, or a tape of spring steel, covered by an insulating layer. Bonded to this is a printed-circuit trace, in the form of a continuous rectangular waveform pattern. The pattern typically has a cyclic pitch of 0.1 inch, 0.2 inch, or 2 millimeters. The slider, about 4 inches long, has two separate but identical printed circuit traces bonded to the surface that faces the scale. These two traces have a waveform pattern with exactly the same cyclic pitch as the waveform on the scale, but one trace is shifted one-quarter of a cycle relative to the other. The slider and the scale remain separated by a small air gap of about 0.007 inch.

Position and Motion Sensors

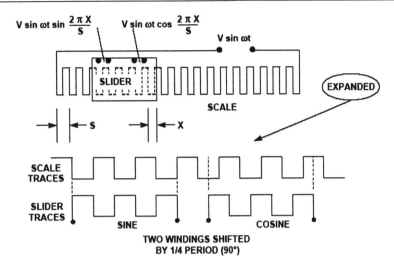

Figure 15.3.16: Linear Inductosyn.

Inductosyn operation resembles that of a resolver. When the scale is energized with a sine wave, this voltage couples to the two slider windings, inducing voltages proportional to the sine and cosine of the slider's spacing within the cyclic pitch of the scale. If S is the distance between pitches, and X is the slider displacement within a pitch, and the scale is energized with a voltage V sinωt, then the slider windings will see terminal voltages of:

V (sine output) = V sinωt sin[2πX/S]

V (cosine output) = V sinωt cos[2πX/S].

As the slider moves the distance of the scale pitch, the voltages produced by the two slider windings are similar to those produced by a resolver rotating through 360°. The absolute orientation of the Inductosyn is determined by counting successive pitches in either direction from an established starting point. Because the Inductosyn consists of a large number of cycles, some form of coarse control is necessary in order to avoid ambiguity. The usual method of providing this is to use a resolver or synchro operated through a rack and pinion or a lead screw.

In contrast to a resolver's highly efficient transformation of 1:1 or 2:1, typical Inductosyns operate with transformation ratios of 100:1. This results in a pair of sinusoidal output signals in the millivolt range which generally require amplification.

Since the slider output signals are derived from an average of several spatial cycles, small errors in conductor spacing have minimal effects. This is an important reason for the Inductosyn's very high accuracy. In combination with 12-bit RDCs, linear Inductosyns readily achieve 25 microinch resolutions.

Rotary inductosyns can be created by printing the scale on a circular rotor and the slider's track pattern on a circular stator. Such rotary devices can achieve very high resolutions. For instance, a typical rotary Inductosyn may have 360 cyclic pitches per rotation, and might use a 12-bit RDC. The converter effectively divides each pitch into 4096 sectors. Multiplying by 360 pitches, the rotary Inductosyn divides the circle into a total of 1,474,560 sectors. This corresponds to an angular resolution of less than 0.9 arc seconds. As in the case of the linear Inductosyn, a means must be provided for counting the individual pitches as the shaft rotates. This may be done with an additional resolver acting as the coarse measurement.

Vector AC Induction Motor Control

Long known for its simplicity of construction, low-cost, high efficiency and long-term dependability, the AC induction motor has been limited by the inability to control its dynamic performance in all but the crudest fashion. This has severely restricted the application of AC induction motors where dynamic control of speed, torque and response to changing load is required. However, recent advances in digital signal processing (DSP) and mixed-signal integrated circuit technology are providing the AC induction motor with performance never before thought possible. Manufacturers anxious to harness the power and economy of vector control can reduce R&D costs and time-to-market for applications ranging from industrial drives to electric automobiles and locomotives with a standard chipset/development system.

It is unlikely that Nikola Tesla (1856–1943), the inventor of the induction motor, could have envisioned that this workhorse of industry could be rejuvenated into a new class of motor that is competitive in most industrial applications.

Before discussing the advantages of vector control it is necessary to have a basic understanding of the fundamental operation of the different types of electric motors in common use.

Until recently, motor applications requiring servo-control tasks such as tuned response to dynamic loads, constant torque and speed control over a wide range were almost exclusively the domain of DC brush and DC permanent magnet synchronous motors. The fundamental reason for this preference was the availability of well understood and proven control schemes. Although easily controlled, DC brush motors suffer from several disadvantages; brushes wear and must be replaced at regular intervals, commutators wear and can be permanently damaged by inadequate brush maintenance, brush/commutator assemblies are a source of particulate contaminants, and the arcing of mechanical commutation can be a serious fire hazard is some environments.

The availability of power inverters capable of controlling high-horsepower motors allowed practical implementation of alternate motor architectures such as the DC permanent magnet synchronous motor (PMSM) in servo control applications. Although eliminating many of the mechanical problems associated with DC brush motors, these motors required more complex control schemes and suffered from several drawbacks of their own. Aside from being costly, DC PMSMs in larger, high-horsepower configurations suffer from high rotor moment-of-inertia as well as limited use in high-speed applications due to mechanical constraints of rotor construction and the need to implement field weakening to exceed baseplate speed.

In the 1960s, advances in control theory, in particular the development of *indirect field-oriented control*, provided the theoretical basis for dynamic control of AC induction motors. Because of the intensive mathematical computations required by indirect field-oriented control, now commonly referred to as *vector control*, practical implementation was not possible for many years. Available hardware could not perform the high-speed precision sensing of rotor position and near real-time computation of dynamic flux vectors. The current availability of precision optical encoders, isolated gate bipolar transistors (IGBTs), high-speed resolver-to-digital converters and high-speed digital signal processors (DSPs) has pushed vector control to the forefront of motor development due to the advantages inherent in the AC induction motor.

A simplified block diagram of an AC induction motor control system is shown in Figure 15.3.17. In this example, a single-chip IC (ADMC300, ADMC330, or ADMC331) performs the control functions. The inputs to the controller chip are the motor currents (normally three-phase) and the motor rotor position and velocity. Hall-effect sensors are often used to monitor the currents, and a resolver and an RDC monitor the rotor position and velocity. The DSP is used to perform the real time vector-type calculations necessary to generate the control outputs to the inverter processors. The transformations required for vector control are also accomplished with the DSP.

The ADMC300 comprises a high performance, 5-channel 16-bit ADC system, a 12-bit 3-phase PWM generation unit, and a flexible encoder interface for position sensor feedback. The ADMC330 includes a 7-channel 12-bit ADC system and a 12-bit 3-phase PWM generator. The ADMC331 includes a 7-channel 12-bit ADC system, and a programmable 16-bit 3-phase PWM generator. It also has additional power factor correction control capabilities. All devices have on-chip DSPs (approximately

20 MHz) based on Analog Device's Modified Harvard Architecture 16-bit DSP core. Third-party DSP software and reference designs are available to facilitate motor control system development using these chips.

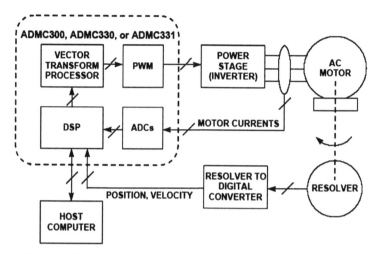

Figure 15.3.17: AC induction motor control application.

Accelerometers

Accelerometers are widely used to measure tilt, inertial forces, shock, and vibration. They find wide usage in automotive, medical, industrial control, and other applications. Modern micromachining techniques allow these accelerometers to be manufactured on CMOS processes at low cost with high reliability. Analog Devices iMEMS® (Integrated Micro Electro Mechanical Systems) accelerometers represent a breakthrough in this technology. A significant advantage of this type of accelerometer over piezoelectric-type charge-output accelerometers is that DC acceleration can be measured (e.g., they can be used in tilt measurements where the acceleration is a constant 1g).

The basic unit cell sensor building block for these accelerometers is shown in Figure 15.3.19. The surface micromachined sensor element is made by depositing polysilicon on a sacrificial oxide layer that is then etched away leaving the suspended sensor element. The actual sensor has tens of unit cells for sensing acceleration, but the diagram shows only one cell for clarity. The electrical basis of the sensor is the differential capacitor (CS1 and CS2) which is formed by a center plate which is part of the moving beam and two fixed outer plates. The two capacitors are equal at rest (no applied acceleration). When acceleration is applied, the mass of the beam causes

Position and Motion Sensors

it to move closer to one of the fixed plates while moving further from the other. This change in differential capacitance forms the electrical basis for the conditioning electronics shown in Figure 15.3.20.

- Tilt or Inclination
- Car Alarms
- Patient Monitors
- Inertial Forces
- Laptop Computer Disc Drive Protection
- Airbag Crash Sensors
- Car Navigation systems
- Elevator Controls
- Shock or Vibration
- Machine Monitoring
- Control of Shaker Tables
- ADI Accelerometer Fullscale g-Range: ±2g to ±100g
- ADI Accelerometer Frequency Range: DC to 1 kHz

Figure 15.3.18: Accelerometer applications.

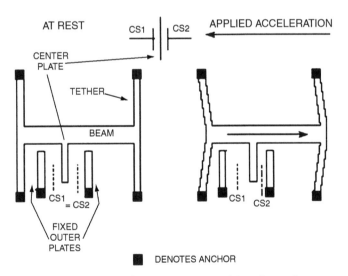

Figure 15.3.19: ADXL-family micromachined accelerometers. (Top view of IC.)

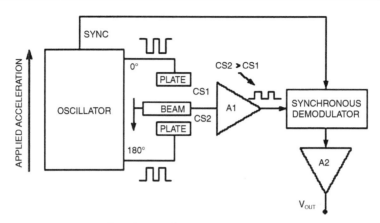

Figure 15.3.20: ADXL-family accelerometers internal signal conditioning.

The sensor's fixed capacitor plates are driven differentially by a 1 MHz square wave: the two square wave amplitudes are equal but are 180° out of phase. When at rest, the values of the two capacitors are the same, and therefore the voltage output at their electrical center (i.e., at the center plate attached to the movable beam) is zero. When the beam begins to move, a mismatch in the capacitance produces an output signal at the center plate. The output amplitude will increase with the acceleration experienced by the sensor. The center plate is buffered by A1 and applied to a synchronous demodulator. The direction of beam motion affects the phase of the signal, and synchronous demodulation is therefore used to extract the amplitude information. The synchronous demodulator output is amplified by A2 which supplies the acceleration output voltage, V_{OUT}.

An interesting application of low-g accelerometers is measuring tilt. Figure 15.3.21 shows the response of an accelerometer to tilt. The accelerometer output on the diagram has been normalized to 1g fullscale. The accelerometer output is proportional to the sine of the tilt angle with respect to the horizon. Note that maximum sensitivity occurs when the accelerometer axis is perpendicular to the acceleration. This scheme allows tilt angles from −90° to +90° (180° of rotation) to be measured. However, in order to measure a full 360° rotation, a dual-axis accelerometer must be used.

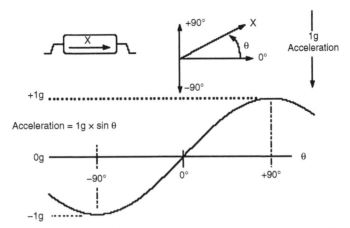

Figure 15.3.21: Using an accelerometer to measure tilt.

References

1. Herman Schaevitz, *The Linear Variable Differential Transformer,* **Proceedings of the SASE**, Volume IV, No. 2, 1946.

2. Dr. Ernest D.D. Schmidt, *Linear Displacement – Linear Variable Differential Transformers – LVDTs*, Schaevitz Sensors, http://www.schaevitz.com.

3. E-Series LVDT Data Sheet, Schaevitz Sensors, http://www.schaevitz.com. Schaevitz Sensors is now a division of Lucas Control Systems, 1000 Lucas Way, Hampton, VA 23666.

4. Ramon Pallas-Areny and John G. Webster, **Sensors and Signal Conditioning**, John Wiley, New York, 1991.

5. Harry L. Trietley, **Transducers in Mechanical and Electronic Design**, Marcel Dekker, Inc., 1986.

6. AD598 and AD698 Data Sheet, Analog Devices, Inc., http://www.analog.com.

7. Bill Travis, *Hall-Effect Sensor ICs Sport Magnetic Personalities*, EDN, April 9, 1998, pp. 81–91.

8. AD22151 Data Sheet, Analog Devices, Inc., http://www.analog.com.

9. Dan Sheingold, **Analog-Digital Conversion Handbook, Third Edition**, Prentice-Hall, 1986.

10. F. P. Flett, *Vector Control Using a Single Vector Rotation Semiconductor for Induction and Permanent Magnet Motors*, **PCIM Conference, Intelligent Motion, September 1992 Proceedings**, available from Analog Devices.

11. F. P. Flett, *Silicon Control Algorithms for Brushless Permanent Magnet Synchronous Machines*, **PCIM Conference, Intelligent Motion, June 1991 Proceedings**, available from Analog Devices.

12. P.J.M. Coussens, et al, *Three Phase Measurements with Vector Rotation Blocks in Mains and Motion Control*, **PCIM Conference, Intelligent Motion, April 1992 Proceedings**, available from Analog Devices.

13. Dennis Fu, *Digital to Synchro and Resolver Conversion with the AC Vector Processor AD2S100*, available from Analog Devices.

14. Dennis Fu, *Circuit Applications of the AD2S90 Resolver-to-Digital Converter*, AN-230, Analog Devices.

15. Aengus Murray and P. Kettle, *Towards a Single Chip DSP Based Motor Control Solution*, **Proceedings PCIM – Intelligent Motion**, May 1996, Nurnberg Germany, pp. 315–326. Also available at http://www.analog.com.

16. D. J. Lucey, P. J. Roche, M. B. Harrington, and J. R. Scannell, *Comparison of Various Space Vector Modulation Strategies*, **Proceedings Irish DSP and Control Colloquium**, July 1994, Dublin, Ireland, pp. 169–175.

17. Niall Lyne, *ADCs Lend Flexibility to Vector Motor Control Applications*, **Electronic Design**, May 1, 1998, pp. 93–100.

18. Frank Goodenough, *Airbags Boom when IC Accelerometer Sees 50g*, **Electronic Design**, August 8, 1991.

15.4 Selecting Position and Displacement Transducers
Tom Anderson, SpaceAge Control, Inc.

As an application development manager for a position transducer supplier, I receive numerous queries on how to solve a broad range of position-measurement challenges.

These inquiries run the gamut from the common (aircraft flight-control surface movement) to the exotic (Formula One racecar suspension travel) to the seemingly impossible (three-dimensional tracking of a golf ball in flight from a fixed position).

These position-measurement challenges usually share one common element. They can be solved using a variety of solutions, but it's not always easy to determine the best one.

There are possibly more options for measuring position than any other type of sensed variable. While there may be more suppliers for pressure transducers, the variety of position transducer types and technologies is unmatched.

The 1997 *Thomas Register* lists 264 suppliers of pressure transducers and 229 suppliers of displacement and position transducers. However, there are 13 categories related to displacement and position measurement, compared to just four categories for pressure measurement.

In this chapter, I introduce you to various position-transducer selection parameters. You'll also find information on position-measurement techniques, technologies, and choices.

Basic Terminology

A brief note on semantics: for ease of communication, this guide refers to transducers and sensors as being the same. While not strictly true, is generally irrelevant whether you are using a position sensor or transducer. The purpose of both is the same—to find out where something is!

Transducers covered here provide position, displacement, and proximity measurements, which are defined as:

- *position* – location of the object's coordinates with respect to a selected reference
- *displacement* – movement from one position to another for a specific distance or angle
- *proximity* – a critical distance signaled by an on/off output

Chapter 15

In this chapter, I focus primarily on transducers for position and displacement measurement. Unless otherwise noted, I use the term "position transducer" to refer to displacement and proximity transducers as well.

The Parameters

On what basis should you select a position transducer? As a starting point, let's look at the laundry list of parameters shown in Figure 15.4.1. While this list is not all-inclusive, it helps you begin to decide what parameters are relevant to your application.

Perhaps the first parameter to address in any application is whether the transducer can physically touch the object being monitored. If your application is sensitive to outside influences, a noncontact transducer may be the most appropriate. Otherwise, a contact sensor might offer advantages not found in a noncontact sensor.

Parameter	Relevant?	Ranking	Choices		
Contact	❏ Yes ❏ No		❏ Contact	❏ Noncontact	
Motion Type	❏ Yes ❏ No		❏ Linear	❏ Rotary	
Dimensions	❏ Yes ❏ No		❏ One-dimensional	❏ Multidimensional	
Measurement Type	❏ Yes ❏ No		❏ Absolute	❏ Incremental	❏ Threshold (Proximity)
Range	❏ Yes ❏ No		❏ Less than 1"	❏ 1–30"	❏ Greater than 30"
Physical Size/Weight	❏ Yes ❏ No		❏ Size Restriction____	❏ Weight Restriction____	
Environmental	❏ Yes ❏ No		❏ Humidity ❏ Moisture	❏ Vibration ❏ Temperature	❏ Corrosion ❏ Other____
Installation/Mounting	❏ Yes ❏ No		❏ Removable	❏ Installation	❏ Time Limit____
Accuracy	❏ Yes ❏ No		❏ Linearity ❏ Hysteresis	❏ Resolution	❏ Repeatability
Lifetime	❏ Yes ❏ No		❏ Cycles____	❏ Hours of Continuous Operation____	
Cost	❏ Yes ❏ No		❏ Less than $50	❏ $50–$500	❏ Greater than $500
Delivery	❏ Yes ❏ No		❏ Less than 1 Week	❏ 1–4 Weeks	❏ Greater than 4 Weeks
Output	❏ Yes ❏ No		❏ Analog Voltage ❏ Sensor Bus____	❏ Analog Current ❏ Visual	❏ Digital ❏ Other____
Frequency Response	❏ Yes ❏ No		❏ Less than 5 Hz	❏ 5–50 Hz	❏ Greater than 50 Hz

Figure 15.4.1: What are your requirements?

At first thought, noncontact transducers may seem like the superior solution for all applications. However, the decision isn't that clear cut. Noncontact products can emit potentially harmful laser- or ultrasonic-based signals. These products also rely on having a clear visual environment to operate in. Frequency response isn't always as high as with a contact sensor, but costs are often higher. Finally, operating-temperature ranges are typically not as broad.

Another parameter to consider early on is whether you need to measure linear or rotary movement. Note that using cable position transducers (like the one shown in Figure 15.4.2), cams, pulleys, levers, electronics, software, and other methods can enable a rotary transducer to measure linear motion, and vice versa. Lack of space, cost, and ease of mounting are a few reasons for doing this.

Figure 15.4.2: Cable position transducers provide extended ranges in small sizes.

Once you decide if you require a contact or noncontact solution and are measuring rotary or linear movement, selecting a transducer technology becomes much easier.

Next, determine if you're monitoring one-dimensional or multidimensional motion. If the motion is multidimensional, determine whether you need to measure in multiple dimensions or if the object is moving in multiple dimensions and you only have to measure one of them.

Often, multidimensional motion is measured with multiple one-dimensional transducers.

Also, think about the type of signal you need to obtain. If you need a signal that specifies a unique position, be sure to specify a transducer with absolute output.

However, if all you need is relative position from a prior position or a simple on/off indicator, then incremental or threshold technology is more appropriate. Figure 15.4.3 gives you a view of some incremental rotary optical encoders.

An important difference between incremental and absolute transducers is that incremental transducers typically need to be reinitialized after powerdown by moving the monitored object to a home position at powerup. This limitation is unacceptable in some applications.

Figure 15.4.3: Incremental rotary optical encoders provide quadrature digital output.

Threshold measurements are on/off in nature and usually involve limit switches or similar devices. As you might guess, absolute devices are usually more expensive than incremental or threshold devices.

Travel, also known as range, varies from microns to hundreds of feet (or more, depending on your definition of transducer). The range of many precision transducers is limited to 10 inches or less.

If your application needs to operate on the Space Station or some other size- and weight-sensitive platform, you need to specify the maximum values for the transducer's dimensions and weight.

The application's operating environment can have a large impact on your technology choice as well. You need to determine what operating and storage temperatures the device will be in and whether you need to meet commercial, industrial, or military environmental requirements.

Also consider whether excessive humidity, moisture, shock, vibration, or EMF will be encountered. Determine whether your environment has other unique aspects, such as high or low pressure or the presence of hazardous or corrosive chemicals.

An often-overlooked parameter is the method and time required for transducer installation and mounting. For testing applications, this parameter may not be so important. However, OEM and large-volume applications often require simple installation and removal to reduce labor costs and enable easy maintenance. See if the transducer can only be mounted with manufacturer-provided special mounting bases or if a variety of mounting techniques can be used. Besides the common threaded fastener approach, some other nonpermanent mounting techniques include suction cups, magnets, industrial adhesives, grooved fittings, and clamping.

In going through the previous parameters, you might have asked yourself, "Hey, what about accuracy?" While accuracy is certainly important and sometimes critical, it's often the last degree of freedom in the selection of a transducer. As you may know from experience, accuracy is not a well-agreed-on term. Typically, various components of accuracy, linearity, repeatability, resolution, and hysteresis are quoted for vendor convenience or per user requirements.

With the availability of software calibration tools today, linearity isn't as important as it once was. For many applications, in fact, repeatability is the most important component.

Accuracy is typically specified in absolute units like mils or microns or in relative units such as percent of full-scale measurement. If you are comparing the accuracy of one device against another, make sure you are comparing apples to apples. For example, see if the accuracies being quoted are at a single temperature or over a temperature range. If you need it, find out if temperature compensation is available.

If you expect to see significant numbers of cycles or if the transducer will be in service for an extended period of time, specify the lifetime and reliability requirements as well. When choosing the transducer, find out what warranties are offered as well as how maintenance and repairs are handled.

A transducer that can be repaired in-house can reduce costs significantly. You should also consider what type of periodic recalibration is recommended and whether calibration procedures are provided.

It's a good idea to ask vendors what type of use their transducers see most often. Common uses include OEM, retrofit, industrial control, commercial, and test and measurement. Hopefully, the transducer has seen previous use in your type of application.

In the early stages of transducer specification, product cost sometimes doesn't even make the list. More often than not, this parameter gains importance as the project moves forward.

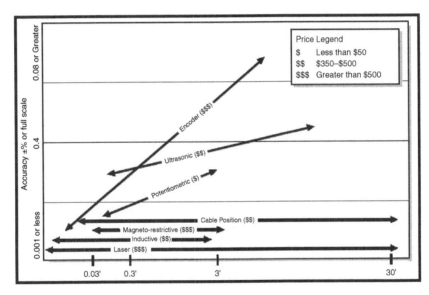

Figure 15.4.4: Selection tradeoffs; typical performance of linear position transducers.

When determining costs, make sure to look at the initial acquisition cost as well as the cost over the product's life. For example, are special signal conditioning electronics, power supplies, electrical connectors, housings, installation tools, or mounting fixtures required? Ask the vendor for typical repair, maintenance, and replacement costs. And, inquire about the cost of the transducer in volume and single-unit quantities. The cost savings (e.g., a cost of $100 in volume but $600 in single quantities) may be an important factor if small-quantity replacement units will be needed in the future.

Another parameter that's occasionally overlooked is the time it takes the product to be delivered to you after you order it. The custom nature of some transducers combined with production processes and manufacturing economics requires lead times of eight weeks or more. This delivery schedule might be acceptable now, but what about in six months when you need extra quantities or a spare part? Evaluate whether or not you can afford to be without a part for an extended period of time.

Obviously, the transducer is going to be a part of a system. So, determine your preferred electrical input and output requirements. Common output choices include analog AC and DC voltage, resistive, current (4–20 mA), digital, and visual (meter).

Increasingly, outputs using sensor bus protocols are being offered. Most position transducers require 50 V or less, and some are self-powered.

Finally, for fast-moving applications, determine the maximum velocity or acceleration that needs to be monitored. Ensure that your data acquisition or control system has an adequate sampling rate to record the resulting data stream.

Check Your Requirements

Now that you're aware of the key parameters, you need to determine which ones are relevant to your application and of these relevant parameters, which are most critical. If you don't prioritize your requirements, it's going to be difficult to make a selection decision. You may come to the conclusion that there is no transducer that can meet your needs. This may be true, but it's more likely that your requirements are too stringent and that you need to make a tradeoff to arrive at the optimum selection.

For example, an engineer recently approached our company looking for a transducer with ±0.0001 inch resolution over 30 inches, and he wanted to keep the cost under $500. He was adamant that all three specifications be met. Our products didn't meet all of his specifications, and we were at a loss as to where we would refer him. After some more discussion, we found out that the resolution requirement was only necessary over a limited portion of the total range and that the cost goal, while important, did have some flexibility.

Hence, in this situation, range was most important, followed by resolution, and then cost.

The moral of this story: focus on your top requirements. Make the best decision you can, given the specifications you need. And keep in mind that you can't have everything, unfortunately.

Next Steps

In this chapter, I've given you some parameters for selecting position transducers. But in case you hadn't noticed, I didn't provide any information on what type of technology you should select for your position transducer. The constant change in transducer technology and the difficulty in generalizing about a particular technology's capabilities and limitations mean there's no way I can cover this area in detail here. Refer to the previous sections of this chapter for more details on various technologies.

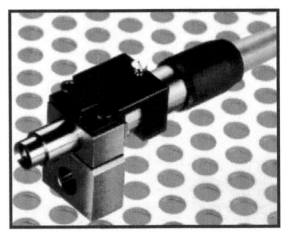

Additionally, choosing the technology should come after determining and prioritizing your requirements. Once your requirements are well known, the choice of technology tends to be self-selecting. For example, just knowing whether you require a contact or noncontact technology can cut your choices almost in half. If you need the latter, a laser position sensor like the one in Figure 15.4.5 may be a good choice.

Figure 15.4.5: Laser position sensors have resolutions of 0.1 μm or better.

To get a feel for the capabilities of some of the more prevalent linear position-measurement technologies, Figure 15.4.4 maps out how these technologies compare against each other based on cost, accuracy, and maximum range. Note that not all technologies are shown.

It may be difficult to clearly define the parameter values you require as well as which parameters are most important in your application. However, it can be even more difficult to obtain these parameters from vendors and then compare one vendor's statements against another's. To get information on products beyond what you see in the vendor's product literature, review transducer-related publications such as *Measurement and Control* and *Sensors* for articles on position-measurement products and technologies.

Also, be sure to ask your colleagues about their experiences and recommendations. They may have a position transducer on hand that you may be able to test for your application.

Of course, in this day and age, make an effort to search Web engines and Internet newsgroups. Numerous engineering, instrumentation, and measurement-oriented newsgroups can be reached via search engines. Extensive sources of position-transducer manufacturers can be found in the *Thomas Register* and the *Sensors Buyer's Guide*.

Contact vendors and request references of similar applications. Ask these references why they selected the product they did and whether they're happy with their decision. Also, find out what other options they considered.

Finally, ask the vendor for product samples or evaluation units that you can use for testing before purchase. If the vendor is hesitant to do this, offer to provide them with a test report summarizing your evaluation. This information may be valuable to them, and they may be more willing to assist you.

Photos 15.4.2, 15.4.3, and 15.4.5 are courtesy of Space Age Control, Oak Grigsby, and Dynamic Control Systems, respectively.

References

[1] J. Fraden, *AIP Handbook of Modern Sensors*, American Institute of Physics, New York, NY, p. 264, 1993, 1996.

Resources

Texts

Schaevitz Engineering, *Handbook of Measurement and Control*, Pennsauken, NJ, 1976.

I. Busch-Vishniac, *Electromechanical Sensors and Actuators*, Springer-Verlag, New York, NY, 1998.

Thomas Register Directory of American Manufacturers, Thomas Publishing Co., New York, NY, 1997.

Internet

Sensors Buyer's Guide, www.sensorsmag.com.

Thomas Register, www.thomasregister.com.

Sources

Sensors

Dynamic Control Systems 7088 Venture St., Ste. 205 Delta, BC Canada V4G 1H5 (604) 940-0141 Fax: (604) 940-0793, www.dynavision.com.

MicroStrain, Inc. 294 N. Winooski Ave. Burlington, VT 05401 (802) 862-6629 Fax: (802) 863-4093, www.microstrain.com.

Midori America 2555 E. Chapman Ave., Ste. 400 Fullerton, CA 92831 (714) 449-0997 Fax: (714) 449-0139, www.thomasregister.com/midori.

OakGrigsby, Inc. 84 N. Dugan Rd. Sugar Grove, IL 60554 (630) 556-4200 Fax: (630) 556-4216 www.oakgrigsby.com.

Senix Corp. 52 Maple St. Bristol, VT 05443 (802) 453-5522 Fax: (802) 453-2549 www.senix.com.

SpaceAge Control, Inc. 38850 20th St. E Palmdale, CA 93550 (661) 273-3000 Fax: (661) 273-4240, www.spaceagecontrol.com.

CHAPTER 16

Pressure Sensors

16.1 Piezoresistive Pressure Sensing

Glenn Harman, Global Product Leader, Honeywell Sensing and Control

Pressure sensors convert input pressures to electrical outputs to measure pressure, force and airflow. These measurements are used to control everything from the water level in your washing machine to the gases emitted by your car's exhaust system. Pressure sensors are used in medical equipment to monitor blood pressure, regulate intravenous infusions, and to detect such things as changes in cranial pressure, hearing problems and glaucoma. People in the manufacturing and process industries rely on pressure sensors to control their machinery and processes. They are essential to the operation of HVAC systems, forklifts, and earth-moving equipment. They measure altitude and turbidity on aircraft and are an important feature of the flight data recorders required on all commercial flights.

Wherever pressure, force or airflow needs to be precisely controlled, there is a potential pressure sensing application. Today's pressure sensors provide a high degree of repeatability, low hysteresis, and long-term stability in applications with input pressures ranging from less than one pound per square inch gauge (psig) to thousands of psig.

Fundamentals of Pressure Sensing Technology

Most pressure, force and airflow sensors are fabricated using silicon-processing techniques common in the semiconductor industry. Therefore, much of the same terminology used in the semiconductor industry also applies to pressure sensor technology. Piezoresistive ion implanted semiconductor technology dominates the component market for pressure sensors for many good reasons. Other approaches, including variable reluctance, variable capacitance, fiber optic, and piezoelectric, are available for niche applications; however, those technologies are not covered in this chapter.

Piezoresistive pressure sensors (strain gage sensors) are often referred to as IC (integrated circuit) sensors, solid-state sensors, monolithic sensors (formed from single-crystal silicon) or just silicon sensors. They are processed in wafer form, where each wafer will contain a few hundred to a few thousand sensor die, depending on the size of the sensor die. A typical sensor chip measures 80 × 80 mils or 2 mm × 2 mm.

Piezoresistive (silicon) pressure sensors contain a sensing element made up of a silicon chip with a thin, circular silicon diaphragm and four piezoresistors. These nearly identical solid-state resistors are buried in the surface of the silicon.

The piezoresistance of a semiconductor refers to the change in resistance caused by strain when pressure or force is applied to the diaphragm. Pressure causes the diaphragm to flex, inducing a stress on the diaphragm and also on the buried resistors.

The resistor values change depending on the amount of pressure applied to the diaphragm. Therefore, a change in pressure (mechanical input) is converted to a change in resistance (electrical output). The sensing element converts or transduces the energy from one form to another, hence the term "pressure transducer."

Pressure sensors are produced first by ion implanting the four piezoresistors into the silicon. Ion implantation is used increasingly to provide improved performance over sensors produced by diffusion.

After the four piezoresistors are formed, the diaphragm is created by chemically etching a controlled shape in the silicon from its backside (on the surface opposite the piezoresistors). The unetched portion of the silicon slice provides a rigid boundary constraint for the diaphragm and a surface for mounting it to some other member.

The thickness of the diaphragm determines the pressure range (sensitivity) of the sensor. However, this relationship is not a linear function. For example, doubling the thickness of the diaphragm decreases the sensitivity by a factor of four. Typical diaphragm thicknesses are 5 to 200 microns (pretty thin stuff), depending on their pressure range. Overpressure is a term used to specify the maximum pressure that may be applied to a sensor's sensing element without causing a permanent change in its output characteristics.

The high sensitivity or gage factor of silicon strain gages is approximately 100 times that of metal strain gages. By implanting the piezoresistors into a homogenous single crystalline silicon medium, they are integrated into the silicon

Pressure Sensors

force-sensing element. Typically, other types of strain gages are bonded to force sensing members of dissimilar material, resulting in thermoelastic strain and complex fabrication processes. Most discrete strain gages are inherently unstable due to bond degradation, temperature sensitivity, and hysteresis caused by thermoelastic strain. Silicon diaphragm pressure sensors are extremely reliable because silicon is an ideal material for receiving the applied force, and the implanted gages are not subject to bonding problems.

As a perfect crystal, silicon does not become permanently stretched but returns to its original shape. Silicon wafers are better than metal for pressure sensing diaphragms because silicon offers extreme elasticity within its operating range. Silicon diaphragms normally fail only by rupturing, usually due to extreme overpressure. Micromachining and laser trimming help manufacturers produce reliable sensors capable of extreme accuracy.

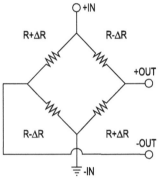

The sensor's resistors can be connected in either a half-bridge or a full "Wheatstone bridge" arrangement, whereby two resistors increase with positive pressure while the other two decrease in resistance. When pressure is applied to the device as shown in Figure 16.1.1, the resistors in the arms of the bridge change by an amount, ΔR. The alignment of the resistor on the silicon determines if the resistor will increase or decrease with applied pressure.

Figure 16.1.1

The resulting differential output voltage VO, is easily shown to be $VO = VB \times \Delta R/R$. Since the change in resistance is directly proportional to pressure, VO can be written as: $VO = (S \times P \times VB) \pm VOS$ where:

VO is the output voltage in mV

S is the sensitivity in mV/V per psi.

P is the pressure in psi.

VB is the bridge voltage in volts.

VOS is the offset error (the differential output voltage when the applied pressure is zero).

The differential output of a "raw" pressure sensor is, however, not precise in terms of calibration and temperature effects. It is partially because of this that sensor manufacturers offer a variety of levels of signal-conditioned sensors from their basic raw state, up through fully calibrated and compensated transmitters with amplified outputs and application-specific integrated circuits (ASIC).

Types of Pressure Measurements

Pressure sensors are categorized, in part, by the type of pressure they measure. Most sensors are offered in a series of products with different listings designed to measure different types of pressure: gauge, differential, absolute, or vacuum gauge.

Most people are accustomed to dealing in gauge pressure—that is, pressure relative to the normal atmospheric pressure that surrounds us. As such, absolute pressure and absolute pressure sensors, which measure pressure relative to a perfect vacuum, can be confusing. Since the zero absolute pressure of a perfect vacuum is impractical to achieve, absolute pressure is much more difficult to measure. It is easier to understand absolute pressure once you have a clear understanding of the more familiar differential and gauge pressures.

Differential pressure is the difference in pressure between two pressure sources—for instance, measuring two sources of pressure to determine the status of a filter. This type of pressure measurement is usually expressed in pounds per square inch differential (psid).

Differential pressure sensors are designed to simultaneously accept two independent pressure sources, so they have two pressure ports. The output is proportional to the pressure difference between the two sources, so we can use this output to determine if the filter needs to be cleaned or replaced. Bidirectional differential pressure sensors are differential pressure sensors that allow the greater input pressure to be applied to either pressure port.

In the differential pressure sensor shown in Figure 16.1.2, measurands (the physical parameters being quantified by the measurements) are applied to both ports.

When one of these two pressure sources is ambient pressure, this is then called gauge pressure. Therefore, gauge pressure is a form of differential pressure measurement in which atmospheric pressure is used as the reference. Measurement of auto tire pressure, where a pressure above atmospheric pressure is needed to maintain tire performance characteristics, is an example. With a gauge device, as shown in Figure 16.1.3, the P1 port is vented to atmospheric pressure, and the measurand is applied to the P2 port.

Pressure Sensors

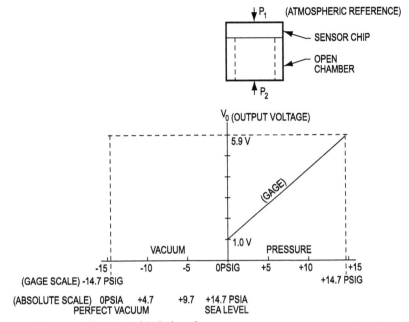

Figure 16.1.3: A high-level gauge pressure sensor output is shown. One-volt output represents ambient pressure.

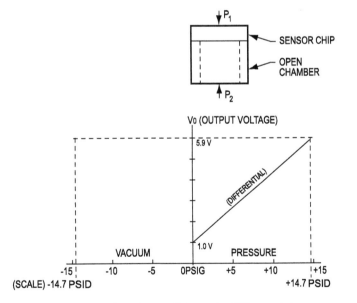

Figure 16.1.2: A signal-conditioned, differential pressure sensor output is shown. One-volt output occurs when pressures are equal on both ports.

Chapter 16

Absolute pressure is measured with respect to a vacuum reference. Absolute pressure sensors are most commonly used to measure changes in barometric pressure or as altimeters. These applications require reference to a fixed pressure, as they cannot simply be referenced to the surrounding ambient pressure.

Absolute pressure sensors must measure input pressure in relation to zero pressure (a total vacuum on one side of the diaphragm). For example, 10 pounds per square inch absolute (psia) would be 10 psi above a perfect vacuum. This is roughly 4.7 psi below the standard atmospheric pressure at sea level of 14.7 psia. Zero psia is then the pressure of a perfect vacuum.

In the absolute pressure sensor shown in Figure 16.1.4, the P2 port is sealed with a vacuum representing a fixed reference. The difference in pressure between the vacuum reference and the measurand applied at the P1 port causes the deflection of the diaphragm, producing the output voltage change.

Note: For illustrative purposes, pounds per square inch (psi) is used as the unit of pressure measure. This unit can obviously be converted to other common pressure units such as in Hg, kPa, bar, and so forth.

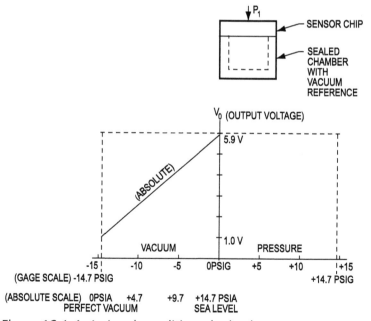

Figure 16.1.4: A signal-conditioned, absolute pressure sensor output is shown. One-volt output represents a perfect vacuum.

Pressure Sensors

Absolute pressure sensors can be made by hermetically sealing a vacuum reference chamber to one side of the sensing element. (See Figure 16.1.5.) Pressures to be measured are then measured relative to this vacuum reference. The actual "vacuum," which is sealed into the sensor, is approximately 0.0005 psia (25 millitorr). If gas becomes trapped in the reference chamber, it will exert pressure during expansion and contraction with temperature in accordance with Boyle's law, so this near absolute vacuum is used as a reference to eliminate any potential thermal errors. One of the advantages of the integrated circuits of piezoresistive sensors is the small volume of trapped vacuum reference, which, in conjunction with a reliable silicon-to-silicon hermetic seal, makes these devices time and temperature stable.

Finally, vacuum gauge pressure is a form of pressure measurement in which vacuum pressures are sensed with reference to ambient pressure. Figure 16.1.6 shows an output from a vacuum gauge sensor.

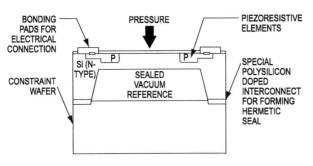

Figure 16.1.5: An absolute pressure sensor with a hermetically sealed vacuum reference chamber on one side of the sensing element.

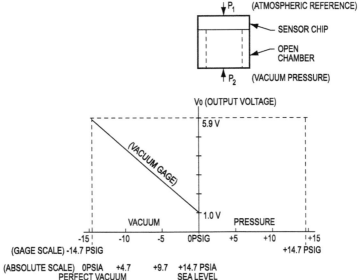

Figure 16.1.6: A high-level vacuum gauge sensor output is shown. One-volt output represents ambient pressure.

Chapter 16

Selecting and Specifying Pressure Sensors

With the type or types of pressure you need to measure in mind, you can begin to narrow your search for the right sensors by considering the performance parameters that are important in your specific application.

Sensor manufacturers want to insure the sensors you purchase are compatible with your application, and most suppliers offer some type of application data sheet to help you gather the information needed to specify the right sensors for your application. For example, point your browser to http://content.honeywell.com/sensing/prodinfo/pressure/technical/c15_7.pdf for a form listing the most relevant specifications for selecting pressure sensors, including pressure, accuracy, electrical packaging and environmental requirements.

What pressure will the transducer measure?

Your first consideration is the maximum pressure of your system. In general, you want to use a transducer that has a maximum pressure range of at least 1.5 times the maximum pressure expected in your system. This extra capacity is advised because many systems, especially hydraulic and process control systems, have pressure spikes or continuous pulsations. The spikes can be five to ten times over the "maximum" pressure. These high-pressure, short-duration spikes can destroy a pressure transducer. Continuous high-pressure pulsations, near or slightly above a transducer's rated maximum, can also limit the life of the transducer. If pulsation frequencies approach the natural (resonance) frequency of the transducer, even low-amplitude pulsations can cause resonance excitation and damage the transducer. However, specifying a higher range transducer is not always a solution because you will sacrifice resolution. You can use a snubber to reduce the spikes, but this is also a trade-off in that it slows the response time of the transducer.

In a sensor's specifications, Span or full-scale output is the algebraic difference between the upper and lower limits of the sensor's pressure range (the difference between the output curve endpoints). Normally these endpoints are null and full scale. Null offset (zero measurand output or ZMO) is the electrical output that occurs when the pressures or forces on both sides of the sensor's diaphragm are equal. "Null" refers to the equal pressure and "null offset" to the output you get when the pressure is at null.

Most manufacturers can tell you how many cycles their products are designed to withstand without loss of performance. For example, Honeywell designs to more than 200 million full-pressure cycles. Consider the system pressure carefully when choosing a transducer. You will be trading off system accuracy versus life of the transducer.

What is the pressure medium?

Another key factor you must consider when selecting a transducer is the medium it will measure. Is it a viscous liquid or a slurry that could plug a pressure port? Is it a solvent or corrosive that could attack the transducer's materials in contact with it, or is it clean, dry air?

These questions will determine whether you need a flush diaphragm device and what materials can be in contact with the medium. Some models have flush diaphragms and others have pressure ports. You can specify stainless steel diaphragms when there is contact with the medium to reduce problems caused by corrosive media. (See Figure 16.1.7.)

Figure 16.1.7: For sensing in harsh or wet environments, you can specify a stainless steel pressure cavity to protect performance.

How accurate does the transducer need to be?

Accuracy is a general term used by many transducer manufacturers to describe the measurement error or uncertainty in the transducer's output. Sources of these errors may include nonlinearity, hysteresis, nonrepeatability, temperature, zero balance, calibration and humidity effects. Most manufacturers specify "accuracy" as the combined effects of nonlinearity, nonrepeatability, and hysteresis. Other errors may be specified separately.

Linearity error is the deviation of the sensor output curve from a specified straight line over a desired pressure range (the degree to which the output of a linear device deviates from ideal performance). It is usually expressed as a percent of full-scale output. One method of computing linearity error is least squares, which mathematically provides a best fit straight line (B.F.S.L.) to the data points (See Figure 16.1.8.) When selecting a pressure transducer, the user must be careful to define the method used to calculate linearity error.

Another method of defining linearity is terminal base linearity (T.B.L.) or end point linearity. (See Figure 16.1.9.) T.B.L. is determined by drawing a straight line (L1) between the end data points on the output curve. Next draw a perpendicular line from line L1 to a data point on the output curve. The data point is chosen to achieve the maximum length of the perpendicular line. The length of the perpendicular line represents terminal base linearity error. T.B.L. is approximately twice the magnitude of B.F.S.L.

Chapter 16

Repeatability error is the deviation in output readings for successive applications of any given input pressure with other conditions remaining constant. (Figure 16.1.10).

Hysteresis error is usually expressed as a combination of mechanical hysteresis and temperature hysteresis. Figure 16.1.11 expresses hysteresis as a combination of the two effects.

Mechanical hysteresis is the output deviation at a certain input pressure when that input is approached first with increasing pressure and then with decreasing pressure.

For many transducers, "accuracy" errors are less than those due to zero balance or temperature. Higher accuracy transducers generally cost more. Does your system truly need this higher accuracy? Using a high-accuracy transducer with a low-resolution measurement instrument creates an ineffective solution.

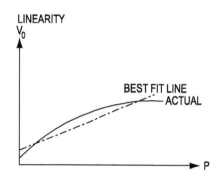

Figure 16.1.8: Best fit straight line linearity.

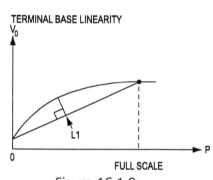

Figure 16.1.9: Terminal base linearity.

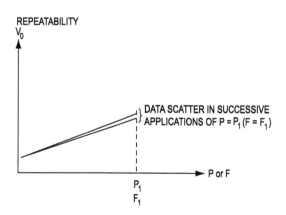

Figure 16.1.10: Repeatability error.

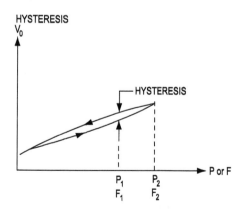

Figure 16.1.11: Hysteresis.

Pressure Sensors

What temperatures will the transducer see?

Pressure transducers, like all physical measurement systems, are subject to error or failure from extreme temperatures. There are usually two temperature ranges specified for a transducer. They are the operating and compensated ranges. The compensated range is a subset of the operating range.

The operating temperature range is that temperature range over which the transducer can be exposed while energized and not suffer damage. However, it will not meet its published performance specifications (temperature coefficients) when subjected to temperatures outside the compensated temperature range.

The compensated temperature range is typically a narrower range within the operating range. Within this band, the transducer is guaranteed to meet its published specifications. Changes in temperature affect the output of the transducer in two ways. They may cause the zero output to change and may also affect the full-scale output. The transducer performance specification should list these temperature errors as:

- $\pm x\%$ of full scale/°C,
- $\pm x\%$ of reading/°C,
- $\pm x\%$ of full scale over the entire compensated temperature range, or
- $\pm x\%$ of reading for the entire compensated temperature range.

Not having these figures available will cause you uncertainty when using a transducer. Is the change in transducer output due to a change in pressure or a change in temperature? Most manufacturers specify the operating and compensated ranges and the temperature coefficients of zero and Span for each product. The temperature effects on a transducer are often the most complicated part of understanding how to use a transducer.

Null temperature shift (thermal zero shift)

Null temperature shift is the change in output at null pressure resulting from a change in temperature. Null temperature shift is not a predictable error because it can shift up or down from unit to unit. A change in temperature causes the entire output curve to shift up or down along the voltage axis. (See Figure 16.1.12.)

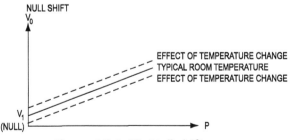

Figure 16.1.12: Null shift error.

Chapter 16

Sensitivity temperature shift is the change in sensitivity due to change in temperature. Change in temperature causes a change in the slope of the sensor output curve. (See Figure 16.1.13.) The more generic error sensitivity shift refers to a change in sensitivity from any environmental change.

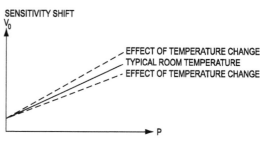

Figure 16.1.13: Sensitivity shift error.

Temperature hysteresis is the output deviation at a certain input, before and after a temperature cycle.

In some applications, some of the manufacturer's published specifications can be reduced or eliminated. For example, if a sensor is used over half the specified temperature range, then you can reduce the specified temperature error by half.

What output signal should I use?

Nearly all transducers are available with a choice of millivolt, amplified voltage, milliampere, or frequency output. The output you select will depend on the distance between the transducer and your system's controller or display, the presence of "noise" or other electrical interference, whether amplification is necessary, and where it is best to place the amplifier. For many OEM products with a short distance between the transducer and the controller, millivolt output is usually adequate and less costly.

If you need to amplify a transducer's output, it may be easier to use a different transducer with a built-in amplifier. For long cable runs, or areas with high electrical noise, a milliampere or frequency output is desirable. For environments with very high levels of radio frequency interference or electromagnetic interference (EEI or EMI), you will need to consider special shielding and filtering in addition to milliampere or frequency outputs.

What is excitation voltage?

The type of output desired might determine the excitation voltage you need. Many amplified transducers have built-in voltage regulators and will operate over a wide range of voltages from an unregulated power source. Some transducers are ratiometric and require a regulated excitation. Ratiometricity implies the sensor output is proportional to the supply voltage with other conditions remaining constant. Ratiometricity error is the change in this proportion and is usually expressed as a percent of Span. The power available may dictate whether you use a regulated or unregulated transducer. Consider the trade-offs between the available excitation and system cost. Many transducer or strain gage amplifiers will provide either a regulated voltage or a regulated current excitation.

Pressure Sensors

Do I need the transducer to be interchangeable?

Is it important that the transducers are interchangeable from system to system, or will you calibrate each unit as part of the system? This is important, especially for an OEM. Once you have shipped your product to your customer, the cost of field calibration can be very high. If the transducers are truly interchangeable, you should be able to replace a transducer in the field and expect the system to stay within specification.

How stable does the transducer need to be over time?

Most transducers will "drift" somewhat over time. It is important to discuss long-term stability with the transducer vendor. This up-front work can pay off in fewer field problems in the future.

How will I connect the transducer to my electrical system?

Will a short cable on the transducer be adequate? Or, in the case of long cable runs, should I have a connector on the transducer? Most pressure transducers can be supplied with cable or optional connectors.

How rugged does the transducer need to be?

One factor that is often overlooked, much to the chagrin of the user, is the ruggedness of the transducer, especially its housing. It is crucial to consider the transducer's prospective operating environment. Is it high in moisture or humidity? How will the transducer be installed? Will there be high levels of shock or vibration? All of these questions should be considered when selecting a housing style.

Because pressure sensors are used in such a wide variety of applications, they are available as light-duty units in cost-effective plastic housings, up through hermetically sealed devices encased in stainless steel housings for extreme or hazardous environments, areas with corrosive chemicals, and applications with sterility concerns, such as food processing or medical equipment manufacturing.

Here's a look at a number of applications where the environment in which the sensor needs to work is important to sensor selection.

1. Autoclaves and sterilizers used in medical, food and process industries

    Sterilizers are used primarily to reduce biohazards by disinfecting instruments or treating biohazard waste before disposal. They can use steam, gas, electricity and UV rays to kill bacteria, but pressure sensors are only used in steam or gas sterilizers. Autoclaves have many uses, including curing and bonding, treating wood and growing chemical crystals. They work exclusively with pressurized steam chambers and may be as small as a microwave or as large as a room.

Chapter 16

Process: Autoclaves, steam and gas sterilizers work using the same principles as a pressure cooker. Air is expelled from the chamber and the pressure is raised to 15 psig. This gives a temperature of 121°C. Sterilizer contents are usually sterile within fifteen minutes.

Sensor functions: Pressure transducers can monitor the chamber pressure performing a dual role. First the pressure is monitored as feedback and control of the P (pressure) in the equation $PV = RT$, assuring variable tolerances are met for certain processes. Secondarily, pressure transducers serve as safety mechanisms to prevent doors from being opened when there is positive chamber pressure.

Recommended environmental specifications:

- Stainless steel housing
- Rated to 150°C with a total error band of less than 1% full scale over the full temperature range
- Shock and vibration protection
- Electrical isolation of media from electronics
- Application-specific integrated circuit with amplified output
- Small size

2. Chillers used in HVAC systems, for process cooling in chemical, food and pharmaceutical manufacturing, cold storage rooms, commercial air cooling, mobile refrigeration for trucks, trains and marine vehicles.

    A chiller is a refrigeration system that cools water, oil or some other fluid. While air conditioners and dehumidifiers condition the air, chillers use the same refrigerating operations to cool liquids.

    Process:

    a. Refrigerant enters the compressor, is compressed, then discharged as a high-temperature, high-pressure, super-heated gas.

    b. Refrigerant travels to an air- or water-cooled condenser where it is converted from a high-temperature gas to a warm temperature liquid.

    c. The thermal expansion valve changes the high-pressure liquid into a low-pressure, cold-saturated gas.

    d. Saturated gas enters the evaporator where it is changed to a cool or dry gas.

    e. The cool, dry gas re-enters the compressor to be pressurized again.

Functions of the sensor: Pressure measurements on the refrigerant are typically made at the compressor inlet (low-pressure gas) and outlet (high-pressure gas), and at other locations in the cycle. The pressures involved range from approximately 100 to 500 psi. Measurements may also be made on the lubricating oil for the compressor or media to be cooled (gas or liquid). Temperatures may range from $-100°C$ to $+250°C$.

Recommended environmental specifications:

- All-wetted design, sealed for dust, moisture and RFI/EMI
- Stainless steel housing
- Temperature range wider than expected temperatures
- Amplified output compatible with relevant controllers

3. Off-road vehicles used in agricultural and construction, and material handling equipment for manufacturing, distribution or disposal applications

   Off-road vehicles have the ability to cross terrain while hauling, lifting, towing or moving solid materials.

   Process: Most dump trucks, aerial lift trucks, loaders, excavators and material handling equipment, such as hoists, lifts and fork lifts, use hydraulics to move material.

   Sensor functions: Off-road vehicle OEMs use pressure transducers to measure the oil pressure in their engines to maximize efficiency and ensure performance. Typical pressure ranges are 150 to 200 psi. Due to the proximity of the sensor to the engine, it must be able to withstand high-end temperature ranges of 1250 to 1400°C.

   Pressure sensors are also used to measure hydraulic pressure, which can be measured both at its source as the hydraulic charge pressure and in the system. Source pressures typically range from 700 to 800 psi, and system pressure ranges can vary from 4,000 to 9,000 psi. The hydraulic pressure provides the vehicle or equipment operator with feedback and balances the power delivered against the load to be lifted.

   Recommended environmental specifications:

   - Stainless steel package designed to withstand extreme environments, including protection from dust, humidity, bumps, shocks, and wash-downs
   - EMC protection, overvoltage and reverse voltage protection
   - -40 to $+150°C$ temp (generally)

4. Industrial air compressors used to supply air for operation of pneumatic controls, actuators or power tools.

   The application determines the type and size of industrial air compressors, which range in size from 2 to 10,000 HP and may be fixed or portable. The most common pressure is 125 psi with the amount of air volume ranging from 1 to 15,000 CFM.

   Process:

   a. Reciprocating compressors take air into a confined space and elevate it to a higher pressure using a piston within a cylinder as the compressing and displacing element.

   b. Rotary compressors use two rotors to take in and compress air within a casing.

   c. Centrifugal compressors depend on the transfer of energy from a rotating impeller to the air. The rotor changes the momentum and pressure of the air, slowing it down.

   Sensor functions: A pressure transducer detects pressure drops across separators used to remove dirt, water and oil from the air. Removal of such contaminants is critical to the efficiency and life of the compressor and neglect can result in costly downtime. Manufacturers typically require filter changes when the pressure drop reaches 10 psi, which is approximately 6 to 12 months of operation. Total pressure drop is 14 psi (new filter elements) and 35 psi (dirty). The amount of energy consumed is also a major consideration.

   Recommended environmental specifications:

   - Stainless steel housing
   - Compensated temperature range with error band of less than 1% FS over the full temperature range
   - Shock and vibration protection
   - Electrical isolation of media from the electronics
   - Application-specific integrated circuit with amplified output
   - Small size

Pressure Sensors

Applicable Standards

ANSI (American National Standards Institute): http://www.ansi.org

FDA (Food and Drug Administration): http://www.fda.gov
Products for food and beverage applications may require FDA and ANSI/NSF agency approvals based on whether the product is exposed to the application medium. The sensor itself would not require FDA approval.

FM (Factory Mutual): http://www.factorymutual.com

IEC (International Electrotechnical Commission): http://www.iec.ch

IEEE (Institute of Electrical and Electronics Engineers): http://ieee.org

ISA (Society for Instrumentation, Systems and Automation): http://www.isa.org

NEMA (National Electrical Manufacturers Association): http://www.nema.org

UL (Underwriter's Laboratories): http://www.ul.com

Interfacing and Design Information

As you design your application, it's important to understand that pressure sensors are not "ideal" devices. Laser trimming on high-level amplified sensors generally reduces null and full-scale errors to approximately one to two percent of Span but does not completely eliminate them. Additional corrective circuitry is sometimes necessary for applications with extremely tight tolerances.

Figure 16.1.14 illustrates the "ideal" pressure sensor. Output drift with time, trimming tolerances, and changes in ambient temperature all contribute to a constant offset error (common-mode error), designated by ΔV_o. Changes in ambient temperature also add another deviation, known as sensitivity shift, which changes the slope of the pressure versus voltage curve.

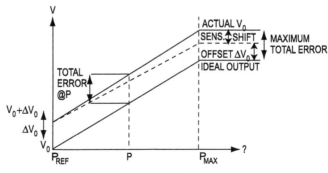

Figure 16.1.14: Sensor errors.

Chapter 16

Methods known as auto-referencing techniques provide a powerful tool to compensate for these errors. System design engineers find these methods attractive since implementation costs are minor in comparison with ultra-stable pressure sensors. Also, device accuracy is substantially increased. Either analog or digital auto-referencing is possible. The digital method is the most cost-effective and easiest to use.

When the accuracy limits of the auto-reference circuit replace errors, the errors can be reduced by as much as 250 times, leaving only the sensitivity shift (normal-mode) error. This is a significant improvement for the added cost involved. In any application where maximizing sensor accuracy is of value, consider an auto-referencing circuit.

Common-mode auto-referencing replaces common-mode error sources. Common-mode errors are present at some reference pressure and contribute the constant offset voltage shown in Figure 16.1.14. These errors are generally larger than the sensitivity shift, especially at pressures close to the reference pressure. Therefore, they allow the greatest accuracy improvement when auto-referenced.

Common-mode errors are easily corrected. Sample the output voltage at reference pressure and compare it to the desired reference voltage. Generate an error correction voltage and subtract it from the output signal at any "measure" pressure. (See Figure 16.1.15.) Common-mode auto-referencing is expressed by the formula:

$$Vcorr = Vout - \Delta Vo.$$

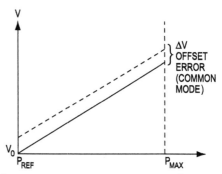

Figure 16.1.15: Common-mode errors.

Vout is any measured output signal; ΔVo is the common-mode error, and Vcorr is the corrected output signal. Note that no slope correction is provided for sensitivity shift error, and the actual output signal will appear as shown in Figure 16.1.16.

The basic functions required to implement common-mode auto-referencing are shown in the block diagram of Figure 16.1.17. They include analog switches, a sample-and-hold, summers, and synchronizing logic for switching between the read and reference cycles on the input and output sides of the pressure sensor. To maintain optimum

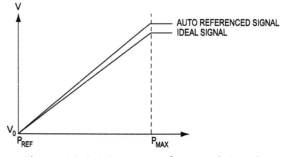

Figure 16.1.16: Auto-referenced signal.

system accuracy, auto-referencing should be used as often as possible in order to eliminate errors due to power supply fluctuations and output drift with time. To assure that the pressure measurements will be the most accurate, they should immediately follow the auto-reference command.

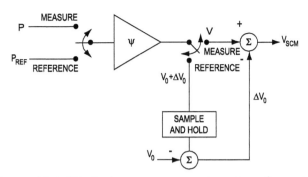

Figure 16.1.17: Basic common-mode auto-referencing.

Although common-mode auto-referencing is almost a universal technique, there are situations where it would be of little value: systems with short measurement cycles where the reference point is read or manually adjusted before cycle startup, or where the sensor is AC coupled and the DC response is ignored.

Certain types of measurement cycles are inherently suited to auto-zeroing (reference pressure is actually zero). Ideally, there is a series of short cycles that can have a quick referencing inserted prior to each cycle. A short measurement cycle preceded by a reference point, followed by a lengthy period of no activity, is also well suited. Many applications are in one of these categories. Many that are not can be converted to the short repeated cycle format with a little design creativity.

Examples of "ideal" auto-zeroing applications are: weighing scale; toilet tanks; washing machines; and pressure reservoirs such as tire pressure, oil pressure, and LP gas tank pressure. The reference condition is applied before the measurement. Other categories are flow measurement and control applications, such as electronic fuel injection systems, sphygmomanometers, and forced air heating systems. Flow rate is zero at some point, usually at system power-up. If an auto-zeroing technique is used, the null-offset and null-shift errors can be eliminated.

The key to an auto-reference circuit is applying the trigger signal to command the reference to take place at the appropriate time. This is called command auto-referencing. There are three levels of sophistication:

1. The simplest auto-referencing method is the manual command. A momentary contact switch initiates the auto-zeroing sequence. This is the most restrictive method, as it requires the user to be present while the system is running in order to periodically reference the sensor. However, it could be done as part of a routine calibration procedure.

2. With semi-automatic command, the user initiates the action. After it is triggered, the system sequences through multiple functions controlled by timers or shift registers. This could include solenoid actuation to switch from measurement to reference pressure, followed by the auto-reference function, then a return to the measurement mode. Figure 16.1.17 illustrates a basic semi-automatic circuit.

3. Using automatic command, the system steps through multiple functions similar to the semi-automatic command. However, on returning to the measurement mode, additional timing circuitry triggers and, after a set measurement time, the sequence is restarted. Depending upon the degree of complexity desired, a small microprocessor-based system and its related software could consolidate the auto-reference circuitry, timing and control logic all into one unit.

Auto-referencing requires an established system reference point. Batch processing and continuous processing are the two main categories of measurement cycle. In a batch process, a reference condition exists at some time, usually at system power-up. For example, a toilet tank has a high water level prior to flushing, corresponding to some reference pressure. When the flushing cycle is complete, the tank is filled to the previous level. The obvious point for auto-referencing is just prior to flushing when the water is at a known level.

In a continuous process, there is no easily accessed reference condition. For example, the volume of fluid in a water tower is being monitored. This is a function of the depth of the water and can be sensed with a pressure sensor. Unlike the toilet, without actually taking a pressure measurement, there is no point in time at which the depth will be known. The sensor/auto-reference/enable system can be used for a simple case when a known reference exists periodically. Also, a reference condition actuator such as a solenoid valve can be used. It can switch the sensor input from the measured pressure to some other reference pressure.

The solenoid can be activated by the user, by some condition such as power-up, or by a timer activated circuit. (See Figure 16.1.18.)

The valve must be activated long enough for the pressure to have a chance to stabilize so a valid reading may be taken. For instance, consider the water tower. A gauge pressure sensor near the bottom senses the water depth. A vent tube to

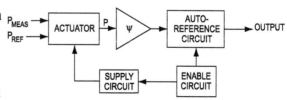

Figure 16.1.18: Auto-reference with reference condition actuators.

the surface serves as a pressure reference. A three-way solenoid valve is the actuator, connecting the water and the vent to the sensor input port. A timer circuit is the enabler. (See Figure 16.1.19.)

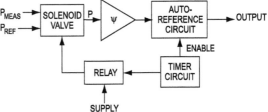

Figure 16.1.19: Timer-actuated circuit, single-port sensor.

Next, suppose the water exits through a single pipe of constant diameter. The velocity can be measured with a differential pressure sensor. A two-way solenoid connected between the two inlet ports serves as the reference actuator as shown in Figure 16.1.20.

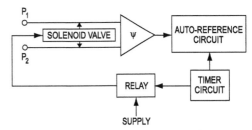

Figure 16.1.20: Timer-activated circuit, dual-port sensor

Latest and Future Developments

Many of the latest developments associated with pressure sensing involve sensors that provide more than pressure measurements. As more and more control systems become CAN-based networks and the systems themselves get "smarter," sensor manufacturers are seizing the opportunity to provide more "information" as opposed to simple output signals.

CAN technology has led to smart sensors that offer a variety of self- and process-related diagnostic functions. They are showing up in "intelligent" appliances, on plant floors, and on-board aircraft and vehicles.

Another trend involves the bundling of different types of sensing elements within the confines of a single chip. Honeywell used this approach when it used a mass airflow sensor to create a microsensor that simultaneously measures ambient temperature, pressure, thermal conductivity and the specific heat of a fluid. Such technology can be applied in chemical plants, chemical storage facilities and on automobiles where, for instance, it enables the engine to adjust itself to changes in fuel properties each time the fuel tank is refilled, resulting in improved gas mileage and cleaner exhaust.

This same approach is being used to develop a high-sensitivity, high-temperature Silicon-On-Insulator (SOI) piezoresistive technology that provides pressure sensing, temperature sensing, and feedback and bias resistor networks all integrated on a single miniature 90-mil-square chip. This technology was designed for applications with both high temperature and high pressure. Its high gage factor is very important in applications where a large "signal-to-noise" ratio is essential to achieving high accuracy performance, particularly over wide temperature ranges such as that required by the turbine engine and down hole oil industries. Sample test results show that the sensor design has successfully eliminated and/or minimized mechanical and thermal hysteresis error sources to levels that challenge the measurement capability of present state-of-the-art instrumentation.

References and Resources

1. "Piezoresistive Technology and Pressure Measurement Types," Honeywell, Inc. http://content.honeywell.com/sensing/prodinfo/pressure/technical/c15_101.pdf
2. "Pressure Sensors Conversion Factors and Chart," Honeywell, Inc. http://content.honeywell.com/sensing/prodinfo/pressure/technical/c15_125.pdf
3. "Pressure Sensors Plumbing and Mounting Considerations," Honeywell, Inc. http://content.honeywell.com/sensing/prodinfo/pressure/technical/c15_121.pdf
4. "Pressure or Force Sensor Switch Circuits," Honeywell, Inc. http://content.honeywell.com/sensing/prodinfo/force/technical/c15_119.pdf
5. "Protecting Pressure Sensor Diaphragm From Rupture Due To Water Hammer," Honeywell, Inc. http://content.honeywell.com/sensing/prodinfo/pressure/technical/c15_120.pdf

16.2 Piezoelectric Pressure Sensors

Roland Sommer and Paul Engeler, Kistler Instrumente AG

The brothers Pierre and Jacques Curie discovered the piezoelectric effect in 1880. They found that some crystalline materials were generating an electrical polarization when subjected to a mechanical load along some crystal directions. Among the materials they investigated were quartz and tourmaline, two crystals which are today still often used in piezoelectric sensors. The first piezoelectric pressure sensor was reported around 1920, but commercial sensors were not available until the 1950s, when electrometer tubes of sufficient quality became available. Today, piezoelectric pressure sensors are widely used in laboratories and in production. The main applications are found in combustion engines, injection molding and ballistics, but they can be used in any field requiring accurate measurements or monitoring of pressure variations. The main advantages of piezoelectric sensors are:

- wide measuring range (span to threshold ratio up to 10^8)
- high rigidity (high natural frequency)
- high linearity between output signal and applied load
- high reproducibility and stability of the properties (when single crystals are used)
- wide operating temperature range
- insensitive to electric and magnetic fields

It is often stated that piezoelectric transducers based on the direct piezoelectric effect can only be used for dynamic measurements. This is partly true, as they react only to a change in the load and hence cannot perform true static measurements. However, a good sensor with a sensing element made of single crystal material, in conjunction with adequate electronics, can be used for accurate measurements down to 0.1 mHz. In other words, quasistatic measurements lasting up to a few hours are possible.

This chapter will give an insight about the design, properties and applications of piezoelectric pressure sensors based on the direct piezoelectric effect (charge generation under mechanical load). These sensors are called active sensors, as they do not need any external power supply. They have a charge output which requires an external charge to voltage converter. Essentially, there are two types of converters, the electrometer and the charge amplifier. The charge amplifier was invented by W.P. Kistler in 1950 and gradually replaced electrometers during the 1960's. The introduction of MOSFET or JFET circuitry and the development of high insulating materials such as Teflon™ and Kapton™ greatly improved performance and propelled the field of piezoelectric measurements into all areas of modern technology.

Chapter 16

Technology Fundamentals

Piezoelectricity

Piezoelectricity is basically defined as a linear electromechanical interaction in a material having no center of symmetry. One distinguishes the direct and the converse piezoelectric effect. In the first case, a mechanical load or deformation of the crystal induces a proportional charge or electrical potential. In the second case, an electric field applied to the crystal induces a mechanical deformation or a load proportional to the field. In this paper, we will focus on pressure sensors based on the direct piezoelectric effect only, which can be described in the following way:

$$D = \varepsilon \cdot E + d \cdot X$$

where D is the induced electric displacement, E the applied electrical field and X the mechanical stress applied to the material. The dielectric constant ε and the piezoelectric coefficient d are describing the materials properties. D and E are vectors, ε is a 2nd rank tensor, d a 3rd rank tensor and X a 4th rank tensor. This means that the piezoelectric properties are anisotropic, the active coefficients for d and ε being determined by the crystal symmetry. The orientation of the crystalline measuring element is therefore critical and determines its properties.

Longitudinal and transverse cuts, volume effect

In a longitudinal cut, the surface A_Q on which the charge Q is induced is the same as the surface A_F on which the load (force) F is applied. The piezoelectric sensitivity depends solely on the longitudinal piezoelectric coefficient d_L. Conversely, in a transverse cut, the induced charge and applied load do not share the same surface. The sensitivity depends both on the transverse piezoelectric coefficient d_T and the surface ratio A_Q/A_F (Figure 16.2.1).

longitudinal cut: $Q = d_L \cdot F$

transverse cut: $Q = d_T \cdot F \cdot A_Q/A_F = d_T \cdot F \cdot l/t$

where l and t are the length and the thickness of the slab.

The advantage of the transverse cut is that the sensitivity can be increased by the geometrical factor l/t, provided the longitudinal and transverse piezoelectric coefficients are the same. This is always the case for quartz and all crystals belonging to the same crystal symmetry as quartz. For tourmaline, lithium niobate, lithium tantalate and piezoceramics, the transverse coefficient is only 1/10th to 1/3rd of the longitudinal coefficient; hence, transverse cuts are seldom used.

Pressure Sensors

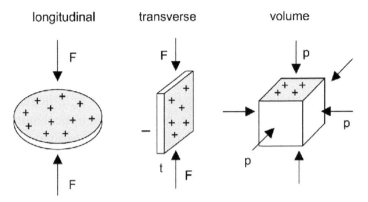

Figure 16.2.1: Longitudinal cut (left): induced charge and applied force share the same surface. Transverse cut (middle): charge is not induced on the same surface on which the force is applied. Volume effect (right): a hydrostatic pressure is applied to the sample, the charge being generated on two opposite surfaces.

Both the longitudinal and the transverse effects are basically uniaxial. The pressure to be measured is converted into an uniaxial force by a diaphragm. The volume effect differs from the uniaxial effect in that the force is applied on all surfaces of the sensing element. The sensitivity is the sum of the longitudinal and the transverse contributions.

Volume effect: $\quad Q = (d_L \cdot p + 2 \cdot d_T \cdot p) \cdot A_Q = d_h \cdot p \cdot A_Q$

where d_h is the hydrostatic piezoelectric coefficient and p the applied pressure. Note that the hydrostatic piezoelectric coefficient is always zero for quartz and all crystals having the same symmetry as quartz. The volume effect is mainly used in shock wave sensors or in hydrophone applications, where the direction of propagation of the pressure wave is not known.

For the longitudinal, transverse and volume effects, the load is applied perpendicular to the piezoelectrically active surface. Shear cuts, where the load is applied parallel to the surface, could in principle be used, but would not significantly improve the properties.

Piezoelectric materials

The sensing element is the heart of the transducer and should hence be selected carefully. Electronic or software compensation for "bad" crystal properties are hardly possible. High sensitivity, high insulation resistance, high mechanical strength, high rigidity, low temperature dependence of the properties over a wide temperature range, low anisotropy, linear relationship between charge and mechanical stress, no aging, no pyroelectricity (insensitive to temperature changes), good machinability, low manufacturing costs—these are a few requirements for a good piezoelectric material. Of course, the ideal material fulfilling all the criteria mentioned above does not exist.

Quartz has been used extensively in the past years and is still the material of choice for most pressure sensors. It is very stable, has a very high mechanical strength, can be used up to 400°C, has outstanding electrical insulation properties, has minimal sensitivity deviation up to 350°C (with special crystal cuts), is not pyroelectric, and is available at low cost. The only disadvantage of quartz is its relatively low sensitivity and its tendency to twin under very high loads.

Tourmaline has a lower sensitivity than quartz, but its temperature range extends to at least 600°C. However, it is pyroelectric and it is only available as natural crystals. Lithium niobate and lithium tantalate have a higher sensitivity, but they are strongly pyroelectric and their insulation resistance is quite low, limiting their practical use to purely dynamic applications.

Crystals of the CGG group (typical crystals are $Ca_3Ga_2Ge_4O_{14}$ and langasite, for instance) have been intensively studied over the last 20 years. Belonging to the same crystal symmetry as quartz, they are not pyroelectric and possess a higher sensitivity. Unlike quartz or gallium orthophosphate, they have no phase transition up to their melting point (above 1300°C), so that their properties remain very stable up to very high temperature and no twinning occurs. However the growth of large crystals is more difficult than for quartz, although langasite crystals up to 4 inches diameter have been grown.

Gallium orthophosphate has the same crystalline structure as quartz. Its sensitivity is twice that of quartz and is practically constant up to 500°C. Its phase transition lies around 970°C (573°C for quartz), extending the useful temperature range to at least 600°C. It is however very difficult to grow (the growth lasts a few months to one year) and is not available as large crystals.

PZT-based piezoceramics and Lead-Metaniobate have a very large piezoelectric sensitivity (up to 100 times that of quartz), but aging (time dependent depolarization), poor linearity and huge pyroelectric effect limit its use to applications for which accuracy is not critical.

High temperature piezoceramics (bismuth titanate based materials) can be used up to 500 or 600°C, but suffer similar (although not as serious) problems as PZT. Their sensitivity is about 5 to 10 times higher than that of quartz. They are used in high temperature applications (e.g., accelerometers operating up to 600°C).

Electronics

As already mentioned in the introduction, piezoelectric transducers are active systems (they do not require any power supply) which have a charge output (high-impedance output). For data acquisition and signal analysis, the charge output must be converted to a voltage, for instance by means of an electrometer or a charge amplifier.

Charge amplifiers

A charge amplifier is basically a high-gain inverting voltage amplifier with a very high input insulation resistance configured as an integrator. A typical measuring chain with a charge amplifier is shown in Figure 16.2.2; examples of laboratory and industrial charge amplifiers are shown in Figure 16.2.3.

The most common output voltage U_{out} is ±10 V. The range capacitors C_r can be switched usually between 10 pF to 100 nF, allowing for measurements over a wide charge range. A time constant can be switched on with the resistor R_t (usually 1 GΩ, 100 GΩ). The Reset/Operate (R/O) switch allows setting the zero point of the charge amplifier.

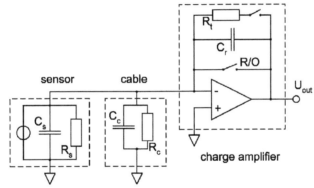

Figure 16.2.2: Measuring chain with charge amplifier. The sensor is modeled as a current source (charge source) in parallel with the sensor capacitance C_s and resistance R_s. The properties of the cable are described by a capacitance C_c and an insulation resistance R_c. The charge amplifier consists of a high-gain amplifier with a range capacitor C_r, a time constant resistor R_t and a reset/operate (R/O) switch.

The output voltage of the **ideal charge amplifier** (infinite open loop gain A, no leakage current, no offset voltage at the input) depends only on the induced charge and the range capacitor, where the input impedance has no influence.

Output signal $\quad U_{out} = -Q/C_r$

Lower frequency limit (–3dB) $\quad f_l = 1/(2\pi \cdot R_t \cdot C_r)$

If no resistor R_t is selected, the charge amplifier operates in DC mode and the steady-state behavior is governed by drift.

These relations are sufficient for most applications. In some extreme cases, the properties of the *real charge amplifier* must be taken into account:

Upper frequency limit $\quad f_u = 200 \ldots 500$ kHz

When operating at frequencies above 100 kHz, the input impedance can no longer be neglected, as the open loop gain of the amplifier depends on frequency.

Drift due to leakage current $\quad I_L < 10$ fA (MOS-FET), $I_L < 100$ fA (J-FET)

Leakage currents cause a drift in the output voltage, which eventually bring the amplifier to saturation. The time dependent charge $Q_L = I_L \cdot t$ generates a time dependent output voltage $U_{out}(t) = -I_L \cdot t / C_r$.

Drift due to offset voltage and low input resistance $\quad U_{off} \approx$ a few mV

The offset voltage at the amplifier input induces a current $I_d = U_{off} / (R_s // R_c)$. As for leakage currents, this current may bring the amplifier into saturation. Some charge amplifiers have a built-in zero adjustment to keep the drift to a very low level.

Should the input resistance be very low (for instance when measuring at high temperature), switching on a time constant resistor R_t or adding a coupling capacitor in series between sensor and amplifier might solve the drift problem. In both cases, the lower limit frequency f_l increases.

High input capacitance (> 1μF) $\quad U_{out} = -\dfrac{Q}{C_r\left(1 + \dfrac{1}{A} + \dfrac{C_s + C_c}{A \cdot C_r}\right)}$

In applications for which very long cables are needed, the cable capacitance cannot be neglected, in particular if the open loop gain of the amplifier is not very high. This results in a decrease of the output signal.

Pressure Sensors

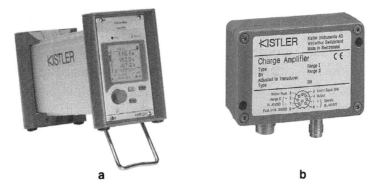

Figure 16.2.3a: Charge amplifier for laboratory applications. Virtually any type of sensors (including force sensors and accelerometers) with charge output can be connected to the input. Figure 16.2.3b: Industrial charge amplifier with two remotely switchable measuring ranges (courtesy of Kistler).

Impedance converter (electrometer)

The impedance converter consists of a MOSFET with high insulation properties (> 100 TΩ) and small leakage currents together with a bipolar transistor with unity gain and a small output resistance (~100 Ω). Unlike charge amplifiers, the output of the impedance converter depends on the total input capacitance (Figure 16.2.4):

$$U_{out} = -\frac{Q}{C_s + C_c + C_r + C_g}, \text{ where } C_g \text{ is the capacitance of the MOSFET}$$

The impedance converter is powered with a constant current (by means of a so-called coupler), generating thus an offset voltage at the output of the converter. Apart from powering the impedance converter, the coupler has to decouple the measuring signal from the offset voltage.

The time constant (lower frequency limit) of the system is given by

$$\tau = R_t \cdot (C_t + C_r + C_c + C_g)$$

The upper frequency limit is determined by the natural frequency of the sensor.

Chapter 16

Impedance converters are very common in sensors with integrated electronics (sensors with voltage output or low impedance sensors). Compared to the charge amplifier, the impedance converter is a cost effective solution and is usually found in industrial applications. Conversely, the charge amplifier offers more flexibility: the output voltage does not depend on the input capacitance, the output voltage is proportional to the applied load, the pressure ranges can easily be switched, allowing any sensor to be connected to the same unit.

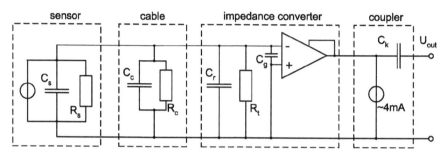

Figure 16.2.4: Measuring chain with an impedance converter.

In Figure 16.2.4, the sensor is modeled as a current source (charge source) in parallel with the sensor capacitance C_s and resistance R_s. The properties of the cable are described by a capacitance C_c and an insulation resistance R_c. The impedance converter consists of an unity gain amplifier with a range capacitor C_r, and a time constant resistor R_t. The coupler powers the impedance converter with a constant current of about 4 mA. The capacitor C_k decouples the measuring signal from the offset voltage.

Noise suppression

Single-ended charge amplifiers are mostly used. They have a common path for one of the signal sides. In order to prevent ground loops, the system (sensor + amplifier) should be grounded at one point only. This solution usually provides a good signal-to-noise ratio. For low level measurements in the presence of a strong electromagnetic field or when very long cables are needed, the differential charge amplifier may be required. Basically, the differential charge amplifier rejects unwanted signals present on both conductors (common mode rejection), as they amplify only the difference signal. This requires sensors with equally balanced capacitances on each signal line.

Pressure Sensors

Sensor Design and Applications

Basic design

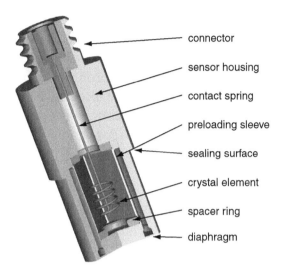

Figure 16.2.5: Basic design of a pressure sensor for general applications. The description of each component is found in the text.

Sensor housing

The sensor housing protects the piezoelectric element against dirt and humidity. The housing also serves as an electrical shield. Further, it provides a means of mounting and sealing the sensor from the pressure media. The sealing area of the sensor is designed to meet the specific applications. All these requirements can only be met using hermetically sealed housings. The housing material is typically a precipitation hardened stainless steel.

Preload sleeve

Preloading the sensing element assures a good linearity and sensitivity stability of the sensor over its complete operating range. The preload must be assured over the entire operating temperature range. The wall of the preload sleeve may measure less than a tenth of a millimeter in thickness to optimize the elasticity and reduce the force shunt. The material of the preload sleeve is usually made of the same material as the sensor housing. Not all sensors use a preload sleeve. Sometimes, the preload is provided by the diaphragm itself.

Chapter 16

Diaphragm

The active area of the diaphragm converts the pressure into a proportional force acting on the element. This force generates a stress into the crystal yielding a proportional charge. Most diaphragms today are hermetically welded to the sensor housing and slightly preloaded or welded to the front end of the sensing element. The diaphragm is the most critical part of the pressure sensor. It determines the durability of the sensor and the accuracy of the measurements. It must be insensitive to brief thermal shocks (i.e., combustion engines) so as not to create measurement errors. In order to achieve superior results, diaphragms must be optimized for various applications.

Connector

The electrical connectors for piezoelectric sensors must have a very high insulation resistance. Depending upon the operating temperature range, the insulator is made of PTFE or aluminum oxide.

Spacer ring

This ring compensates for the difference between the thermal expansion of the crystal and preload sleeve material. By appropriate dimensioning and material selection, the ring effectively reduces temperature induced output drifts.

Crystal element

The shape and size of the piezoelectric crystals vary depending upon sensor design and application. The examples shown in Figures 16.2.5 and 16.2.6b use a design with three crystal elements using the transverse effect. The mechanically unloaded surfaces are metal vacuum coated and insulated from each other to form electrodes for collecting the electrical charge. A helical-shaped spring contacts the flat surfaces and carries the charge to the connector. The cylindrical side contacts the preloading sleeve.

Design of the piezoelectric sensing element

Today, sensing element designs can vary significantly to meet specific requirements such as pressure range, operating temperature and environmental considerations.

Crystal disks cut for the longitudinal effect (Figure 16.2.6a)

Sensing elements using these plates can consist of one or several crystal plates. The sensitivity of these elements is directly proportional to the number of plates used. Further, these plates do not require metalized surfaces. The charge is collected directly from the loaded surfaces using electrodes. Such elements are used for high frequency (short rise time) or high temperature pressure sensors.

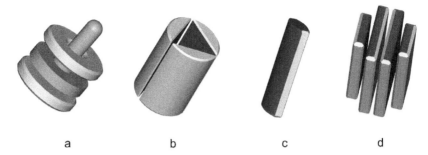

Figure 16.2.6: Different crystal element design. a) is based on longitudinal cuts. b), c) and d) are based on transverse cuts.

Set of crystals cut for the transverse effect (Figure 16.2.6b)

In the transverse effect, the force is acting on the end faces and the electrical charge appears on the mechanically unloaded faces of the crystal. These faces are metal coated and insulated from each other. Depending on their size, a set of three crystals generates a charge equal to 5 to 15 piezoelectric disks of the longitudinal effect. These elements are ideal for small to medium sized pressure or force sensors and yield a high sensitivity.

Crystal rod cut for the transverse effect (Figure 16.2.6c)

In this configuration, the metal coating of each unloaded surfaces extends to one of the end faces. The smaller sized crystal generates a charge equal to 4 to 6 piezoelectric disks of the longitudinal effect. These crystals allow for very small sensing element designs.

Set of crystal plates cut for the transverse effect (Figure 16.2.6d)

Each of the four plates is similar to the crystal rod previously described. The set has a sensitivity of approximately 7 to 10 piezoelectric disks of the longitudinal effect. However, with a plate thickness of only a few tenths of a millimeter, several plates are used in parallel to increase the load bearing capacity. This configuration is ideal for very small sensors.

Chapter 16

Standard sensors for general pressure applications

Sensor with longitudinal crystal cuts   **Sensor with transverse crystal cuts**

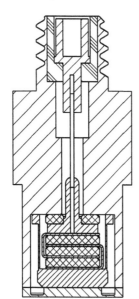 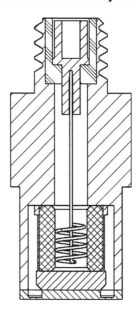

Figure 16.2.7: Sensors for general pressure applications. Left: Design with longitudinal cuts (crystal discs). Right: Design with transverse cuts (crystal slabs).

Both sensors shown in Figure 16.2.7 have the same sensitivity. While the sensor on the left is using crystals in the longitudinal effect and has a higher resonant frequency (approx. 300 kHz), the sensor on the right is using crystals in the transverse effect, and is therefore essentially insensitive to the twinning effect in quartz. The crystals can be cut in certain ways to optimize the temperature coefficient of sensitivity. Sensors using the longitudinal effect of quartz should not be used for high-temperature applications.

Sensor manufacturers offer a large selection of optional mounting hardware such as: adaptors, connecting nipples, etc. This allows a small number of sensor types to be used effectively for many applications. In situations where space is limited, sensors can be installed directly without the need for an adaptor.

Sensors are available in various sizes. Smaller versions have an outer diameter of 6 mm, the larger ones 11 mm. While the smaller sensors have a sensitivity of 16 pC/bar, the larger produce 80 pC/bar with measuring ranges from 250 bar to 1,000 bar. It is worth noting that piezoelectric sensors have a very large dynamic measuring range.**

Acceleration compensation

All pressure sensors are sensitive to acceleration, especially in the axial direction. In high vibration environments with small pressures levels, the acceleration induced signal can reach a few mbar/g. While in most applications the acceleration error can be ignored, it can disturb measurements where the sensor is subjected to strong vibrations while measuring small pressure. In such applications (e.g., acoustic or sound level detection), acceleration compensated sensors should be used.

Acceleration compensated pressure sensors may use longitudinal or transverse effect crystals.

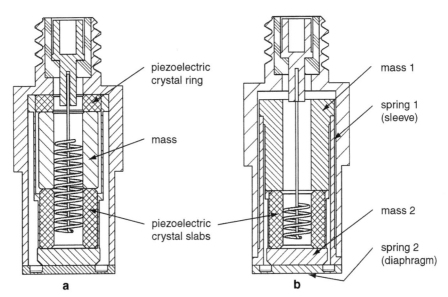

Figure 16.2.8: Design for acceleration-compensated sensors.
a): traditional design, b): new design.

** *The same sensor can be used for a partial measuring range of 1 bar as well as for a full range of 1000 bar.*

Chapter 16

Traditional design (Figure 16.2.8a):

The helical-shaped spring connects the charge from the set of crystal slabs to the mass. The crystal ring is installed with opposite polarity to the transverse effect crystals, thereby reducing the pressure sensitivity of the sensor. The compensating mass is dimensioned so that the crystal ring produces the same, but opposite signal acting on the transverse effect crystals. Under acceleration, the signals originating from the two sets of crystals cancel each other. Acceleration compensation adds complexity to the sensor, but a recent design achieves acceleration compensation without the crystal ring.

New design (Figure 16.2.8b):

In this new patented design, the two spring-mass systems are adjusted to be equal eliminating any acceleration-induced forces to act on the crystals. The vibration sensitivity of these designs is typically smaller than 100 µbar/g.

Ground insulated pressure sensors

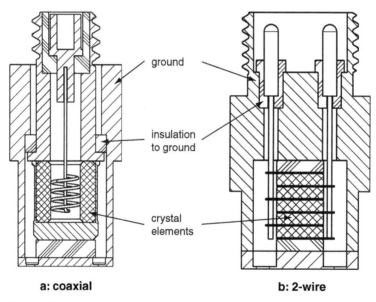

Figure 16.2.9: Ground insulated sensors. a): coaxial design. b): 2-wire design.

Ground isolated sensors reduce ground loop hum which arises when big potential differences build up between the sensor and the measuring chain (electronics). This may happen when electronics and sensors are connected with very long cables. In the coaxial design (Figure 16.2.9a), the outer shield of the coaxial cable is not electrically connected to the sensor housing. The two-wire system (Figure 16.2.9b) requires true differential design of the piezoelectric measuring elements and also needs a true differential charge amplifier. These sensors are primarily used in high to very high temperature applications in gas turbines, aeronautic and aerospace applications. Most of these sensors are also acceleration compensated.

Sensors for ballistics (high pressure)

Piezoelectric pressure, force and acceleration sensors have been used for many years for ballistics measurements. About fifty years ago, the first piezoelectric pressure sensors used for ballistics had a mechanical piston instead of a diaphragm. These sensors had to be disassembled and cleaned after every shot. Today, piezoelectric high-pressure sensors are used worldwide for acceptance testing of ammunition and gun powder. These tests assure that the ammunitions pass the stringent pressure profile requirements specified by the gun manufacturers.

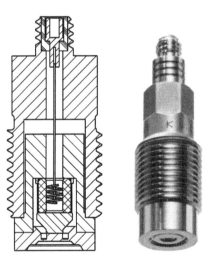

Figure 16.2.10: Typical design for a ballistics sensor. The thread is separated from the crystal element (Antistrain design). Available with M10 to M12 thread (courtesy of Kistler).

Today, front sealing sensors have replaced the shoulder sealing sensors. Figure 16.2.10 shows a front sealing "Antistrain" pressure sensor. This design eliminates measurement errors caused by the mechanically induced mounting strain. Further, the Antistrain design also eliminates sensitivity variations due to varying mounting torque. The measurement range for ballistic high-pressure sensors range from 5,000 bar to 10,000 bar with a linearity of less than ±0.5 % full scale with excellent long-term sensitivity stability. The life expectancy is several thousands of pressure cycles. As with any piezoelectric sensors, the same ballistic pressure sensor is capable of measuring a few 100 bar up to 10,000 bar.

Older designs required the user to apply periodically a protective coating of grease to the diaphragm to protect it from the hot gases. New sensor designs are capable of measuring accurately with an excellent life expectancy without the need for protective coatings.

Cavity pressure sensors

Measuring the cavity pressure in molds of plastics injection machines improves the quality and reduces the production time cycles. Optimizing this process assures that the parts have a constant weight and remain dimensionally stable. Continuous monitoring of the pressure profiles during molding improves yield resulting in reduced production cost.

Special pressure sensors were developed for these applications. Contrary to conventional sensors, they do not have a diaphragm (Figure 16.2.11). As the hot plastics material contacts the relatively colder cavity of the mold, the plastic immediately forms a skin acting as a diaphragm, hence preventing the hot material from penetrating the less than 0.03-mm wide gap between the measuring element and the mold. This eliminates the need for a diaphragm that would limit the life expectancy of these pressure sensors. Further, the front end of these sensors can be shaped to conform to the surface of the mold.

Most sensors are mounted directly into the mold without the addition of a protective sleeve. This reduces the size of the sensor, making it suitable in molds used for very small parts. Direct mount sensors have typically diameters from 4 to 8 mm (Figure 16.2.11a), the smallest has a diameter of only 1 mm.

Cavity pressure sensors operate reliably in temperatures up to 200°C. Although the mold temperature limits the functionality of the sensors, the actual melt temperature of the material can be higher (400°C). Most of the sensors include an integrated O-ring to reduce contamination that often lodges between the sensor element and the mold, but this is not essential for the proper operation of the sensors.

Pressure Sensors

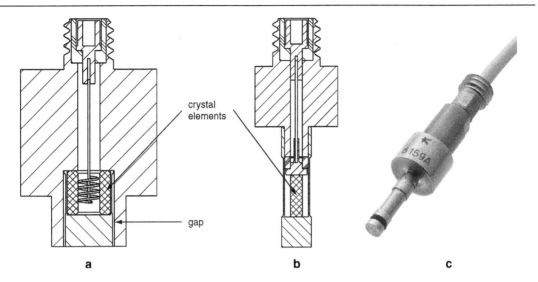

Figure 16.2.11: Sensors for cavity pressure measurements. a): front diameter 6 mm. b) and c): front diameter 2.5 mm (courtesy of Kistler).

Sensors for engines

Piezoelectric pressure sensors were used early on to measure combustion pressure in engines. Today, most piezoelectric pressure sensors are used for the development of automotive engines. New sensors are being developed continuously to meet the higher demands for smaller engines and higher accuracy.

The typical operating temperature varies between 150 to 350°C depending on measuring location and test conditions. However, the diaphragm temperature can be considerably higher. The diaphragm is exposed directly to the hot cyclic combustion gases of approximately 2,500°C. Special diaphragm designs and materials minimize thermally induced measurement errors and assure high life cycle durability. Meeting these requirements pose special challenges to the diaphragm design and demands that the sensors must be optimized for specific applications.

Today's combustion engines are becoming smaller and have four or more valves per cylinder. This limits the space available for the combustion pressure sensor. Sensors with an M5 mounting thread and a front diameter of 4 mm make these installations possible (Figure 16.2.12a). For further space restricted applications, sensors can be integrated directly into specially designed spark plugs or glow plugs (Figure 16.2.12b).

Water-cooled sensors improve zero stability (Figure 16.2.13a). Initially, water cooling of the sensors was required to protect the sensors from the high temperatures of the engines. Modern sensors can be used without water-cooling, hence will not be damaged if for some reason the cooling is interrupted. The smallest water-cooled sensor today has an M8 mounting thread.

To monitor large diesel and gas engines for power stations and ships, robust, uncooled pressure sensors are used. These sensors are designed for continuous operation and feature a long lifetime (Figure 16.2.13b). Often, the measurement systems include an integrated charge amplifier customized for specific applications.

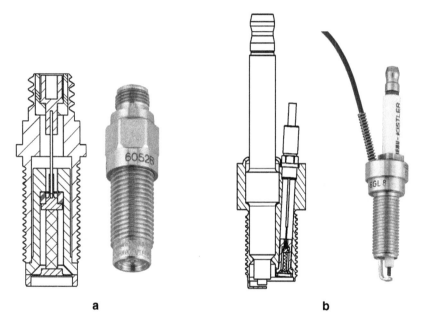

Figure 16.2.12a: Combustion pressure sensors with an M5 mounting thread.
(Courtesy of Kistler.)

Figure 16.2.12b: Small pressure sensor integrated in a M12 spark plug.
(Courtesy of Kistler.)

Pressure Sensors

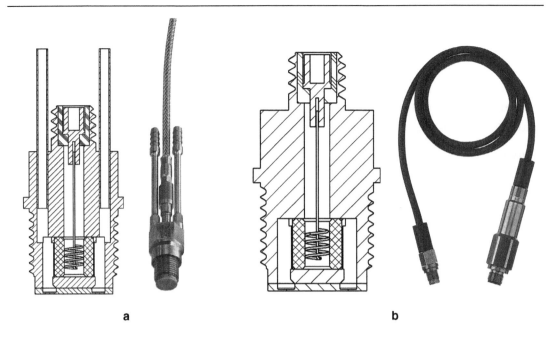

Figure 16.2.13a: Water-cooled pressure sensors M8 for engine applications.
(Courtesy of Kistler.)

Figure 16.2.13b: Robust uncooled sensor M10 for engine monitoring.
(Courtesy of Kistler.)

High-temperature pressure sensors

The gas turbine manufacturers need high-temperature pressure sensors able to operate at temperatures higher than 400°C in order to monitor small fluctuations of the combustion pressure. Sensors operating up to 650°C with a pressure range of 0 .. 250 bar are available (Figure 16.2.14). To withstand these high temperatures, the sensing part of the sensor is made of tourmaline discs and the sensor is fitted with a two-wire mineral insulated cable. The sensor design is similar to Figure 16.2.9b.

Figure 16.2.14: Transducer for operating temperatures up to 650°C.
(Courtesy of Vibro-Meter SA.)

Chapter 16

Sensor Selection

The sensor manufacturers can help you find the most appropriate sensor for your application. Data sheets are nowadays available on the internet and together with search functions or application oriented tools greatly facilitate the search for the right sensor. Hereafter is a list of the most relevant sensor and electronic properties you need to know before starting a search:

Sensor

Pressure range – the maximum pressure the sensor will have to measure.

Calibrated partial range – useful for measurements in a smaller pressure range. Note that piezoelectric sensors offer several decades of measuring range without loss of accuracy.

Overload – the maximum pressure the sensor will have to withstand.

Natural frequency – This defines the upper frequency limit of the dynamic pressure signal.

Sensitivity – the sensitivity of the sensor (usually in pC/bar or mV/bar).

Thermal sensitivity shift – the variation of sensitivity as a function of temperature.

Linearity – usually in ±%FS (full scale). Deviation from a true linear relationship between output signal and pressure.

Hysteresis – usually in %FS (full scale). Output difference when the same increasing and then decreasing pressure is applied consecutively.

Insulation resistance – a sensor with good electrical insulation properties will allow measurements down to the mHz range. Note that the electrical insulation of the sensor decreases as the temperature increases.

Temperature range – the minimum and maximum operating temperatures. If the temperature exceeds 300°C, special attention is needed for the connector and the cables.

Sensor dimensions – the maximum allowed dimensions of the sensor (mounting hole).

Shock resistance – the maximum shock resistance (usually in g).

Acceleration sensitivity – relevant if the sensor is mounted on a vibrating structure (engine for instance). One distinguishes radial and axial acceleration sensitivities.

Thermal shock – output signal (usually in bar) due to a thermal shock on the sensor. Important in engine measurements.

Environment – usually the sensors are hermetically sealed and made of stainless steel. Special attention is however required if you intend to use the sensor in a very corrosive or abrasive medium, or if the sensor will be submitted to nuclear radiations.

Connector and cables – low-noise coaxial cables (charge output), standard coaxial cable (voltage output). High-temperature applications (> 300°C) require special cables and connectors.

Sensor output – charge or voltage output. Voltage output is achieved by integrating the electronics in the sensor. The advantage is that you do not need any additional electronics (charge amplifier), but the flexibility is greatly reduced, as only one pressure range and one time constant are available. The temperature is limited to about 150°C.

The lower frequency limit depends basically on the range capacitor and the range resistor. Let us assume that you have the following configuration:

Sensitivity of the pressure sensor: 20 pC/bar
Pressure range: 200 bar
Full scale output: 10 V

First you have to select the range capacitor. The output sensitivity (sensitivity of the charge amplifier) is given by:

20 pC/bar · 200 bar / 10 V = 400 pC/V

This is also the ideal value of the range capacitor C_r = 400 pF. Of course, a real charge amplifier is not configured with thousands of range capacitors, but only with 4 or 5 very precise decade capacitors (100 pF, 1, 10, 100 nF, for instance). In our example, the effective range capacitor is the next higher capacitor, i.e., 1 nF.

In AC mode, the time constant, τ, depends on the product $R_t \cdot C_r$. Assuming R_t = 100 GΩ, the time constant is now

$\tau = R_t \cdot C_r = 100$ s

and the lower frequency limit (3 dB amplitude decrease of a sine wave) is computed according to

$f_l = 1 / (2\pi \cdot \tau) = 1.6$ mHz

In DC mode, there is no resistor R_t and the long time behavior is governed by drift. Assuming an offset voltage U_{off} of 5 mV at the input stage of the charge amplifier and a sensor insulation of 10 TΩ, the resulting drift is

5 mV / 10^{13} Ω / 1 nF = 0.5 µV/s, which is negligible

At higher temperatures, the drift is no longer negligible. At 400°C, the sensor insulation drops down to 10^7 Ω, generating a drift of about 500 mV/s at the output. Without switching a time constant resistor, the charge amplifier would saturate in less than 20 s! An alternative way to eliminate the drift is to include a very high capacitor C_D in series between the sensor and the charge amplifier (coupling capacitor). In this case, no drift occurs and the time constant is now given by $R_S \cdot C_D$.

Electronics

Integrated electronics – provides a cost effective solution for applications requiring little changes in the transducer's setting. Limited to temperature below 150°C.

Charge amplifier – provides the best flexibility, as the full pressure range of the sensor can be exploited and time constants can be adjusted. Furthermore, the output does not depend on the input capacitance.

Impedance converter – cost effective alternative to the charge amplifier. The output depends however on the input capacitance (cable capacitance) and they offer fewer signal conditioning capabilities.

Differential charge amplifier – for applications with small signals in a very noisy environment.

Manufacturer's Links

Kistler Instrumente AG	www.kistler.com
Endevco Corporation	www.endevco.com
AVL List GmbH	www.avl.com
Vibro-Meter SA	www.vibrometer.ch
PCB Piezotronics Inc	www.pcb.com
Columbia Research Lab, Inc.	www.columbiaresearchlab.com
Dytran Instruments, Inc.	www.dytran.com

Latest and Future Developments

Today's development is driven mainly by two factors: miniaturization and high temperature. These two factors have been pushing the research for new piezoelectric materials with better properties: higher sensitivity and good stability at high temperature (>300°C). As a result, new materials belonging to the CGG group are replacing the standard materials (quartz and tourmaline) in miniature or high temperature sensors.

Miniaturization

Sensors have to be smaller. This is a consequence of the trend observed in many application areas. For instance, engines are always more compact, requiring sensors of diameters M5 or even smaller. In injection molding of plastics, sensors with diameter 2.5 mm for pressure measurements up to 2,000 bar are now commonly used. Reducing the size of sensors is a challenging process. Apart from the difficulty of manufacturing small sensor components, the sensitivity decreases proportionally to the diaphragm area. As the diameter decreases, the force acting on the piezoelectric element decreases with the square of the diameter. Accordingly, reducing the diameter of the sensor (diaphragm) by a factor of two results in a sensitivity drop by a factor of four. This may be overcome by cutting very thin slabs of crystals and using the geometrical factor in transverse crystal designs to enhance the sensitivity or/and by using materials with higher sensitivity.

High temperature

Standard pressure sensors can be used today up to 350°C. Above this temperature, many problems have to be overcome: the piezoelectric element has to be stable and still maintain high insulation, corrosion can be avoided by using nickel-based alloys (but they are difficult to machine or to weld), conventional cables have to be replaced by integrated metal cables, etc. Of course, the sensor must withstand these temperatures often over thousands of hours, as the machines (turbines for instance) cannot be stopped at will to replace a defective sensor. Some manufacturers offer sensors operating up to 600 or 700°C, based on tourmaline, lithium niobate or high-temperature piezoceramics.

Chapter 16

Sensor identification

Each sensor is calibrated by the manufacturer and comes with a so-called calibration sheet. Before using the sensor, the operator has to enter the correct sensor sensitivity in the charge amplifier. For applications requiring many sensors, this may be a tedious task and a mistake is possible. Sensor identification, coupled with a database containing all the needed information (ID, sensitivity, sensitivity shift under temperature, etc.) has now become available and greatly facilitates the implementation of large numbers of sensors. One way is to integrate a TEDS (Transducer Electronic Data Sheet, according to IEEE 1451.4 Standard) digital chip in the sensor or its connector. When plugging the sensor to the charge amplifier, the data is read from the TEDS and the charge amplifier automatically adjusts its parameters according to the sensor sensitivity. TEDS are rewritable, allowing subsequent modifications or adaptations of the sensor properties. Another way is to integrate an ID chip (for instance a coded surface acoustic wave (SAW) tag) in the sensor. The charge amplifier recognizes the sensor and can get the relevant information in a database, provided it is connected to a PC.

References and Resources

1. Bill B., "Messen mit Kristallen," Die Bibliothek der Technik, Band 227 (2002) or "Measuring with crystals," Kistler V900-335e.

2. Cavalloni C., & Sommer R., "PiezoStar Crystals: A New Dimension in Sensor technology," Kistler Special Print 920-240e-07.03 (2003).

3. Gautschi G., "Piezoelectric Sensorics," Springer (2002) and references therein.

4. IEEE Standard on Piezoelectricity (ANSI/IEEE Std 176-1987).

5. Ikeda T., "Fundamental of Piezoelectricity," Oxford University Press (1996).

6. Krempl P.W., Schleinzer G. & Wallnöfer W., "Gallium Phosphate $GaPO_4$: A New Piezoelectric Crystal Material for High-Temperature Sensorics," Sensors and Actuators A61, 361-363 (1997).

7. Tichy J. & Gautschi G., "Piezoelektrische Messtechnik," Springer (1980).

8. Wilson J., "Noise Suppression and Prevention in Piezoelectric Transducer Systems," Endevco TP 270.

CHAPTER **17**

Sensors for Mechanical Shock

Anthony Chu, Endevco

At the first Shock and Vibration Symposium in 1947, mechanical shock was defined as "a sudden and violent change in the state of motion of the component parts or particles of a body or medium resulting from the sudden application of a relatively large external force, such as a blow or impact." Since then the specific words used have changed somewhat but the meaning remains the same. Most analysts treat shock as a transient vibration. No matter how it is described or what source produced it, the effects of mechanical shock on structures and equipment create major design problems for a wide variety of systems.

17.1 Technology Fundamentals

Shock measurement is usually accomplished by measuring the acceleration, velocity, or displacement response of the body. Shock measurement is important in studying the effectiveness of protective packaging design, earthquakes, effects of explosive events (pyroshock), effects of handling and or dropping items, transportation environments, many military applications, automotive crash testing and ballistic effects.

Shock Measurement

Shock measurement usually requires good high frequency response, good linearity and a wide dynamic range. Frequency content can reach 100 kHz and higher, and amplitudes may exceed 100 kg ($\approx 10^6$ m/s²). Long duration transients may also require good low-frequency response. Two basic categories of shocks may have to be measured, velocity shocks and oscillatory shocks.

Velocity Shocks

Velocity shock has two components: intensity, usually measured in g's (1 standard g = 9.80665 m/s²), and duration, measured in milliseconds. A drop from table-top height onto a hard floor can result in shock in excess of 1,000 g's over a period of about 3 milliseconds. A component in an artillery shell experiences about 16,000 g's for 12 milliseconds. Higher shock levels usually have shorter duration, perhaps fractions of milliseconds, and lower shock levels can have a duration as long as 20 milliseconds.

Shocks resulting in crushing or bending of impact surfaces may last hundreds of milliseconds.

Oscillatory Shocks

Oscillatory shocks, such as pyroshock and ballistic shock are shocks that do not cause a significant change of velocity of the object experiencing the shock. Instead, they cause an oscillatory response or ringing. For example, hitting a large bell with a hammer will cause the bell to oscillate (ring) but will not significantly change the bell's velocity.

High Amplitudes

Many shocks involve large forces and result in high amplitude acceleration response. Accurate measurement requires a wide dynamic range and good linearity.

Low level (less than 100 g's) can usually be measured using general-purpose accelerometers. Higher peak acceleration levels will usually require accelerometers that have been designed specifically for shock measurements.

Shock accelerometers have lower sensitivities and higher resonance frequencies than general purpose accelerometers. And, of course, they are designed to survive many applications of the high internal stresses caused by high amplitude shocks.

High Frequencies, Short Rise Times

Shock transients often have very short rise times requiring good high-frequency response and minimal ringing. Pyroshock measurements may require physical mitigation of extremely high frequencies that excite accelerometer resonance frequency. If the accelerometer survives the mechanical stresses of the resonance, it will produce an excessive output signal level which may overload signal conditioning causing clipping and loss of data. Mechanical filtering combined with electrical filtering can minimize these problems, but the characteristics of the mechanical and electrical filters need to be well-matched to prevent distortion in the frequency range of interest (usually up to 10 kHz).

High Sensor Stresses

The sensor measuring the shock must not only be able to survive it, but also to maintain its accuracy and linearity for many repetitions of similar shocks. Internal stresses in the sensor may cause shifts in performance characteristics, cumulative damage, fatigue failures, nonlinearities, or other problems. Some of the symptoms of such damage are significant changes of resistance and/or capacitance, excess noise, change of zero measurand output, and change of sensitivity.

17.2 Sensor Types, Advantages and Disadvantages

Shock is commonly measured with an accelerometer or velocity sensor measuring the response of the device under test as it responds to the transient force input. Non-contact measurements may be made using optical techniques such as laser Doppler velocimetry. Occasionally, strain gages are used to directly measure the strain(s) induced in a structure by the shock transient.

Piezoelectric Accelerometers

Piezoelectric accelerometers, described elsewhere in this handbook, have proven to be very popular for many shock measurement applications because of their inherent ruggedness and wide variety of characteristics available. Although mechanical structures vary widely (compression, annular shear, flat plate shear, etc.), they can be divided into two basic types, charge mode (high output impedance) and internal electronic (low output impedance).

Charge Mode (PE)

Charge mode accelerometers are used with an external charge amplifier or in-line impedance converter that senses the charge generated by the strain of the PE crystal. Their output is measured in picocoulombs and must be very carefully transmitted using well-shielded low noise coaxial cable. If the cable is not treated for triboelectric noise or not well-shielded, the cable noise or electromagnetic interference noise may seriously corrupt the signal.

Charge mode accelerometers are inherently the most rugged and most reliable because of their simplicity of design and construction.

Internal Electronic PE (IEPE)

Internal electronic piezolectric (IEPE) accelerometers incorporate a preamplifier in the accelerometer case. Their output is pre-amplified and has low output impedance. It is, therefore, less susceptible to environmental noise and cable noise. However, including electronic components, with multiple electrical connections, inside the accelerometer reduces their inherent reliability. Also, the internal electronics has a fixed gain which determines the dynamic range of the sensor. This limits the flexibility of application.

Piezoresistive Accelerometers (PR)

Modern piezoresistive (PR) accelerometers are manufactured using MEMS (micro electro mechanical systems) technology. Those designed for shock measurement typically have semiconductor strain gages implanted into complex spring-mass structures. Their primary advantage over other types of accelerometers is that they have frequency response down to DC (steady state, infinite low-frequency time constant). Therefore they can be used to measure very long duration shocks with no degradation of fidelity. When this steady state responding characteristic is combined with high resonance frequency, low sensitivity capability, an ideal shock measurement sensor results. MEMS technology enables this combination of characteristics to be realized in a relatively small physical size sensor.

The biggest disadvantage of PR accelerometers is their relative fragility if excited to resonance. Since their seismic systems have mechanical amplification factors up to 100 at resonance response, even a small amount of excitation energy near their natural frequency will cause excessive response with high stresses and likely breakage. Therefore, PR accelerometers must be very carefully applied in shock measurement applications.

Laser Doppler Velocimeters

Laser Doppler velocimeters measure the Doppler shift in the frequency of a modulated laser beam reflected from a target surface. The reflected beam is compared with the incident beam and the frequency shift is used to determine the velocity of the target. Since this is a non-contact measurement, it can sometimes be used where a sensor could not be mounted. It also eliminates the possible errors caused by mounting and mechanical effects in a contact sensor.

Disadvantages are that the instrument is relatively large and expensive and the measurement requires a line of sight be established between the velocimeter and the target. Some velocimeters have specific stand-off requirements restricting the geometry of a test setup.

Strain Gages

Because the objective of many shock tests is to try to determine stresses in the subject structure, strain gages are often used instead of, or in addition to, other sensors. They can be located and oriented to directly measure strain in critical areas. However, they do not characterize the shock transient or the structural response of the device under test (DUT). Strain measurements may be affected by local flaws or discontinuities in the structure.

17.3 Selecting and Specifying

Applications such as package drop-testing, automotive crash-testing, and pyroshock/simulation require accelerometers with special capabilities. An accidental drop of a cellular phone from standing height can produce peak acceleration levels well over 10,000 g's. Many novices in shock testing make the wrong assumption that the shock measurements of an object can be approximated using a rigid body model, and completely forget about the localized material responses. In a high-g shock test where structural responses are often nonlinear and difficult to characterize, choosing the right accelerometers can be critical. In addition to the characteristics of the shock to be measured, other environmental effects must also be considered. Basic accelerometer selection considerations for shock measurement follow.

Expected Amplitude

Range – Usable output range should not be confused with survivability. An accelerometer may only have to measure 100g full scale, but it may be required to survive an initial shock of 10,000g preceding the lower level event. Another distinction should be made between the maximum expected level from a Shock Response Spectrum and the actual input spectrum the accelerometer is likely to experience.

Low-pass filtering – Low-pass filters can be used at the input of an amplifier to prevent overload condition due to unexpected input spikes.

Zeroshift – Zeroshift or DC-offset, a sudden change of baseline level during a high-g event (see Figure 17.3.1), is a type of error usually associated with, but not limited to, PE/ISOTRON accelerometers. There is no universal standard for this parameter, so it cannot be found in the Performance Specifications. PR devices designed for high-g shock are typically immune from zeroshift. Accelerometers with a built-in mechanical filter, such as Endevco Models 7255A and 7270AM6, are designed to eliminate this problem. Figure 17.3.1 illustrates the principle of a built-in mechanical filter. There have been several papers written on this subject; readers are advised to refer to Endevco Technical Paper TP290 and TP308.

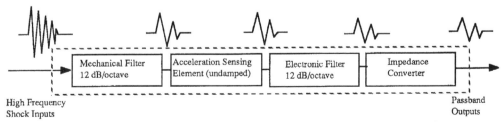

Figure 17.3.1: Model for accelerometer with built-in mechanical filter.

Chapter 17

Survivability – In lower level shock applications, most accelerometers can survive the environment without causing internal damage. But in high-g testing, physical damage to the sensor is often a reality. It is suggested to overestimate the maximum shock level when selecting the range of a shock accelerometer. A general rule-of-thumb: the closer the accelerometers are to the source (explosive or impact), the higher the input g level. Survivability extends to cables and connections. In high-g shock, a small amount of unrestrained mass would translate into large force, causing connector failures (or bad contacts) and generating triboelectric noise with ordinary coaxial cable. Solder terminals (see Figure 17.3.2) and ribbon wires are recommended in high-g applications because of their light weight, but extra care is required in installing and handling these delicate connections.

Figure 17.3.2:
Shock accelerometer with solder terminals.

Expected Frequency Content

As will be shown, the frequency content of the transient to be measured can vary widely. Especially when measuring shock, frequency extremes—both low and high—should be considered.

Low Frequency

When measuring long duration shock or measuring rigid body motion of the structure (i.e., ship shock), accelerometers with DC response are required to capture the low frequency information accurately. If the acceleration data is to be integrated to yield velocity or displacement information, DC response is an absolute necessity. Low frequencies may be contaminated by thermal transient response, discussed later in this chapter.

High Frequency

The input spectrum of high g-shock, be it mechanical or pyrotechnic, has always been underestimated by practitioners in the measurement industry, leading to improper test equipment selections. Furthermore, most transducer manufacturers have very limited experience in high-g shock test, and it is reflected in the design approaches of many so called "shock" accelerometers. This author suggests that, in close range (near-field) high-g shock measurement, the accelerometer must be protected from all ultra-high-frequency input energy in order to avoid sensor resonance, which is the root cause of many problems in high-g shock measurement.

All undamped, spring-mass type accelerometers have a finite seismic resonance. When the resonance of such a device is excited, integrity of the output signals is suspected. To ensure linear response and minimize error, spectrum of the acceleration input must stay within the transducer's recommended bandwidth. As a general rule-of-thumb, the maximum usable bandwidth for an undamped accelerometer is to be less than one-fifth of the transducer resonance. This rule is generally well observed in the vibration test community, but not so well observed in shock measurements.

Unfortunately, the term "maximum usable bandwidth" is often mistaken for the available bandwidth of a Shock Response Spectrum by many test engineers. Since most Shock Response Spectra stop at 10 kHz or 20 kHz, accelerometers with resonance in the neighborhood of 100 kHz are considered adequate for high g-shock applications, ignoring the fact that there is much energy beyond 20 kHz. The problem is further complicated by the issue of damage potential of high frequency. It is a well established fact that shock energy above 10 kHz seldom causes any damage to the test article, and it is routinely overlooked in most data analysis. These high frequency components, although posing no danger to the article, seriously affect the linear operation of any spring-mass type accelerometer.

It has been demonstrated that the input spectrum of most high-g shock measurements contains frequency components for above 100 kHz. These high-frequency components tend to cause resonance of almost any real structure, including accelerometers. A few papers and articles have been published concerning the effect of ultra-high frequency impulses on shock measurements. This out-of-band transient phenomenon is referred to in the papers as a "Pre-Pulse" stress wave that approximates a true-impulse.

Chapter 17

Two types of shock simulations are capable of generating near true-impulses

a) *Close-Range (Near-Field) Pyrotechnic Shock*

In pyrotechnic shock, the process of explosion involves chemical reactions in a substance which convert the explosive material into its gaseous state at very high temperature and pressure. Most explosives, such as Flexible Linear Shaped Charge and pyrotechnic bolts, do not contain as much energy as ordinary fuel, but generate an extremely high rate of energy release during explosion. The response of the structure near the immediate region can actually approach a true impulse due to the instantaneous velocity change at the explosive interface. As a result, measuring the data in the area surrounding a pyrotechnic explosion has always been a nightmare for engineers and scientists.

Depending on the explosive location and the point of measurement, the amount of high-frequency energy reaching the transducer is inversely proportional to the distance between them. In a remote sensing location where the shock wave has to propagate through a long path or many joints of dissimilar materials to reach the transducer, high-frequency components can be significantly attenuated.

b) *Close-Range (Near-Field) Metal-to-Metal Impact*

Most shock simulation devices, such as drop towers and pneumatic hammers, rely on high velocity metal-to-metal impact to generate the required shock spectrum. When the point of impact allows very little material deformation (like in all reusable shock machines), the acceleration response of the structure can also approach a true impulse. Again, the input spectrum is highly dependent upon the accelerometer location relative to the point of impact.

Failure Modes

Although these common methods of shock simulation present a formidable challenge for the entire measurement system, from sensor to data capture, the accelerometer is by far the most vulnerable under such conditions. There are two types of widely used shock accelerometers: piezoresistive and piezoelectric devices. Each reacts differently under the attack of near true-impulses. Three common failure modes are observed:

Sensors for Mechanical Shock

a) *Sensor Failure*

Recent new designs in piezoresistive accelerometers have tremendously improved their usable bandwidth and rigidity. One piezoresistive sensor (Endevco Model 7270A) exhibits seismic resonance above 1 MHz, leaving quite a large margin of safety for the general rule-of-thumb. Under the attack of true-impulses, however, the sensor can still be set into resonance (at 1 MHz) due to the nature of the input signals. Since the gauge mechanism is practically undamped, displacement of the elements goes out of control at resonance and eventually causes permanent gauge damage. The result of this type of failure is complete loss of data.

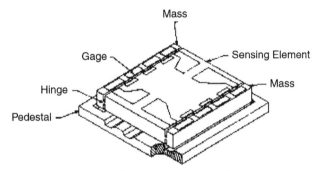

Figure 17.3.3: Sensing element of high-resonance frequency PR accelerometer.

Piezoelectric sensors are more robust under the same conditions, but they fail in other fashions:

b) *Zeroshift*

This subject has been well examined in many technical papers. A piezoresistive accelerometer generally does not exhibit zeroshift until the gauge mechanism has been damaged or, is in the process of deterioration. Piezoelectric sensors, on the other hand, account for most of the zeroshift phenomena associated with transducers.

When a piezoelectric element is set into resonance, three things can happen:

1. Relative displacement of the sensing element at its resonance can exceed 100 times of the input. Internal stress at the molecular level is therefore unusually high. This overstress condition produces spurious charge outputs due to domain switching, a characteristic common in polycrystalline materials. The result of this type of phenomenon is DC offset in the time history, as shown in Figure 17.3.4.

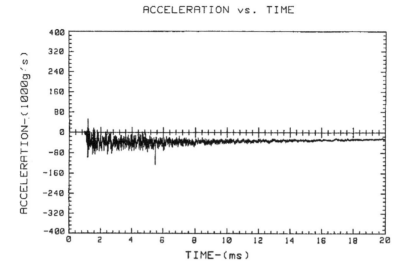

Figure 17.3.4: Zero shift caused by shock transient.

2. Crystal elements that have monocrystalline structure do not exhibit domain switching phenomenon, but they produce zeroshift in another fashion. Most monocrystal (such as quartz) shock accelerometers are compression type design, as depicted in Figure 17.3.5. In this type of design, the sensor assembly is held together by a preloaded screw. When the transducer is excited into resonance, the amount of relative displacement between the components can actually result in shifting of their original positions. These physical movements of sensor parts cause a sudden change of the preload condition and manifest themselves as hysteresis effect—zeroshift at the output.

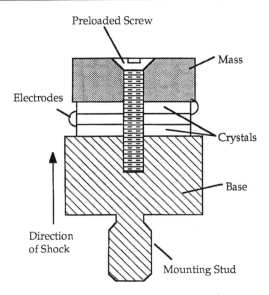

Figure 17.3.5: Compression type accelerometer.

3. The crystal material is not overstressed, and no physical shifting of parts occurred, but a huge amount of charge is generated simply due to sensor resonance. This unexpected amount of electrical signal can saturate, or in many instances, damage the subsequent signal conditioners. The result of this type of malfunction is loss of data or gross DC offset in the time history.

   Slight amount of zeroshift in the time history can yield unrealistic velocity and displacement during data reduction. The real danger remains that, although data with gross DC offsets are generally discarded, the minor offsets in the acceleration data (mostly unnoticeable by naked eyes) are accepted as good measurements.

c) *Nonlinearity*

   The output of a transducer at resonance is sometimes nonlinear and not repeatable. The response of a saturated charge converter is also nonlinear and not repeatable. The result of this type of malfunction is poor repeatability in SRS, leading to incorrect definition of the shock environment.

Chapter 17

Structural Resonance

Shock inputs with fast rise times (near true impulse) will often excite several resonant modes of the device under test (DUT) as well as the natural frequency of the accelerometer. If resonant modes of the structure are at frequencies near the natural frequency of the accelerometer, a multiplicative effect occurs. The structure multiplies the input by the Q of the DUT and that multiplied response becomes the input to the accelerometer. Thus the Q of the DUT is multiplied by the Q of the accelerometer. Although this is the true response of the DUT (at that location), the double amplification may be sufficient to damage the accelerometer or force it into a nonlinearity.

Environmental Effects

As with any measurement, shock sensors respond to their total environment. Although sensor designers attempt to minimize response to non-shock inputs, there will always be some environmental effects on the measurement. The most common environmental effects are caused by temperature, thermal transients, transverse motion and electromagnetic interference.

Temperature

The sensitivity of an accelerometer varies as a function of temperature. Normal calibration data is applicable at +20°C. If an accelerometer is to be used at other temperatures, its expected responses should be understood. Since most laboratory shock test measurements are made at or near room temperature, this is often not of any consequence. However, if field shock measurements are to be made at temperatures significantly different from "room temperature," temperature sensitivity may be of concern.

For piezoelectric accelerometers, sensitivity variation with temperature is primarily a function of the change in properties of the piezoelectric material. A design is usually specified for operation over as wide a temperature span as stability and structural integrity permit. Over this range it will have a characteristic response, and individual variations from that will depend on material parameters, assembly and part variations. A typical sensitivity plot for a temperature-compensated accelerometer is shown in Figure 17.3.6.

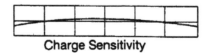

Figure 17.3.6: Typical temperature response curve of compensated accelerometer.

Sensors for Mechanical Shock

For piezoresistive accelerometers, sensitivity variation is a function of the piezoresistive coefficients of the material used, the element resistance, and any compensating resistors used within the bridge circuit. Typically, series resistors are placed in the input leads (when using constant voltage power supplies) to rotate the curve shape for minimum sensitivity change.

Sensitivity variations of high-sensitivity piezoresistive accelerometers can often be measured by resting the units in a temperature chamber and simply inverting the unit in the gravity field (a static measurement). However, the low sensitivity of shock accelerometers makes this impractical.

Because piezoelectrics cannot measure static phenomena, their sensitivity variation is usually tested by vibrating them in a temperature chamber. The unit is placed on the end of a rod with low thermal conductivity that protrudes into a temperature chamber from a vibration exciter on which it is mounted. A standard accelerometer is placed on the exciter outside the chamber (at room temperature). The test is performed at a low frequency, usually 100 to 300 Hz, to avoid any rod resonances. Low-sensitivity shock accelerometers may have such low output under these conditions that this test is impractical.

Where the need for accuracy is so great that the unit-to-unit variations from typical curve becomes unacceptable, special vibration calibrations can be performed on accelerometers, at any temperature within their rated range, on special order.

Piezoresistive, variable capacitance and other dc-responding accelerometers usually exhibit a small output with zero input (zero measurand output, ZMO). When this zero offset changes with temperature, it causes a thermal zero shift. Without temperature compensation, PR accelerometers typically have such large thermal zero shift that they are unusable. Optimum temperature compensation minimizes this effect. Typical thermal zero shift of a standard production PR accelerometer is on the order of a few millivolts, with a maximum specification limit of 15 or 20 mV.

Thermal Transients

Piezoelectric accelerometers may produce an output while their temperature is changing. This thermal transient response is commonly called pyroelectric effect. In almost all testing applications temperature changes in accelerometers occur gradually over a period of several seconds or minutes. As a result, thermal transient response may not be detected because most amplifiers have an input filter to produce a low frequency cut-off to ensure that slowly varying thermal transient output is not transmitted. The effect of thermal transients on shock measurements is usually very small because most shock events are completed in such a short time. However, these outputs must be considered if the amplifier passes these low frequencies (usually less than 1 Hz) or if the outputs are sufficiently large that they would overload the amplifier. The amplitude and frequency content of thermal transient response is a function of the magnitude and rate of change of temperature. The pyroelectric characteristics of piezoelectric crystals are known and the output for any particular accelerometer-amplifier combination can be experimentally determined under specified temperature transient conditions.

There are three mechanisms that cause thermal transient outputs; they are called primary, secondary, and tertiary pyroelectric effects. The thermal transient response of an accelerometer is the resultant of all three.

Primary pyroelectric effect is the output caused by uniform temperature change in a constrained crystal. It occurs on surfaces perpendicular to the axis of polarization. Some natural piezoelectric crystals (e.g., quartz) do not produce primary pyroelectric output. Compression designs using ferroelectric ceramics have a large primary output. On the other hand, ferroelectric shear accelerometers do not produce a primary pyroelectric response because their electrode surfaces are parallel to the axis of polarization. Their pyroelectric response is comparable to natural crystals.

Secondary pyroelectric response is caused by thermal deformation of the crystal from uniform heating. Some natural crystals (e.g., tourmaline) have a large secondary output, but still small compared to ferroelectrics.

Tertiary pyroelectric response is produced by all accelerometers. It is caused by a temperature gradient across a crystal. The tertiary component is dependent on mechanical design, polarization axis, and electrode orientation, not on the specific crystal material. In shear designs, the crystal elements are usually well isolated from the case structure so that thermally induced case strains do not produce an output. This, plus the absence of primary pyroelectric response, makes shear type accelerometers less affected by transient temperature changes than compression types. The Endevco Isobase design isolates the crystal from the accelerometer base, and decreases the thermal conductivity from the environment to the crystal. The Isobase design reduces the pyroelectric output by approximately one order of magnitude. The use of low-frequency roll-off in amplifiers will reduce the amplitude of slowly varying pyroelectric outputs to an insignificant level and block any steady state outputs. In most applications, amplifiers with a low-frequency cut-off of 3 Hz or above will have no significant output errors caused by pyroelectric effects. However, amplifiers with extended low-frequency response can pass some transient pyroelectric signals.

ANSI recommends a pyroelectric test procedure for the complete transducer-amplifier combination. In this procedure, the transducer is mounted on a test block whose mass is much greater than the transducer mass. The test block and the transducer are then subjected to a sudden 50°F ambient temperature change while the amplifier output (Figure 17.3.7) is monitored.

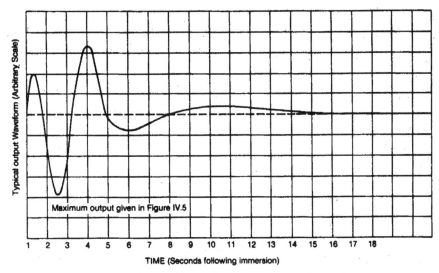

Figure 17.3.7: Thermal transient response of accelerometer.

The results of such a test on various transducers are presented in Table 17.3.1. Two different methods are compared in Table 17.3.1. In the older method, only the mounting block was immersed; in the newer method the block and the accelerometer are both immersed.

Table 17.3.1: Thermal transient response of several accelerometers

Model	Type	Sensitivity PC/g	Total Immersion	Immerse Block Only
2215E	Compression	1650	51	0.05
2220D	Shear	2.80	18	negligible
2222C	Shear	1.20	20	negligible
2275	Isobase	11.0	12	negligible
2221D	Shear	17.2	0.6	negligible
7701-50	Isoshear	50	0.2	negligible

In obtaining these data, the transducer and mounting block were stabilized at 82°F and then plunged into an ice bath at 32°F. The transient output following immersion was a series of two to five low-frequency polarity reversals of approximately sinusoidal shape within a time span of 1 to 10 seconds. The results are indicative of the maximum pyroelectric sensitivity of various transducers only under the described test conditions. Actual pyroelectric outputs for any particular accelerometer-amplifier combination should be experimentally determined under the temperature conditions present in the measurement.

A word of caution: Compression accelerometers (built with P-8 or P-10 crystals) which have been subjected to changing ambient temperatures while disconnected should not be connected to the amplifier without first shorting the transducer terminals. Damaging (not lethal, but shocking) potentials of several hundred volts have been observed in open-circuited accelerometers due to pyroelectricity.

Transverse Motion

For any single-axis accelerometer, one axis provides maximum response to an acceleration input. In an ideal accelerometer there would be no output signal for acceleration inputs along any other axes. In practice it is very difficult to produce this ideal transducer due to small manufacturing tolerances. However, a quality accelerometer will have minimal transverse sensitivity. Typical practical values are 3% or less in the axis of maximum transverse sensitivity.

Almost all Endevco accelerometers are tested for transverse sensitivity as a normal inspection step in their manufacture. The input motion must be unidirectional. Most electrodynamic vibration exciters do not have adequate low transverse motion to test high quality accelerometers. Because of this, Endevco developed a special mechanical, large displacement shaker which operates at 12 Hz and 7 g. Direction of motion is well-controlled, and the moving assembly permits rotating the accelerometer while it is being vibrated. Thus, the transverse sensitivity can be measured to produce a plot similar to Figure 17.3.8 and the maximum transverse sensitivity noted on the accelerometer calibration card. The value supplied is only a measure of the misalignment of the accelerometer. Further errors can be caused by improper surface preparation and mounting (flatness, smoothness, perpendicularity).

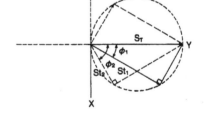

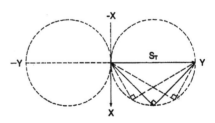

Figure 17.3.8:
Typical transverse sensitivity plot.

Sensors for Mechanical Shock

Electromagnetic Interference

Radio frequency and magnetic fields have no effect on piezoelectric elements. However, if an accelerometer includes magnetic materials, a spurious output may be observed when it is vibrated in a high magnetic field or subjected to high intensity changing magnetic flux. Adequate isolation must be provided against RF ground loops and stray signal pick-up. Insulated mounting studs can be used to electrically isolate accelerometers from ground. High intensity RF or magnetic fields may require special shielding of the accelerometer, cable, and amplifier.

Other types of accelerometers will have varying sensitivities to EMI/RFI environments depending on their construction, internal shielding, and connecting cables.

Some shock environments also include a high intensity electromagnetic pulse. Since this causes a rapidly changing magnetic field around the accelerometer, it can generate a high voltage pulse in the accelerometer and connecting cable. The induced voltage may exceed the shock signal, and may be high enough to damage signal conditioning electronics. Although it cannot prevent damage, a placebo sensor can sometimes be used to minimize the distortion of the acceleration data.

A placebo accelerometer is identical to the measurement accelerometer in all respects except it has zero sensitivity to acceleration. Therefore, any signal from the placebo must be noise induced by the environment. By placing the placebo near the measurement unit, it is exposed to the same environment. Thus the signal from the placebo can be subtracted from the signal from the measurement unit, and the result is the acceleration signal. This is not a foolproof technique, though, as some environmental effects may change the performance of the measurement unit without producing noise (change sensitivity, cause zero offset, etc.).

17.4 Applicable Standards

IEST

Institute of Environmental Standards and Technology, 5005 Newport Drive, Suite 506, Rolling Meadows, IL 60008-3841, USA, 1-847-255-1561; www.iest.org; IEST RP-DTE011.1, 2003

ISA

ISA, The Instrumentation, Systems and Automation Society, 67 Alexander Drive, Research Triangle Park, NC 27709, USA, 1-919-990-9314; www.isa.org; ISA-RP37.2-1982, ISA-RP37.5-1975, ISA-Dtr37.14.01 (draft in process).

17.5 Interfacing Information

The fast rise time, high frequency content and high amplitude of most shock pulses requires special attention to the mounting of the accelerometer. Electrical connections and signal conditioning must be capable of accurately processing the transient signal without introducing extraneous noise.

Mechanical Interface, Mounting

For specimens having small cross-sectional dimensions, the attachment method, as well as the size and mass of the accelerometer, can alter the stiffness of the specimen. Ideally, the dimensions of the accelerometer should be small compared to the dimensions of the structure in the local area where the accelerometer is attached. If the accelerometer dimensions are too large, the local stiffness of the structure increases and the resonance frequency and amplitude of vibration are correspondingly changed.

Similarly, the use of a fixture or accelerometer mounting stud may produce these stiffening effects. Choose microminiature accelerometers and use cement mounting for these small structures.

In addition to the accelerometer's effect on the dynamics of the mechanical system, an error source can also be introduced if the accelerometer is not securely attached to the structure. As frequency content increases, one must take special steps to attach the accelerometer. Accelerometers that are designed for stud mounting will ideally be mounted using the provided mounting stud and properly torquing them into the specified tapped hole.

Accelerometers may be cemented directly to the test surface with several types of epoxies and quick-set cements. The strength of the bond should be evaluated, particularly if severe shock amplitudes are expected. If strains may be present on the surface, cements should be chosen with appropriate elastic properties. To avoid degradation, bonds must be thin, for example, 0.1 mm or thinner. Cyanoacrylate adhesives provide extremely thin bonds and minimal response change with miniature designs.

Threaded studs are used to mount most accelerometers. Several important considerations are:

1. The surface condition of both the accelerometer and test specimen must be flat, smooth and clean. For most applications, the surface should approximate a 0.0003 inch TIR flatness and less than 32 microinch rms roughness.

2. Coating all mating surfaces with a thin film of oil or acoustic couplant improves coupling at high frequencies and is recommended when frequency components exceed 2000 Hz.

3. The manufacturer's recommendations for mounting torque should always be followed.

4. Insulated mounting studs may be needed for electrical isolation in some applications. When used, the resonant frequency is reduced and the effect on frequency response can be significant. For example, the resonant frequency of a 30 kHz, 1-ounce accelerometer typically reduces to 25–26 kHz when an insulated stud is used. Below 5 kHz the difference is less than 1%.

5. If the accelerometer cannot be mounted to a rigid block or boss, an accelerometer with low base strain sensitivity should be used.

Figure 17.5.1: An assortment of special mounting blocks and adapters.

The use of mounting blocks and fixtures almost always degrades frequency response above 1000–2000 Hz for all but the microminiature accelerometers. When tests are performed at higher frequencies, a frequency response calibration of the accelerometer with fixturing is recommended. If biaxial or triaxial measurements are required, biaxial or triaxial accelerometers provide better frequency response than uniaxial accelerometers on a fixture or block. Figure 17.5.2 shows typical deviations in frequency response for a 30-gram accelerometer having a 30-kHz resonance frequency when mounted by different techniques.

Chapter 17

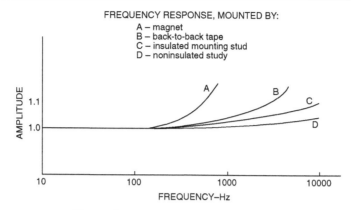

Figure 17.5.2: Effect of attachment means on frequency response of a 30-gram accelerometer.

Endevco Technical Paper Number 312, by James Mathews, explores the effect of various mounting methods on lighter weight accelerometers at different temperatures. Generally, the best adhesive is a cyanoacrylate such as Aron Alpha, or Super Glue. They cure very quickly at room temperature, and provide a broad frequency range and good temperature range. Disadvantages are the need for a solvent to break down the glue bond for removal, time-consuming removal, and difficulty getting a good bond on a rough surface.

Dental cement is also useful in many applications. It provides good transmissibility and high strength. However, like the cyanoacrylates, it is difficult to remove from the test structure and from the accelerometer.

Not Recommended for shock measurement:

Waxes, such as petro-wax and beeswax, double-sided tape, hot glue and magnetic mounts are not recommended for accelerometer mounting for shock measurements.

Petro-wax is often a convenient alternative, but has rather strict limitations. It is very easy to use, but requires a thin (less than 0.1 in.) wax layer. It is also very quick to use and easy to remove. Because of limited bonding strength, Petro-wax should not be used at peak vibration levels above about 20 g, frequencies above 2 kHz, or temperatures above 200°F.

Double-sided tape is also quick and convenient, and has a temperature range similar to cyanoacrylates. However, it has low bonding strength limiting the amplitude range. Cable motion of some top connector or high profile accelerometers can cause transverse forces that may break the bond.

Hot glue from a glue gun may be useful, is better than waxes or tape, but is problematic for shock measurements. Hot glue mounting is appropriate for temperatures from −18°C to +93°C. Above 93°C, hot glue loses its stiffness and frequency response decreases rapidly. It is very convenient and provides quick cure time. The quick cure time is also one of its greatest disadvantages, requiring the user to mount the accelerometer as soon as the glue is applied. Maintaining a thin bond line for maximum transmissibility is difficult with hot glue applications.

Magnetic mounts are not recommended for shock measurement. For vibration measurements, if the mounting surface is a magnetic material, accelerometers are often mounted using a magnetic mounting attachment. They are often used on industrial machinery that has cast iron or steel structures. They offer ease of use, quick mounting, broad temperature range, good holding strength, and are available in a variety of sizes. Disadvantages are size and weight (may increase mass loading effects), reduced bandwidth and the need for careful application. If a user "slaps" the accelerometer/magnet onto a hard surface, the high frequency, high amplitude shock can cause catastrophic damage to an accelerometer.

TIP: If possible, it is advisable to have the accelerometer calibrated for frequency response, at the expected usage level and temperature, mounted as it will be mounted in service. Ideally, shock measurement sensors should be calibrated using a shock input.

Electrical Interface, Signal Conditioning

Signal conditioning electronics must be properly matched to the accelerometer used. Shock measurement may be accomplished by almost any type of sensor, so the user must be sure to match the signal conditioner to the type of sensor. Charge mode ("High Impedance") accelerometers require a charge amplifier; IEPE ("Low Impedance") accelerometers require a signal conditioner that provides constant current excitation and a coupling capacitor; strain gage (piezoresistive or metal strain gage) and variable capacitance accelerometers require a bridge amplifier (usually an "instrumentation amplifier") with excitation voltage source; velocity sensors require an instrumentation amplifier but do not need any excitation voltage source.

Regardless of the type of sensor, the signal conditioner must have adequate low- and high-frequency response and any filtering must be designed for transients.

17.6 Design Techniques and Tips, with Examples

High Mechanical Resonance Frequency

Mechanical resonance of a shock sensor should be at least three times (five times is recommended) the highest frequency in the transient being measured. Frequency response at one-fifth of the resonance frequency, for most shock sensors, is up by about four percent; at one-third the resonance frequency, response will be up about 12%.

Rugged

High frequency or short rise time of the shock pulse may excite the resonance of the sensor, so the sensor should be able to withstand occasional resonance excitation without permanent damage. Even relatively low level excitation at frequencies near the resonance frequency can cause excessive response because of the high Q resonance of most shock sensors. Piezoresistive (silicon strain gage) accelerometers are often broken by this kind of excitation. Although piezoelectric accelerometers may survive resonance excitation, they may be damaged such that their sensitivity decreases. Repetitive resonance excitation of piezoelectric accelerometers may cause low cycle fatigue failure of the crystal or other internal components.

Damped Resonant Response

Some piezoresistive and variable capacitance accelerometer designs incorporate mechanical damping to flatten their frequency response curves and reduce the Q of their resonance responses. The ideal damping ratio is 0.707 times critical to provide minimal phase distortion and ideal frequency response.

Mechanical Filtering

There are two shock accelerometer designs (Endevco 7255A and 7270AM6) that incorporate mechanical filtering between the mounting surface and the sensing mechanism. This provides mechanical low-pass filtering to protect the sensing mechanism and to flatten frequency response. It is combined with an integral matched electronic filter to provide flat response up to 10 kHz and then produce rapid rolloff at higher frequencies. External mechanical filter adapters have also been utilized in some special applications. Figure 17.6.1 shows a mechanical schematic of an internal mechanical filter.

Sensors for Mechanical Shock

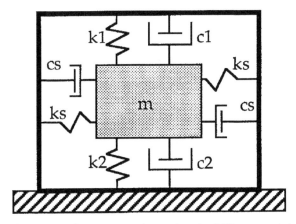

Figure 17.6.1: Internal mechanical filter model.

Electronic Filtering

Electronic low-pass filtering is often used in shock measurement systems. It can be incorporated into the sensor design, included in the signal conditioning, provided for anti-aliasing prior to an analog to digital converter, or performed with digital signal processing as part of the data post processing. In most cases, the earlier in the instrumentation chain the low-pass filtering is performed, the better the results and the less the likelihood of distortion, clipping, or overloading. Figure 17.6.2 shows a schematic of an accelerometer with both mechanical and electronic filtering built in.

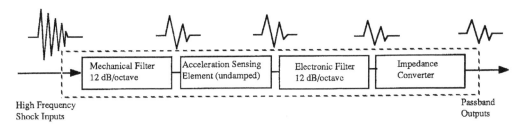

Figure 17.6.2: Integral electronic filter schematic.

17.7 Latest and Future Developments

MEMS

Micro-Electro-Mechanical Systems (MEMS) are fabricated by chemically etching silicon or other similar materials to precisely controlled shapes. If a silicon wafer, sliced from a silicon crystal, is photolithographically masked and the proper etchant applied, geometrically precise shapes can be produced. Depending on the orientation of the wafer cut, the etchant, and the time of exposure, deep rectangular or other precise shapes can be etched into the silicon. This, combined with doping and thin film processes, can be used to produce very small, precisely controlled mechanical structures, which may also incorporate electronic components.

MEMS is used to fabricate small, lightweight, extremely robust silicon sensing elements that are used in some shock sensors.

References

1. Ed. Jon S. Wilson, *Shock and Vibration Measurement Technology*, Publ. Endevco (Part No. 29005), San Juan Capistrano, CA, 2002.

2. *IEST RP DTE-011.1, Sensor Selection for Shock and Vibration Measurement*, Publ. Institute of Environmental Sciences and Technology, Chicago, 2003.

3. *IEST RP DTE-012.1, Handbook of Dynamic Data Acquisition and Analysis*, Publ. Institute of Environmental Sciences and Technology, Chicago, 1993.

4. Anthony Chu, *Endevco Technical Paper 290, Zeroshift of Piezoelectric Accelerometers in Pyroshock Measurements,* Endevco, San Juan Capistrano, CA, 1987.

5. Anthony Chu, *Endevco Technical paper 308, Built-In Mechanical Filter in a Shock Accelerometer,* Endevco, San Juan Capistrano, CA.

6. James Mathews, *Endevco Technical Paper 312, Guide to Adhesively Mounting Accelerometers,* Endevco, San Juan Capistrano, CA, 1998.

CHAPTER 18

Test and Measurement Microphones

Dr. John Carey, Larson Davis, Inc.

A microphone provides an analog output signal which is proportional to the variation in acoustic pressure acting upon a flexible diaphragm. This electrical signal is then used to transmit, record or measure the characteristics of the acoustic signal. The most common applications are related to audio broadcasting, recording and reproduction, where the frequencies of importance are in the human hearing range, 20 Hz–20 kHz. These microphones are often highly directional. This is a desirable feature when the intention is to reproduce the sound produced by a particular singer or instrument, particularly in the presence of background noise. Also, their ability to represent the measured sound level with precision is less important than their ability to reproduce the sound electrically in terms of parameters important to audio technicians, such as color, depth, warmth, etc.

Measurement microphones differ from those used for audio applications in that their primary role is to electrically reproduce the sound waveform without distortion and with a linear relationship between the voltage out and the pressure sensed by the microphone diaphragm. This precision must be maintained over a wide range of frequencies and amplitudes when measuring sound waves arriving from different angles. Furthermore, they are expected to maintain this degree of precision over a range of temperature and barometric pressure variations.

18.1 Measurement Microphone Characteristics

The numerical relationship between the output voltage and the acoustic pressure sensed by the diaphragm is the *sensitivity*, which is expressed in units of mV/Pa. Pa is the abbreviation for Pascal, a metric unit of pressure expressed in Newton/m^2. The magnitude of the sensitivity is important in relation to the inherent electrical noise of the measurement system, since one cannot properly measure a voltage that is near the voltage noise floor of the instrument itself. Thus, for a given microphone and measurement system, the magnitude of the sensitivity will establish the minimum sound pressure, which can be accurately measured.

Any variation of the sensitivity would produce a distortion in the output signal compared to the acoustic signal. Thus, the following parameters are essential in a measurement microphone.

- The sensitivity should be nearly constant over the range of frequencies to be measured. Thus, a quality measurement microphone should have a "flat" frequency response.

- The sensitivity should be nearly constant over a wide range of sound pressure levels. This is expressed as *linearity*, since a constant sensitivity would produce a straight-line graph of output voltage versus sound pressure.

- The sensitivity should be nearly constant over a wide range of temperature and barometric pressure.

18.2 Common Microphone Types

Dynamic microphones, in which an electrical coil connected to the diaphragm is moved through an electrical field, generating a voltage proportional to the velocity of the moving element. Dynamic microphones have many characteristics that are desirable for audio applications, but they have a high sensitivity to vibration, limited dynamic range and their frequency response is generally not flat enough for measurement applications.

Piezoelectric Microphones, in which the force of the sound pressure acting on the diaphragm is transmitted to a piezoelectric element, generating a charge proportional to the applied force. These microphones are often used in very high pressure situations, such as measuring explosive blasts. Also, because of their rugged design, they find use when trying to measure small dynamic pressure fluctuations riding on top of large static pressures. In general, the noise floor of these microphones limit their use for precision sound measurement applications.

Condenser (capacitive) microphones, in which the movement of the diaphragm relative to a fixed backplane produces a variation in capacitance proportional to the deflection of the diaphragm. An electrical circuit provided in the microphone preamplifier converts the capacitance variation to a voltage variation. The characteristics of condenser microphones—high sensitivity, wide dynamic range, flat frequency response, low internal noise, low distortion and high stability—make them the design of choice for measurement microphones. Less precision versions of the condenser microphone, much less expensive to manufacture than measurement quality designs, are also commonly used in audio applications such as broadcasting and portable telephones.

18.3 Traditional Condenser Microphone Design

Figure 18.3.1 presents a sectional view of a typical condenser microphone showing details of the mechanical elements.

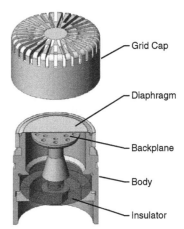

Figure 18.3.1: Measurement microphone, cut-away view.

The diaphragm, a thin metal sheet stretched and clamped over the body of the microphone, deflects with the variation of air pressure about the atmospheric value. A stainless steel backplane makes up the other plate of a capacitor, whose capacitance varies as the diaphragm is displaced in response to sound pressure variations. The precision required to produce a microphone of measurement quality is extreme; the diaphragm thickness will be in the range 1–2 μinch and the space between the diaphragm and the backplane will be on the order of 0.001 inch.

In the conventional condenser microphone, a highly stable DC polarization voltage, also referred to as the bias voltage, is applied between the diaphragm and the backplane which generates charges of opposite polarity on them. Variations in the spacing between the diaphragm and the backplane resulting from the action of the acoustic field result in proportional variations of the difference in voltage between them. In practice the microphone is threaded onto a cylindrical preamplifier of the same diameter containing an electrical circuit that has an ultra-high input impedance. The polarization voltage is typically 100–200 volts, although lower levels in the 25–35 volt range are sometimes used to reduce the sensitivity of the microphone or to minimize the potential for electrical breakdown in the air gap when working in high humidity environments.

The dynamics of the condenser microphone can be modeled as a spring-mass-damper system, where the diaphragm is the mass, the tension in the diaphragm the spring and the damping is provided by air friction in the space between the diaphragm and backplane. The frequency response is therefore flat up to a relatively high frequency where the effects of the resonance of this mass-spring-damper system begin to be seen. The lower frequency is controlled by the size of the vent opening, which is necessary to equalize the interior of the microphone to variations in atmospheric pressure. The majority of applications of condenser microphones are at frequencies well above the low frequency –3 dB point, which is typically a few hertz.

18.4 Prepolarized (or Electret) Microphone Design

An alternative design for a condenser microphone is the prepolarized microphone, in which the electric field between the diaphragm and backplane is generated by surface charges fixed on the backplane, called an "electret." This is usually accomplished by coating the backplane with a polymer material and charging it electrically. Operationally, the prepolarized microphone will have the same performance specifications as a similar condenser microphone of traditional design. The major advantage of the prepolarized is that no DC polarization voltage is required, greatly simplifying the design and manufacturing cost of the preamplifier used to drive it. This allows for cost effective coaxial cables and power supplies to be utilized.

Prepolarized microphones, having no free charges, can also be used in very high humidity environments where electric discharge between the diaphragm and backplane might otherwise occur which would generate noise and possibly damage the microphone. This is one of the drawbacks of the traditional condenser microphone with its high polarization voltage.

18.5 Frequency Response

The frequency response is measured using an electrostatic actuator, a device that is attached in place of the usual protective grid cap. This places a metal electrode in close proximity to the diaphragm, to which a high DC voltage (~800 volts) and a lower voltage AC signal are applied from an external generator. Due to electrostatic attraction, the alternating voltage exerts an oscillating force on the diaphragm, causing the diaphragm to displace in the same manner as would a variation of pressure on its exterior surface. By sweeping the frequency of the generator over a range of frequencies, the frequency response curve can be determined and plotted. The lower curve in Figure 18.5.1 shows the electrostatic actuator frequency response of a ½" microphone. This technique is not effective at lower frequencies, so this method is used primarily to determine the variation of the frequency response from the flat region up to and beyond the resonance frequency.

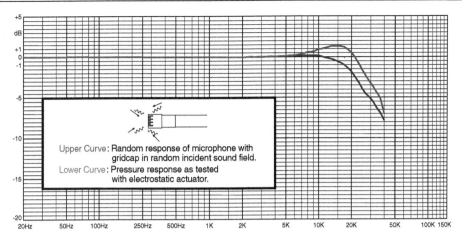

Figure 18.5.1: Frequency response of ½" microphone using an electrostatic actuator (lower curve).

Effect of Angle of Incidence of Sound Wave

The electrostatic actuator simulates the action of a pure pressure variation uniformly distributed across the surface of the diaphragm. This is called the *pressure response* of the microphone. In practice, one situation in which the diaphragm might be excited by a uniform pressure is when the microphone is sensing the pressure within a cavity whose dimensions are much smaller than a wavelength of the exciting frequency. In this case, there is no wave motion and the pressure varies uniformly throughout the cavity as a function of time. However, in most sound measurement situations, the pressure driving the diaphragm is the result of one or more sound waves impinging on its surface from various directions.

An acoustic wave generates both pressure variations and local particle motions in the fluid medium. In certain circumstances, the particle velocity is in the same direction as the propagation of the wave and also in-phase with the pressure variation. One of these situations is when a wave is propagating within a tube whose diameter is much less than the wavelength of the wave. Another is when the wave is generated by a single source in a free and empty space (no reflections) and the measurement is being made at a distance far from the source. We refer to these as *plane waves* and we can use them to illustrate the effect of angle of incidence of the response of the microphones.

Chapter 18

Pressure Microphones

Consider the situation in the acoustic free field where a sound wave from a single source is approaching the microphone in a direction parallel to the surface of the diaphragm. The angle of incidence is defined as relative to the direction normal to the surface of the diaphragm, so this situation is referred to as 90° incidence. The wave moves smoothly across the diaphragm without reflection, so the diaphragm is driven only by the pressure associated with the traveling wave. For this reason, the microphone design we have been describing is referred to as a *pressure microphone*. At low frequencies, where the wavelength of the sound is much larger than the diameter of the microphone, the pressure across the surface will be essentially uniform. However, at higher frequencies, as the wavelength of the sound decreases, the pressure across the diaphragm becomes non-uniform, eventually leading to with areas of positive and negative pressure at very high frequencies, as shown in Figure 18.5.2.

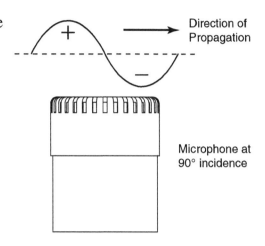

Figure 18.5.2: Wave traversing microphone diaphragm, 90° incidence.

The overall effect of this is a diminishing force on the diaphragm. As a result, the frequency response of the microphone will drop-off at higher frequencies, as shown by the 90° incidence curve on the right in Figure 18.5.3, limiting the upper frequency limit of the flat response portion of the curve, and thus the useful measurement range of the microphone. Since this effect is based on the diameter of the microphone as a function of the wavelength of the sound, the upper limit of the useful frequency range of this type of microphone is higher for smaller diameters.

If we rotate the microphone so that the wave strikes the surface of the diaphragm as shown by the 0° incidence diagram on the left of Figure 18.5.3, this effect will not occur because the pressure will be uniform across the face of the diaphragm at all frequencies. However, because the diaphragm is solid, it is necessary that the particle velocity at the diaphragm surface be zero, which is accomplished by the presence of a reflected wave propagating back in the direction towards the source (this phenomenon only occurs at higher frequencies). The result is an increase in the pressure seen by the diaphragm. The curve labeled 0° in Figure 18.5.3 illustrates how this reflection affects the shape of the upper end of the flat region of the frequency response of the microphone as a function of frequency. The effect of this reflection for other angles of incidence is less pronounced, as shown in Figure 18.5.3.

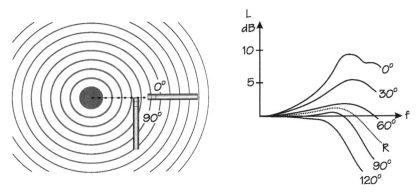

Figure 18.5.3: Free field corrections.

For most microphone designs, this 0° *free field correction* can be calculated mathematically by modeling the microphone as a cylinder whose axis is aligned with the direction of propagation of the incident sound wave. To determine the frequency response of a particular microphone for waves approaching from any angle, the free field correction for that angle is added to the frequency response obtained using an electrostatic actuator. We refer to the frequency response of the microphone for 90° incidence as the *pressure response*, and for 0° incidence as the *free field response*.

Free Field Microphones

Note that the effect of the reflection from the diaphragm at 0° degree incidence is to increase the frequency at which the curve drops off, although the departure from the flat response makes this frequency region unusable for measurement purposes. By modifying the microphone design to increase the damping, it is possible to reduce the amplitude of the peak associated with the refection, thereby increasing the upper limiting frequency of the flat region of the curve. This creates a new type of microphone, the *free field* microphone, which measures best when oriented such that the wave is approaching with 0° incidence. The microphone is designed to compensate for its presence in the sound field.

In Figure 18.5.4, we see the typical pressure (electrostatic actuator curve) and free field response of a free field microphone.

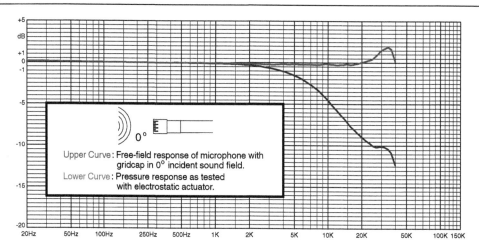

Figure 18.5.4: Free field microphone frequency response.

Figure 18.5.5 shows the directional characteristics of a free field microphone in a polar format, along with the same data for a sound level meter using this microphone. The effect of the body of the instrument can be clearly seen for sound waves approaching from angles greater than 90°.

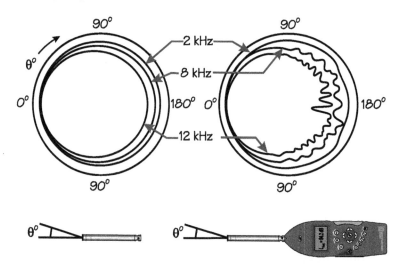

Figure 18.5.5: Directionality of a ½" free field microphone, alone and mounted on a sound level meter.

From this data, we can see that when using a free field microphone to measure the sound generated by a single source in a free field, the best measurement is obtained by aiming it towards the source, at 0° incidence. When using a pressure microphone, however, the best measurement is obtained by orienting the microphone axis perpendicular to the direction of wave propagation, at 90° incidence.

The main situations where a measurement is made of a single sound source in a free field are as follows:

1. In an open space outdoors, far from other noise sources.

2. Inside an anechoic chamber, a specially designed test room whose surfaces are covered with long, acoustically absorbent wedges designed to trap and absorb incident acoustic waves.

Random Incidence Microphones

In most acoustic measurement situations the microphone is exposed to sound waves approaching it from many directions. This could be due to the presence of multiple sound sources, moving sound sources, or the result of waves reflected from nearby hard surfaces as is common with indoor measurements. When using a microphone under such conditions, where impinging sound waves could be arriving from a variety of directions that might be varying with time, we define a *random incidence correction* in a similar manner to the free field correction. To determine this, measurements are performed in an anechoic environment using a single sound source to determine the frequency response of the microphone using angles of incidence over a 360° range, in steps of approximately 5°. Assuming that the waves arrive with equal probability over this range, a mathematical calculation can produce the frequency response which would have been obtained had it been possible to actually simulate such a test. The procedure for determining the random incidence response of a microphone is described in the international standard *IEC 1183 (1994) Electroacoustics-Random-incidence and diffuse-field calibration of sound level meters*. The result is a frequency response curve lying between the pressure and free field response curves, as indicated by the curve labeled R in Figure 18.5.3. For a pressure microphone, the random incidence response will be fairly close to the pressure response so these are often used in random incident measurement situations. The upper curve in Figure 18.5.1 represents the random incidence frequency response of a ½" pressure microphone.

The design of a pressure microphone can be modified to provide a frequency response that is optimized for random incidence measurements. Free field microphones, however, do not generally have an acceptable random incidence response.

Thus, a particular measurement microphone will be designed for use under one of the following acoustic field conditions:

- pressure
- free field
- random incidence

In many cases pressure microphones having diameters of ½" or less are designated as both pressure and random incidence microphones since their frequency response characteristics are satisfactory for use in both types of acoustic fields.

18.6 Limitations on Measurement Range

Lower Level Limits

In addition to the limitations on useful frequency range described above, there are also limitations on the sound pressure level, in dB, over which an accurate measurement can be obtained. The lower measurement limit of the microphone is established by its cartridge thermal noise. This is the dB level that would be read by a measurement instrument connected to the microphone output when there is no acoustic pressure applied to the microphone. There are two sources of thermal noise, air damping and preamplifier circuitry. The air damping causes a white noise that is a property of the microphone. The preamplifier has low frequency noise which is inversely proportional to frequency as well as white noise. In general, larger microphones have lower thermal noise, so they are often used for measurements of low noise levels.

Upper Level Limits

Mathematical modeling of the dynamics of the condenser microphone shows that the dynamic response in terms of output signal to capacitance variation is slightly nonlinear.

This produces distortion in the output signal that increases at higher amplitudes. The upper measurement limit is established as the amplitude, in dB, for which the total harmonic distortion of the output signal exceeds 3%, a level that can be calculated analytically for a particular microphone design. The actual limit is not only dependent upon it's physical limitations, but it is effected by both the sensitivity of the microphone (mV/Pa) as well as the output voltage of the mated preamplifier. Lower sensitivity microphones can typically measure higher decibel levels.

Effect of Diaphragm Tension

The dynamics of a microphone are strongly influenced by the tension applied to the diaphragm during manufacture. A looser diaphragm will deflect further for a given pressure excitation, producing a greater sensitivity. However, it will also reduce the upper frequency limit and the upper amplitude limit. Most manufacturers of condenser microphones produce two versions of ½" diameter microphones, the one with the looser diaphragm being referred to as a *high sensitivity microphone*.

18.7 Effect of Environmental Conditions

The sensitivity of a condenser microphone can be influenced to some degree by variations in temperature, humidity and barometric pressure. Microphone standards require that the manufacturer specify these variations. More significantly, as will be explained in more detail later, sound level meter standards establish maximum permitted variations in performance over specified ranges of temperature, humidity and barometric pressure. Since the microphone is the component having the major influence on these variations, this in effect establishes required performance specifications in terms of temperature, humidity and barometric pressure on the microphone that is to be used with the sound level meter.

Typical Performance Specifications

The most common types of microphones commercially available have diameters of 1", ½" and ¼". These are generally available in pressure, free field or random incidence designs. And, as mentioned above, ½" microphones are often available in standard and high sensitivity versions. Typical performance specifications are as follows:

Pressure and Random Incidence Microphones				
Diameter, Inch	¼	½	½, high sensitivity	1"
Type	P and RI	P and RI	P and RI	Pressure
Sensitivity, mV/Pa	1.3	12	50	50
Low Frequency, Hz*	4	4	3	3
High Frequency, kHz*	70	25	10	8
Thermal Noise, dB	31	18	15	10
3% Distortion Limit, dB	170	160	145	145

* *Based on frequency response variation within ±2%.*

Chapter 18

Free Field Microphones				
Diameter, in.	¼	½	½, high sensitivity	1"
Sensitivity, mV/Pa	4	12	50	50
Low Frequency, Hz*	4	4	3	3
High Frequency, kHz*	80	40	20	18
Thermal Noise, dB	30	20	15	10
3% Distortion Limit, dB	160	160	145	145

Based on frequency response variation within ±2%.

18.8 Microphone Standards

Microphones used for test and measurement purposes are governed by the international standard *IEC 1094-4, measurement microphones-specifications for working standard microphones* (IEC stands for International Electrotechnical Commission). This standard establishes the parameters that must be measured and provided by the manufacturer. In addition to those presented above, this standard requires the reporting of the following specifications:

- Effective front volume
- Linearity range
- Static pressure coefficient
- Temperature coefficient
- Relative humidity coefficient
- Pressure equalizing time constant
- Long-term stability coefficient
- Short-term stability coefficient

Sound Level Meter Standards

One of the main uses of measurement microphones is as the sensing element on precision sound level meters. The two standard organizations whose standards dominate the design and utilization of sound level meters are:

- International Electrotechnical Commission (IEC)
- American National Standards Institute (ANSI)

These standards are the following:

- IEC 61672-1 (2002-05) Sound Level Meters-Part 1: Specifications
- IEC 61672-2 (2002-05) Sound Level Meters-Part 2: Pattern Evaluation Tests
- ANSI S1.4-1983 (R2001) with Amd. S1.4A 1995 Specifications for Sound Level Meters
- ANSI S1.43-1997 (R2002) Specifications for Integrating-Averaging Sound Level Meters

Test and Measurement Microphones

While many items in these standards deal with meter characteristics, such as detection and averaging of signals, certain others such as acoustic frequency response and directionality specifications involve the microphone directly, since it is the element of the meter which has the predominant influence on them. The ANSI standards are largely followed in North America while Europe and much of the rest of the world follow the IEC standards. In many respects, particularly those relating to signal processing, the specifications are very similar. However, there is a great divergence concerning acoustic frequency response specifications, which are mainly determined by the microphone alone.

The specifications of IEC 61672-1 require the sound level meter to have a very flat frequency response over a specified range of frequencies for sound waves impinging on the microphone diaphragm at 0° incidence, but they are not too demanding on the variation from that response for waves approaching the microphone from other directions. Conversely, ANSI S1.4-1983 (R2001) and ANSI S1.43-1997 (R2002) require the frequency response averaged over all angles of incidence to be quite flat while they are less demanding then IEC 61672-1 on the response associated with 0° incidence.

As a result of these differences, for a particular sound level meter to meet the IEC standards, it should use a free field microphone but to meet the ANSI specifications it should use a random incidence microphone. It also means that a sound level meter equipped to meet the IEC standard will measure most accurately when measuring a single source with the sound level meter "pointed" towards the source and less accurately when measuring in a space having multiple sources or reflections of waves from room surfaces.

For obvious reasons, the sound level meter equipped to meet the ANSI standard will measure more accurately in multiple source or diffuse field situations than one equipped to meet IEC standards.

When there is only one source, the ANSI meter will measure less accurately than the IEC meter, but an issue is how to orient the random incidence microphone to measure as accurately as possible. Examination of the frequency response as a function of incident angle indicates that the meter should be oriented such that the sound waves from the source strike the microphone diaphragm at a angle in the range 70°–80° in order to provide the best measurement.

Chapter 18

In practice, the selection of a microphone is as simple as deciding whether the best measurement is desired, or the ability to state that the measurement was made using a sound level meter that meets the specifications of a particular standard. In some cases an application standard may require that the measurement device meet an IEC or ANSI standard. Otherwise, select the microphone type based on its functional specifications and orient the microphone appropriately.

These standards also define several classes of sound level meters, two of which are Class or Type 1 (precision) and Class or Type 2 (general purpose) based on accuracy (at this moment ANSI uses the designation Type but IEC switched from Type to Class in the most recent update of the standard). This has led to situations where users refer to microphones as Type 1 or Type 2 when in fact this designation refers to sound level meters, not microphones. This can essentially be taken as a way of defining the accuracy of the microphone in question in terms of whether it could be used on a sound level meter that is to meet the accuracy requirements of Type 1 or 2.

18.9 Specialized Microphone Types

Sound Intensity Microphones

A single microphone can measure only the sound pressure at its surface. A technique called sound intensity measurement permits the measurement of the energy flow in a defined direction. This is accomplished by using a pair of closely spaced measurement quality microphones and outputting their signals to a dual channel analyzer capable of measuring the cross-spectral characteristics of the two signals. Figure 18.9.1 shows a typical sound intensity probe used for these measurements.

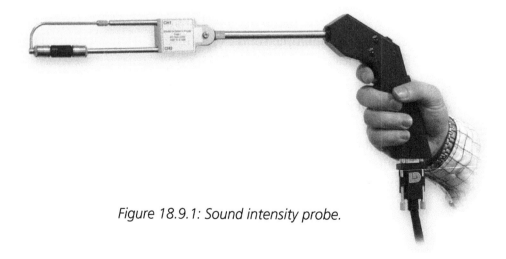

Figure 18.9.1: Sound intensity probe.

A detailed description of the technique is beyond the scope of this chapter, but it is appropriate to comment on the microphones used for this measurement. The measurement of the phase between the signals is one of the most important parameters in the determination of sound intensity. To obtain good results, it is essential that there be a very low phase difference between them. For this reason, sound intensity microphones are generally sold in phase-matched pairs. Another feature is that their protective grid caps incorporate a small threaded stud in the center to permit them to be connected to both ends of a spacer, which holds them securely apart at a known distance.

Array Microphones

There is a growing interest in the use of measurement techniques such as acoustic holography in which simultaneous time recordings are made of the signals from a large number (16+) of microphones arranged in a fixed array. Obviously cost becomes an important issue when using such a large number of microphones, yet precision in frequency response and phase-match must be maintained to provide sufficient accuracy.

Small size is also important to minimize the interference of the microphones on the sound field being measured. Usually the microphone capsule and the preamplifier are integrated into a single unit that can easily be fixed into a support frame defining the physical arrangement of the microphones in space. The microphones are of a prepolarized type to eliminate the need for a bias voltage and the internal electronics of the preamplifier are ICP®-powered, a technology commonly used for accelerometers with integrated circuits where an external current source of 2–4 mA provides the power.

Figure 18.9.2: Array microphone.

Modern versions offer TEDS (Transducer Electronic Data Sheet) technology, in which technical specifications including sensitivity, manufacturer, model and serial numbers and calibration information are stored internally in digital format. This permits measuring instruments having the capability to read TEDS data to utilize such information automatically in creating a measurement setup, making it unnecessary to manually enter the sensitivity for each microphone, for example.

Probe Microphones

A probe microphone is essentially a small diameter condenser microphone having a rigid cap instead of a diaphragm with a threaded opening into the space between the diaphragm and the end cap permitting the attachment of a small diameter hollow probe tube.

The tubes have a diameter of approximately 1.25 mm and can have a variety of lengths, typically in the range 25–200 mm. The long, narrow tube is used for measurements in small enclosed spaces, in hard-to-access areas and in the acoustic near field where the small size and high acoustic impedance of the probe have a minimum effect on the sound field. Since the tubes are usually made of stainless steel, measurements can be made in areas where the temperature is as high as 800°C (as long as the temperature inside the microphone and preamplifier housing is maintained below specified maximum limits).

Figure 18.9.3: Probe microphone. (Courtesy of G.R.A.S. Sound and Vibration.)

Although the presence of the tube causes the frequency response to begin rolling off above 300 Hz, the roll-off follows a smooth curve up to 20 kHz. This permits the use of correction curves, either supplied by the manufacturer or determined by a calibration procedure, to calculate the sound pressure level at the probe tip based on that measured by the microphone element within the body of the probe microphone assembly.

18.10 Calibration

Sound Level Calibrators

A very useful device when working with microphones is the battery powered sound level calibrator, such as shown in Figure 18.10.1.

Figure 18.10.1:
Sound level calibrator.

It has an internal speaker, which is electrically driven at a fixed frequency, typically 250 Hz or 1 kHz, at a feedback-controlled voltage level such that a microphone inserted into the opening in the calibrator will be exposed to a constant sound pressure level, typically 94 or 114 dB.

Pistonphone Calibrator

A pistonphone is a mechanical device in which a vibrating piston generates an acoustic pressure field within a cavity at a particular frequency, typically 250 Hz, having a fixed sound pressure level, typically 124 dB. When a microphone is partially inserted into an opening in this cavity, its diaphragm will be exposed to this same sound pressure level. The acoustic output level generated by a pistonphone is a function of the barometric pressure, so the value is specified for measurements made at sea level. When using a pistonphone, it therefore necessary to measure the barometric pressure at the measurement location and correct the specified output level of the pistonphone to compensate for the effect of atmospheric pressure variations from the sea level value.

Chapter 18

Insert Voltage Calibration

Most manufacturers utilize a calibration procedure referred to as *insert voltage calibration* to determine the open circuit sensitivity of the microphones they produce. In this procedure, an electrical generator is connected between ground and the diaphragm material, a digital voltmeter is connected between the backplane and ground, and a sound level calibrator is placed over the microphone. Two measurements are made. In the first, the sound level calibrator is switched on to apply an acoustic signal of known amplitude and frequency to the microphone and the output voltage is measured. Then, with the calibrator still in place over the microphone, but switched off, an electrical signal is applied to the microphone and the output adjusted until the voltmeter reads the same voltage as had been read when the calibrator had been exciting the microphone acoustically. These measurements permit the calculation of the *open circuit sensitivity* of the microphone.

Field Calibration

The open circuit sensitivity is the sensitivity that would on be obtained if the microphone were used in conjunction with a preamplifier having infinite input impedance. In practice, however, the input impedances of microphone preamplifiers are not infinite and vary between different types and models. As a result, the sensitivity of the combined microphone and preamplifier may differ from the open circuit sensitivity by as much as 0.25 dB. In order to obtain an accurate measurement using a particular microphone and preamplifier, it has become standard practice before each measurement session to use a sound level calibrator or pistonphone and adjust the measurement system, or sound level meter, to read the dB output level specified for the calibrator. With modern sound level meters and measurement systems, a corresponding value of sensitivity can be determined and tracked over time to identify changes in the sensitivity of a microphone which can be indicative of problems.

Reciprocity Calibration

A condenser microphone displays reciprocal dynamic behavior, meaning that a time-varying electrical signal applied between the diaphragm and the backplane will cause the diaphragm to vibrate and generate an acoustic output having a similar waveform, just as a time-varying acoustic pressure generates an electrical output having the same waveform.

Based on this, it is possible using three microphones, two of which are reciprocal, to determine the absolute free field and random incidence sensitivities of all three based on pair-wise testing of them using one as a sound source and the other as a receiver. The details are beyond the scope of this chapter. For additional information, the reader is referred to the international standards IEC 61094-2 and 3. Complete systems to perform microphone reciprocity microphone calibration are commercially available.

18.11 Major Manufacturers of Test and Measurement Microphones

ACO Pacific
2604 Read Avenue
Belmont, CA 94002
U.S.A.
www.acopacific.com

Brüel & Kjær Sound & Vibration
Skodsborgvej 307
2850 Nærum
Denmark
www.bksv.com

G.R.A.S. Sound & Vibration
Staktoften 22D
2950 Vedbæk
Denmark
www.gras.dk

Microtech Gefell GmbH
Muehlberg 18, D-07926
Gefall, Germany
www.microtechgefell.como

Norsonic AS
PO Box 24
N-3421 Lierskogen
Norway
www.norsonic.com

Ono Sokki
1-16-1 Hakusan, Midori-ku
Yokohama 226-8507
Japan
www.onosokki.co.jp

PCB Piezotronics, Inc.
3425 Walden Ave.
Depew, NY 14043
www.pcb.com

References and Resources

1. C. M. Harris (ed), Handbook of Acoustical Measurements and Noise Control, 3rd ed., McGraw-Hill, New York, NY 10020, 1988.

2. Leo L. Beranek (ed.), Noise and Vibration Control, McGraw-Hill, New York, NY 10020, 1971.

3. Allan D. Pierce, Acoustics, An Introduction to Its Physical Principles and Applications McGraw-Hill, New York, NY 10020, 1981.

4. F. J. Fahy, Sound Intensity, Elsevier Applied Science, London and New York, 1989.

CHAPTER 19

Strain Gages

19.1 Introduction to Strain Gages

Dr. Thomas Kenny, Department of Mechanical Engineering, Stanford University

Strain gages are used in many types of sensors. They provide a convenient way to convert a displacement (strain) into an electrical signal. Their "output" is actually a change of resistance. It can be converted to a voltage signal by connecting the strain gage(s) in a bridge configuration. A few sensors use only a single strain gage element in the bridge, along with three fixed resistors. Others use two strain gages and two fixed resistors, and most recent designs use four strain gages. The gages may be almost any material, but some materials are much more efficient strain gages than others. Proprietary metal alloys and semiconductor silicon are the most commonly used materials.

Piezoresistance

A piezoresistor is a device that exhibits a change in resistance when it is strained. There are two components of the piezoresistive effect in most materials—the geometric component and the resistive component.

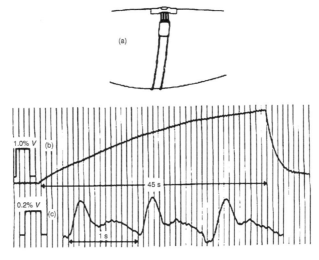

Figure 19.1.1: Liquid strain gage: mercury tube.

Chapter 19

The geometric component of piezoresistivity comes from the fact that a strained element undergoes a change in dimension. These changes in cross-sectional area and length affect the resistance of the device.

A good example of the geometric effect of piezoresistivity is the liquid strain gage. A great many of these devices were in use years ago. Imagine an elastic tube filled with a conductive fluid, such as mercury. The resistance of the mercury in the tube can be measured with a pair of metal electrodes, one at each end, as shown in Figure 19.1.1. Since mercury is essentially incompressible, forces applied along the length of the tube stretch it, and also cause the diameter of the tube to be reduced, with the net effect of having the volume remain constant. The resistance of the strain gage is given by

R = (resistivity of mercury)(length of tube)/(cross-sectional area of tube)

Since

$$R = \frac{\rho L}{A} = \frac{\rho L^2}{V},$$

then

$$\frac{dR}{dL} = \frac{2\rho L}{V} = \frac{2R}{L}$$

We define a quantity called the gage factor K as:

$$K = \frac{dR/R}{dL/L}$$

Since

$$\frac{dR}{dL} = \frac{2R}{L},$$

we have $K = 2$ for a liquid strain gage.

This means that the fractional change in resistance is twice the fractional change in length. In other words, if a liquid strain gage is stretched by 1%, its resistance increases by 2%. This is true for all liquid strain gages, since all that is needed is that the medium be incompressible.

Liquid strain gages were in use in hospitals for measurements of fluctuations in blood pressure. A rubber hose filled with mercury was stretched around a human limb, and the fluctuations in pressure were recorded on a strip-chart recorder, and the shape of the pressure pulses were used to diagnose the condition of the arteries. Such devices have been replaced by solid-state strain gage instruments in modern hospitals, but this example is still interesting from an introductory standpoint.

Metal wires can also be used as strain gages. As is true for the liquid strain gage, stretching of the wire changes its geometry in a way that acts to increase the resistance. For a metal wire, we can calculate the gage factor as we did for the liquid gage, except that we can't assume the metal is incompressible, and we can't assume the resistivity is a constant:

$$R = \frac{\rho L}{A} = \frac{\rho L}{\pi r^2} = \frac{4\rho L}{\pi D^2}$$

$$dR = \frac{4L}{\pi D^2} d\rho + \frac{\rho}{\pi D^2} dL - \frac{8\rho L}{\pi D^3} dD$$

$$\frac{dR}{R} = \frac{d\rho}{\rho} + \frac{dL}{L} - \frac{2dD}{D}$$

Then

$$K = \frac{dR/R}{dL/L} = \frac{d\rho/\rho}{dL/L} + 1 - \frac{2dD/D}{dL/L}$$

Since

$$-\frac{dD/D}{dL/L}$$

is defined as Poisson's ratio, v, we have

$$K = 1 + 2v + \frac{d\rho/\rho}{dL/L}$$

For different metals, this quantity depends on the material properties, and on the details of the conduction mechanism. In general, metals have gage factors between 2 and 4.

Now, since the stress times the area is equal to the force, and the fractional change in resistance is equal to the gage factor times the fractional change in length (the strain), and stress is Young's modulus times the strain, we have

$$F = \sigma A = EA\frac{dL}{L} = \frac{EA}{K}\frac{dR}{R}$$

or

$$\frac{dR}{R} = \frac{FK}{EA}$$

So the fractional change in resistance of a strain gage is proportional to the applied force and is proportional to the gage factor divided by Young's modulus for the material. Clearly, we would prefer to have a large change in resistance to simplify the design of the rest of a sensing instrument, so we generally try to choose small diameters, small Young's modulus, and large gage factors when possible. The elastic limits of most materials are below 1%, so we are generally talking about resistance changes in the 1%–0.001% range. Clearly, the measurement of such resistances is not trivial, and we often see resistance bridges designed to produce voltages that can be fed into amplification circuits.

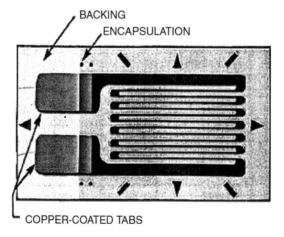

Figure 19.1.2: Thin film strain gage.
(Courtesy of Vishay.)

Thin Film Strain Gages

For many years, there has been an industry associated with the fabrication and marketing of thin metal film strain gages and the necessary tools and equipment for attaching these gages and the wires to various mechanical structures. A photograph of a thin film strain gage is shown in Figure 19.1.2. This particular strain gage consists of a metal wire patterned so that it is primarily sensitive to elongation in one direction. Strain gages are available from several vendors, and literally hundreds of patterns of the metal film can be selected, with different patterns providing sensitivity to strain in particular directions. In recent years, much use has been made of the fact that doped silicon is a conductor that exhibits a gage factor as large as 200, depending on the amount of doping. This creates an opportunity to make strain gages from silicon, and to use them to produce more sensitive devices than would be easy to make in any other material.

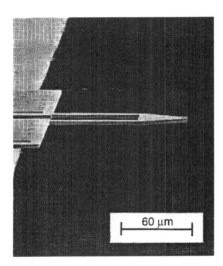

Figure 19.1.3: Silicon strain gage.
(Dr. Marco Tortonese, PhD thesis, Stanford University, 1992.)

Microdevices

Another aspect of the utility of silicon is that recent years have seen the development of a family of etching techniques that allow the fabrication of micromechanical structures from silicon wafers. Generally referred to as silicon micromachining, these techniques use the patterning and processing techniques of the electronics industry to define and produce micromechanical structures.

Chapter 19

Micromachining can be used to fabricate piezoresistive cantilevers for a wide variety of applications. Recent research (Ben Chui at Stanford and John Mamin at IBM Almaden) has focused on the development of piezoresistive cantilevers for data storage applications. In this design, a 100 micron-long piezoresistive cantilever is dragged along a polycarbonate disk at 10 mm/s, bouncing up and down as it passes over sub-micron indentations in the surface of the disk. This idea is essentially a high-performance phonograph needle. The devices shown in Figure 19.1.4 illustrate cantilevers developed for this data storage application. Since 2000, researchers at IBM Zurich, led by Vettiger, have been making large 2-D arrays of piezoresistive cantilevers suitable for a high-density data storage system based on this approach.

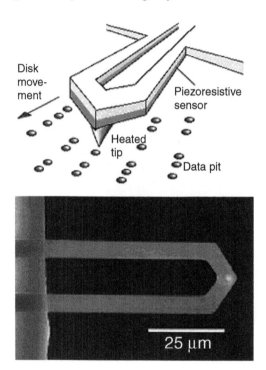

Figure 19.1.4: AFM thermo-mechanical data storage.
(Thesis of Ben Chui, PhD, published in 1998.)

A great deal more could be said about these techniques, but for now we simply state that these techniques are capable of producing diaphragms and cantilevers of silicon with thickness of microns and lateral dimensions of hundreds of microns up to millimeters (see Figure 19.1.3). The mechanical properties of these structures are exactly what we would expect from the bulk mechanical characteristics of silicon.

Since these microstructures can have sensitive strain gages embedded in them, it is easy to see that a number of useful sensing devices can be built. Particular examples include strain gage-based pressure sensors, where an array of strain gages can be positioned around the perimeter of a thin diaphragm and connected into a bridge configuration to automatically cancel out other noise and drift signals from the gages.

Strain Gage Accuracy

Another issue associated with strain gages is the accuracy of the resistance measurement. Generally, accuracy is improved by using larger currents and producing larger voltage changes. However, the practical limit to the amount of current that can be used comes about due to power dissipation in the resistive element. For this reason, the technologies for bonding thin film strain gages have been optimized to maximize the thermal conduction from the thin film to the substrate. Improving the thermal conductance enables the use of more current in the measurement.

Many strain gages, and particularly doped silicon strain gages, are sensitive to temperature changes. In some cases, this is a useful effect—especially if the application also needs to measure temperature. Generally, this is not the case, so it is necessary to compensate for this sensitivity. The easiest way to do this is to fabricate reference resistors from the same material, and locate them so that they do not sense the strain signal. A bridge configuration can be easily arranged to retain the strain sensitivity while canceling the temperature sensitivity of an array of strain gages. Such arrangements are very important and easily produced, so they are very common.

Applications

The applications of strain gages are in sensors where medium-to-large amounts of strain are expected to occur (0.001%–1%), where very low-cost devices are needed, where miniature silicon devices are necessary, and where signals are expected at frequencies from DC to a few kHz. The frequency limitation comes about because the bonding configuration of these devices generally leads to large stray capacitance, which tends to filter out rapidly varying signals.

Chapter 19

Example Calculation: Piezoresistive Cantilever

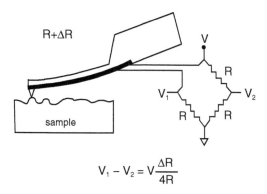

$$V_1 - V_2 = V\frac{\Delta R}{4R}$$

Figure 19.1.5: Piezoresistive cantilever.
(Courtesy of the PhD thesis of Marco Tortonese [Stanford, 1998].)

This example calculation and figure are taken from the thesis of Dr. Marco Tortonese, in which fabrication and operation of an AFM based on piezoresistive cantilevers are described in detail. The calculation of the sensitivity of a piezoresistive cantilever is presented here to provide an example of the strain gage calculations.

As shown in Figure 19.1.5, we use a piezoresistive cantilever to sense variations in the shape of a surface which is passed beneath. This technique has been demonstrated as the Atomic Force Microscope (AFM) by several graduate students in the group of Cal Quate at Stanford. In AFM, attractive forces between a sharp tip and a sample surface cause slight cantilever deflections. If the cantilever is thin enough, forces associated with atomic interactions between individual atoms can be measured.

The load-deflection relationship for a simple cantilever beam is

$$Z = \frac{L^3 F}{3EI}$$

where

$$I = \frac{\omega T^3}{12}$$

Here, L is the length, T is the thickness, and w is the width. Since $F = kZ$, we have stiffness:

$$k = \frac{E\omega T^3}{4L^3}$$

For a deflection Z, the cantilever has an angle of deflection of approximately

$$\theta = \frac{Z}{L},$$

and, therefore, a radius of curvature of approximately

$$R = \frac{L}{\theta} = \frac{L^2}{Z}$$

The strain in the upper surface of the cantilever is caused by the difference in arc length for the upper and lower surfaces.

$$\Delta L = L_{upper} - L_{lower} = (R+T)\theta - R\theta = T\theta = \frac{TZ}{L}$$

The strain is given by

$$\varepsilon = \frac{\Delta L}{L} = \frac{TZ}{L^2}$$

$$\varepsilon = \frac{TF}{L^2 k} = \frac{TF}{L^2 \dfrac{E\omega T^3}{4L^3}} = \frac{4LF}{E\omega T^2}$$

For a typical AFM cantilever (as shown in Figure 19.1.4), we have parameters $T = 4$ μm, $L = 100$ μm,
$w = 4$ μm, $E = 2 \times 10^{11}$ N/m2, and $F = 10^{-7}$ N.

Therefore,

$$\varepsilon = \frac{4(100 \times 10^{-6} m)(10^{-7} N)}{(2 \times 10^{11} N/m^2)(4 \times 10^{-6} m)(4 \times 10^{-6} m)^2} = 3 \times 10^{-6}$$

Since doped silicon has a gage factor of about 100, we would expect a change in resistance dR/R of 0.03% for this example.

Chapter 19

In fact, the cantilever does not take on a circular deflection, and the strain is largely concentrated at the base. If we place our strain gage at the base, we can expect a strain enhancement of order 5–10 times, thereby increasing the resistance change.

With a good circuit it is possible to measure resistance changes as small as one part in 10^6, so this is indeed a reasonable measurement. It is not simple, but it is possible.

In many cases in AFM, forces as small as 10^{-10} N are measured, which requires a careful electrical circuit design.

19.2 Strain-Gage Based Measurements

Analog Devices Technical Staff
Walt Kester, Editor

The most popular electrical elements used in force measurements include the resistance strain gage, the semiconductor strain gage, and piezoelectric transducers. The strain gage measures force indirectly by measuring the deflection it produces in a calibrated carrier. Pressure can be converted into a force using an appropriate transducer, and strain gage techniques can then be used to measure pressure. Flow rates can be measured using differential pressure measurements which also make use of strain gage technology.

- **Strain: Strain Gage, Piezoelectric Transducers**
- **Force: Load Cell**
- **Pressure: Diaphragm to Force to Strain Gage**
- **Flow: Differential Pressure Techniques**

Figure 19.2.1: Strain-gage based measurements.

The resistance strain gage is a resistive element which changes in length, hence resistance, as the force applied to the base on which it is mounted causes stretching or compression. It is perhaps the most well-known transducer for converting force into an electrical variable.

Unbonded strain gages consist of a wire stretched between two points as shown in Figure 19.2.2. Force acting on the wire (area = A, length = L, resistivity = p) will cause the wire to elongate or shorten, which will cause the resistance to increase or decrease proportionally according to:

$$R = pL/A$$

and
$$\Delta R/R = GF \Delta L/L,$$

where GF = Gage factor (2.0 to 4.5 for metals, and more than 150 for semiconductors).

The dimensionless quantity $\Delta L/L$ is a measure of the force applied to the wire and is expressed in *microstrains* ($1\mu e = 10^{-6}$ cm/cm) which is the same as parts-per-million (ppm). From this equation, note that larger gage factors result in proportionally larger resistance changes—hence, more sensitivity.

Excerpted from *Practical Design Techniques for Sensor Signal Conditioning*, Analog Devices, Inc., www.analog.com.

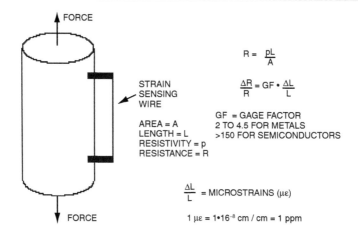

Figure 19.2.2: Unbonded wire strain gage.

Bonded strain gages consist of a thin wire or conducting film arranged in a coplanar pattern and cemented to a base or carrier. The gage is normally mounted so that as much as possible of the length of the conductor is aligned in the direction of the stress that is being measured. Lead wires are attached to the base and brought out for interconnection. Bonded devices are considerably more practical and are in much wider use than unbonded devices.

Perhaps the most popular version is the foil-type gage, produced by photo-etching techniques, and using similar metals to the wire types (alloys of copper-nickel (Constantan), nickel-chromium (Nichrome), nickel-iron, platinum-tungsten, etc. (See Figure 19.2.4). Gages having wire sensing elements present a small surface area to the specimen; this reduces leakage currents at high temperatures and permits higher isolation potentials between the sensing element and the specimen. Foil sensing elements, on the other hand, have a large ratio of surface area to cross-sectional area and are more stable under extremes of temperature and prolonged loading. The large surface area and thin cross section also permit the device to follow the specimen temperature and facilitate the dissipation of self-induced heat.

Strain Gages

Figure 19.2.3:
Bonded wire strain gage.

- SMALL SURFACE AREA
- LOW LEAKAGE
- HIGH ISOLATION

Figure 19.2.4:
Metal foil strain gage.

- PHOTO ETCHING TECHNIQUE
- LARGE AREA
- STABLE OVER TEMPERATURE
- THIN CROSS SECTION
- GOOD HEAD DISSIPATION

Chapter 19

Semiconductor strain gages make use of the piezoresistive effect in certain semiconductor materials such as silicon and germanium in order to obtain greater sensitivity and higher-level output. Semiconductor gages can be produced to have either positive or negative changes when strained. They can be made physically small while still maintaining a high nominal resistance. Semiconductor strain gage bridges may have 30 times the sensitivity of bridges employing metal films, but are temperature sensitive and difficult to compensate. Their change in resistance with strain is also nonlinear. They are not in as widespread use as the more stable metal film devices for precision work; however, where sensitivity is important and temperature variations are small, they may have some advantage. Instrumentation is similar to that for metal-film bridges but is less critical because of the higher signal levels and decreased transducer accuracy.

PARAMETER	METAL STRAIN GAGE	SEMICONDUCTOR STRAIN GAGE
Measurement Range	0.1 to 40,000 µc	0.001 to 3000 µc
Gage Factor	2.0 to 4.5	50 to 200
Resistance, Ω	120, 350, 600, ..., 5000	1000 to 5000
Resistance Tolerance	0.1% to 0.2%	1% to 2%
Size, mm	0.4 to 150 Standard: 3 to 6	1 to 5

Figure 19.2.5: Comparison between metal and semiconductor strain gages.

Strain gages can be used to measure force, as in Figure 19.2.6 where a cantilever beam is slightly deflected by the applied force. Four strain gages are used to measure the flex of the beam, two on the top side, and two on the bottom side. The gages are connected in an all-element bridge configuration. This configuration gives maximum sensitivity and is inherently linear. This configuration also offers first-order correction for temperature drift in the individual strain gages.

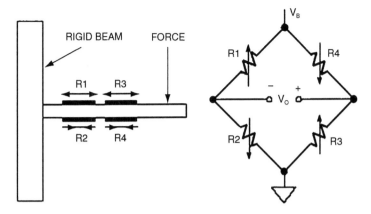

Figure 19.2.6: Strain gage beam force sensor.

Strain gages are low-impedance devices; they require significant excitation power to obtain reasonable levels of output voltage. A typical strain-gage based load cell bridge will have (typically) a 350 Ω impedance and is specified as having a sensitivity in terms of millivolts full scale per volt of excitation. The load cell is composed of four individual strain gages arranged as a bridge as shown in Figure 19.2.7. For a 10 V bridge excitation voltage with a rating of 3 mV/V, 30 millivolts of signal will be available at full scale loading. The output can be increased by increasing the drive to the bridge, but self-heating effects are a significant limitation to this approach: they can cause erroneous readings or even device destruction. Many load cells have "sense" connections to allow the signal conditioning electronics to compensate for DC drops in the wires. Some load cells have additional internal resistors which are selected for temperature compensation.

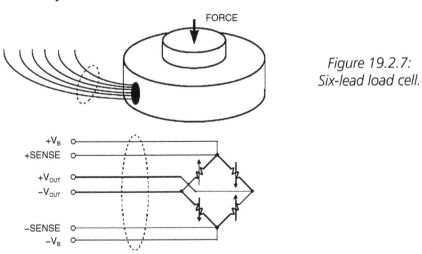

Figure 19.2.7: Six-lead load cell.

Chapter 19

Pressure Sensors

Pressures in liquids and gases are measured electrically by a variety of pressure transducers. A variety of mechanical converters (including diaphragms, capsules, bellows, manometer tubes, and Bourdon tubes) are used to measure pressure by measuring an associated length, distance, or displacement, and to measure pressure changes by the motion produced.

The output of this mechanical interface is then applied to an electrical converter such as a strain gage or piezoelectric transducer. Unlike strain gages, piezoelectric pressure transducers are typically used for high-frequency pressure measurements (such as sonar applications or crystal microphones).

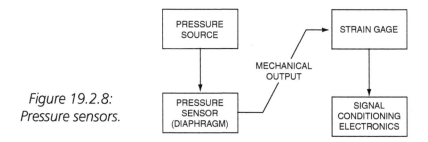

Figure 19.2.8: Pressure sensors.

There are many ways of defining flow (mass flow, volume flow, laminar flow, turbulent flow). Usually the *amount* of a substance flowing (mass flow) is the most important, and if the fluid's density is constant, a volume flow measurement is a useful substitute that is generally easier to perform. One commonly used class of transducers, which measures flow rate indirectly, involves the measurement of pressure. Figure 19.2.9 shows a bending vane with an attached strain gage placed in the flow to measure flow rate.

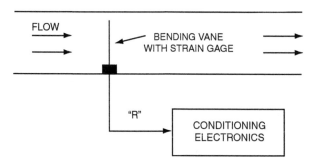

Figure 19.2.9: Bending vane with strain gage used to measure flow rate.

Strain Gages

Bridge Signal Conditioning Circuits

An example of an all-element varying bridge circuit is a fatigue monitoring strain sensing circuit as shown in Figure 19.2.10. The full bridge is an integrated unit that can be attached to the surface on which the strain or flex is to be measured. In order to facilitate remote sensing, current excitation is used. The OP177 servos the bridge current to 10 mA around a reference voltage of 1.235 V. The strain gauge produces an output of 10.25 mV/1000 µe. The signal is amplified by the AD620 instrumentation amplifier which is configured for a gain of 100. Full-scale strain voltage may be set by adjusting the 100 Ω gain potentiometer such that, for a strain of –3500 µE, the output reads –3.500 V; and for a strain of +5000 µE, the output registers +5.000 V. The measurement may then be digitized with an ADC which has a 10 V full-scale input range. The 0.1 µF capacitor across the AD620 input pins serves as an EMI/RFI filter in conjunction with the bridge resistance of 1 kΩ. The corner frequency of the filter is approximately 1.6 kHz.

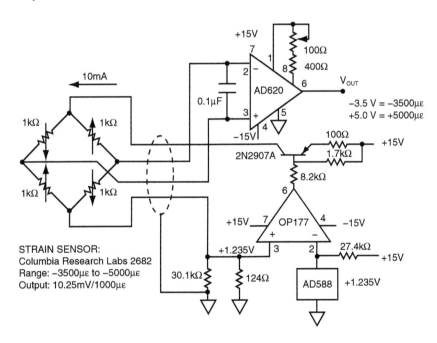

Figure 19.2.10:
Precision strain gage sensor amplifier.

Chapter 19

Another example is a load cell amplifier circuit shown in Figure 19.2.11. A typical load cell has a bridge resistance of 350 Ω. A 10.000 V bridge excitation is derived from an AD588 precision voltage reference with an OP177 and 2N2219A used as a buffer. The 2N2219A is within the OP177 feedback loop and supplies the necessary bridge drive current (28.57 mA). To ensure this linearity is preserved, an instrumentation amplifier is used. This design has a minimum number of critical resistors and amplifiers, making the entire implementation accurate, stable, and cost effective. The only requirement is that the 475 Ω resistor and the 100 Ω potentiometer have low temperature coefficients so that the amplifier gain does not drift over temperature.

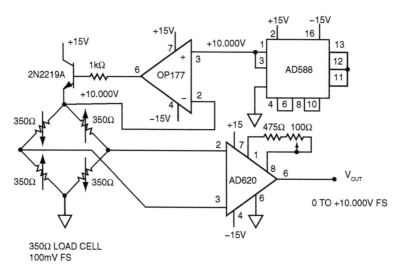

Figure 19.2.11: Precision load cell amplifier.

As has been previously shown, a precision load cell is usually configured as a 350 Ω, bridge. Figure 19.2.12 shows a precision load-cell amplifier that is powered from a single supply. The excitation voltage to the bridge must be precise and stable, otherwise it introduces an error in the measurement. In this circuit, a precision REF195 5 V reference is used as the bridge drive. The REF195 reference can supply more than 30mA to a load, so it can drive the 350Ω bridge without the need of a buffer. The dual OP213 is configured as a two op amp in-amp with a gain of 100. The resistor network sets the gain according to the formula:

$$G = 1 + \frac{10k\Omega}{1k\Omega} + \frac{20k\Omega}{196\Omega + 28.7\Omega} = 100$$

For optimum common-mode rejection, the resistor ratios must be precise. High tolerance resistors (±0.5% or better) should be used.

For a zero volt bridge output signal, the amplifier will swing to within 2.5 mV of 0 V. This is the minimum output limit of the OP213. Therefore, if an offset adjustment is required, the adjustment should start from a positive voltage at V_{REF} and adjust V_{REF} downward until the output (V_{OUT}) stops changing. This is the point where the amplifier limits the swing. Because of the single supply design, the amplifier cannot sense signals which have negative polarity. If linearity at zero volts input is required, or if negative polarity signals must be processed, the V_{REF} connection can be connected to a voltage which is mid-supply (2.5 V) rather than ground. Note that when V_{REF} is not at ground, the output must be referenced to V_{REF}.

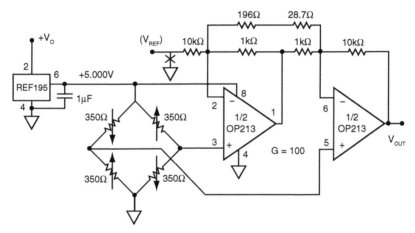

Figure 19.2.12: Single supply load cell amplifier.

The AD7730 24-bit sigma-delta ADC is ideal for direct conditioning of bridge outputs and requires no interface circuitry. The simplified connection diagram is shown in Figure 19.2.13. The entire circuit operates on a single +5 V supply which also serves as the bridge excitation voltage. Note that the measurement is ratiometric because the sensed bridge excitation voltage is also used as the ADC reference. Variations in the +5 V supply do not affect the accuracy of the measurement.

Chapter 19

The AD7730 has an internal programmable gain amplifier which allows a full-scale bridge output of ±10mV to be digitized to 16-bit accuracy. The AD7730 has self and system calibration features which allow offset and gain errors to be minimized with periodic recalibrations. A "chop" mode option minimizes the offset voltage and drift and operates similarly to a chopper-stabilized amplifier. The effective input voltage noise RTI is approximately 40 nV rms, or 264 nV peak-to-peak. This corresponds to a resolution of 13 ppm, or approximately 16.5-bits. Gain linearity is also approximately 16-bits.

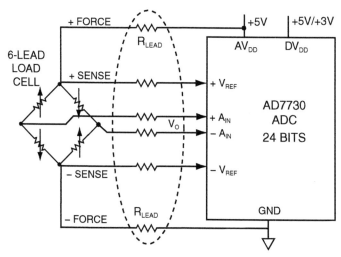

Figure 19.2.13: Load cell application using the AD7730 ADC.

- **Assume:**
  - ◆ Full-scale Bridge Output of ±10 mV, +5 V Excitation
  - ◆ "Chop Mode" Activated
  - ◆ System Calibration Performed: Zero and Full-scale
- **Performance:**
  - ◆ Noise RTI: 40 nV rms, 264 nV p-p
  - ◆ Noise-Free Resolution: = = 80,000 Counts (16.5 bits)
  - ◆ Gain Nonlinearity: 18ppm
  - ◆ Gain Accuracy: < 1 µV
  - ◆ Offset Voltage: <1 µV
  - ◆ Offset Drift: 0.5 µV/°C
  - ◆ Gain Drift: 2 ppm/°C
  - ◆ Note: Gain and Offset Drift Removable with System Recalibration

Figure 19.2.14: Performance of AD7730 load cell ADC.

References

1. Ramon Pallas-Areny and John G. Webster, Sensors and Signal Conditioning, John Wiley, New York, 1991.

2. Dan Sheingold, Editor, Transducer Interfacing Handbook, Analog Devices, Inc., 1980.

3. Walt Kester, Editor, 1992 Amplifier Applications Guide, Section 2, 3, Analog Devices, Inc., 1992.

4. Walt Kester, Editor, System Applications Guide, Section 1, 6, Analog Devices, Inc., 1993.

5. Harry L. Trietley, Transducers in Mechanical and Electronic Design, Marcel Dekker, Inc., 1986.

6. Jacob Fraden, Handbook of Modern Sensors, Second Edition, Springer-Verlag, New York, NY, 1996.

7. The Pressure, Strain, and Force Handbook, Vol. 29, Omega Engineering, One Omega Drive, P.O. Box 4047, Stamford CT, 06907-0047, 1995. (http://www.omega.com)

8. The Flow and Level Handbook, Vol. 29, Omega Engineering, One Omega Drive, P.O. Box 4047, Stamford CT, 06907-0047, 1995. (http://www.omega.com)

9. Ernest O. Doebelin, Measurement Systems Applications and Design, Fourth Edition, McGraw-Hill, 1990.

10. AD7730 Data Sheet, Analog Devices, http://www.analog.com.

19.3 Strain Gage Sensor Installations

George C. Low, HITEC Corporation

The various types of strain gages covered elsewhere in this book can be installed by a variety of different methods. This section will attempt to provide some detail on each of the more popular methods of installation using the most popular strain gage types (bonded foil resistance strain gages and free filament strain gages). The reader is encouraged to study more about the methods that are of interest, as not all of the intricacies of the installation techniques can be covered in this brief section. Please be aware that there is a definite art to installing strain gages as it is primarily a manual process, and particularly with the esoteric, free filament strain gage installations, the quality of the installation is dependent to some degree on the installer's experience and is not entirely based on whether or not the proper steps were followed. There are various modifications to the following installation techniques based on exact requirements, environment, etc., but these can be considered general installation technique guidelines.

There are methods of installing strain gages in which the strain gage is actually created during the installation process. This type of installation is commonly referred to as vacuum depositing or sputtering. This particular installation technique is not considered part of the general strain gage installation methods and is only covered briefly at the end of this section.

We will break down the strain gage installations into three broad categories: General Stress Analysis, Precision Transducer Installations, and Elevated Temperature Installations. The final section will be on specialty installations and will briefly make mention of other types of installations.

General Stress Analysis Installation (Bonded Foil Strain Gage)

For general stress analysis, the user is primarily interested in obtaining stress/strain data as fast as possible and as accurately as possible. Examples of this are FEA model validation, general design validation, structural failure analysis, simple accelerated life cycle testing, etc. These types of installations usually do not warrant the same type of manufacturing methods as a high performance and high accuracy transducer (for example, a post cure operation).

The most common method of installation for a bonded foil resistance strain gage in this category is using a room temperature cure cyanoacrylate adhesive consisting of a catalyst and an adhesive. This type of installation requires a minimum amount of tools and equipment and also a minimum amount of experience. The basic procedure is as follows:

a. Surface preparation, either a chemical cleaning process or a combination of fine grit abrasive and chemical cleaning. It should be noted that in some types of tests, the surface being measured must not be altered, which dictates a chemical cleaning only.

b. Gage location centerlines or layout lines are burnished on the part using a method that will not cause bumps or other anomalies under the gage grid after installation. Examples of this can be simply a pencil, a scribe tool with a brass tip, etc. In some cases the layout lines can be applied using laser marking.

c. The gage is now carefully placed in position and held in place using a piece of Mylar® tape. It is sometimes desirable not to seal the strain gage coupon on all four sides with the tape. This is to allow "squeeze out" of the adhesive and provide a uniform bond line with no bumps or air pockets. Carefully lift back the Mylar tape with gage adhered to the tape and fold it back over itself exposing the bottom or bondable surface of the strain gage.

d. The catalyst is now applied to the gage backing, while the adhesive is applied to the component surface.

e. Fold the gage back in place, and using thumb pressure, press and hold the gage per the manufacturer's recommended guidelines, usually at least one minute.

At this point, the strain gage is bonded to the component and is ready for the next step, which is lead wire attachment. The lead wire length and material is selected based on the user's test and instrumentation requirements. A suitable coating must be installed as well in order to seal the installation from the environment and also to provide some mechanical protection.

This is the most basic installation technique, and care must be taken when choosing the strain gage itself. Parameters to consider include: grid length and type, backing and foil type, resistance, self-temperature compensation rating, etc. General stress analysis applications can also utilize the higher quality and higher capability of heat cured epoxy adhesive systems that are more commonly utilized in transducer applications.

Chapter 19

Precision Transducer Installations

These installations cover a wide range of transducers, from torque transducers to shear beams to pressure transducers. Also included in this section would be the ever-popular component "transducerization" which includes the in-situ components that are slightly modified to accept strain gages and they become the transducer. An example of this is the suspension "pushrods" on an open wheel racecar. In this case, the actual suspension pushrods have a couple pockets milled into it in which strain gages are installed and the pushrod itself becomes a transducer that the team can use to determine optimum suspension setup prior to a race.

These types of installations require better adhesives and routinely utilize heat cure epoxy adhesives. Various types are available depending on operating temperature of the transducer, surface porosity of the base transducer material, and so forth.

The general procedure, for bonded foil resistance strain gages, follows some similar steps as the general stress analysis installations:

a. Surface preparation, usually a fine grit abrasive blast is used. Chemical cleaning is also required prior to the actual strain gage application in order to ensure a contamination-free surface.

b. Gage location centerlines are burnished on the part using a method that will not cause bumps or other anomalies under the gage grid after installation. Examples of this can be simply a pencil, a scribe tool with a brass tip, etc. In some cases the layout lines can be applied using laser marking.

c. The gage is now carefully placed in position and held in place using a piece of Mylar tape. It is sometimes desirable to not seal the strain gage coupon on all four sides with the tape. This is to allow "squeeze out" of the adhesive and provide a uniform bond line with no bumps or air pockets. Carefully lift back the Mylar tape with gage adhered to the tape and fold it back over itself exposing the bottom or bondable surface of the strain gage.

d. The adhesive is now applied to the gage backing and the component. Allow to air dry per the manufacturer's recommended guidelines.

e. Fold the gage back in place, place a piece of Teflon® film over the Mylar tape, and place a suitably sized rubber pad over the Teflon film.

f. A critical part of the installation is using the correct clamping pressure to clamp the gage. Each adhesive has a recommended clamping pressure for precision transducer applications. It should also be noted that some applications require more than the recommended pressure. This author is aware of an application that required more than twice the manufacturer's recommended clamp pressure in order to meet certain specifications such as creep performance. The clamp should also be calibrated and of a suitable design that allows for efficient clamping and unclamping if used for production type work. The clamped transducer is then placed in an oven and allowed to ramp up at a controlled rate to the desired cure temperature. A common type of installation on a steel transducer body, for instance, would require a cure of 2 hours at 350°F.*

g. The next step in the procedure is the post cure. This is important for long-term stable transducer operation. Allow the transducer to cool after the cure operation and remove the clamp. Place the transducer back in the oven and post cure the transducer (with no clamp) at 50°F over either the cure temperature or the maximum operating temperature, whichever is higher.

h. After installation, the gages are wired as appropriate, usually in a Wheatstone bridge configuration. Further transducer manufacturing steps occur but are outside of the scope of this section.

Elevated Temperature Installations

Installations under this category cover installations for use over about 700°F. For that reason they require the use of free filament wire strain gages. It should also be noted that at these higher temperatures, essentially the only measurements that can be made with any certainty are dynamic measurements, as opposed to static. Some static measurements are said to have been made up to 1200°F, but this author does not know the accuracies and repeatability of those measurements.

The following are general procedures to be followed for free filament strain gage applications. As in the other installation categories covered, there can be variations to this based on material requirements, experience, etc. For this type of installation in particular, the installer's experience and skill are critical to a quality installation that will survive the rigors of a jet engine spin pit test, for instance.

Chapter 19

These strain gages can be installed using two different methods, either using ceramic cement, or via a ROKIDE® flame spray process. Ceramic cement is usually utilized for applications below 1200°F, where the use of the flame spray process would provide unwanted reinforcement to a thin specimen, and also where the installer cannot spray due to space constraints. The ROKIDE process provides better erosion characteristics over ceramic cement but does not perform as well in fatigue as ceramic cements do. ROKIDE is essentially aluminum oxide and comes in various purity levels.

The basic procedures are as follows:

Ceramic Cement Process

This process utilizes ceramic cement which, when applied in the appropriate steps, ultimately encapsulates the free filament strain gage grid in ceramic cement, which protects the strain gage from the harsh elevated temperature environment.

a. Where possible, pre-bake the component to eliminate any surface oils, etc.

b. Carefully burnish the surface of the component extending the lines beyond the area to be grit blasted. This area would include the entire area where the strain gage grid is applied as well as the lead wire routing areas that require ceramic cement application.

c. Mask the component with tape for the gage locations and lead wire paths. Grit blast using a pressure blaster using new grit of suitable size for the particular component.

d. Remove all tape and inspect for contaminants.

e. Mask the outline of the gage location and lead wire routing areas with Mylar tape. Apply a pre-coat of ceramic cement per manufacturer's guidelines to both areas. After the cement has air dried for a nominal length of time, remove the Mylar tape. Oven cure the pre-coat per manufacturer's guidelines.

f. The strain gages themselves come from the manufacturer on a slide with integral mastic, which holds the gage shape. Carefully remove the gage from the manufacturer's slide and position the gage over the correct gage location.

g. Carefully press the mastic into contact with the pre-coat using fine-tipped tweezers or other suitable tool.

Strain Gages

    h. Using a clean brush, apply ceramic cement in thin layers over the exposed grid and lead wires in between the tape bars. Allow to air dry and place in an oven for cure per the manufacturer's guidelines.

    i. Verify the gage resistance and remove the remaining tape bars.

    j. Apply a thin layer of ceramic cement over the newly exposed areas of the grid and lead wires. Allow to air dry and place in an oven for cure per the manufacturer's guidelines.

    k. Once all lead wires have been attached, perform final electrical inspection by checking the resistance of the circuit, as well as the insulation resistance.

    l. A typical ceramic cement installation of this type is between 0.007" to 0.008" thick.

ROKIDE Flame Spray Process

This process utilizes a special spray gun which, using oxygen and acetylene and the appropriate grade of ROKIDE rod, sprays a molten ceramic onto the desired surface. This ultimately encapsulates the strain gage grid and protects it from the harsh elevated temperature environment.

    a. Where possible, pre-bake the component to eliminate any surface oils, etc.

    b. Carefully burnish the surface of the component extending the lines beyond the area to be grit blasted. This area would include the entire area where the strain gage grid is applied as well as the lead wire routing areas that require ROKIDE flame spray application.

    c. Mask the component with suitable tape for the gage locations and lead wire paths. Grit blast using a pressure blaster using new grit of suitable size for the particular component. Clean with dry, contamination-free air.

    d. Apply a thin nickel aluminide base coat. The nickel aluminide retards oxidation and also provides a better mechanical bond for the aluminum oxide (ROKIDE). Clean with dry, contamination-free air.

    e. Apply a thin aluminum oxide pre-coat, which will electrically insulate the free filament strain gage from the component surface. Clean with dry, contamination-free air.

    f. The strain gages themselves come from the manufacturer on a slide with integral mastic, which holds the gage shape. Carefully remove the gage from the manufacturer's slide and position the gage over the correct gage location.

g. Carefully press the mastic into contact with the pre-coat using fine-tipped tweezers or other suitable tool. Ensure the gage grid is not pressed too hard so as to conform it to the aluminum oxide pre-coat.

h. Box in the gage grid by placing tape around the perimeter of the grid using suitable high temperature tape. This ensures that only a minimum amount of surface area will be covered by the aluminum oxide, which does cause a reinforcing effect.

i. Apply a light aluminum oxide tack-coat to the exposed gage grid and gage leads. Always take resistance readings throughout the process to ensure the gage grid has not become damaged.

j. Remove the perimeter tape first, and then the tape bars which were originally holding the strain gage grid to the aluminum oxide pre-coat.

k. Box in the installation again using the same tape as in the previous operation. This second application of perimeter tape should be positioned about 1/32" beyond the first tape application. This will provide a layering effect, which will minimize the sharp edge of the final aluminum oxide installation.

l. Spray the final coat of aluminum oxide over the entire strain gage and strain gage leads as
appropriate.

m. Remove all tape and inspect for circuit resistance and insulation resistance.

n. A typical ROKIDE installation of this type is generally about 0.012" thick.

Other Installation Methods

One unique method of strain gage installation requires specialized manufacturing equipment and technical knowledge. This is referred to as sputtering or vacuum deposition. During this process the strain gage itself is created during the installation process. It is beyond this scope of this section to address this specialized method of installation.

Another specialized method is called thick film, which again is outside of the scope of this section.

These two methods are considered specialized and have been covered in the relevant section of the strain sensor chapter. In general they do not cover nearly as broad a range of applications as the three sections broken out above; therefore more attention has been given to the more common installations.

Another strain gage installation technique lends itself to higher volume applications (usually 10,000 pieces and higher). Assuming the flexure to be gaged is essentially flat, the flexures can be arranged in an array. This can be either as part of the manufacturing process as in chemical milling of thin cantilever/bending beams in a sheet, or can be separate flexures arranged in an array via appropriate manufacturing fixtures. The actual strain gage application is performed by bonding an entire sheet of gages (strain gages also in an array form that have not yet been separated into individual pieces) to the flexures. The array is then subject to clamping pressures using a press type setup. The gaged flexures are then trimmed and separated from each other and the result is an efficiently gaged batch of beams.

A twist to this process is when the strain gage foil is bonded to the backing, which is in turn bonded to the flexure all in one operation. The composite is then etched at the same time; i.e., the strain gage pattern as well as the flexure is created during the same operation. This, of course, only lends itself to the thin (<0.1") beam type flexures.

A weldable strain gage installation is yet another form of installation. This type of gage comes from the manufacturer as a complete assembly consisting of a metal shim, a strain gage, lead wires, and potting/coating, completely bonded and assembled. The potting/coating is an appropriate compound suitable for the environment. Weldable strain gages come in both high temperature and room temperature versions. These gages are applied to the surface of the component to be tested using spot-welding techniques. Weldable strain gages are essentially only used in the field in areas where more standard strain gages cannot be installed due to either installer's skill or in locations where it is impossible to perform any other type of strain gage installation.

The general techniques and processes listed in this section should not be considered the final word on strain gage installations. As mentioned previously, there are many different variations to these processes based on operating environment, size of the component or transducer, materials being used, and so forth. In all cases, it is imperative the installer read all instructions for the installation materials being utilized. The precision transducer processes, for example, can be used for installing semiconductor strain gages, although some steps need to be altered such as clamping pressure, etc. The basic process, however, is very similar.

CHAPTER **20**

Temperature Sensors

John Fontes, Senior Applications Engineer, Honeywell Sensing and Control

Because temperature can have such a significant effect on materials and processes at the molecular level, it is the most widely sensed of all variables. Temperature is defined as a specific degree of hotness or coldness as referenced to a specific scale. It can also be defined as the amount of heat energy in an object or system. Heat energy is directly related to molecular energy (vibration, friction and oscillation of particles within a molecule): the higher the heat energy, the greater the molecular energy.

Temperature sensors detect a change in a physical parameter such as resistance or output voltage that corresponds to a temperature change. There are two basic types of temperature sensing:

- **Contact** temperature sensing requires the sensor to be in direct physical contact with the media or object being sensed. It can be used to monitor the temperature of solids, liquids or gases over an extremely wide temperature range.

- **Non-contact** measurement interprets the radiant energy of a heat source in the form of energy emitted in the infrared portion of the electromagnetic spectrum. This method can be used to monitor non-reflective solids and liquids but is not effective with gases due to their natural transparency.

20.1 Sensor Types and Technologies

Temperature sensors comprise three families: electro-mechanical, electronic, and resistive. The following sections discuss how each sensor type is constructed and used to measure temperature and humidity.

Electro-mechanical

Bi-metal thermostats are exactly what the name implies: two different metals bonded together under heat and pressure to form a single strip of material. By employing the different expansion rates of the two materials, thermal energy can be converted into electro-mechanical motion.

There are two basic bi-metal thermostat technologies: snap-action and creeper. The snap-action device uses a formed bi-metal disc to provide a near instantaneous change of state (open to close and close to open). The creeper style uses a bi-metal strip to slowly open and close the contacts. The opening speed is determined by the bi-metal selected and the rate of temperature change of the application.

Bi-metal thermostats are also available in adjustable versions. By turning a screw, a change in internal geometry takes place that changes the temperature setpoint.

Bulb and capillary thermostats make use of the capillary action of expanding or contracting fluid to make or break a set of electrical contacts. The fluid is encapsulated in a reservoir tube that can be located 150mm to 2000mm from the switch. This allows for slightly higher operating temperatures than most electro-mechanical devices. Due to the technology involved, the switching action of these devices is slow in comparison to snap-action devices.

Electronic

Silicon sensors make use of the bulk electrical resistance properties of semiconductor materials, rather than the junction of two differently doped areas. Especially at low temperatures, silicon sensors provide a nearly linear increase in resistance versus temperature or a positive temperature coefficient (PTC). IC-type devices can provide a direct, digital temperature reading, so there's no need for an A/D converter.

Infrared (IR) pyrometry. All objects emit infrared energy provided their temperature is above absolute zero (0 Kelvin). There is a direct correlation between the infrared energy an object emits and its temperature.

IR sensors measure the infrared energy emitted from an object in the 4–20 micron wavelength and convert the reading to a voltage. Typical IR technology uses a lens to concentrate radiated energy onto a thermopile. The resulting voltage output is amplified and conditioned to provide a temperature reading.

Factors that affect the accuracy of IR sensing are the reflectivity (the measure of a material's ability to reflect infrared energy), transmissivity (the measure of a material's ability to transmit or pass infrared energy), and emissivity (the ratio of the energy radiated by an object to the energy radiated by a perfect radiator of the surface being measured).

An object that has an emissivity of 0.0 is a perfect reflector, while an object with an emissivity of 1.0 emits (or absorbs) 100% of the infrared energy applied to it. (An emissivity of 1.0 is called a "blackbody" and does not exist in the real world.)

Temperature Sensing

Thermocouples are formed when two electrical conductors of dissimilar metals or alloys are joined at one end of a circuit. Thermocouples do not have sensing elements, so they are less limited than resistive temperature devices (RTDs) in terms of materials used and can handle much higher temperatures. Typically, they are built around bare conductors and insulated by ceramic powder or formed ceramic.

All thermocouples have what are referred to as a "hot" (or measurement) junction and a "cold" (or reference) junction. One end of the conductor (the measurement junction) is exposed to the process temperature, while the other end is maintained at a known reference temperature. (See Figure 20.1.1.) The cold junction can be either a reference junction that is maintained at 0°C (32°F) or at the electronically compensated meter interface.

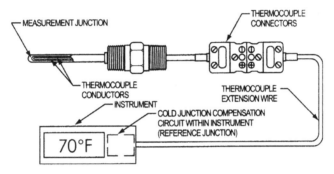

Figure 20.1.1: Thermocoupler.
Source: Desmarais, Ron and Jim Breuer. "How to Select and Use the Right Temperature Sensor." Sensors Online. January 2001. http://www.sensorsmag.com/articles/0101/24/index.htm

When the ends are subjected to different temperatures, a current will flow in the wires proportional to their temperature difference. Temperature at the measurement junction is determined by knowing the type of thermocouple used, the magnitude of the millivolt potential, and the temperature of the reference junction.

Thermocouples are classified by calibration type due to their differing voltage or EMF (electromotive force) vs. temperature response. The millivolt potential is a function of the material composition and conductor metallurgical structure. Instead of being assigned a value at a specific temperature, thermocouples are given standard or special limits of error covering a range of temperature.

Chapter 20

Resistive Devices

Thermistors (or thermally sensitive resistors) are devices that change their electrical resistance in relation to their temperature. They typically consist of a combination of two or three metal oxides that are sintered in a ceramic base material and have lead wires soldered to a semiconductor wafer or chip, which are covered with epoxy or glass.

Thermistors are available in two different types: positive temperature coefficient (PTC) and negative temperature coefficient (NTC). PTC devices exhibit a positive change or increase in resistance as temperature rises, while NTC devices exhibit a negative change or decrease in resistance when temperature increases. The change in resistance of NTC devices is typically quite large, providing a high degree of sensitivity. They also have the advantage of being available in extremely small configurations for extremely rapid thermal response.

In addition to metal oxide technology, PTC devices can also be produced using conductive polymers. These devices make use of a phase change in the material to provide a rapid increase in electrical resistance. This allows for their use in protection against excessive electrical current as well as excessive temperature.

Like RTDs, thermistors' resistance value is specified with a plus-or-minus tolerance at a particular temperature. Thermistors are usually specified at 25°C.

Thermistors' resistance can be made virtually linear using support circuitry such as a Wheatstone bridge. The resistance can then be interpreted using look-up tables to perform a switching function or to drive a meter. They can also be used in liquid level sensing applications.

RTDs (resistive temperature devices), like thermistors, employ a change in electrical resistance to measure or control temperature. RTDs consist of a sensing element, connection wires between the element and measurement instrument, and a support for positioning the element in the process.

The metal sensing element is an electrical resistor that changes resistance with temperature. The element usually contains a coil of wire or conductive film with conductors etched or cut into it. It is usually housed in ceramic and sealed with ceramic cement or glass. (See Figure 20.1.2.)

The sensing element should be positioned where it can reach process temperature quickly. Wire wound devices should be adequately secured in high vibration and shock applications. Extension wires between the element and instrument allow resistance to be measured from great distances.

Temperature Sensing

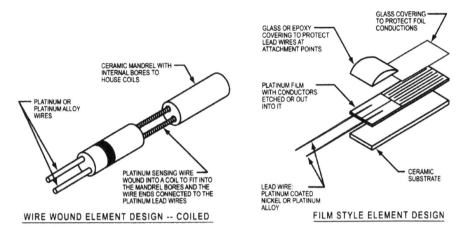

Figure 20.1.2: Sensing element designs.
Source: Desmarais, Ron and Jim Breuer. "How to Select and Use the Right Temperature Sensor." Sensors Online. January 2001. http://www.sensorsmag.com/articles/0101/24/index.htm

Flexible wire wound and etched foil RTDs are available in various standard configurations. Typically a Kapton®, silicone rubber, Mylar or clear polyester dielectric material is used for electrical insulation. They can be mounted on curved or irregular surfaces using pressure sensitive adhesives, thermally conductive glues, silicone tape, or mechanical clamps. This type of configuration is far superior for monitoring a large area such as the outside diameter of a pipe or tank. They can also be integrated into a flexible heater circuit for optimum control.

20.2 Selecting and Specifying Temperature Sensors

The following sections address what differentiates each sensor from one another, including temperature, accuracy, and interchangeability. The advantages and disadvantages of each sensor type are also identified.

Selecting Temperature Sensors

General Considerations

How to select the best temperature sensor? In general, all sensor types are useful temperature measurement options, but each has its advantages and disadvantages. For example:

- Thermistors provide high resolution, have the widest range of applications, are the most sensitive, and are low cost, but are nonlinear and have limited temperature range.

- Thermocouples have the highest temperature region and are durable for high-vibration and high-shock applications, but require special extension wire.
- RTDs are nearly linear and are highly accurate and stable, but they are large and expensive.
- Silicon types are low cost and nearly linear, but have a limited temperature range.

An important consideration in selecting thermal sensors are the materials used, which have temperature limitations. Tolerance, accuracy, and interchangeability are also important. Tolerance is a specific requirement, usually plus or minus a particular temperature. Accuracy is the sensor's ability to measure the temperature's true value over a temperature range.

Regardless of the sensor technology selected, user safety should be the primary concern. Never select a device solely because it has the lowest cost. Choose the device that offers the best performance for its price and always adhere to the manufacturer's guidelines and recommendations.

Each temperature sensing application can present its own unique set of requirements and problems and needs to be evaluated on an individual basis. Here are some questions to consider.

Does the application require contact or non-contact sensing?

If the application is moving or if physical contact is not practical due to contamination or hazardous material issues, infrared is the technology of choice.

What temperature range is the sensor required to control or monitor?

Thermocouples have the broadest temperature range, –200°C to 2315°C. (Some devices within this range do not have ANSI calibration types established.) Depending on design and material, thermsistors have a usable range of –100 to 500°C. Bi-metal thermostats can handle temperatures from –85 to 371°C.

For cryogenic temperatures, RTDs and some silicon-based devices are capable of approaching absolute zero (0K). Maximum temperatures range from 150°C to 200°C. Support circuitry must be thermally isolated from the sensor so as not to exceed its capabilities.

For non-contact (infrared) devices, temperatures below –18°C or above 538°C would require a custom unit.

With all of these devices, it is possible to exceed these ranges through the use of thermowells or by placing the device in a location relative to the heat source. However, this type of approach can affect the accuracy and response of the system.

What is the rate of temperature change of the application?

For applications where the rate of temperature change is rapid (>1.0°C/minute), the mass of the sensor may become an issue. The thermal inertia of the sensor is based on its mass. For extremely rapid changes, sensor mass should be kept to a minimum to allow it to more accurately track the change of the application. This includes the mass and thermal conductivity of the thermowell or other protective material.

For applications where the sensor will be remotely located due to environmental or other issues, design verification testing should be performed. This involves using two or more sensors to monitor the temperature of the application, while another sensor monitors the temperature at the proposed sensor location. In this way, sensor location can be optimized.

How tightly do you need to control or monitor the temperature?

For certain medical applications or processes involving chemical reactions, tolerances of ±0.1°C or less may be required. For any application requiring tolerances of less than ±1.7°C, an electronic system will be required. Silicon, RTD, thermocouple or thermistor-based systems can all be designed to maintain these extremely tight tolerances. Typically RTDs will provide the greatest overall accuracy.

Remember, in control applications, component accuracy and system accuracy may be totally different. If your system accuracy is no better than + 3°C, it does not make sense to buy the most expensive sensor. You may be able to use a bi-metal thermostat and achieve the same system accuracy at a significantly lower total cost.

Are agency approvals such as UL, CSA, FDA, etc. required at the system or component level?

When agency approvals are required, component selection is essentially limited to electro-mechanical devices. Application electrical loads must be reviewed with the manufacturer for conformance to agency requirements. When loads do not conform, it is still possible to have the component reviewed for approval in the specific application.

In some medical and other applications, electronic-based control systems may be not acceptable without the use of a snap-action thermostat as a safety device. Since the failure mode of the electronic system cannot be guaranteed, the thermostat is designed to open the circuit and prevent an over-temperature condition.

Chapter 20

How important is total system cost in the selection of the sensor?

In high-end applications costing thousands of dollars, the cost of the temperature sensor is typically insignificant. For this reason, the selection is based on required system accuracy. If an accuracy of ±3°C is acceptable, a low cost electro-mechanical device or thermistor-based circuit may be more than sufficient. An accuracy of ±0.1°C will require a more sophisticated (and expensive) alternative.

When dealing with low-cost items like consumer goods (drip coffee makers, popcorn poppers, etc.), the sensor can become a much larger percentage of the total cost. In these applications, the accuracy of the sensor becomes much less important and cost and reliability become more significant. In high volumes, a commercial grade thermostat can do the job reliably for less than $0.50 total cost. With the cost of assembly, circuit board, components, wave soldering, etc., the total cost of a thermocouple or thermistor-based circuit may be higher, but the reliability may actually be lower due to the additional number of components and solder joints.

In the end, the ultimate question is, how much accuracy do you need and how much can you afford to pay? You also need to consider the environmental conditions in which the sensor will be used and what must be done to ensure its survival.

Selecting Electro-mechanical Sensors

Device selection is of extreme importance when using electro-mechanical devices. In high moisture or corrosive environments, the use of a hermetically sealed device should be strongly considered. Non-hermetic devices may be used for these types of applications based on the manufacturer's input. However, they should be sealed with epoxy or some type of overmold.

For high vibration or shock applications, the design of the device is critical. Typical commercial grade thermostats rely on only armature spring pressure to maintain contact closure. This can cause chattering under high vibration levels and premature contact failure. Consult with the manufacturer during the design phase of this type of application.

Devices that will be subjected to temperatures below −17.8°C (0°F) should employ an inert dry gas internally to minimize moisture condensation or icing on the contact surface.

Application electrical loads should strictly adhere to the manufacturer's advertised limits. Using a device above rated limits can lead to premature failure of the device and the application.

Device selection should be based on environmental exposure temperature, not operating temperatures. A device may be required to operate in a range that conforms to its material capabilities, but be in close proximity to significantly higher temperatures. Over time, this will lead to deterioration of the device.

Bi-metal thermostats have been used for over fifty years in applications as varied as drip coffee makers and the Space Shuttle. While they can be used as a low-cost control solution for the appliance market, they can also provide a highly reliable device for long-term use in military and spaceflight applications.

Devices are available in either open or close on temperature rise variants. Depending on the type of device selected, setpoints can range from –85 to +371°C (–120 to +700°F).

Some devices employ an internal heater to provide protection against both excessive temperature and current.

Bi-metal devices are available in many different sizes, configurations, and capabilities. Since they usually carry the actual application load, they do not require any additional circuitry to perform their function. Current carrying capacities range from dry circuit loads to as high as 25 Amps. Standard production accuracies are available to ±1.7°C (±3°F).

Thermostats are typically rated in life cycles at a specific electrical load and can vary significantly depending on whether they are used in a control or monitoring application.

Advantages

- Direct interface with application for fast response
- No additional circuitry/components required
- Available in both hermetic and non-hermetically sealed designs
- High current carrying capacity
- Wide operating temperature range
- Application/market-based pricing
- NASA qualified high reliability and military versions available

Disadvantages

- Less accurate than most electronic-based systems
- Larger size than electronic-based systems
- Creepage-type device cannot interface with electronic components
- Can fail "closed" at end of life

Bulb and capillary thermostats are available in single pole, single throw (SPST), single pole, double throw (SPDT) and adjustable setpoint designs in a temperature range of –35 to +400°C (–31 to 752°F). They are also available in manual re-set designs.

Due to the high electrical loads associated with the typical end applications and their relatively slow switching action, extremely hard contact materials are used. This results in a high initial contact resistance that precludes their use in low current applications.

Advantages

- Control can be located at a significant distance from application being sensed
- Built-in overtemperature systems available
- Broad operating temperature range
- High current carrying capability

Disadvantages

- Large size
- Relatively expensive
- Limited number of potential applications
- User programmable

Selecting Electronic Sensors

Silicon sensors are available in a wide variety of designs, outputs, and costs. Temperature ranges are available from the cryogenic (1.4K) to 200°C. With high sensitivity and a nearly linear resistance curve, they are ideal for many applications.

IC versions are available with on-chip signal conditioning for direct voltage or current output to controllers or meters. Because they have memory, IC-types can be very accurately calibrated. They work effectively in multi-sensor environments such as communications networks.

The output value of most IC sensors is proportional with temperature over a specific range. Standard accuracy is usually assigned, but can often be calibrated at a specific temperature. Along with basic temperature control and indication, various temperature compensation functions are often directly incorporated in printed circuits.

Depending on the application, silicon sensors can be designed as an element into probe assemblies or incorporated directly onto printed circuit boards in a surface-mount configuration. Care must be taken in the design of circuits employing silicon-based technology as excess current can cause self-heating of the sensing element. This can significantly reduce system accuracy.

Some manufacturers have developed IC designs that can be used in place of thermostats in some applications. They feature factory programmed or user programmable setpoints or hysteresis. They are available in standard Joint Electron Device Engineering Council (JEDEC) configurations.

The operating parameters of the user programmable types are set either through the use of external resistors or are digitally programmed through a two-wire interface with a processor.

Advantages

- Less expensive than RTDs
- More linear than thermistors
- Easier to use than RTDs or thermocouples due to higher output
- IC types feature on-chip signal conditioning
- Many IC types include communication protocols with bus-type data acquisition systems

Disadvantages

- Not as linear as RTDs
- Less accurate than other electronic-based systems

- More expensive than thermistors or thermocouples
- Limited temperature range
- Slower thermal response than other electronic-based systems
- Typically larger than RTDs and thermistors
- Require larger package sizes for immersion
- Additional components/circuitry required to control application loads

Infrared (IR) pyrometry. Most IR devices are portable, battery powered, hand-held units that provide a digital read-out of the application temperature. They are also available as fixed-mount devices that can be used with fiber-optic cables for remote sensing. Outputs can be used to drive a display or control loop.

It is important to use a device with the correct field of view for the application to be measured. To ensure an accurate reading, the object being measured must completely fill the field of view of the measurement system. The measurement system will determine the average temperature of all devices within the field of view including the background.

Several factors can have an impact on both device and system performance. Dust in the atmosphere between the sensor and the target will absorb or deflect some the radiated energy and cause large variations in measurements. Fiber-optic cables may be used to reduce the distance between the sensor and target to minimize error, but the environmental temperature capabilities of the cable itself must now be considered.

Advantages

- Allows for non-contact measurement of moving objects or hazardous materials
- Can be used in conjunction with fiber optics for remote sensing
- Typical temperature range –18 to +538°C (0 to 1000°F)
- Accuracy to ±1%

Disadvantages

- Accuracy can be affected by surface finish
- Field of view must be matched to target size
- Ambient temperature can affect readings

- Wavelength filter must be matched to the application
- Higher cost ($200+) can be even higher if control circuitry is required
- Calibration can be difficult and costly
- Additional components/circuitry required to control application loads
- Dust, gas, or other vapors in the environment can effect the accuracy of the system

Thermocouples have the widest temperature range of all sensor technologies, −200 to +2315°C (−328 to +4200°F), and can be used in a wide variety of environments. (See Table 21.2.1.) Their inherently simple design allows them to withstand high levels of mechanical shock and vibration. Their small size provides nearly immediate response to small temperature changes.

Table 21.2.1: Thermocouple application information.

	Application Information
E	Recommended for continuously oxidizing or inert atmospheres. Sub-zero limits of error not established. Highest thermoelectric output of the common thermocouple types.
J	Suitable for vacuum, reducing or inert atmospheres, oxidizing atmospheres with reduced life. Iron oxidizes rapidly above 1000°F (538°C), so only heavy-gauge wire is recommended for high temperature. Bare elements should not be exposed to sulfurous atmospheres above 1000°F (538°C).
K	Recommended for continuous oxidizing or neutral atmospheres. Mostly used above 1000°F (538°C). Subject to failure if exposed to sulfur. Preferential oxidation of chromium in positive leg at certain low oxygen concentrations causes "green rot" and large negative calibration drifts most serious in the 1500°F–1900°F (816°C–1038°C) range. Ventilation or inert sealing of the protection tube can prevent this.
N	Can be used in applications where Type K elements have shorter life and stability problems due to oxidation and the development of "green rot."
T	Usable in oxidizing, reducing, or inert atmospheres as well as vacuum. Not subject to corrosion in moist atmospheres. Limits of error published for sub-zero temperature ranges.
R & S	Recommended for high temperature. Must be protected in a nonmetallic tube and ceramic insulators. Continued high-temperature use causes grain growth that can lead to mechanical failure. Negative calibration drift caused by rhodium diffusion to the pure leg of platinum as well as from rhodium volatilization. Type R is used in industry and Type S in the laboratory.
B	Same as R & S but has a lower output. Also has a higher maximum temperature and is less susceptible to grain growth.

Source: Desmarais, Ron and Jim Breuer. "How to Select and Use the Right Temperature Sensor." Sensors Online. January 2001. http://www.sensorsmag.com/articles/0101/24/index.htm

Since different combinations of materials produce higher voltage outputs at different temperatures, they are matched to produce a nearly linear voltage curve. This makes thermocouples much simpler to interface with a meter or controller.

When exposed to corrosive or oxidizing environments, the thermocouple element should be protected by a thermowell, stainless steel probe of mineral-filled cable.

Thermocouples are available in three different junction styles:

- **Exposed junction.** The sensing tip consists of two dissimilar wires joined together by welding, soldering, crimping, or brazing to form a "hot junction." It has the fastest response time of the three types.
- **Grounded junction.** The hot junction is welded to the inside of a protective metal sheath. This protects the junction but affects thermal response. It also makes the device more susceptible to electromagnetic interference (EMI).
- **Ungrounded junction.** The hot junction is electrically insulated from its protective metal sheath by a thermally conductive material. This increases thermal lag but isolates the junction from EMI.

Exposed hot junctions are susceptible to oxidation and corrosion. They can be protected through the use of thermowells or stainless steel sheaths in both grounded and un-grounded styles. However, this can significantly affect system accuracy.

A grounded hot junction will protect the sensor. However it will increase thermal response times and make the sensor more susceptible to EMI. It also increases conduction and radiation errors.

An ungrounded hot junction will also protect the sensor. However, because the sensor is electrically isolated from the sheath or thermowell, the influence of EMI is much less. The penalty of this design is even larger conduction and radiation errors and much slower response.

Advantages

- Small size provides rapid temperature response
- Relatively inexpensive
- Wide temperature range
- More durable than RTDs for use in high-vibration and high-shock applications
- ANSI established calibration types

Disadvantages

- Must be protected from corrosive environments
- Smaller gage wire sizes are less stable and have a shorter operating life
- Use of plated-copper instrumentation wire results in errors when ambient temperatures change
- Special extension wires are required
- Reference junction compensation is required
- Less stable than RTDs in moderate or high temperatures
- Should be tested to verify performance under controlled conditions for critical applications
- Additional components/circuitry required to control application loads

Selecting Resistive Sensors

Thermistors' construction makes them by far the most sensitive of any sensors to temperature changes. Because they do not contain materials such as platinum, they are relatively inexpensive in comparison to wire wound RTDs. Their small size lends them to a variety of applications. Plus, they can be easily molded into protective packages for durability.

However, because of the materials used, their operating temperature range (–100°C to 300°C) is narrower than RTDs or thermocouples. And because their resistance vs. temperature characteristics are nonlinear, they are typically used to measure narrow temperature ranges to minimize nonlinearity. PTC-type thermistors have a much smaller useful temperature range than NTC types.

Another drawback associated with thermistors is that they can fail in a "closed" mode. This could potentially produce a resistance that is interpreted by the system as a temperature reading and not as a component failure.

As with RTDs, thermistors are powered devices. They require electrical input to function. In applications where the power budget is critical or an IC interface is required, a battery back-up might be required.

It may also be necessary to take into account the mass and self-heating of the device in the application. Since these are resistance devices, they generate their own heat in addition to the heat they are measuring. As the temperature of the application increases, the resistance of the device decreases, increasing the self-heating effect. If the mass and thermal conductivity are sufficient, this effect will be negligible. However, based on the system accuracy required, it should be considered.

Thermistors do not have standard resistance vs. temperature characteristics, so interchangeability can pose a problem. This can require costly system re-design when a change in manufacturer is contemplated. Most manufacturers have their own proprietary resistance curves, usually published as ratios based on a resistance of 25°C.

Thermistors can be either very robust or extremely fragile. Bead thermistors typically have extremely thin wire leads that must be adequately secured in high vibration or shock environments. The bead itself must also be rigidly attached to the application in these environments.

High temperature exposure can have an effect on the long-term stability of the sensor. Some ceramic materials, particularly those chosen for lower impedance values, exhibit a tendency to vary from the initial resistance curve.

In critical applications, the manufacturer should be consulted during the design phase to ensure selection of the correct device. In corrosive environments, epoxy-coated beads can degrade in a relatively short period of time. For these types of applications, a glass encapsulated device or probe assembly should be used even though the cost is slightly higher.

Advantages

- Low component cost
- Fast thermal response
- Large change in resistance vs. temperature for more resolution
- Extremely small size means faster reaction to change in temperature and ability to use in variety of assemblies
- Linearized resistance types available
- High resistance values so no lead wire compensation necessary

Disadvantages
- Limited temperature range
- Lower temperature exposures than RTDs or thermocouples
- No established resistance standards
- Self heating can affect accuracy
- Non-linear resistance change requires additional components for accurate interpretation
- Increased component count decreases system reliability
- Additional components/circuitry required to control application loads

RTDs are used for a variety of consumer applications including automotive, thermostats, small appliances, ovens, refrigerators, air conditioners, furnaces, and instant water heaters. They are also popular in industry applications such as process control, electronic circuits and assemblies, printers, laptops, computers, power supplies, battery packs, motor and bearing temperature, HVAC, instruments, and environmental chambers. Medical applications include respiratory, culture, incubator, and disposables. RTDs are very stable over time, and due to the simplicity of their design, are not typically affected by environmental conditions.

RTDs are typically manufactured from materials having a positive temperature coefficient (a usable resistance vs. temperature change). The most common materials are copper, nickel, nickel/iron alloy, tungsten, and platinum.

However, platinum can be exposed to temperatures up to 1200°F and is recognized as the most accurate, stable, and predictable as a resistor. Its useful temperature range is also higher than nickel, copper or nickel/iron alloy. (See Table 20.2.2.) Plus, it provides the most linear resistance change, allowing for the easiest interpretation.

Table 20.2.2: Sensing element materials and temperature limits.

Material	Usable Temperature Range
Platinum	−450°F to 1200°F
Nickel	−150°F to 600°F
Copper	−100°F to 300°F
Nickel/Iron	32°F to 400°F

Source: Desmarais, Ron and Jim Breuer. "How to Select and Use the Right Temperature Sensor." Sensors Online. January 2001. http://www.sensorsmag.com/articles/0101/24/index.htm

Because of the low natural electrical resistance of platinum, it is necessary to wind a long, thin wire around a ceramic core to achieve a 100 Ω resistance. This is a standard value for the type of RTD. Lower values are available, but they make signal interpretation more difficult.

Other materials such as copper are used in the manufacture of low cost RTDs. Copper's change in resistance is actually more linear than platinum. However, its temperature range is more limited and it is susceptible to corrosion, which platinum is not. Copper and nickel/iron alloy devices can be adversely affected by corrosive or high moisture environments and should be isolated from these types of conditions.

The wires that connect the sensing element to the measuring instrument are usually made of nickel, nickel alloys, tinned copper, sliver-plated copper, or nickel-plated copper, and are insulated with PVC, FEP Teflon®, TFE Teflon®, and fiberglass. The materials selected also influence the temperatures in which the RTD can be used.

Manufacturers usually offer a low-temperature and high-temperature construction. Low-temperature configurations have a range of 400°F to 500°F, and are comprised of Teflon®-insulated nickel or silver-plated copper wires with an epoxy seal. High temperature configurations have a range of 900°F to 1200°F and use fiberglass-insulated, nickel-plated copper wire with a ceramic cement.

RTDs can be mounted with various materials. Most commonly, the sensing element and wires are inserted into a closed-end stainless steel tube, which is packed with vibration damping or heat transfer material such as ceramic powder.

The sensing element in the RTD is manufactured to a specific electrical resistance at a specific temperature. Resistance curves for RTDs have been standardized by various agencies, making RTDs easily interchangeable among manufacturers.

When integrating RTDs into circuits, the resistance of the interconnections must be taken into account. For every increase of 0.33 Ω caused by lead wires, etc., a 1°C error will be introduced.

It may also be necessary to take into account the mass and self-heating of the device in the application. Since these are resistance devices, they generate their own heat in addition to the heat they are measuring. As the temperature of the application increases, the resistance of the device decreases, increasing the self-heating effect. If the mass and thermal conductivity are sufficient, this effect will be negligible. However, based on the system accuracy required, it should be considered.

Advantages

- Very accurate and repeatable
- Wide temperature range −200 to +650°C (−328 to +1202°F) depending on type
- Extremely stable over time: >0.1°C/year drift
- Larger voltage output than thermocouples
- Excellent resistance linearity
- Resistance can be determined in the laboratory and will not vary significantly over time
- Area or point sensing
- Low variation for better interchangeability
- Can use standard instrumentation cable to connect to control equipment
- Established industry accepted resistance curves

Disadvantages

- Higher cost than thermistors or thermocouples
- Self heating of the RTD can affect overall system accuracy
- Larger size than thermistors or thermocouples
- Not as durable as thermocouples in high-vibration and high-shock environments

20.3 Applicable Standards

Standards Bodies

American National Standards Institute (ANSI): http://www.ansi.org
A private, non-profit organization responsible for administering the U.S. voluntary standardization and conformity assessment system.

American Society of Testing and Materials (ASTM): http://www.astm.org
One of the largest voluntary standards development organizations in the world. Develops and publishes voluntary consensus standards for materials, products, systems, and services.

Canadian Standards Association (CSA): http://www.csa.ca
: Not-for-profit membership-based association serving business, industry, government, and consumers in Canada and around the world. Develops standards for enhancing public safety and health, advancing the quality of life, helping to preserve the environment, and facilitating trade.

Instrumentation, Systems, and Automation Society (ISA): http://www.isa.org
: Helps advance the theory, design, manufacture, and use of sensors, instruments, computers, and systems for measurement and control in a variety of applications.

International Electrotechnical Commission (IEC): http://www.iec.ch
: Prepares and publishes international standards for all electrical, electronic, and related technologies.

International Organization for Standardization (ISO): http://www.iso.ch/iso/en/ISOOnline.openerpage
: A network of national standards institutes from 146 countries working in partnership with international organizations, governments, industry, business and consumer representatives.

Japanese Standards Association (JSA): http://www.jsa.or.jp/default_english.asp
: Objective is "to educate the public regarding the standardization and unification of industrial standards, and thereby to contribute to the improvement of technology and the enhancement of production efficiency."

National Institute of Standards and Technology (NIST): http://www.nist.gov
: Founded in 1901, NIST is a non-regulatory federal agency within the U.S. Commerce Department's Technology Administration. Its mission is to develop and promote measurement, standards, and technology to enhance productivity, facilitate trade, and improve the quality of life.

Note: All thermistor testing and calibration baths are measured to the "International Temperature Scale of 1990" (http://www.its-90.com) using instrumentation from Hart Scientific (http://www.hartscientific.com/).

Industry Organizations

American Society for Quality (ASQ): http://www.asq.org/
> Purpose is to improve workplace and communities by advancing learning, quality improvement, and knowledge exchange. Advises the U.S. Congress, government agencies, state legislatures, and other groups and individuals on quality-related topics.

International Measurement Confederation (IMEKO): http://www.mit.tut.fi/imeko/
> Non-governmental federation of 36 member organizations. Promotes international interchange of scientific and technical information in the field of measurement and instrumentation and the international cooperation among scientists and engineers from research and industry.

National Conference of Standards Laboratories International (NCSL International): http://www.ncsli.org/
> A professional association for individuals involved in all aspects of measurement science.

Underwriter's Laboratories (UL): http://www.ul.com
> An independent, not-for-profit product-safety testing and certification organization.

Applicable Standards and Specifications

Electro-mechanical Devices

MIL-PRF-24236
> Switches, (Bi-metallic and Metallic), General Specification for
> Issued by: Defense Supply Center (DOD)

ANSI Z21.21
> Thermostats, gas appliance
> Issued by: American National Standards Institute

UL873
> Standard for Temperature Indicating and Regulating Equipment
> Issued by: Underwriters Laboratory

CAN/CSA 22.2 No. 24-1993
> Temperature Indicating and Regulating Equipment
> Issued by: Canadian Standards Association

Chapter 20

Thermistors

MIL-PRF-23648
Resistor, Thermal (Thermistor) Insulated, General Specification for
Issued by: Defense Supply Center (DOD)

ANSI/EIA 337
Glass Coated Thermistor Beads and Thermistor Beads in Glass Probes and Glass Rods (NTC) General Specification for
Issued by: American National Standards Institute

ANSI/EIA 275
Thermistor Definitions and Test Methods
Issued by: American National Standards Institute

Thermocouples

ANSI MC96.1
Thermocouples, General Specification for
Issued by: Instrument Society of America

MIL-T-24388
Resistive Temperature Devices (RTDs) and Thermocouples for Shipboard Use, General Specification for
Issued by: U.S. Naval Sea Systems Command

RTDs

IEC-751
Resistance Standards for Resistive Temperature Devices (RTDs)
Issued by: International Electrotechnical Commission

JIS C 1604
Resistance Standards for Resistive Temperature Devices (RTDs)
Issued by: Japanese Standards Association

DIN 43760
Resistance Standards for Nickel Resistive Temperature Devices (RTDs)
Issued by: Deutsches Institut fur Normung

BS 1904
Resistance Standards for Resistive Temperature Devices (Same as IEC 751)
Issued by: British Standards Association

SAMA RC21-4-1966
Resistance Standards for Resistive Temperature Devices (RTDs)
Issued by: Scientific Apparatus Makers Association

MIL-T-24388
Resistive Temperature Devices (RTDs) and Thermocouples for Shipboard Use, General Specification for
Issued by: U.S. Naval Sea Systems Command

20.4 Interfacing and Design Information

The most important consideration with any type of sensing technology is sensor location.

In a control application, where the rate of temperature change is fairly slow, the sensor should be located as close to the heat source as possible. In this way, the thermal lag is minimal. The heat source will cycle more frequently; however, it will eliminate potential undershoot or overshoot of the application.

When the rate of temperature change is rapid due to the thermal conductivity of the material or because of frequent changes in the mass being heated, the sensor should be located as close to the material as possible. This will cause the heat source cycle to be longer and increase fluctuations of the workload. With an electronic-based system, these fluctuations can be minimized with a PID controller.

In all circumstances, the distance between the heat source, sensor and mass to be heated should be as short as possible. This will minimize thermal lag, workload temperature fluctuations and power usage.

Bi-metallic and Bulb and Capillary Thermostats

Electro-mechanical sensors are typically the simplest components to interface with their applications. Since they are capable of either opening or closing with increasing temperature, they are capable of interrupting a power circuit to control or shut down a circuit or of closing a circuit to sound an alarm, turn on a fan, etc.

In most circumstances, thermostats are connected to one leg of the power source. When the application temperature is reached, the device will function to either make or break the circuit.

When the electrical load required by the application exceeds the capabilities of the thermostat, the thermostat can be used in conjunction with a relay, contactor or some other type of power handling component.

Resistance and Accuracy

Sensor accuracy is a function of production tolerance and any additional calibration that the sensor may get. Calibration can improve the accuracy of an RTD by 10 times over production tolerance. The accuracy values in Table 20.4.1 apply to production tolerance tight trim RTDs with ice point tolerances of R_0 ±0.1%. The thin film values in Table 20.4.2 are for tight trim platinum RTDs. Both thin film and wire-wound tight trim RTDs with 0.00385 alpha values meet IEC 751 Class B.

In qualifying volumes, RTDs can be laser trimmed for tight resistance interchangeability at any temperature between 0°C and 150°C or to an ice point resistance other than 100 Ω or 1000 Ω. Laser trimming also allows matching the resistance of RTDs with different alpha values at a target temperature.

Table 20.4.1: Accuracy* vs. temperature.

Ice Point, Alpha Value	1000Ω 0.00375	100Ω 0.00385	100Ω 0.003902
Temperature °C	±ΔResistance (Ω)		
−200	5.1	0.5	0.5
−100	2.4	0.3	0.3
0	1.0	0.1	0.1
100	2.2	0.2	0.2
200	4.3	0.4	0.4
300	6.2	0.6	0.6
400	8.3	0.8	0.8
500	9.6	1.0	1.0
600	10.4	1.2	1.2
Temperature °C	±ΔTemperature (°C)		
−200	1.2	1.2	1.2
−100	0.6	0.6	0.6
0	0.3	0.3	0.3
100	0.6	0.6	0.6
200	1.2	1.2	1.2
300	1.8	1.8	1.8
400	2.5	2.5	2.5
500	3.0	3.0	3.0
600	3.3	3.6	3.6

* Figures are for production tolerance tight trim RTDs.

Table 20.4.2: Platinum RTD resistance vs. temperature.

Ice Point, Alpha Value & RTD Type	1000Ω 0.00375 Pt Thin Film	100Ω 0.00385 Pt Thin Film	100Ω 0.00385 Pt WW	100Ω 0.003902 Pt WW
Temperature °C	Resistance (Ω)			
−200	199.49	18.10	18.10	19.76
−180	284.87	26.81	26.81	28.01
−160	368.57	35.35	35.35	36.17
−140	450.83	43.75	43.75	44.27
−120	531.83	52.04	52.04	52.31
−100	611.76	60.21	60.21	60.31
−80	690.78	68.30	68.30	68.27
−60	769.01	76.32	76.32	76.22
−40	846.58	84.27	84.27	84.15
−20	923.55	92.16	92.16	92.08
0	1000.00	100.00	100.00	100.00
20	1075.96	107.79	107.79	107.92
40	1151.44	115.54	115.54	115.84
60	1226.44	123.24	123.24	123.76
80	1300.96	130.89	130.89	131.69
100	1375.00	138.50	138.50	139.61
120	1448.56	146.06	146.06	147.53
140	1521.63	153.57	153.57	155.45
160	1594.22	161.04	161.04	163.37
180	1666.33	168.46	168.46	171.29
200	1737.96	175.83	175.83	179.21
220	1809.11	183.16	183.16	187.14
240	1879.78	190.43	190.43	195.06
260	1949.96	197.67	197.67	202.98
280	2019.67	204.85	204.85	210.90
300	2088.89	211.99	211.99	218.82
320	2157.63	219.08	219.08	226.74
340	2225.89	226.12	226.12	234.66
360	2293.66	233.12	233.12	242.59
380	2360.96	240.07	240.07	250.51
400	2427.78	246.98	246.98	258.43
420	2494.11	253.83	253.83	266.35
440	2559.96	260.65	260.65	274.27
460	2625.33	267.41	267.41	282.19
480	2690.22	274.13	274.13	290.11
500	2754.63	280.80	280.80	298.04
520	2818.55	287.42	287.42	305.96
540	2881.99	294.00	294.00	313.88
560	2944.96	300.53	300.53	321.80

Chapter 20

Table 20.4.2: Platinum RTD resistance vs. temperature (continued).

Ice Point, Alpha Value & RTD Type	1000Ω 0.00375 Pt Thin Film	100Ω 0.00385 Pt Thin Film	100Ω 0.00385 Pt WW	100Ω 0.003902 Pt WW
Temperature °C	Resistance (Ω)			
580	3007.44	307.01		
600	3069.44	313.44		
620	3130.96	319.83		
640	3191.99	326.18		
660	3252.55	332.47		
680	3312.62	338.72		
700	3372.21	344.92		
720	3431.32	351.08		
740	3489.95	357.18		
750	3519.09	360.22		

Temperature Circuits

Two-wire circuit: A Wheatstone bridge is the most common approach for measuring an RTD. As R_T increases or decreases with temperature, V_{out} also increases or decreases. Use an op-amp to observe V_{out}. Lead wire resistance, L_1 and L_2 directly adds to the RTD leg of the bridge. (See Figure 21.4.1.)

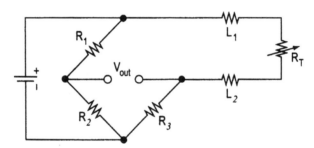

Figure 20.4.1: Two-wire temperature circuit.

Three-wire circuit: In this approach, L_1 and L_3 carry the bridge current. When the bridge is in balance, no current flows through L_2, so no L_2 lead resistance is observed. The bridge becomes unbalanced as R_T changes. Use an op-amp to observe V_{out} and prevent current flow in L_2. The effects of L_1 and L_3 cancel when $L_1 = L_3$ since they are in separate arms of the bridge. (See Figure 20.4.2.)

Temperature Sensing

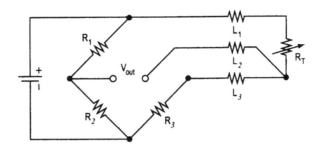

Figure 20.4.2: Three-wire temperature circuit.

Four-wire circuit: A four-wire approach uses a constant current source to cancel lead wire effects even when $L_1 \neq L_4$. Use an op-amp to observe V_{out} and prevent current flow in L_2 and L_3. (See Figure 20.4.3.)

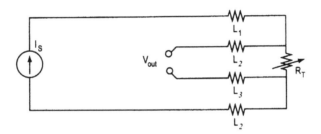

Figure 20.4.3: Four-wire temperature circuit.

Temperature switch. The following circuit causes an output voltage to rail whenever the temperature of the RTD rises above a fixed value T_1. The open-collector output simplifies the interfacing of this circuit with additional electronics. (See Figure 20.4.4.)

Chapter 20

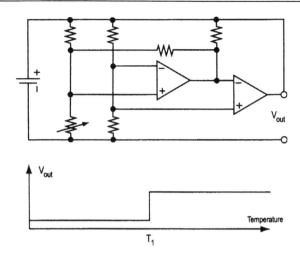

Figure 20.4.4: Temperature switch circuit.

Temperature switch with hysteresis. The following circuit uses positive feedback from the output to self heat the RTD enough to develop a hysteresis in the behavior of the switch. Once on, the temperature must drop low enough to offset the self heating before the switch will disable. (See Figure 20.4.5.)

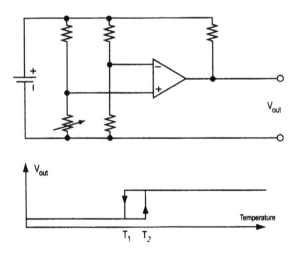

Figure 20.4.5: Temperature switch with hysteresis circuit.

Temperature Sensing

Heat Conduction Equation and RTD Self Heating

Heat flow. The Heat Conduction Equation (Figure 20.4.6) governs time response and other heat flow phenomena. Solutions to the Heat Conduction Equation consist of a time-independent final temperature distribution and a series sum of exponentially damped orthogonal functions that describe the evolution of the temperature distribution from the initial condition *f(x)* to the final condition.

$$\text{Heat Conduction Equation: } \alpha^2 \nabla^2 u = \frac{\partial u}{\partial t}$$

$$\alpha^2 = \frac{K}{ps} \text{ Thermal Diffusivity } (m^2/s)$$

$$K = \text{Thermal Conductivity } (J/s \bullet m \bullet °C)$$

$$p = \text{Density } (kg/m^3)$$

$$s = \text{Specific Heat } (J/°C)$$

Figure 20.4.6: Heat conduction equation.

Note: Do not confuse the alpha used in the equation in Figure 20.4.6 with the alpha used to describe an RTD's R-T curve.

Apply the Heat Conduction Equation to a thin film RTD mounted to a very thermally conductive—i.e., metal—surface. Since the RTD is very thin, approximate the problem as one-dimensional in x with a general solution u(x, t) as shown in Figure 20.4.7.

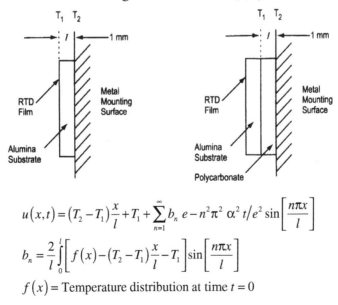

$$u(x,t) = (T_2 - T_1)\frac{x}{l} + T_1 + \sum_{n=1}^{\infty} b_n e - n^2 \pi^2 \alpha^2 t/e^2 \sin\left[\frac{n\pi x}{l}\right]$$

$$b_n = \frac{2}{l} \int_0^l \left[f(x) - (T_2 - T_1)\frac{x}{l} - T_1 \right] \sin\left[\frac{n\pi x}{l}\right]$$

$$f(x) = \text{Temperature distribution at time } t = 0$$

Figure 20.4.7: Applying the heat conduction equation.

559

Chapter 20

Self-heating: Once heat is introduced into the RTD by resistive heating, the equation in Figure 20.4.8, which defines thermal conductivity, must be satisfied:

$$j_u = -K \frac{\partial u(T)}{\partial x}$$

Figure 20.4.8: Thermal conductivity equation.

Applying the conductivity equation as a boundary condition on the general solution for the RTD-on-a-surface example, $u(x, t)$ results in the self-heating relationship shown in Figure 20.4.9.

$$\frac{P}{A} = \alpha^2 u'(0)$$

P = Thermal power dissipated in the RTD = $v_+^2 / R(T)$

A = Surface area of the RTD

Yielding our result:

$$\frac{P}{A} = \alpha^2 \frac{(T_2 - T_1)}{l} \text{ or } T_1 = \frac{lP}{\alpha^2 A} + T_2 = \frac{lV_+^2}{\alpha^2 A\, R(T)} + T_2$$

Figure 20.4.9: Self-heating relationship.

Example 1: Applying the result to a low thermal impedance situation, examine an HEL-700 at 0°C, with 0.254 mm (0.010 in) thick alumina substrate (diffusivity k J 38 W/m°C) and 1000 Ω ice point resistance. Here the self-heating error calculated from a 2.3 mA current is negligible, less than 0.02°C.

Example 2: Examining a high thermal impedance situation, use the same RTD, encapsulated in a plastic or epoxy package such as a TO-92. Approximating this as an intervening 1 mm thick layer of polycarbonate with diffusivity of 0.199 W/m°C, the 2.3 mA current now generates a 12.4°C offset.

A plastic encapsulated RTD will exhibit significantly greater temperature offset error than the same un-encapsulated RTD when both are mounted to a surface (or environment) with good thermal conductivity. However, for air measurement, the opposite occurs as in Table 20.4.3.

Table 20.4.3: Temperature offset in still air.

RTD Current	Ceramic SIP	Encapsulated
0.1mA	<0.02°C	<0.02°C
1.0mA	0.83°C	0.50°C

Conclusion: When the thermal conductivity of the sensor packaging is lower than the thermal conductivity of the environment being measured, then the sensor packaging can increase self heating. More importantly, lower operating currents always reduce or eliminate self-heating errors.

20.5 Latest and Future Developments

Sensor manufacturers are currently working on temperature sensors that can withstand higher and higher temperatures for longer periods of time. For example, Honeywell is working with a turbo charger manufacturer to develop both RTDs and thermistors to meet application temperatures as high as 1100°C.

Increasing demand for more fuel-efficient vehicles is driving this development. With the implementation of EURO 4 emission standard in 2005 as well as LEV standard in the US, engine manufacturers are trying to squeeze more and more power out of smaller and smaller engines. The current direction is toward the increased use of turbo chargers combined with smaller engine size. High-temperature sensors allow the control systems to monitor gas consumption engine performance and overheating of the turbo charger.

References and Resources

Bakker, A. "CMOS Smart Temperature Sensors: An Overview." Proceedings of IEEE Sensors 2002. Piscataway, NJ: IEEE, 2002.

Bakker, A. and Jonah H. Huijsing. *High Accuracy CMOS Smart Temperature Sensors*. Boston: Kluwer Academic Publishers, 2000.

Desmarais, Ron and Jim Breuer. "How to Select and Use the Right Temperature Sensor." *Sensors* 18 (2001):24-36.

Honeywell web site, temperature sensor information: http://content.honeywell.com/sensing/prodinfo/temperature/#technical

Honeywell web site, thermistor information: http://content.honeywell.com/sensing/hss/thermal/product/thermisters.asp

Mathews, David. "Choosing and Using a Temperature Sensor." *Sensors* 17 (2000):54-57.

Measurements Science Conference (MSC): http://www.msc-conf.com/

Quelch, D. "Humidity Sensors for Industrial Applications." International Conference on Sensors and Transducers, Vol. 1. Tavistock, UK: Trident Exhibitions, 2001.

Scolio, Jay. "Temperature Sensor: ICs Simplify Designs." *Sensors* 17 (2000): 48-53.

CHAPTER **21**

Nanotechnology-Enabled Sensors:
Possibilities, Realities, and Applications

Sharon Smith, Lockheed Martin Corporation
David J. Nagel, The George Washington University

This article is reprinted from *Sensors* magazine, November 2003. Used with permission.

If you make or use sensors, your business will likely feel the impact of current and future developments in nanotechnology, a very promising new branch of small-scale technology named for the unit of measure at which it operates: the nanometer, or 0.001 micron. Nanotechnology enables us to create functional materials, devices, and systems by controlling matter at the atomic and molecular scales, and to exploit novel properties and phenomena [1]. Consider that most chemical and biological sensors, as well as many physical sensors, depend on interactions occurring at these levels and you'll get an idea of the effect nanotechnology will have on the sensor world.

The trend toward the small began with the miniaturization of macro techniques, which led to the now well-established field of microtechnology. Electronic, optical, and mechanical microtechnologies have all profited from the smaller, smarter, and less costly sensors that resulted from work with ICs, fiber optics, other micro-optics, and MEMS (microelectromechanical systems). As we continue to work with these minuscule building blocks, there will be a convergence of nanotechnology, biotechnology, and information technology, among others, with benefits for each discipline. Substantially smaller size, lower weight, more modest power requirements, greater sensitivity, and better specificity are just a few of the improvements we'll see in sensor design.

Nanosensors and nano-enabled sensors have applications in many industries, among them transportation, communications, building and facilities, medicine, safety, and national security, including both homeland defense and military operations. Consider nanowire sensors that detect chemicals and biologics [2], nanosensors placed in blood cells to detect early radiation damage in astronauts [3], and nanoshells that detect and destroy tumors [4]. Many start-up companies are already at work developing these devices in an effort to get in at the beginning. Funding for nanotechnology increased by more than a factor of 5 between 1997 and 2003, and is still on the rise [5]. So this is a good time to examine the possibilities—and the limitations—of this small new world.

Chapter 21

21.1 Possibilities

The current global enthusiasm for nanotechnology is an offshoot of several late 20th century advances. Of particular importance was the ability to manipulate individual atoms in a controlled fashion—a sort of atomic bricklaying—by techniques such as scanning probe microscopy. Initial successes in producing significant amounts of silver and gold nanoparticles helped to draw even more attention, as did the discovery that materials and devices on the atomic and molecular scales have new and useful properties due in part to surface and quantum effects.

Another major contributor was the creation of carbon nanotubes (CNTs), extremely narrow, hollow cylinders made of carbon atoms. Both single-and multi-walled CNTs could, for example, be functionalized at their ends to act as biosensors for DNA or proteins. The single-walled versions can have different geometries (see Figure 21.1.1). Depending on the exact orientation of the carbon atoms, a CNT can exhibit either conducting (metallic) or semiconducting properties. This characteristic, and the ability to grow CNTs at specific locations and manipulated afterward, make it likely that the tubes will be important for electronics and sensors. For instance, they can be used in the fabrication of nanoscale field-effect transistors for electronics or as biological probes for sensors, either singly or as an array.

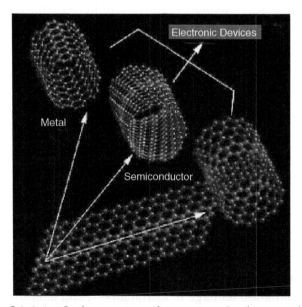

Figure 21.1.1: Carbon nanotubes can exist in a variety of forms and can be either metallic or semiconducting in nature, depending on their atomic structure. (Courtesy of NASA Ames Research Center, Moffett Field, CA.)

Increasingly Integrated Technologies. The technologies associated with materials, devices, and systems were once relatively separate, but integration has become the ideal. First, transistors were made into ICs. Next came the integration of micro-optics and micromechanics into devices that were packaged individually and mounted on PCBs. The use of flip chips (where the chip is the package), and placement of passive components within PCBs, are blurring the distinction between devices and systems. The high levels of integration made possible by nanotechnology has made the (very smart) material essentially the device and possibly also the system. Larry Bock, chief executive for Nanosys, recently noted that "nanotech takes the complexity out of the system and puts it in the material" [6].

We can now seriously contemplate sensing the interaction of a small number of molecules, processing and transmitting the data with a small number of electrons, and storing the information in nanometer-scale structures. Fluorescence and other means of single-molecule detection are being developed. IBM and others are working on data storage systems that use proximal probes to make and read nanometer-scale indentations in polymers. These systems promise read/write densities near 1×10^{12} bits/sq. in., far in excess of current magnetic storage capabilities [7]. Although presenting a significant challenge, integration of nano-scale technologies could lead to tiny, low-power, smart sensors that could be manufactured cheaply in large numbers. Their service areas could include in situ sensing of structural materials, sensor redundancy in systems, and size- and weight-constrained structures such as satellites and space platforms.

Nanomaterials and nanostructures are other promising application areas. Two functions often separated in many sensors, especially those for chemicals and biological substances, are recognition of the molecule or other object of interest and transduction of that recognition event into a useful signal. Nanotechnology will enable us to design sensors that are much smaller, less power hungry, and more sensitive than current micro- or macrosensors. Sensing applications will thus enjoy benefits far beyond those offered by MEMS and other microsensors.

Manufacturing Advances. Recent advances in top-down manufacturing processes have spurred both micro- and nanotechnologies. Makers of leading-edge ICs use lithography, etching, and deposition to sculpt a substrate such as silicon and build structures on it. Conventional microelectronics has approached the nanometer scale—line widths on chips are near the 100 nm level and are continuing to shrink. MEMS devices are constructed in a similar top-down process. As these processes begin working on smaller and smaller dimensions, they can be used to make a variety of nanotechnology components, much as a large lathe can be used to make small parts in a machine shop.

Chapter 21

In the nano arena, various bottom-up methods use individual atoms and molecules to build useful structures. Under the right conditions, the atoms, molecules, and larger units can self-assemble [8]. Alternatively, directed assembly can be used [9].

In either case, the combination of nano-scale top-down and bottom-up processes gives materials and device designers a wide variety of old and new tools. Designers can also combine micro- and nanotechnologies to develop new sensor systems.

Computational Design. Recently developed experimental tools, notably synchrotron X-radiation and nuclear magnetic resonance, have revealed the atomic structures of many complex molecules. But this knowledge is not enough; we need to understand the interactions of atoms and molecules in the recognition and sometimes the transduction stages of sensing. The availability of powerful computers and algorithms for simulating nano-scale interactions means that we can design nanosensors computationally, and not just experimentally, by using the molecular dynamics codes and calculations that are already fundamental tools in nanotechnology.

21.2 Realities

Although the excitement over nanotechnology and its prospective uses is generally well founded, the development and integration of nanosensors must take into account the realities imposed by physics, chemistry, biology, engineering, and commerce. For example, as nanotechnologies are integrated into macro-sized systems, we'll have to provide for and control the flow of matter, energy, and information between the nano and macro scales.

The Usual Design Problems—Intensified. Many of the design considerations for nanosensors are similar to those for microsensors, notably interface requirements, heat dissipation, and the need to deal with interference and noise, both electrical and mechanical. Each interface in a microsystem is subject to unwanted transmission of electrical, mechanical, thermal, and, possibly, chemical, acoustical, and optical fluxes. Dealing with unwanted molecules and signals in very small systems often requires ancillary equipment and low-temperature operation to reduce noise. Flow control is especially critical in chemical and biological sensors into which gaseous or liquid analytes are brought and from which they are expelled. Furthermore, the very sensitive, tailored surfaces of these sensors are prone to degradation from the effects of foreign substances, heat, and cold. But the ability to install hundreds of sensors in a small space allows malfunctioning devices to be ignored in favor of good ones, thus prolonging a system's useful lifetime.

Risk and Economics. The path from research to engineering to products to revenues to profits to sustained commercial operations, difficult for technologies of any scale, is particularly challenging for nanotechnologies. One major impediment to their adoption is the common reluctance to specify new technologies for high-value systems. Another is that at present most nano-scale materials are hard to produce in large volumes, so unit prices are high and markets are limited. Costs will decrease over time, but small companies may have a struggle making their profit goals quickly enough to survive.

21.3 Applications

Few sensors today are based on pure nanoscience, and the development of nano-enabled sensors is in the early stages; yet we can already foresee some of the possible devices and applications. Sensors for physical properties were the focus of some early development efforts, but nanotechnology will contribute most heavily to realizing the potential of chemical and biosensors for safety, medical, and other purposes. Vo-Dinh, Cullum, and Stokes recently provided an overview of nanosensors and biochips for the detection of biomolecules [10].

Physical Sensors. Researchers at the Georgia Institute of Technology led by Walter de Heer devised the world's smallest "balance" (see Figure 21.3.1) by taking advantage of the unique electrical and mechanical properties of carbon nanotubes [11,12]. They mounted a single particle on the end of a CNT and applied an electrical charge to it. Acting much like a strong, flexible spring, the CNT oscillated, without breaking, and the mass of the particle was calculated from changes in the resonance vibrational frequency with and without the particle. This approach may allow the mass of individual biomolecules to be measured.

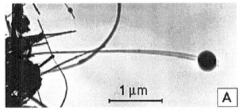

Figure 21.3.1: The mass of a carbon sphere shifts the resonance frequency of the carbon nanotube to which it is attached. (Courtesy of Walter de Heer, Georgia Institute of Technology, Atlanta, GA.)

Chapter 21

Electrometers. Cleland and Roukes at the California Institute of Technology reported the fabrication and characterization of a working, submicron mechanical electrometer [13].

This device (see Figure 21.3.2) has demonstrated charge sensitivity below a single electron charge per unit bandwidth (~0.1 electrons/Hz at 2.61 MHz), better than that of state-of-the-art semiconductor devices.

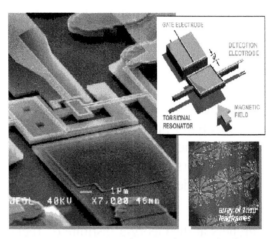

Figure 21.3.2: A nanometer-scale mechanical electrometer consists of a torsional mechanical resonator, a detection electrode, and a gate electrode used to couple charge to the mechanical element. A schematic and micrographs of a single element and an array of elements are shown. (Reprinted with copyright permission from Nature Publishing Group).

Chemical Sensors. Various nanotube-based gas sensors have been described in the past few years. Modi et al. have developed a miniaturized gas ionization detector based on CNTs [14]. The sensor could be used for gas chromatography. Titania nanotube hydrogen sensors [15] have been incorporated in a wireless sensor network to detect hydrogen concentrations in the atmosphere. And Kong et al. have developed a chemical sensor for gaseous molecules such as NO_2 and NH_3 that is based on nanotube molecular wires [16].

Datskos and Thundat used a focused ion beam technique to fabricate nanocantilevers (see Figure 21.3.3) and have developed an electron transfer transduction approach to measure cantilever motion [17]. The results might be sensitive enough to detect single chemical and biological molecules. Structurally modified semiconducting nanobelts of ZnO have also been demonstrated applicable to nanocantilever sensors [18].

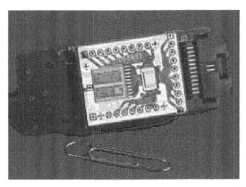

Figure 21.3.3: This nano-array incorporates capacitive readout cantilevers and electronics for signal analysis. (Courtesy of Thomas G. Thundat, Ph.D., Oak Ridge National Laboratory, Oak Ridge, TN.)

Biosensors. Nanotechnology will also enable the very selective, sensitive detection of a broad range of biomolecules. By using the sequential electrochemical reduction of the metal ions onto an alumina template, we can now create cylindrical rods made up of metal sections 50 nm to 5 microns long [19,20]. These particles, trademarked Nanobarcodes (see Figure 21.3.4), can be coated with analyte-specific entities such as antibodies for selective detection of complex molecules. DNA detection with these nano-scale coded particles has also been demonstrated.

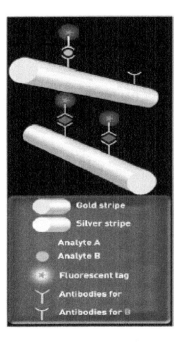

Figure 21.3.4: DNA and other biomaterials can be sensed using encoded antibodies of Nanobarcodes particles.

Chapter 21

Researchers at NASA Ames Research Center have taken a different route [21]. They cover the surface of a chip with millions of vertically mounted CNTs 30–50 nm in dia. (see Figure 21.3.5). When the DNA molecules attached to the ends of the nanotubes are placed in a liquid containing DNA molecules of interest, the DNA on the chip attaches to the target and increases its electrical conductivity. This technique, expected to reach the sensitivity of fluorescence-based detection systems, may find application in the development of a portable sensor.

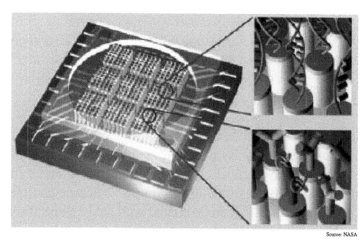

Figure 21.3.5: Vertical carbon nanotubes are grown on a silicon chip. DNA molecules attached at the ends of the tubes detect specific types of DNA in an analyte. (Courtesy of NASA Ames Research Ctr., Moffett Field, CA.)

Deployable Nanosensors. The SnifferSTAR, a lightweight, portable chemical detection system (see Figure 21.3.6), is a good example of nanotechnology's potential for field applications [22]. This unique system combines a nanomaterial for sample collection and concentration with a MEMs-based chemical lab-on-a-chip detector. SnifferSTAR will likely find work in defense and homeland security and is ideal for deployment on unmanned systems such as micro unmanned aerial vehicles.

Nanotechnology-Enabled Sensors: Possibilities, Realities, and Applications

Figure 21.3.6: The SnifferSTAR is a nano-enabled chemical sensor integrated into a micro unmanned aerial vehicle. (Courtesy of Sandia National Laboratories, Albuquerque NM, and Lockheed Martin Corp.)

And More. Other areas we expect to benefit from nanotechnology-based sensors include transportation (land, sea, air, and space); communications (wired and wireless, optical, and RF); buildings and facilities (homes, offices, factories); humans (especially for health and medical monitoring); and robotics of all types. We'll also see nano-enabled sensors increasingly integrated into commercial and military products. Many new companies will make nano materials and some will make sensors based on them. The URLs of some of these companies are given after the Reference listings.

23.4 Summary

Nanotechnology is certain to improve existing sensors and be a strong force in developing new ones. The field is progressing, but considerable work must be done before we see its full impact. Among the obvious challenges are reducing the cost of materials and devices, improving reliability, and packaging the devices into useful products. Nevertheless, we are beginning to see nano-scale materials and devices being integrated into real-world systems, and the future looks very bright indeed for technology on a tiny scale.

Chapter 21

References

1. "Small Wonders, Endless Frontiers: A Review of the National Nanotechnology Initiative," National Academy Press, 2002.

2. Cui, Y., et al., "Nanowire Nanosensors for Highly Sensitive and Selective Detection of Biological and Chemical Species," Science, Vol. 293, Aug. 17, 2001, pp. 1289–1292.

3. "Space Mission for Nanosensors," The Futurist, Nov./Dec. 2002, p. 13.

4. Cassell, J.A., "DoD grants $3M to Study Nanoshells for Early Detection, Treatment of Breast Cancer," NanoBiotech News, Vol. 1, No. 3, Aug. 13, 2003.

5. Moore, S.K., "U.S. Nanotech Funding Heads for $1 Billion Horizon," IEEE Spectrum Online.

6. Bock, L, in "Nano's Veteran Entrepreneur," Smalltimes, July/Aug. 2003, pp. 27–32.

7. IBM Research News.

8. Bernt, I., et al., "Molecular Self-Assembly: Organic Versus Inorganic Approaches," Springer Verlag, M. Fujita, ed., 1st Ed., May 15, 2000.

9. Zyvex Capabilities.

10. Vo-Dinh, T., B.M. Cullum, and D.L. Stokes, "Nanosensors and Biochips: Frontiers in Biomolecular Diagnosis," Sensors and Actuators B, 74 (2001) pp. 2–11.

11. Poncharal, P., et al., "Electrostatic Deflections and Electromechanical Resonances of Carbon Nanotubes," Science, 283:1513–1516 (1999), pp. 1513–1516.

12. Toon, J., "Weighing the Very Small: 'Nanobalance' Based on Carbon Nanotubes Shows New Application for Nanomechanics," Georgia Tech Research News, March 4, 1999.

13. Cleland, A.N., and M.L. Roukes, "A Nanometre-scale Mechanical Electrometer," Nature, Vol. 392, March 12, 1998.

14. Modi, A., et al., "Miniaturized Gas Ionization Sensors using Carbon Nanotubes," Nature, Vol. 424, Jul. 10, 2003, pp. 171–174.

15. Grimes, C.A., et al., "A Sentinel Sensor Network for Hydrogen Sensing," Sensors/MDPI, 2003, pp. 69–82.

16. Kong, J., et al., "Nanotube Molecular Wires as Chemical Sensors," Science, Vol. 287, Jan. 28, 2000, pp. 622–625.

17. Datskos, P.G., and T. Thundat, "Nanocantilever Signal Transduction by Electron Transfer," J Nanosci Nanotech, Vol. 2, 2002, pp. 369–372.

18. Hughes, W.L., and Z.L. Wang, "Nanobelts as nanocantilevers," Applied Physics Letters, Vol. 82, No. 17, April 28, 2003, pp. 2886–2888.

19. Nicewarner-Pena, R., et al., "Submicrometer Metallic Barcodes," Science, Vol. 294, 2001, p. 137.

20. Freemantle, F., "Nano Bar Coding for Bioanalysis," C&EN: News of the Week, Science, Vol. 79, No. 41, Oct. 8, 2001, p. 13.

21. Smalley, E., "Chip Senses Trace DNA," Technology Research News, Jul. 30/Aug. 6, 2003.

22. "Ultralight device analyzes gases immediately. Flying SnifferSTAR may aid civilians and U.S. Military," Sandia National Laboratories, Press Release, Jan. 23, 2003.

For Further Reading

Integrated Nano-Technologies, Henrietta, NY.

Materials Modification, Inc., Fairfax, VA.

Molecular Nanosystems, Inc., Palo Alto, CA.

Nanomix Inc., Emeryville, CA.

Nanoplex Technologies, Inc., Mountain View, CA.

Nanosphere, Inc., Northbrook, IL.

Nanosys, Inc., Palo Alto, CA.

Chapter 21

Sharon Smith, Ph.D., is Director of Technology, Lockheed Martin Corp., Bethesda, MD; 301-897-6267, sharon.smith@lmco.com.

David J. Nagel, Ph.D., is a Research Professor, Electrical and Computer Engineering Department, The George Washington University, Washington, DC; 202-994-5293, nagel@gwu.edu.

CHAPTER 22

Wireless Sensor Networks:
Principles and Applications

Chris Townsend, Steven Arms, MicroStrain, Inc.

22.1 Introduction to Wireless Sensor Networks

Sensors integrated into structures, machinery, and the environment, coupled with the efficient delivery of sensed information, could provide tremendous benefits to society. Potential benefits include: fewer catastrophic failures, conservation of natural resources, improved manufacturing productivity, improved emergency response, and enhanced homeland security [1]. However, barriers to the widespread use of sensors in structures and machines remain. Bundles of lead wires and fiber optic "tails" are subject to breakage and connector failures. Long wire bundles represent a significant installation and long term maintenance cost, limiting the number of sensors that may be deployed, and therefore reducing the overall quality of the data reported. Wireless sensing networks can eliminate these costs, easing installation and eliminating connectors.

The ideal wireless sensor is networked and scaleable, consumes very little power, is smart and software programmable, capable of fast data acquisition, reliable and accurate over the long term, costs little to purchase and install, and requires no real maintenance.

Selecting the optimum sensors and wireless communications link requires knowledge of the application and problem definition. Battery life, sensor update rates, and size are all major design considerations. Examples of low data rate sensors include temperature, humidity, and peak strain captured passively. Examples of high data rate sensors include strain, acceleration, and vibration.

Recent advances have resulted in the ability to integrate sensors, radio communications, and digital electronics into a single integrated circuit (IC) package. This capability is enabling networks of very low cost sensors that are able to communicate with each other using low power wireless data routing protocols. A wireless sensor network (WSN) generally consists of a basestation (or "gateway") that can communicate with a number of wireless sensors via a radio link. Data is collected at the wireless sensor node, compressed, and transmitted to the gateway directly or, if required, uses other wireless sensor nodes to forward data to the gateway. The transmitted data is then presented to the system by the gateway connection. The purpose of this chapter is to provide a brief technical introduction to wireless sensor networks and present a few applications in which wireless sensor networks are enabling.

22.2 Individual Wireless Sensor Node Architecture

A functional block diagram of a versatile wireless sensing node is provided in Figure 22.2.1. A modular design approach provides a flexible and versatile platform to address the needs of a wide variety of applications [2]. For example, depending on the sensors to be deployed, the signal conditioning block can be re-programmed or replaced. This allows for a wide variety of different sensors to be used with the wireless sensing node. Similarly, the radio link may be swapped out as required for a given applications' wireless range requirement and the need for bidirectional communications. The use of flash memory allows the remote nodes to acquire data on command from a basestation, or by an event sensed by one or more inputs to the node. Furthermore, the embedded firmware can be upgraded through the wireless network in the field.

The microprocessor has a number of functions including:

1) managing data collection from the sensors
2) performing power management functions
3) interfacing the sensor data to the physical radio layer
4) managing the radio network protocol

A key feature of any wireless sensing node is to minimize the power consumed by the system. Generally, the radio subsystem requires the largest amount of power. Therefore, it is advantageous to send data over the radio network only when required. This sensor event-driven data collection model requires an algorithm to be loaded into the node to determine when to send data based on the sensed event. Additionally, it is important to minimize the power consumed by the sensor itself. Therefore, the hardware should be designed to allow the microprocessor to judiciously control power to the radio, sensor, and sensor signal conditioner.

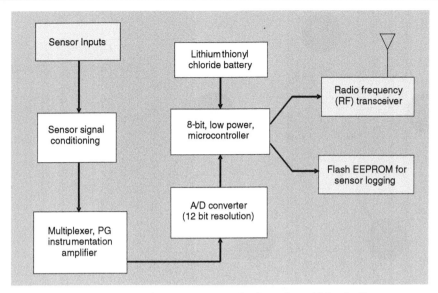

Figure 22.2.1: Wireless sensor node functional block diagram.

22.3 Wireless Sensor Networks Architecture

There are a number of different topologies for radio communications networks. A brief discussion of the network topologies that apply to wireless sensor networks are outlined below.

Star Network (Single Point-to-Multipoint)

A star network (Figure 22.3.1) is a communications topology where a single basestation can send and/or receive a message to a number of remote nodes. The remote nodes can only send or receive a message from the single basestation, they are not permitted to send messages to each other. The advantage of this type of network for wireless sensor networks is in its simplicity and the ability to keep the remote node's power consumption to a minimum. It also allows for low latency communications between the remote node and the basestation. The disadvantage of such a network is that the basestation must be within radio transmission range of all the individual nodes and is not as robust as other networks due to its dependency on a single node to manage the network.

Chapter 22

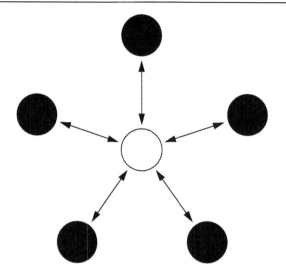

*Figure 22.3.1:
Star network topology.*

Mesh Network

A mesh network allows for any node in the network to transmit to any other node in the network that is within its radio transmission range. This allows for what is known as multihop communications; that is, if a node wants to send a message to another node that is out of radio communications range, it can use an intermediate node to forward the message to the desired node. This network topology has the advantage of redundancy and scalability. If an individual node fails, a remote node still can communicate to any other node in its range, which in turn, can forward the message to the desired location. In addition, the range of the network is not necessarily limited by the range in between single nodes, it can simply be extended by adding more nodes to the system. The disadvantage of this type of network is in power consumption for the nodes that implement the multihop communications are generally higher than for the nodes that don't have this capability, often limiting the battery life. Additionally, as the number of communication hops to a destination increases, the time to deliver the message also increases, especially if low power operation of the nodes is a requirement.

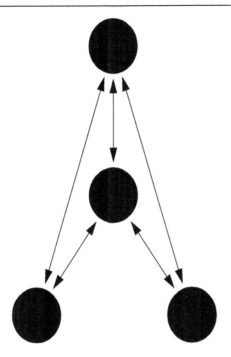

*Figure 22.3.2:
Mesh network topology.*

Hybrid Star – Mesh Network

A hybrid between the star and mesh network provides for a robust and versatile communications network, while maintaining the ability to keep the wireless sensor nodes power consumption to a minimum. In this network topology, the lowest power sensor nodes are not enabled with the ability to forward messages. This allows for minimal power consumption to be maintained. However, other nodes on the network are enabled with multihop capability, allowing them to forward messages from the low power nodes to other nodes on the network. Generally, the nodes with the multi-hop capability are higher power, and if possible, are often plugged into the electrical mains line. This is the topology implemented by the up and coming mesh networking standard known as ZigBee.

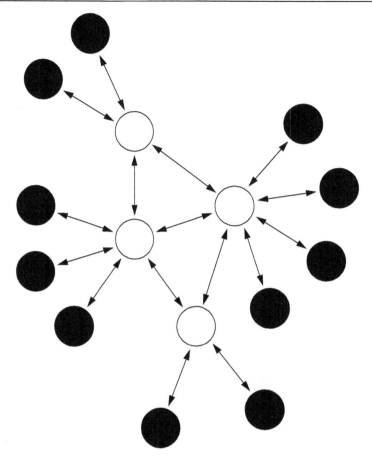

Figure 22.3.3: Hybrid star-mesh network topology.

22.4 Radio Options for the Physical Layer in Wireless Sensor Networks

The physical radio layer defines the operating frequency, modulation scheme, and hardware interface of the radio to the system. There are many low power proprietary radio integrated circuits that are appropriate choices for the radio layer in wireless sensor networks, including those from companies such as Atmel, MicroChip, Micrel, Melexis, and ChipCon. If possible, it is advantageous to use a radio interface that is standards based. This allows for interoperability among multiple companies networks. A discussion of existing radio standards and how they may or may not apply to wireless sensor networks is given next.

IEEE802.11x

IEEE802.11 is a standard that is meant for local area networking for relatively high bandwidth data transfer between computers or other devices. The data transfer rate ranges from as low as 1 Mbps to over 50 Mbps. Typical transmission range is 300 feet with a standard antenna; the range can be greatly improved with use of a directional high gain antenna. Both frequency hopping and direct sequence spread spectrum modulation schemes are available. While the data rates are certainly high enough for wireless sensor applications, the power requirements generally preclude its use in wireless sensor applications.

Bluetooth (IEEE802.15.1 and .2)

Bluetooth is a personal area network (PAN) standard that is lower power than 802.11. It was originally specified to serve applications such as data transfer from personal computers to peripheral devices such as cell phones or personal digital assistants. Bluetooth uses a star network topology that supports up to seven remote nodes communicating with a single basestation. While some companies have built wireless sensors based on Bluetooth, they have not been met with wide acceptance due to limitations of the Bluetooth protocol including:

1) Relatively high power for a short transmission range.

2) Nodes take a long time to synchronize to network when returning from sleep mode,
   which increases average system power.

3) Low number of nodes per network (<=7 nodes per piconet).

4) Medium access controller (MAC) layer is overly complex when compared to that required for wireless sensor applications.

IEEE 802.15.4

The 802.15.4 standard was specifically designed for the requirements of wireless sensing applications. The standard is very flexible, as it specifies multiple data rates and multiple transmission frequencies. The power requirements are moderately low; however, the hardware is designed to allow for the radio to be put to sleep, which reduces the power to a minimal amount. Additionally, when the node wakes up from sleep mode, rapid synchronization to the network can be achieved. This capability allows for very low average power supply current when the radio can be periodically turned off. The standard supports the following characteristics:

1) Transmission frequencies, 868 MHz/902–928 MHz/2.48–2.5 GHz.

2) Data rates of 20 Kbps (868 MHz Band) 40 Kbps (902 MHz band) and 250 Kbps (2.4 GHz band).

3) Supports star and peer-to-peer (mesh) network connections.

4) Standard specifies optional use of AES-128 security for encryption of transmitted data.

5) Link quality indication, which is useful for multi-hop mesh networking algorithms.

6) Uses direct sequence spread spectrum (DSSS) for robust data communications.

It is expected that of the three aforementioned standards, the IEEE 802.15.4 will become most widely accepted for wireless sensing applications. The 2.4-GHz band will be widely used, as it is essentially a worldwide license-free band. The high data rates accommodated by the 2.4-GHz specification will allow for lower system power due to the lower amount of radio transmission time to transfer data as compared to the lower frequency bands.

ZigBee

The ZigBee™ Alliance is an association of companies working together to enable reliable, cost-effective, low-power, wirelessly networked monitoring and control products based on an open global standard. The ZigBee alliance specifies the IEEE 802.15.4 as the physical and MAC layer and is seeking to standardize higher level applications such as lighting control and HVAC monitoring. It also serves as the compliance arm to IEEE802.15.4 much as the Wi-Fi alliance served the IEEE802.11 specification. The ZigBee network specification, to be ratified in 2004, will support both star network and hybrid star mesh networks. As can been seen in Figure 22.4.1, the ZigBee alliance encompasses the IEEE802.15.4 specification and expands on the network specification and the application interface.

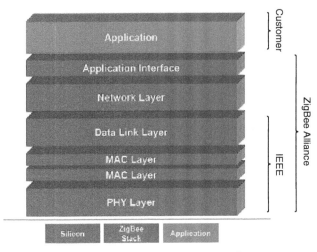

Figure 22.4.1: ZigBee stack.

IEEE1451.5

While the IEEE802.15.4 standard specifies a communication architecture that is appropriate for wireless sensor networks, it stops short of defining specifics about the sensor interface. The IEEE1451.5 wireless sensor working group aims to build on the efforts of previous IEEE1451 smart sensor working groups to standardize the interface of sensors to a wireless network. Currently, the IEEE802.15.4 physical layer has been chosen as the wireless networking communications interface, and at the time of this writing the group is in the process of defining the sensor interface.

22.5 Power Consideration in Wireless Sensor Networks

The single most important consideration for a wireless sensor network is power consumption. While the concept of wireless sensor networks looks practical and exciting on paper, if batteries are going to have to be changed constantly, widespread adoption will not occur. Therefore, when the sensor node is designed power consumption must be minimized. Figure 22.5.1 shows a chart outlining the major contributors to power consumption in a typical 5000-ohm wireless strain gage sensor node versus transmitted data update rate. Note that by far, the largest power consumption is attributable to the radio link itself.

Chapter 22

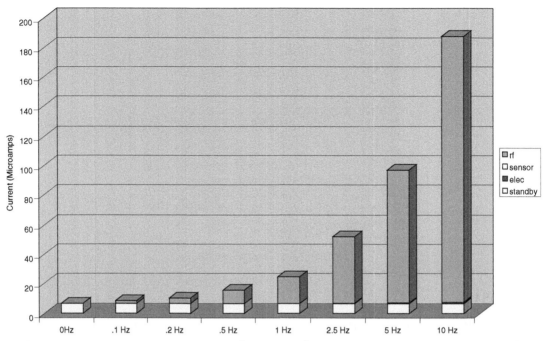

Figure 22.5.1: Power consumption of a 5000-ohm strain gage wireless sensor node.

There are a number of strategies that can be used to reduce the average supply current of the radio, including:

- Reduce the amount of data transmitted through data compression and reduction.
- Lower the transceiver duty cycle and frequency of data transmissions.
- Reduce the frame overhead.
- Implement strict power management mechanisms (power-down and sleep modes).
- Implement an event-driven transmission strategy; only transmit data when a sensor event occurs.

Power reduction strategies for the sensor itself include:

- Turn power on to sensor only when sampling.
- Turn power on to signal conditioning only when sampling sensor.
- Only sample sensor when an event occurs.
- Lower sensor sample rate to the minimum required by the application.

22.6 Applications of Wireless Sensor Networks

Structural Health Monitoring – Smart Structures

Sensors embedded into machines and structures enable condition-based maintenance of these assets [3]. Typically, structures or machines are inspected at regular time intervals, and components may be repaired or replaced based on their hours in service, rather than on their working conditions. This method is expensive if the components are in good working order, and in some cases, scheduled maintenance will not protect the asset if it was damaged in between the inspection intervals. Wireless sensing will allow assets to be inspected when the sensors indicate that there may be a problem, reducing the cost of maintenance and preventing catastrophic failure in the event that damage is detected. Additionally, the use of wireless reduces the initial deployment costs, as the cost of installing long cable runs is often prohibitive.

In some cases, wireless sensing applications demand the elimination of not only lead wires, but the elimination of batteries as well, due to the inherent nature of the machine, structure, or materials under test. These applications include sensors mounted on continuously rotating parts [4], within concrete and composite materials [5], and within medical implants [6,7].

Industrial Automation

In addition to being expensive, leadwires can be constraining, especially when moving parts are involved. The use of wireless sensors allows for rapid installation of sensing equipment and allows access to locations that would not be practical if cables were attached. An example of such an application on a production line is shown in Figure 22.6.1. In this application, typically ten or more sensors are used to measure gaps where rubber seals are to be placed. Previously, the use of wired sensors was too cumbersome to be implemented in a production line environment. The use of wireless sensors in this application is enabling, allowing a measurement to be made that was not previously practical [8].

Chapter 22

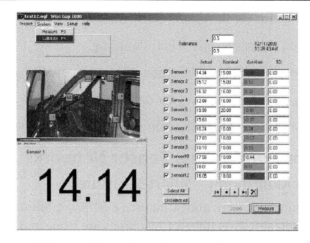

Figure 22.6.1: Industrial application of wireless sensors.

Other applications include energy control systems, security, wind turbine health monitoring, environmental monitoring, location-based services for logistics, and health care.

Application Highlight – Civil Structure Monitoring

One of the most recent applications of today's smarter, energy-aware sensor networks is structural health monitoring of large civil structures, such as the Ben Franklin Bridge (Figure 22.6.2), which spans the Delaware River, linking Philadelphia and Camden, N.J [9,10]. The bridge carries automobile, train and pedestrian traffic. Bridge officials wanted to monitor the strains on the structure as high-speed commuter trains crossed over the bridge.

Figure 22.6.2: Ben Franklin Bridge.

A star network of ten strain sensors was deployed on the tracks of the commuter rail train. The wireless sensing nodes were packaged in environmentally sealed NEMA rated enclosures. The strain gages were also suitably sealed from the environment and were spot welded to the surface of the bridge steel support structure. Transmission range of the sensors on this star network was approximately 100 meters.

The sensors operate in a low-power sampling mode where they check for presence of a train by sampling the strain sensors at a low sampling rate of approximately 6 Hz. When a train is present the strain increases on the rail, which is detected by the sensors. Once detected, the system starts sampling at a much higher sample rate. The strain waveform is logged into local Flash memory on the wireless sensor nodes. Periodically, the waveforms are downloaded from the wireless sensors to the basestation. The basestation has a cell phone attached to it which allows for the collected data to be transferred via the cell network to the engineers' office for data analysis.

This low-power event-driven data collection method reduces the power required for continuous operation from 30 mA if the sensors were on all the time to less than 1 mA continuous. This enables a lithium battery to provide more than a year of continuous operation.

Resolution of the collected strain data was typically less than 1 microstrain. A typical waveform downloaded from the node is shown in Figure 22.6.3. Other performance specifications for these wireless strain sensing nodes have been provided in an earlier work [11].

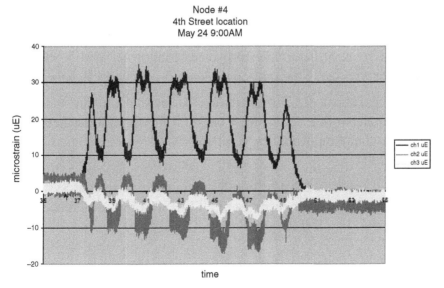

Figure 22.6.3: Bridge strain data.

22.7 Future Developments

The most general and versatile deployments of wireless sensing networks demand that batteries be deployed. Future work is being performed on systems that exploit piezoelectric materials to harvest ambient strain energy for energy storage in capacitors and/or rechargeable batteries. By combining smart, energy saving electronics with advanced thin film battery chemistries that permit infinite recharge cycles, these systems could provide a long term, maintenance free, wireless monitoring solution [12].

Conclusion

Wireless sensor networks are enabling applications that previously were not practical. As new standards-based networks are released and low power systems are continually developed, we will start to see the widespread deployment of wireless sensor networks.

Acknowledgment

The authors gratefully acknowledge the support of the National Science Foundation (NSF) through its SBIR programs. This chapter does not reflect the views or opinions of the NSF or its staff.

References

1. Lewis, F.L., "Wireless Sensor Networks," Smart Environments: Technologies, Protocols, and Applications, ed. D.J. Cook and S.K. Das, John Wiley, New York, 2004.
2. Townsend C.P, Hamel M.J., Arms S.W. (2001): Telemetered Sensors for Dynamic Activity & Structural Performance Monitoring, SPIE's 8th Annual Int'l Conference on Smart Structures and Materials, Newport Beach, CA.
3. A. Tiwari, A., Lewis, F.L., Shuzhi S-G.; "Design & Implementation of Wireless Sensor Network for Machine Condition Based Maintenance," Int'l Conf. Control, Automation, Robotics, & Vision (ICARV), Kunming, China, 6–9 Dec. 2004.
4. Arms, S.A., Townsend, C.P.; "Wireless Strain Measurement Systems – Applications & Solutions," Proceedings of NSF-ESF Joint Conference on Structural Health Monitoring, Strasbourg, France, Oct 3–5, 2003.

5. Arms, S.W., Townsend, C.P., Hamel, M.J.; "Validation of Remotely Powered and Interrogated Sensing Networks for Composite Cure Monitoring," paper presented at the 8th International Conference on Composites Engineering (ICCE/8), Tenerife, Spain, August 7–11, 2001.

6. Townsend, C.P., and Arms, S.W., Hamel, M.J.; "Remotely Powered, Multichannel, Microprocessor-Based Telemetry systems for Smart Implantable Devices and Smart Structures," SPIE's 6th Annual Int'l Conference on Smart Structures and Materials, Newport Beach, CA, Mar 1–5 1999.

7. Morris, B.A., D'Lima, D.D., Slamin, J., Kovacevic, N., Townsend, C.P., Arms, S.W., Colwell, C.W., e-Knee: The Evolution of the Electronic Knee Prosthesis: Telemetry Technology Development, Supplement to Am. Journal of Bone & Joint Surgery, January 2002.

8. Kohlstrand, K.M, Danowski, C, Schmadel, I, Arms, S.W; "Mind The Gap: Using Wireless Sensors to Measure Gaps Efficiently," *Sensors* Magazine, October 2003.

9. Galbreath, J.H, Townsend, C.P., Mundell, S.W., Hamel M.J., Esser B., Huston, D., Arms, S.W. (2003): Civil Structure Strain Monitoring with Power-Efficient High-Speed Wireless Sensor Networks, Proceedings International Workshop for Structural Health Monitoring, Stanford, CA.

10. Arms, S.W., Newhard, A.T., Galbreath, J.H., Townsend, C.P., "Remotely Reprogrammable Wireless Sensor Networks for Structural Health Monitoring Applications," ICCES International Conference on Computational and Experimental Engineering and Sciences, Medeira, Portugal, July 2004.

11. Arms, S.W., Townsend, C.P., Galbreath, J.H., Newhard, A.T.; "Wireless Strain Sensing Networks," Proceedings 2nd European Workshop on Structural Health Monitoring, Munich, Germany, July 7–9, 2004.

12. Churchill, D.L., Hamel, M.J., Townsend, C.P., Arms, S.W., "Strain Energy Harvesting for Wireless Sensor Networks," Proc. SPIE's 10th Int'l Symposium on Smart Structures & Materials, San Diego, CA. Paper presented March, 2003.

APPENDIX A

Lifetime Cost of Sensor Ownership

Analyzing It, Calculating It

Overview

> *"There is nothing in the world that some man cannot make a little worse and sell a little cheaper, and he who considers price only is that man's lawful prey."*
>
> —John Ruskin (1819–1900)

The lifetime cost of a sensor or transducer involves more than the initial cost for the item itself. By looking at the total cost of ownership, an optimum purchase decision can be made specific to your application.

Introduction

If you purchase a car, the initial purchase price may only be 60% of the total lifetime cost of the vehicle. Gas, oil, repairs, insurance, maintenance, taxes, license fees, and other costs can exceed the initial purchase price over a 5- to 10-year typical vehicle lifetime.

If your company purchases a PC, the initial purchase price may only be 10% of the total lifetime cost of the computer. Installation, support, training, upgrades, and repairs usually dwarf the initial outlay.

Have you looked at the total cost of ownership for the sensors and transducers you are using? Do you look at these costs before making a specification?

Typical "Initial Cost" Purchase Analysis

When someone asks you how much did something "cost," you typically state a figure based on what was shown on the quote, invoice, or receipt. In the case of a transducer, this is often only the cost of the transducer and possibly an amount for shipping, taxes, and related transaction costs.

This cost accounting may make the boss and the finance department happy. It can also reduce effectiveness and profitability.

What's Missing

You may say to yourself, "I'm only buying a simple transducer. What other costs could there be?"

I'm glad you asked. Below is a list of other costs that you may incur in the purchasing, maintaining, installing, and use of your transducer. These costs, in total, can become much larger than the initial "invoice" purchase price.

Installation. Does the transducer design require you to make a special mounting plate or is flexible mounting inherent in the product? How long does installation take? Can installation be performed by a lower-skilled employee or must a higher-skilled technician or engineer perform the task?

Cabling, Connectors, and Signal Conditioning. Does the sensor require the purchase of additional electrical cable, electrical connectors, signal conditioning, and related instrumentation?

Reliability. What is the stated lifetime of the product? Does it have an MTBF (mean time before failure) rating? Does the vendor have reliability statistics of the product being used in an environment similar to your own? Unscheduled downtime costs can be huge in factory automation, aviation, and capital-intensive applications.

Scheduled Down Time. Is calibration or scheduled maintenance required? How will this downtime affect your operations? Will alternate sensors need to be installed? Can this work be done during other maintenance periods?

Repairs. Is the product repairable or is it discarded at the end of its lifetime? Are there costs associated with its disposal? Can the repair be performed on site or must the item be returned to the manufacturer?

Calibrations. Can calibrations be performed on site or is factory return required? How often are calibrations required and what is the cost?

Usability. Is the signal from the product easy to work with or does it require specialized power supplies, amplifiers, and related equipment that must be procured, installed, learned, and configured?

Lead Time. Longer lead times require you to spend more time scheduling and may require you to stock sensors to avoid stock out situations.

On-time Performance. Does the sensor get delivered on time? If you planned for receiving the item in 7 days but the shipment does not show up for 21 days, you will spend valuable time re-scheduling resources and nagging the vendor to get the product to you.

Environmental Rating. Unintended uses can often make environmental protection an important feature. A misplaced cup of coffee or an inadvertent blow from a steel-toed shoe can wreak havoc on your "office environment" sensor. And increase your costs. And if you plan to add environmental protection yourself, remember to add this cost to the solution's total cost.

Shipping. It may not seem like much to pay a flat small fee for shipping. But add that flat small fee over spare parts, factory calibrations, repairs, and replacements and the amount can become substantial.

If shipping is based on weight and volume, look at the products you are considering specifying. Are there any size or weight differences? Are there are tariff differences related to the products originating from different countries?

Stocking Requirements. Lead time, reliability, repairability, ontime shipping and other factors influence the stocking (inventory) levels required for the transducer.

A rule of thumb is that annual inventory carrying costs are 25% with ranges from 18% to 75%. Your carrying costs may be higher than 25% based on this analysis:

Cost of Money	6 to 12%
Taxes	2 to 6%
Insurance	1 to 3%
Warehouse Expenses	2 to 5%
Physical Handling	2 to 5%
Inventory Control	3 to 6%
Obsolescence	6 to 12%
Deterioration and Pilferage	3 to 6%
Recalibration	5 to 10%
Total	30 to 65%

To reiterate, the above are *annual* carrying costs that will continue as long as you hold the products in your inventory.

Warranty. What is the length of warranty? What are the terms of the warranty? Are extended warranties available? What are the warranty restrictions?

Training. Are there extraordinary education or training requirements to use the sensor and related instrumentation? Is calibration straightforward is a course required?

Documentation. Are adequate user manuals and application notes available? Do users need to spend valuable time learning and documenting the product?

Customer Service. Is customer service readily available? What are the hours of operation? How responsive is customer service to your inquiries regarding pricing, shipping information, and repairs? Is Web site pricing and ordering available?

Technical Support. Is technical support available 24/7/365? Are there fees associated with technical support? Is the information provided complete, accurate, and timely?

Still Not Convinced?

Do you believe total cost of ownership is not relevant in your application? Consider the experience of an airline who went with "an affordable" choice only to find out 15 months later that the transducers were surviving for only 12 months on average and needed to be replaced annually. The replacement transducer selected did cost 20% more but was available off-the-shelf and was previously qualified for aircraft use. There have been no failures with the replacement transducers and no replacements have been required after 36 months of continuous use.

The Bottom Line

To do an interactive comparison of sensor and transducer total lifetime costs, use the Total Cost of Ownership Calculator at http://spaceagecontrol.com/calctco.htm.

Conclusion

The selection of the proper sensor or transducer for a given application includes an evaluation of the costs of the sensor or transducer. Initial purchase costs can be less than 20% of the product's lifetime costs.

Only by considering the lifetime costs can you ensure you are specifying an optimum solution.

References and Resources:

1. For an analysis on downtime costs, see *The Hidden Cost of Downtime* (SmartSignal Corporation).

2. Richardson, Helen: Transportation & Distribution, "Control Your Costs Then Cut Them," December 1995.

APPENDIX B

Smart Sensors and TEDS FAQ

1. What is IEEE 1451?

IEEE 1451 is a set of standards that were established to address smart sensor systems and to develop a comprehensive set of sensor and software protocols. It was hoped that the standard would pave the way for seamless connection of many parts of smart sensors and associated hardware and software.

2. What is IEEE 1451.2?

The first of these standards was IEEE 1451.2. The standard was designed to gain a standard way to specify the device operation and calibration, a standard physical interface between the sensor and the communications device, and the ability to use standardized off-the-shelf components to build smart sensors. IEEE 1451.2 had considerable challenges that ultimately led to virtually no adoption of it in practical applications.

3. What is IEEE 1451.4?

IEEE 1451.4 directs its attention to only the TEDS part of the sensor and signal conditioning system. IEEE 1451.4 adopts a valuable approach by taking a much more simple approach to other smart sensor concepts by simply focusing on the self-identification aspects of a sensor. IEEE1451.4 does this by specifying a table of self-identifying parameters that are stored in the sensor in the form of a TEDS (Transducer Electronic Datasheet).

4. What is P1451.4?

The IEEE1451.4 committee have only published a draft specification. All indications are that the standard will change slightly before it is published in its final format. National Instruments and its sensor partners have decided not to wait for the final version and are suggesting that sensor vendors start producing sensors based on the draft specification. The draft specification is defined as P1451.4 for (P)reliminary release, to separate it from the eventual 1451.4 specification.

Appendix B

5. What is TEDS?

TEDS is an acronym for Transducer Electronic Datasheet. It is a table of parameters that identify the transducer and are held in the transducer on a EEprom for interrogation by external electronics. The definition of the table is still not fully defined by the IEEE committee so carries the designation IEEE P 1451.4 for preliminary. Note: TEDS is the data contained on a sensor that is defined by IEEE1451.4. Honeywell Sensotec has for the last 8 years used the TEDS concept in its SIG CAL or SIG MOD. This TEDS in the SIG CAL Plug and Play technology is defined by Honeywell Sensotec rather than the IEEE1451.4 standard.

6. What data is carried in 1451.4 TEDS?

There are four areas of TEDS data. One is the basic data that identifies the transducer. The EXTENDED TEDS is where all the electrical and physical properties are stored. The USER area is where a sensor user can store data regarding the sensors location, next calibration date etc. The TEMPLATES section has yet to be fully defined by the IEEE1451.4 committee but will likely contain additional data that is distinct for each class of sensor and will be compiled by the manufacturer. An example of this might be the calibration curve for a ASTME74 load cell.

Transducer Electronic Data Sheet
TEDS

Basic TEDS	Manufacturer ID	Sensotec
	Model Number	41
	Serial Number	462992
	Version Letter	53e
Standard and Extended TEDS	Calibration Date	April 22, 2002
	Temperature effect on span	0.0045
	Temperature effect on offset	0.0045
	Min Operating Temperature	-53
	Max Operating Temperature	121
	Response Time	0.0005
	Min Electrical Output	-2
	Max Electrical Output	+2
	Sensitivity	1.998
	Bridge Impedance	350
	Excitation Nominal	10
	Excitation Maximum	15
	Excitation Minimum	3
	Max Current Draw	30
User Area	Sensor Location	23 right dyno
	Calibration Due Date	April 21, 2003
Templates	Special Calibration Date	12.3-0 175x-0.00
	Wiring Code	Wiring Code #15

7. What is the output of a TEDS sensor?

There are two types of TEDS sensor outputs: four wire outputs or two wire outputs/ two wire ICP outputs (Integrated Charge Pump) used on accelerometers. In the case of four wire systems the standard calls for two additional wires to carry the digital TEDS data. On two wire systems the Digital TEDS data in digital format share the same wires as the analog signal.

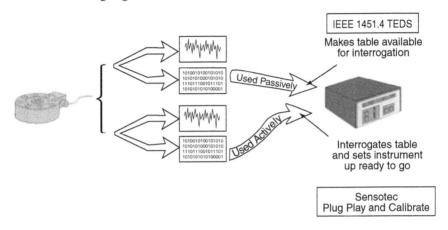

8. What is Plug and Play?

Plug and Play is the name adopted for the technology surrounding IEEE1451.4. Plug and Play, however, suggests that the user can plug the sensor into the signal conditioning and everything is automatically configured and is ready to take measurements. The IEEE1451.4 standard does not explicitly ensure that this is the case. For example, the IEEE1451.4 specification does not, as yet, address the hardware connection by specifying a connector and wiring code nor does it address what to do with the TEDS data once the signal conditioning has read the information off the EEprom. It is up to sensor manufacturers to use the dot 4 standard either passively or actively. Honeywell Sensotec has chosen to use the IEEE1451.4 table actively and provides 'Plug, Play and Calibrate' of the sensor and signal conditioning system.

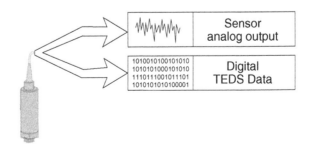

Appendix B

9. Honeywell Sensotec 'Sig Cal' and IEEE 1451.4

The IEEE1451.4 standard was written around the same EErom as utilized by the Honeywell Sensotec 'Sig Cal'. Conversion of a Sig Cal sensor to IEEE1451.4 requires a software change only. The IEEE 1451.4 defines a passive system only where the table of data that resides in the sensor. 'Sig Cal' on the other hand uses a Honeywell Sensotec defined table of data and then uses that data to set up the SC2000 so that it is ready to run. Sig Cal is true 'Plug, Play and Calibrate'

APPENDIX C

Units and Conversions

Table of Basic and Derived SI Units

BASIC UNITS

Dimension	Unit	Symbol
Length	meter	m
Mass	kilogram	kg
Time	second	s
Electric current	ampere	A
Temperature, thermodynamic	kelvin	K
Amount of matter	mole	mol
Angle	radian	rad
Solid angle	steradian	sr
Luminous intensity	candela	cd

SOME DERIVED SI UNITS

Quantity	Units	Symbol	Basic Units
Area	square meter		m^2
Volume	cubic meter		m^3
Frequency	Hertz, cycles per second	Hz	$1/s$
Density	kilogram per cubic meter		kg/m^3
Velocity	meter/sec		m/s
Angular velocity	radian/sec		rad/s
Acceleration	meter/second squared		m/s^2
Angular acceleration	radians per second square		rad/s^2
Volumetric flow rate	cubic meter per second		m^3/s
Mass flow rate	kg per second		kg/s
Force	Newton	N	$kg\text{-}m/s^2$
Surface Tension	Newton per meter	N/m	kg/s^2
Pressure, stress	Pascal (Newton per square meter)	Pa (N/m2)	$kg/m\text{-}s2$
Dynamic viscosity	Newton-second per square meter	$N\text{-}s/m^2$	$kg/m\text{-}s^2$
Kinematic viscosity	meter squared per second		m^2/s
Work, energy	joule, newton-meter, watt-second	J, N-m, W-s	$kg\text{-}m2/s2$
Power	watt, joule per second	W, J/s	$kg\text{-}m^2/s^3$

Appendix C

SOME DERIVED SI UNITS (continued)

Quantity	Units	Symbol	Basic Units
Specific heat, gas constant	joule per kilogram degree	J/kg-K	m^2/s^2-K
Enthalpy	joule per kilogram	J/kg	m^2/s^2
Entropy	joule per kilogram degree	J/kg-K	m^2/s^2-K
Thermal conductivity	watt per meter degree	W/m-K	kg-m/s^3-K
Diffusion coefficient	meter squared per second		m^2/s
Electrical charge	coulomb	C	A-s
Electromotive force	volt	V	kg-m^2/A-s^3
Electric field strength	volt per meter	V/m	kg-m/A-s^3
Electric resistance	ohm	ohm	kg-m^2/A^2-s^3
Electric Conductivity	amperes per volt meter	A/V-m	A^2-s^3/kg-m^3
Electric capacitance	farad	F	A^2-s^4/kg-m^2
magnetic flux	Weber	Wb	kg-m^2/s^2-A
Inductance	henry	H	kg-m^2/s^2-A^2
Magnetic flux density	tesla	T	kg/s^2-A

Units and Conversions

Conversion Factors

This table provides conversion factors to convert various units to their SI equivalents.

Name	To convert from	to	multiply by	divide by
Acceleration	ft/sec^2	m/s^2	0.3048	3.2810
Area	acre	m^2	4047	2.471E-04
Area	ft^2	m^2	9.294E-02	10.7600
Area	hectare	m^2	1.000E+04	1.000E-04
Area	in^2	m^2	6.452E-04	1550
Density	g/cm^3	kg/m^3	1000	1.000E-03
Density	lbm/ft^3	kg/m^3	16.02	6.243E-02
Density	lbm/in^3	kg/m^3	2.767E+04	3.614E-05
Density	lb*s^2/in^4	kg/m^3	1.069E+07	9.357E-08
Density	slug/ft^3	kg/m^3	515.40	1.940E-03
Energy	BTU	J	1055	9.478E-04
Energy	cal	J	4.1859	0.2389
Energy	erg	J	1.000E-07	1.000E+07
Energy	ev	J	1.602E-19	6.242E+18
Energy	ft*lbf	J	1.3557	0.7376
Energy	kiloton TNT	J	4.187E+12	2.388E-13
Energy	kW*hr	J	3.600E+06	2.778E-07
Energy	megaton TNT	J	4.187E+15	2.388E-16
Force	dyne	N	1.000E-05	1.000E+05
Force	lbf	N	4.4484	0.2248
Force	ozf	N	0.2780	3.5968
Heat capacity	BTU/lbm*°F	J/kg*°C	4188	2.388E-04
Heat transfer coefficient	BTU/hr*ft^2*°F	W/m^2*°C	5.6786	0.1761
Length	AU	m	1.496E+11	6.685E-12
Length	ft	m	0.3048	3.2810
Length	in	m	2.540E-02	39.3700
Length	mile	m	1609	6.214E-04
Length	Nautical mile	m	1853	5.397E-04
Length	parsec	m	3.085E+16	3.241E-17
Mass	amu	kg	1.661E-27	6.022E+26
Mass	lbm	kg	0.4535	2.2050
Mass	lb*s^2/in	kg	1200.00	5.711E-03
Mass	slug	kg	14.59	6.853E-02
Mass flow rate	lbm/hr	kg/s	1.260E-04	7937
Mass flow rate	lbm/sec	kg/s	0.4535	2.2050
Moment of inertia	ft*lb*s^2	kg*m^2	1.3557	0.7376
Moment of inertia	in*lb*s^2	kg*m^2	0.1130	8.8510
Moment of inertia	oz*in*s^2	kg*m^2	7.062E-03	141.60
Power	BTU/hr	W	0.2931	3.4120
Power	hp	W	745.71	1.341E-03
Power	tons of refrigeration	W	3516	2.844E-04
Pressure	bar	Pa	1.000E+05	1.000E-05
Pressure	dyne/cm^2	Pa	0.1000	10.0000
Pressure	in. mercury	Pa	3377	2.961E-04
Pressure	in. water	Pa	248.82	4.019E-03
Pressure	kgf/cm^2	Pa	9.807E+04	1.020E-05
Pressure	lbf/ft^2	Pa	47.89	2.088E-02

Appendix C

(continued)

Name	To convert from	to	multiply by	divide by
Pressure	lbf/in^2	Pa	6897	1.450E-04
Pressure	mbar	Pa	100.00	1.000E-02
Pressure	microns mercury	Pa	0.1333	7.501
Pressure	mm mercury	Pa	133.3	7.501E-03
Pressure	std atm	Pa	1.013E+05	9.869E-06
Specific heat	BTU/lbm*°F	J/kg*°C	4186	2.389E-04
Specific heat	cal/g*°C	J/kg*°C	4186	2.389E-04
Temperature	°F	°C	0.5556	1.8000
Thermal conductivity	BTU/hr*ft*°F	W/m*°C	1.7307	0.5778
Thermal conductivity	BTU*in/hr*ft^2*°F	W/m*°C	0.1442	6.9340
Thermal conductivity	cal/cm*s*°C	W/m*°C	418.60	2.389E-03
Thermal conductivity	cal/ft*hr*°F	W/m*°C	6.867E-03	145.62
Time	day	s	8.640E+04	1.157E-05
Time	sidereal year	s	3.156E+07	3.169E-08
Torque	ft*lbf	N*m	1.3557	0.7376
Torque	in*lbf	N*m	0.1130	8.8504
Torque	in*ozf	N*m	7.062E-03	141.61
Velocity	ft/min	m/s	5.079E-03	196.90
Velocity	ft/s	m/s	0.3048	3.2810
Velocity	km/hr	m/s	0.2778	3.6000
Velocity	miles/hr	m/s	0.4470	2.2370
Viscosity - absolute	centipose	N*s/m^2	1.000E-03	1000
Viscosity - absolute	g/cm*s	N*s/m^2	0.1000	10
Viscosity - absolute	lbf/ft^2*s	N*s/m^2	47.87	2.089E-02
Viscosity - absolute	lbm/ft*s	N*s/m^2	1.4881	0.6720
Viscosity - kinematic	centistoke	m^2/s	1.000E-06	1.000E+06
Viscosity - kinematic	ft^2/sec	m^2/s	9.294E-02	10.7600
Volume	ft^3	m^3	2.831E-02	35.3200
Volume	in^3	m^3	1.639E-05	6.102E+04
Volume	liters	m^3	1.000E-03	1000
Volume	U.S. gallons	m^3	3.785E-03	264.20
Volume flow rate	ft^3/min	m^3/s	4.719E-04	2119
Volume flow rate	U.S. gallons/min	m^3/s	6.309E-05	1.585E+04

Electrical Units

Name (Symbol)	Base Unit	Unit Symbol
Angular Frequency	radians/second	ω
Admittance (Y)	siemens	S
Capacitance (C)	farad	F
Capacitive Reactance (X_C)	ohm	Ω
Center Frequency (F_c)	hertz	Hz
Charge	coulomb	C
Conductance (G) [℧]	siemen [mho]	S (℧) [1/Ω]
Conductivity (σ)	siemens/meter	S/m
Current (I)	ampere	A
Dipole Moment	coulomb·meter	p
Electric Charge	coulomb	Q
Electric Flux	coulomb	Ψ
Electric Flux Density	coulomb/meter^2	D
Electric Susceptibility	1	χ_e
Energy (E)	joule	J
Frequency	hertz	Hz
Impedance (Z)	ohm	Ω
Inductance (L)	henry	H
Inductive Reactance (X_L)	ohm	Ω
Joule	watt·second	J
Linear Current Density	amperes/meter^2	α
Magnetic Field	tesla	T
Magnetic Flux	weber	Φ
Magnetic Flux Linkage	weber	Λ
Magnetic Flux Density	tesla	B
Magnetic Susceptibility	1	χ_m
Magnetic Vector Potential	weber/meter	A
Magnetization	ampere/meter	M
Magnetomotive Force	weber	Wb
Mutual Inductance	henry	$L_{i,i}$ $M_{i,i}$
Permeability	henrys/meter	μ
Permittivity	farads/meter	ε
Phase Angle	radian	ϕ
Phase Coefficient	radian	β
Potential (V)	volt	V
Power (P)	watt	W
Reluctance (R)	1/henrys	1/H
Reluctivity	meters/henry	ν
Resistance (R)	ohm	Ω
Resistivity (ρ)	ohm·meter	Ω*m
Resonant Frequency (F_r)	hertz	Hz
Surface Density of Charge	coulombs/meter^2	σ
Susceptance (B)	siemen	S
Volume Density of Charge	coulombs/meter^3	ρ
Volume Resistivity	ohms/meter^3	ρ
Wavelength	meter	λ

Appendix C

Numerical Prefixes

Prefix	Symbol	Multiplier
vendeko	v	10^{-33}
xenno	x	10^{-27}
yocto	y	10^{-24}
zepto	z	10^{-21}
atto	a	10^{-18}
femto	f	10^{-15}
pico	p	10^{-12}
nano	n	10^{-9}
micro	μ	10^{-6}
milli	m	10^{-3}
centi	c	10^{-2}
deci	d	10^{-1}
deka	da	10^{1}
hecto	h	10^{2}
kilo	k	10^{3}
mega	M	10^{6}
giga	G	10^{9}
tera	T	10^{12}
peta	P	10^{15}
exa	E	10^{18}
zetta	Z	10^{21}
yotta	Y	10^{24}
xenna	X	10^{27}
vendeka	V	10^{33}

APPENDIX D

Physical Constants

From: http://physics.nist.gov/constants

Fundamental Physical Constants — Extensive Listing

Quantity	Symbol	Value	Unit	Relative std. uncert. u_r
UNIVERSAL				
speed of light in vacuum	c, c_0	299 792 458	m s^{-1}	(exact)
magnetic constant	μ_0	$4\pi \times 10^{-7}$	N A^{-2}	
		$= 12.566\,370\,614... \times 10^{-7}$	N A^{-2}	(exact)
electric constant $1/\mu_0 c^2$	ε_0	$8.854\,187\,817... \times 10^{-12}$	F m^{-1}	(exact)
characteristic impedance of vacuum $\sqrt{\mu_0/\varepsilon_0} = \mu_0 c$	Z_0	376.730 313 461...	Ω	(exact)
Newtonian constant of gravitation	G	$6.6742(10) \times 10^{-11}$	m^3 kg^{-1} s^{-2}	1.5×10^{-4}
	$G/\hbar c$	$6.7087(10) \times 10^{-39}$	(GeV/c^2)$^{-2}$	1.5×10^{-4}
Planck constant	h	$6.626\,0693(11) \times 10^{-34}$	J s	1.7×10^{-7}
in eV s		$4.135\,667\,43(35) \times 10^{-15}$	eV s	8.5×10^{-8}
$h/2\pi$	$\hbar$	$1.054\,571\,68(18) \times 10^{-34}$	J s	1.7×10^{-7}
in eV s		$6.582\,119\,15(56) \times 10^{-16}$	eV s	8.5×10^{-8}
$\hbar c$ in MeV fm		197.326 968(17)	MeV fm	8.5×10^{-8}
Planck mass $(\hbar c/G)^{1/2}$	m_P	$2.176\,45(16) \times 10^{-8}$	kg	7.5×10^{-5}
Planck temperature $(\hbar c^5/G)^{1/2}/k$	T_P	$1.416\,79(11) \times 10^{32}$	K	7.5×10^{-5}
Planck length $\hbar/m_P c = (\hbar G/c^3)^{1/2}$	l_P	$1.616\,24(12) \times 10^{-35}$	m	7.5×10^{-5}
Planck time $l_P/c = (\hbar G/c^5)^{1/2}$	t_P	$5.391\,21(40) \times 10^{-44}$	s	7.5×10^{-5}
ELECTROMAGNETIC				
elementary charge	e	$1.602\,176\,53(14) \times 10^{-19}$	C	8.5×10^{-8}
	e/h	$2.417\,989\,40(21) \times 10^{14}$	A J^{-1}	8.5×10^{-8}
magnetic flux quantum $h/2e$	Φ_0	$2.067\,833\,72(18) \times 10^{-15}$	Wb	8.5×10^{-8}
conductance quantum $2e^2/h$	G_0	$7.748\,091\,733(26) \times 10^{-5}$	S	3.3×10^{-9}
inverse of conductance quantum	G_0^{-1}	12 906.403 725(43)	Ω	3.3×10^{-9}
Josephson constant[1] $2e/h$	K_J	$483\,597.879(41) \times 10^9$	Hz V^{-1}	8.5×10^{-8}
von Klitzing constant[2] $h/e^2 = \mu_0 c/2\alpha$	R_K	25 812.807 449(86)	Ω	3.3×10^{-9}
Bohr magneton $e\hbar/2m_e$	μ_B	$927.400\,949(80) \times 10^{-26}$	J T^{-1}	8.6×10^{-8}
in eV T^{-1}		$5.788\,381\,804(39) \times 10^{-5}$	eV T^{-1}	6.7×10^{-9}
	μ_B/h	$13.996\,2458(12) \times 10^9$	Hz T^{-1}	8.6×10^{-8}
	μ_B/hc	46.686 4507(40)	m^{-1} T^{-1}	8.6×10^{-8}
	μ_B/k	0.671 7131(12)	K T^{-1}	1.8×10^{-6}
nuclear magneton $e\hbar/2m_p$	μ_N	$5.050\,783\,43(43) \times 10^{-27}$	J T^{-1}	8.6×10^{-8}
in eV T^{-1}		$3.152\,451\,259(21) \times 10^{-8}$	eV T^{-1}	6.7×10^{-9}
	μ_N/h	7.622 593 71(65)	MHz T^{-1}	8.6×10^{-8}
	μ_N/hc	$2.542\,623\,58(22) \times 10^{-2}$	m^{-1} T^{-1}	8.6×10^{-8}
	μ_N/k	$3.658\,2637(64) \times 10^{-4}$	K T^{-1}	1.8×10^{-6}
ATOMIC AND NUCLEAR				
General				

Appendix D

From: http://physics.nist.gov/constants

Fundamental Physical Constants — Extensive Listing

Quantity	Symbol	Value	Unit	Relative std. uncert. u_r
fine-structure constant $e^2/4\pi\varepsilon_0\hbar c$	α	$7.297\,352\,568(24) \times 10^{-3}$		3.3×10^{-9}
inverse fine-structure constant	α^{-1}	$137.035\,999\,11(46)$		3.3×10^{-9}
Rydberg constant $\alpha^2 m_e c/2h$	R_∞	$10\,973\,731.568\,525(73)$	m^{-1}	6.6×10^{-12}
	$R_\infty c$	$3.289\,841\,960\,360(22) \times 10^{15}$	Hz	6.6×10^{-12}
	$R_\infty hc$	$2.179\,872\,09(37) \times 10^{-18}$	J	1.7×10^{-7}
$R_\infty hc$ in eV		$13.605\,6923(12)$	eV	8.5×10^{-8}
Bohr radius $\alpha/4\pi R_\infty = 4\pi\varepsilon_0\hbar^2/m_e e^2$	a_0	$0.529\,177\,2108(18) \times 10^{-10}$	m	3.3×10^{-9}
Hartree energy $e^2/4\pi\varepsilon_0 a_0 = 2R_\infty hc$				
$= \alpha^2 m_e c^2$	E_h	$4.359\,744\,17(75) \times 10^{-18}$	J	1.7×10^{-7}
in eV		$27.211\,3845(23)$	eV	8.5×10^{-8}
quantum of circulation	$h/2m_e$	$3.636\,947\,550(24) \times 10^{-4}$	$m^2\,s^{-1}$	6.7×10^{-9}
	h/m_e	$7.273\,895\,101(48) \times 10^{-4}$	$m^2\,s^{-1}$	6.7×10^{-9}
Electroweak				
Fermi coupling constant[3]	$G_F/(\hbar c)^3$	$1.166\,39(1) \times 10^{-5}$	GeV^{-2}	8.6×10^{-6}
weak mixing angle[4] θ_W (on-shell scheme)				
$\sin^2\theta_W = s_W^2 \equiv 1 - (m_W/m_Z)^2$	$\sin^2\theta_W$	$0.222\,15(76)$		3.4×10^{-3}
Electron, e^-				
electron mass	m_e	$9.109\,3826(16) \times 10^{-31}$	kg	1.7×10^{-7}
in u, $m_e = A_r(e)$ u (electron relative atomic mass times u)		$5.485\,799\,0945(24) \times 10^{-4}$	u	4.4×10^{-10}
energy equivalent	$m_e c^2$	$8.187\,1047(14) \times 10^{-14}$	J	1.7×10^{-7}
in MeV		$0.510\,998\,918(44)$	MeV	8.6×10^{-8}
electron-muon mass ratio	m_e/m_μ	$4.836\,331\,67(13) \times 10^{-3}$		2.6×10^{-8}
electron-tau mass ratio	m_e/m_τ	$2.875\,64(47) \times 10^{-4}$		1.6×10^{-4}
electron-proton mass ratio	m_e/m_p	$5.446\,170\,2173(25) \times 10^{-4}$		4.6×10^{-10}
electron-neutron mass ratio	m_e/m_n	$5.438\,673\,4481(38) \times 10^{-4}$		7.0×10^{-10}
electron-deuteron mass ratio	m_e/m_d	$2.724\,437\,1095(13) \times 10^{-4}$		4.8×10^{-10}
electron to alpha particle mass ratio	m_e/m_α	$1.370\,933\,555\,75(61) \times 10^{-4}$		4.4×10^{-10}
electron charge to mass quotient	$-e/m_e$	$-1.758\,820\,12(15) \times 10^{11}$	$C\,kg^{-1}$	8.6×10^{-8}
electron molar mass $N_A m_e$	$M(e), M_e$	$5.485\,799\,0945(24) \times 10^{-7}$	$kg\,mol^{-1}$	4.4×10^{-10}
Compton wavelength $h/m_e c$	λ_C	$2.426\,310\,238(16) \times 10^{-12}$	m	6.7×10^{-9}
$\lambda_C/2\pi = \alpha a_0 = \alpha^2/4\pi R_\infty$	$\bar{\lambda}_C$	$386.159\,2678(26) \times 10^{-15}$	m	6.7×10^{-9}
classical electron radius $\alpha^2 a_0$	r_e	$2.817\,940\,325(28) \times 10^{-15}$	m	1.0×10^{-8}
Thomson cross section $(8\pi/3)r_e^2$	σ_e	$0.665\,245\,873(13) \times 10^{-28}$	m^2	2.0×10^{-8}
electron magnetic moment	μ_e	$-928.476\,412(80) \times 10^{-26}$	$J\,T^{-1}$	8.6×10^{-8}
to Bohr magneton ratio	μ_e/μ_B	$-1.001\,159\,652\,1859(38)$		3.8×10^{-12}
to nuclear magneton ratio	μ_e/μ_N	$-1838.281\,971\,07(85)$		4.6×10^{-10}
electron magnetic moment anomaly $|\mu_e|/\mu_B - 1$	a_e	$1.159\,652\,1859(38) \times 10^{-3}$		3.2×10^{-9}
electron g-factor $-2(1 + a_e)$	g_e	$-2.002\,319\,304\,3718(75)$		3.8×10^{-12}

Physical Constants

From: http://physics.nist.gov/constants

Fundamental Physical Constants — Extensive Listing

Quantity	Symbol	Value	Unit	Relative std. uncert. u_r
electron-muon magnetic moment ratio	μ_e/μ_μ	206.766 9894(54)		2.6×10^{-8}
electron-proton magnetic moment ratio	μ_e/μ_p	−658.210 6862(66)		1.0×10^{-8}
electron to shielded proton magnetic moment ratio (H_2O, sphere, 25 °C)	μ_e/μ'_p	−658.227 5956(71)		1.1×10^{-8}
electron-neutron magnetic moment ratio	μ_e/μ_n	960.920 50(23)		2.4×10^{-7}
electron-deuteron magnetic moment ratio	μ_e/μ_d	−2143.923 493(23)		1.1×10^{-8}
electron to shielded helion[5] magnetic moment ratio (gas, sphere, 25 °C)	μ_e/μ'_h	864.058 255(10)		1.2×10^{-8}
electron gyromagnetic ratio $2|\mu_e|/\hbar$	γ_e	$1.760 859 74(15) \times 10^{11}$	$s^{-1} T^{-1}$	8.6×10^{-8}
	$\gamma_e/2\pi$	28 024.9532(24)	$MHz\ T^{-1}$	8.6×10^{-8}

Muon, μ^-

Quantity	Symbol	Value	Unit	Relative std. uncert. u_r
muon mass	m_μ	$1.883 531 40(33) \times 10^{-28}$	kg	1.7×10^{-7}
in u, $m_\mu = A_r(\mu)$ u (muon relative atomic mass times u)		0.113 428 9264(30)	u	2.6×10^{-8}
energy equivalent	$m_\mu c^2$	$1.692 833 60(29) \times 10^{-11}$	J	1.7×10^{-7}
in MeV		105.658 3692(94)	MeV	8.9×10^{-8}
muon-electron mass ratio	m_μ/m_e	206.768 2838(54)		2.6×10^{-8}
muon-tau mass ratio	m_μ/m_τ	$5.945 92(97) \times 10^{-2}$		1.6×10^{-4}
muon-proton mass ratio	m_μ/m_p	0.112 609 5269(29)		2.6×10^{-8}
muon-neutron mass ratio	m_μ/m_n	0.112 454 5175(29)		2.6×10^{-8}
muon molar mass $N_A m_\mu$	$M(\mu), M_\mu$	$0.113 428 9264(30) \times 10^{-3}$	$kg\ mol^{-1}$	2.6×10^{-8}
muon Compton wavelength $h/m_\mu c$	$\lambda_{C,\mu}$	$11.734 441 05(30) \times 10^{-15}$	m	2.5×10^{-8}
$\lambda_{C,\mu}/2\pi$	$\lambdabar_{C,\mu}$	$1.867 594 298(47) \times 10^{-15}$	m	2.5×10^{-8}
muon magnetic moment	μ_μ	$-4.490 447 99(40) \times 10^{-26}$	$J\ T^{-1}$	8.9×10^{-8}
to Bohr magneton ratio	μ_μ/μ_B	$-4.841 970 45(13) \times 10^{-3}$		2.6×10^{-8}
to nuclear magneton ratio	μ_μ/μ_N	−8.890 596 98(23)		2.6×10^{-8}
muon magnetic moment anomaly $|\mu_\mu|/(e\hbar/2m_\mu) - 1$	a_μ	$1.165 919 81(62) \times 10^{-3}$		5.3×10^{-7}
muon g-factor $-2(1 + a_\mu)$	g_μ	−2.002 331 8396(12)		6.2×10^{-10}
muon-proton magnetic moment ratio	μ_μ/μ_p	−3.183 345 118(89)		2.8×10^{-8}

Tau, τ^-

Quantity	Symbol	Value	Unit	Relative std. uncert. u_r
tau mass[6]	m_τ	$3.167 77(52) \times 10^{-27}$	kg	1.6×10^{-4}
in u, $m_\tau = A_r(\tau)$ u (tau relative atomic mass times u)		1.907 68(31)	u	1.6×10^{-4}
energy equivalent	$m_\tau c^2$	$2.847 05(46) \times 10^{-10}$	J	1.6×10^{-4}

Appendix D

From: http://physics.nist.gov/constants

Fundamental Physical Constants — Extensive Listing

Quantity	Symbol	Value	Unit	Relative std. uncert. u_r
in MeV		1776.99(29)	MeV	1.6×10^{-4}
tau-electron mass ratio	m_τ/m_e	3477.48(57)		1.6×10^{-4}
tau-muon mass ratio	m_τ/m_μ	16.8183(27)		1.6×10^{-4}
tau-proton mass ratio	m_τ/m_p	1.89390(31)		1.6×10^{-4}
tau-neutron mass ratio	m_τ/m_n	1.89129(31)		1.6×10^{-4}
tau molar mass $N_A m_\tau$	$M(\tau), M_\tau$	$1.90768(31) \times 10^{-3}$	kg mol^{-1}	1.6×10^{-4}
tau Compton wavelength $h/m_\tau c$	$\lambda_{C,\tau}$	$0.69772(11) \times 10^{-15}$	m	1.6×10^{-4}
$\lambda_{C,\tau}/2\pi$	$\bar\lambda_{C,\tau}$	$0.111046(18) \times 10^{-15}$	m	1.6×10^{-4}
Proton, p				
proton mass	m_p	$1.67262171(29) \times 10^{-27}$	kg	1.7×10^{-7}
in u, $m_p = A_r(p)$ u (proton relative atomic mass times u)		1.00727646688(13)	u	1.3×10^{-10}
energy equivalent	$m_p c^2$	$1.50327743(26) \times 10^{-10}$	J	1.7×10^{-7}
in MeV		938.272029(80)	MeV	8.6×10^{-8}
proton-electron mass ratio	m_p/m_e	1836.15267261(85)		4.6×10^{-10}
proton-muon mass ratio	m_p/m_μ	8.88024333(23)		2.6×10^{-8}
proton-tau mass ratio	m_p/m_τ	0.528012(86)		1.6×10^{-4}
proton-neutron mass ratio	m_p/m_n	0.99862347872(58)		5.8×10^{-10}
proton charge to mass quotient	e/m_p	$9.57883376(82) \times 10^{7}$	C kg^{-1}	8.6×10^{-8}
proton molar mass $N_A m_p$	$M(p), M_p$	$1.00727646688(13) \times 10^{-3}$	kg mol^{-1}	1.3×10^{-10}
proton Compton wavelength $h/m_p c$	$\lambda_{C,p}$	$1.3214098555(88) \times 10^{-15}$	m	6.7×10^{-9}
$\lambda_{C,p}/2\pi$	$\bar\lambda_{C,p}$	$0.2103089104(14) \times 10^{-15}$	m	6.7×10^{-9}
proton rms charge radius	R_p	$0.8750(68) \times 10^{-15}$	m	7.8×10^{-3}
proton magnetic moment	μ_p	$1.41060671(12) \times 10^{-26}$	J T^{-1}	8.7×10^{-8}
to Bohr magneton ratio	μ_p/μ_B	$1.521032206(15) \times 10^{-3}$		1.0×10^{-8}
to nuclear magneton ratio	μ_p/μ_N	2.792847351(28)		1.0×10^{-8}
proton g-factor $2\mu_p/\mu_N$	g_p	5.585694701(56)		1.0×10^{-8}
proton-neutron magnetic moment ratio	μ_p/μ_n	$-1.45989805(34)$		2.4×10^{-7}
shielded proton magnetic moment (H$_2$O, sphere, 25 °C)	μ'_p	$1.41057047(12) \times 10^{-26}$	J T^{-1}	8.7×10^{-8}
to Bohr magneton ratio	μ'_p/μ_B	$1.520993132(16) \times 10^{-3}$		1.1×10^{-8}
to nuclear magneton ratio	μ'_p/μ_N	2.792775604(30)		1.1×10^{-8}
proton magnetic shielding correction $1 - \mu'_p/\mu_p$ (H$_2$O, sphere, 25 °C)	σ'_p	$25.689(15) \times 10^{-6}$		5.7×10^{-4}
proton gyromagnetic ratio $2\mu_p/\hbar$	γ_p	$2.67522205(23) \times 10^{8}$	s^{-1} T^{-1}	8.6×10^{-8}
	$\gamma_p/2\pi$	42.5774813(37)	MHz T^{-1}	8.6×10^{-8}
shielded proton gyromagnetic ratio $2\mu'_p/\hbar$ (H$_2$O, sphere, 25 °C)	γ'_p	$2.67515333(23) \times 10^{8}$	s^{-1} T^{-1}	8.6×10^{-8}

Fundamental Physical Constants — Extensive Listing

Quantity	Symbol	Value	Unit	Relative std. uncert. u_r
	$\gamma'_p/2\pi$	42.576 3875(37)	MHz T^{-1}	8.6×10^{-8}
Neutron, n				
neutron mass	m_n	1.674 927 28(29) $\times 10^{-27}$	kg	1.7×10^{-7}
in u, $m_n = A_r(n)$ u (neutron relative atomic mass times u)		1.008 664 915 60(55)	u	5.5×10^{-10}
energy equivalent	$m_n c^2$	1.505 349 57(26) $\times 10^{-10}$	J	1.7×10^{-7}
in MeV		939.565 360(81)	MeV	8.6×10^{-8}
neutron-electron mass ratio	m_n/m_e	1838.683 6598(13)		7.0×10^{-10}
neutron-muon mass ratio	m_n/m_μ	8.892 484 02(23)		2.6×10^{-8}
neutron-tau mass ratio	m_n/m_τ	0.528 740(86)		1.6×10^{-4}
neutron-proton mass ratio	m_n/m_p	1.001 378 418 70(58)		5.8×10^{-10}
neutron molar mass $N_A m_n$	$M(n), M_n$	1.008 664 915 60(55) $\times 10^{-3}$	kg mol^{-1}	5.5×10^{-10}
neutron Compton wavelength $h/m_n c$	$\lambda_{C,n}$	1.319 590 9067(88) $\times 10^{-15}$	m	6.7×10^{-9}
$\lambda_{C,n}/2\pi$	$\bar\lambda_{C,n}$	0.210 019 4157(14) $\times 10^{-15}$	m	6.7×10^{-9}
neutron magnetic moment	μ_n	$-0.966\,236\,45(24) \times 10^{-26}$	J T^{-1}	2.5×10^{-7}
to Bohr magneton ratio	μ_n/μ_B	$-1.041\,875\,63(25) \times 10^{-3}$		2.4×10^{-7}
to nuclear magneton ratio	μ_n/μ_N	$-1.913\,042\,73(45)$		2.4×10^{-7}
neutron g-factor $2\mu_n/\mu_N$	g_n	$-3.826\,085\,46(90)$		2.4×10^{-7}
neutron-electron magnetic moment ratio	μ_n/μ_e	1.040 668 82(25) $\times 10^{-3}$		2.4×10^{-7}
neutron-proton magnetic moment ratio	μ_n/μ_p	$-0.684\,979\,34(16)$		2.4×10^{-7}
neutron to shielded proton magnetic moment ratio (H_2O, sphere, 25 °C)	μ_n/μ'_p	$-0.684\,996\,94(16)$		2.4×10^{-7}
neutron gyromagnetic ratio $2|\mu_n|/\hbar$	γ_n	1.832 471 83(46) $\times 10^8$	s^{-1} T^{-1}	2.5×10^{-7}
	$\gamma_n/2\pi$	29.164 6950(73)	MHz T^{-1}	2.5×10^{-7}
Deuteron, d				
deuteron mass	m_d	3.343 583 35(57) $\times 10^{-27}$	kg	1.7×10^{-7}
in u, $m_d = A_r(d)$ u (deuteron relative atomic mass times u)		2.013 553 212 70(35)	u	1.7×10^{-10}
energy equivalent	$m_d c^2$	3.005 062 85(51) $\times 10^{-10}$	J	1.7×10^{-7}
in MeV		1875.612 82(16)	MeV	8.6×10^{-8}
deuteron-electron mass ratio	m_d/m_e	3670.482 9652(18)		4.8×10^{-10}
deuteron-proton mass ratio	m_d/m_p	1.999 007 500 82(41)		2.0×10^{-10}
deuteron molar mass $N_A m_d$	$M(d), M_d$	2.013 553 212 70(35) $\times 10^{-3}$	kg mol^{-1}	1.7×10^{-10}
deuteron rms charge radius	R_d	2.1394(28) $\times 10^{-15}$	m	1.3×10^{-3}
deuteron magnetic moment	μ_d	0.433 073 482(38) $\times 10^{-26}$	J T^{-1}	8.7×10^{-8}
to Bohr magneton ratio	μ_d/μ_B	0.466 975 4567(50) $\times 10^{-3}$		1.1×10^{-8}
to nuclear magneton ratio	μ_d/μ_N	0.857 438 2329(92)		1.1×10^{-8}

Appendix D

From: http://physics.nist.gov/constants

Fundamental Physical Constants — Extensive Listing

Quantity	Symbol	Value	Unit	Relative std. uncert. u_r
deuteron-electron magnetic moment ratio	μ_d/μ_e	$-4.664\,345\,548(50) \times 10^{-4}$		1.1×10^{-8}
deuteron-proton magnetic moment ratio	μ_d/μ_p	$0.307\,012\,2084(45)$		1.5×10^{-8}
deuteron-neutron magnetic moment ratio	μ_d/μ_n	$-0.448\,206\,52(11)$		2.4×10^{-7}
Helion, h				
helion mass[5]	m_h	$5.006\,412\,14(86) \times 10^{-27}$	kg	1.7×10^{-7}
in u, $m_h = A_r(h)$ u (helion relative atomic mass times u)		$3.014\,932\,2434(58)$	u	1.9×10^{-9}
energy equivalent	$m_h c^2$	$4.499\,538\,84(77) \times 10^{-10}$	J	1.7×10^{-7}
in MeV		$2808.391\,42(24)$	MeV	8.6×10^{-8}
helion-electron mass ratio	m_h/m_e	$5495.885\,269(11)$		2.0×10^{-9}
helion-proton mass ratio	m_h/m_p	$2.993\,152\,6671(58)$		1.9×10^{-9}
helion molar mass $N_A m_h$	$M(h), M_h$	$3.014\,932\,2434(58) \times 10^{-3}$	kg mol^{-1}	1.9×10^{-9}
shielded helion magnetic moment (gas, sphere, 25 °C)	μ'_h	$-1.074\,553\,024(93) \times 10^{-26}$	J T^{-1}	8.7×10^{-8}
to Bohr magneton ratio	μ'_h/μ_B	$-1.158\,671\,474(14) \times 10^{-3}$		1.2×10^{-8}
to nuclear magneton ratio	μ'_h/μ_N	$-2.127\,497\,723(25)$		1.2×10^{-8}
shielded helion to proton magnetic moment ratio (gas, sphere, 25 °C)	μ'_h/μ_p	$-0.761\,766\,562(12)$		1.5×10^{-8}
shielded helion to shielded proton magnetic moment ratio (gas/H$_2$O, spheres, 25 °C)	μ'_h/μ'_p	$-0.761\,786\,1313(33)$		4.3×10^{-9}
shielded helion gyromagnetic ratio $2|\mu'_h|/\hbar$ (gas, sphere, 25 °C)	γ'_h	$2.037\,894\,70(18) \times 10^{8}$	s^{-1} T^{-1}	8.7×10^{-8}
	$\gamma'_h/2\pi$	$32.434\,1015(28)$	MHz T^{-1}	8.7×10^{-8}
Alpha particle, α				
alpha particle mass	m_α	$6.644\,6565(11) \times 10^{-27}$	kg	1.7×10^{-7}
in u, $m_\alpha = A_r(\alpha)$ u (alpha particle relative atomic mass times u)		$4.001\,506\,179\,149(56)$	u	1.4×10^{-11}
energy equivalent	$m_\alpha c^2$	$5.971\,9194(10) \times 10^{-10}$	J	1.7×10^{-7}
in MeV		$3727.379\,17(32)$	MeV	8.6×10^{-8}
alpha particle to electron mass ratio	m_α/m_e	$7294.299\,5363(32)$		4.4×10^{-10}
alpha particle to proton mass ratio	m_α/m_p	$3.972\,599\,689\,07(52)$		1.3×10^{-10}
alpha particle molar mass $N_A m_\alpha$	$M(\alpha), M_\alpha$	$4.001\,506\,179\,149(56) \times 10^{-3}$	kg mol^{-1}	1.4×10^{-11}
PHYSICO-CHEMICAL				
Avogadro constant	N_A, L	$6.022\,1415(10) \times 10^{23}$	mol^{-1}	1.7×10^{-7}
atomic mass constant $m_u = \frac{1}{12}m(^{12}C) = 1$ u	m_u	$1.660\,538\,86(28) \times 10^{-27}$	kg	1.7×10^{-7}

Physical Constants

From: http://physics.nist.gov/constants

Fundamental Physical Constants — Extensive Listing

Quantity	Symbol	Value	Unit	Relative std. uncert. u_r
$= 10^{-3}$ kg mol$^{-1}/N_A$				
energy equivalent	$m_u c^2$	$1.49241790(26) \times 10^{-10}$	J	1.7×10^{-7}
in MeV		$931.494043(80)$	MeV	8.6×10^{-8}
Faraday constant[7] $N_A e$	F	$96485.3383(83)$	C mol^{-1}	8.6×10^{-8}
molar Planck constant	$N_A h$	$3.990312716(27) \times 10^{-10}$	J s mol^{-1}	6.7×10^{-9}
	$N_A hc$	$0.11962656572(80)$	J m mol^{-1}	6.7×10^{-9}
molar gas constant	R	$8.314472(15)$	J mol^{-1} K^{-1}	1.7×10^{-6}
Boltzmann constant R/N_A	k	$1.3806505(24) \times 10^{-23}$	J K^{-1}	1.8×10^{-6}
in eV K^{-1}		$8.617343(15) \times 10^{-5}$	eV K^{-1}	1.8×10^{-6}
	k/h	$2.0836644(36) \times 10^{10}$	Hz K^{-1}	1.7×10^{-6}
	k/hc	$69.50356(12)$	m^{-1} K^{-1}	1.7×10^{-6}
molar volume of ideal gas RT/p				
$T = 273.15$ K, $p = 101.325$ kPa	V_m	$22.413996(39) \times 10^{-3}$	m^3 mol^{-1}	1.7×10^{-6}
Loschmidt constant N_A/V_m	n_0	$2.6867773(47) \times 10^{25}$	m^{-3}	1.8×10^{-6}
$T = 273.15$ K, $p = 100$ kPa	V_m	$22.710981(40) \times 10^{-3}$	m^3 mol^{-1}	1.7×10^{-6}
Sackur-Tetrode constant				
(absolute entropy constant)[8]				
$\frac{5}{2} + \ln[(2\pi m_u k T_1/h^2)^{3/2} kT_1/p_0]$				
$T_1 = 1$ K, $p_0 = 100$ kPa	S_0/R	$-1.1517047(44)$		3.8×10^{-6}
$T_1 = 1$ K, $p_0 = 101.325$ kPa		$-1.1648677(44)$		3.8×10^{-6}
Stefan-Boltzmann constant				
$(\pi^2/60) k^4/\hbar^3 c^2$	σ	$5.670400(40) \times 10^{-8}$	W m^{-2} K^{-4}	7.0×10^{-6}
first radiation constant $2\pi h c^2$	c_1	$3.74177138(64) \times 10^{-16}$	W m^2	1.7×10^{-7}
first radiation constant for spectral radiance $2hc^2$	c_{1L}	$1.19104282(20) \times 10^{-16}$	W m^2 sr^{-1}	1.7×10^{-7}
second radiation constant hc/k	c_2	$1.4387752(25) \times 10^{-2}$	m K	1.7×10^{-6}
Wien displacement law constant				
$b = \lambda_{max} T = c_2/4.965114231...$	b	$2.8977685(51) \times 10^{-3}$	m K	1.7×10^{-6}

[1] See the "Adopted values" table for the conventional value adopted internationally for realizing representations of the volt using the Josephson effect.

[2] See the "Adopted values" table for the conventional value adopted internationally for realizing representations of the ohm using the quantum Hall effect.

[3] Value recommended by the Particle Data Group (Hagiwara, et al., 2002).

[4] Based on the ratio of the masses of the W and Z bosons m_W/m_Z recommended by the Particle Data Group (Hagiwara, et al., 2002). The value for $\sin^2 \theta_W$ they recommend, which is based on a particular variant of the modified minimal subtraction ($\overline{MS}$) scheme, is $\sin^2 \hat{\theta}_W(M_Z) = 0.23124(24)$.

[5] The helion, symbol h, is the nucleus of the ^{3}He atom.

[6] This and all other values involving m_τ are based on the value of $m_\tau c^2$ in MeV recommended by the Particle Data Group, (Hagiwara, et al., 2002), but with a standard uncertainty of 0.29 MeV rather than the quoted uncertainty of -0.26 MeV, $+0.29$ MeV.

[7] The numerical value of F to be used in coulometric chemical measurements is $96485.336(16)$ $[1.7 \times 10^{-7}]$ when the relevant current is measured in terms of representations of the volt and ohm based on the Josephson and quantum Hall effects and the internationally adopted conventional values of the Josephson and von Klitzing constants K_{J-90} and R_{K-90} given in the "Adopted values" table.

[8] The entropy of an ideal monoatomic gas of relative atomic mass A_r is given by $S = S_0 + \frac{3}{2} R \ln A_r - R \ln(p/p_0) + \frac{5}{2} R \ln(T/K)$. [9] The relative atomic mass $A_r(X)$ of particle X with mass $m(X)$ is defined by $A_r(X) = m(X)/m_u$, where $m_u = m(^{12}C)/12 = M_u/N_A = 1$ u is the atomic mass constant, N_A is the Avogadro constant, and u is the atomic mass unit. Thus the mass of particle X in u is $m(X) = A_r(X)$ u and the molar mass of X is $M(X) = A_r(X) M_u$.

Appendix D

From: **http://physics.nist.gov/constants**

[10] This is the value adopted internationally for realizing representations of the volt using the Josephson effect.

[11] This is the value adopted internationally for realizing representations of the ohm using the quantum Hall effect. [a] This is the lattice parameter (unit cell edge length) of an ideal single crystal of naturally occurring Si free of impurities and imperfections, and is deduced from measurements on extremely pure and nearly perfect single crystals of Si by correcting for the effects of impurities.

APPENDIX E

Dielectric Constants

Dielectric Constants and Strengths

Values are relative dielectric constants (relative permittivities). As indicated by $e_r = 1.00000$ for a vacuum, all values are relative to a vacuum.

[Multiply by $e_0 = 8.8542 \times 10^{-12}$ F/m (permittivity of free space) to obtain absolute permittivity.]

Substance	Dielectric Constant k (relative)	Dielectric Strength (V/mil)	Max Temp (°F)
ABS (plastic)	2.4 - 3.8	410	140
Air	1.00054	30 - 70	
Alumina	8.1 - 9.5		
Aluminum Silicate	5.3 - 5.5		
Bakelite	3.7		
Bakelite (mica filled)	4.7	325 - 375	
Beeswax (yellow)	2.7		
Beryllium oxide	6.7		
Butyl Rubber	2.4		
Diamond	5.5 - 10		
Delrin (acetyl resin)	3.7	500	180
Enamel	5.1	450	
Epoxy glass PCB	5.2	700	
Formica XX	4.00		
Fused quartz	3.8		
Fused silica (glass)	3.8		
Germanium	16		
Glass	4 - 10		
Gutta-percha	2.6		
Halowax oil	4.8		
Kapton® Type 100	3.9	7400	500
Type 150	2.9	4400	
Kel-F	2.6		
Lexan®	2.96	400	275
Lucite	2.8		
Mica	4.5 - 8.0	3800 -5600	
Mica, Ruby	5.4		
Micarta 254	3.4 - 5.4		
Mylar®	3.2	7000	250

Appendix E

(continued)

Substance	Dielectric Constant k (relative)	Dielectric Strength (V/mil)	Max Temp (°F)
Neoprene	6 - 9	600	
Neoprene rubber	6.7		
Nomex®		800	450
Nylon	3.2 - 5	400	280
Oil (mineral, squibb)	2.7	200	
Paper (bond)	3.0	200	
Paraffin	2-3		
Phenolica (glass-filled)	5 - 7		
Phenolics (cellulose-filled)	4 - 15		
Phenolics (mica-filled)	4.7 - 7.5		
Plexiglass®	2.2 - 3.4	450 - 990	
Polyethylene LDPE/HDPE	2.3	450 - 1200	170
Polyamide	2.5 - 2.6		
Polypropylene	2.2	500	250
Polystyrene	2.5 - 2.6	500	
Polyvinylchloride (PVC)	3	725	140
Porcelain	5.1 - 5.9	40 -280	
Pyrex glass (Corning 7740)	5.1	335	
Quartz (fused)	4.2	150 - 200	
RT/Duroid 5880 (go to Rogers)	2.20		
Rubber (silicone)	3.2	150 - 500	170
Ruby	11.3		
Silicon	11.7 - 12.9	100 - 700	300
Silicone oil	2.5		
Silicone RTV	3.6	550	
Steatite	5.3-6.5		
Strontium titanate	233		
Teflon® (PTFE)	2.0 - 2.1	1000	480
Tenite	2.9 - 4.5		
Transformer oil	4.5		
Vacuum (free space)	1.00000		
Valox®		1560	400
Vaseline	2.16		
Vinyl	2.8 - 4.5		
Water (32°F) (68°F) (212°F)	88.0 80.4 55.3	80	
Water (distilled)	76.7 - 78.2		
Wood	1.2 - 2.1		

APPENDIX F

Index of Refraction

Material	Index
Vacuum	1.00000
Air at STP	1.00029
Ice	1.31
Water at 20 C	1.33
Acetone	1.36
Ethyl alcohol	1.36
Sugar solution(30%)	1.38
Fluorite	1.433
Fused quartz	1.46
Glycerine	1.473
Sugar solution (80%)	1.49
Typical crown glass	1.52
Crown glasses	1.52-1.62
Spectacle crown, C-1	1.523
Sodium chloride	1.54
Polystyrene	1.55-1.59
Carbon disulfide	1.63
Flint glasses	1.57-1.75
Heavy flint glass	1.65
Extra dense flint, EDF-3	1.7200
Methylene iodide	1.74
Sapphire	1.77
Rare earth flint	1.7-1.84
Lanthanum flint	1.82-1.98
Arsenic trisulfide glass	2.04
Diamond	2.417

APPENDIX G

Engineering Material Properties

This table gives various engineering material properties listed alphabetically. The units are SI.

Density	Acrylic	1400	kg/m^3
Density	Air (2800 m)	0.9800	kg/m^3
Density	Air (STP)	1.2930	kg/m^3
Density	Aluminum 2024-T3	2770	kg/m^3
Density	Aluminum 3003	2700	kg/m^3
Density	Aluminum 6061-T6	2700	kg/m^3
Density	Aluminum 7079-T6	2740	kg/m^3
Density	Ammonia - liquid	682.10	kg/m^3
Density	Argon - liquid	1390	kg/m^3
Density	Beryllium QMV	1850	kg/m^3
Density	Borosilicate Ohara E6	2180	kg/m^3
Density	Borosilicate Tempax	2230	kg/m^3
Density	Concrete	2242	kg/m^3
Density	Copper - pure	8900	kg/m^3
Density	Dow Corning 200 (350cSt)	968.00	kg/m^3
Density	Fused silica	2200	kg/m^3
Density	Glass wool	64.00	kg/m^3
Density	Gold - pure	1.932E+04	kg/m^3
Density	Helium - liquid	125.00	kg/m^3
Density	Hydrogen - liquid	70.00	kg/m^3
Density	Iron	7830	kg/m^3
Density	Lead - pure	1.134E+04	kg/m^3
Density	Magnesium AZ31B-H24	1770	kg/m^3
Density	Magnesium HK31A-H24	1790	kg/m^3
Density	Methane - liquid	424.00	kg/m^3
Density	Molybdenum - wrought	1.030E+04	kg/m^3
Density	Neon - liquid	1200	kg/m^3
Density	Nickel - pure	8900	kg/m^3
Density	Nitrogen - liquid	804.00	kg/m^3
Density	Nylon	1700	kg/m^3
Density	Platinum	2.145E+04	kg/m^3
Density	Polycarbonate	1300	kg/m^3
Density	Polyethylene	2300	kg/m^3
Density	PTFE	1200	kg/m^3
Density	SiC Alpha	2975	kg/m^3
Density	SiC sintered KT	2975	kg/m^3
Density	Silver - pure	1.050E+04	kg/m^3

Appendix G

(continued)

Property	Material	Value	Units
Density	Steel AISI 304	8030	kg/m³
Density	Steel AISI C1020	7850	kg/m³
Density	Tantalum	1.660E+04	kg/m³
Density	Titanium B 120VCA	4850	kg/m³
Density	Tungsten	1.930E+04	kg/m³
Density	Water (4 C)	999.97	kg/m³
Density	White pine	513.00	kg/m³
Elastic modulus	Aluminum 2024-T3	7.310E+10	Pa
Elastic modulus	Aluminum 6061-T6	7.310E+10	Pa
Elastic modulus	Aluminum 7079-T6	7.172E+10	Pa
Elastic modulus	Beryllium QMV	2.897E+11	Pa
Elastic modulus	Borosilicate Ohara E6	5.743E+10	Pa
Elastic modulus	Borosilicate Tempax	6.200E+10	Pa
Elastic modulus	Copper - pure	1.172E+11	Pa
Elastic modulus	Gold - pure	7.448E+10	Pa
Elastic modulus	Lead - pure	1.379E+10	Pa
Elastic modulus	Magnesium AZ31B-H24	4.483E+10	Pa
Elastic modulus	Magnesium HK31A-H24	4.414E+10	Pa
Elastic modulus	Molybdenum - wrought	2.759E+11	Pa
Elastic modulus	Nickel - pure	2.207E+11	Pa
Elastic modulus	Platinum	1.469E+11	Pa
Elastic modulus	SiC Alpha	4.760E+11	Pa
Elastic modulus	SiC sintered KT	3.320E+11	Pa
Elastic modulus	Silver - pure	7.241E+10	Pa
Elastic modulus	Steel AISI 304	1.931E+11	Pa
Elastic modulus	Steel AISI C1020	2.034E+11	Pa
Elastic modulus	Tantalum	1.862E+11	Pa
Elastic modulus	Titanium B 120VCA	1.021E+11	Pa
Elastic modulus	Tungsten	3.448E+11	Pa
Electrical resistivity	Aluminum 2017	4.000E-08	ohm*m
Electrical resistivity	Aluminum 3003	4.000E-08	ohm*m
Electrical resistivity	Aluminum 99.996%	2.655E-08	ohm*m
Electrical resistivity	Copper	1.673E-08	ohm*m
Electrical resistivity	Nickel ASTM B160	1.000E-07	ohm*m
Electrical resistivity	Steel AISI 304	7.200E-07	ohm*m
Electrical resistivity	Steel AISI C1020	1.000E-07	ohm*m
Heat capacity	Air	1006	J/kg*°C
Heat capacity	Aluminum 2024-T3	963.00	J/kg*°C
Heat capacity	Aluminum 6061-T6	963.00	J/kg*°C
Heat capacity	Aluminum 7079-T6	963.00	J/kg*°C
Heat capacity	Beryllium QMV	1884	J/kg*°C
Heat capacity	Borosilicate glass	710.00	J/kg*°C
Heat capacity	Concrete	1000	J/kg*°C
Heat capacity	Copper - pure	385.00	J/kg*°C
Heat capacity	Dow Corning 200 (350 cST)	1465	J/kg*°C
Heat capacity	Ethanol (25°C)	2453	J/kg*°C
Heat capacity	Gold - pure	130.00	J/kg*°C
Heat capacity	Ice	2093	J/kg*°C
Heat capacity	Iron	440.00	J/kg*°C
Heat capacity	Lead - pure	130.00	J/kg*°C

Engineering Material Properties

(continued)

Heat capacity	Magnesium AZ31B-H24	1047	J/kg*°C
Heat capacity	Magnesium HK31A-H24	544.00	J/kg*°C
Heat capacity	Methanol (25°C)	2547	J/kg*°C
Heat capacity	Molybdenum - wrought	293.00	J/kg*°C
Heat capacity	Nickel - pure	461.00	J/kg*°C
Heat capacity	Platinum	130.00	J/kg*°C
Heat capacity	SiC Alpha	1300	J/kg*°C
Heat capacity	SiC sintered KT	1340	J/kg*°C
Heat capacity	Silica (0°C)	937.00	J/kg*°C
Heat capacity	Silver - pure	235.00	J/kg*°C
Heat capacity	Steel AISI 304	503.00	J/kg*°C
Heat capacity	Steel AISI C1020	419.00	J/kg*°C
Heat capacity	Tantalum	126.00	J/kg*°C
Heat capacity	Titanium B 120VCA	544.00	J/kg*°C
Heat capacity	Tungsten	138.00	J/kg*°C
Heat capacity	Water	4216	J/kg*°C
Heat of combustion	Methane	55.70	MJ/kg
Heat of combustion	Octane	47.70	MJ/kg
Heat of fusion	Nitrogen	25.50	kJ/kg
Heat of fusion	Water	334.00	kJ/kg
Heat of vaporization	Ammonia	1368	kJ/kg
Heat of vaporization	Argon	162.76	kJ/kg
Heat of vaporization	Helium	23.93	kJ/kg
Heat of vaporization	Hydrogen	451.90	kJ/kg
Heat of vaporization	Methane	577.40	kJ/kg
Heat of vaporization	Neon	87.03	kJ/kg
Heat of vaporization	Nitrogen	199.20	kJ/kg
Heat of vaporization	Water (100°C)	2258	kJ/kg
Heat transfer coef.	Air v = 0 m/s	5.6000	W/°C*m^2
Heat transfer coef.	Air v = 3.4 m/s	18.90	W/°C*m^2
Heat transfer coef.	Air v = 6.7 m/s	38.00	W/°C*m^2
Kinematic viscosity	Air (101 kPa)	1.800E-05	m^2/s
Kinematic viscosity	Dow Corning 200 (350cSt)	1.121E-07	m^2/s
Kinematic viscosity	Water (0°C)	1.753E-06	m^2/s
Poisson's ratio	Aluminum 2024-T3	0.3300	
Poisson's ratio	Aluminum 6061-T6	0.3300	
Poisson's ratio	Aluminum 7079-&6	0.3300	
Poisson's ratio	Beryllium QMV	0.0300	
Poisson's ratio	Borosilicate Ohara E6	0.1950	
Poisson's ratio	Borosilicate Tempax	0.2200	
Poisson's ratio	Copper - pure		
Poisson's ratio	Gold - pure	0.4200	
Poisson's ratio	Lead - pure	0.4200	
Poisson's ratio	Magnesium AZ31B-H24	0.3500	
Poisson's ratio	Magnesium HK31A-H24	0.3500	
Poisson's ratio	Molybdenum - wrought	0.3200	
Poisson's ratio	Nickel - pure		
Poisson's ratio	Platinum	0.3900	
Poisson's ratio	Silver - pure	0.3700	
Poisson's ratio	Steel AISI 304	0.2900	

Appendix G

(continued)

Property	Material	Value	Units
Poisson's ratio	Steel AISI C1020	0.2900	
Poisson's ratio	Tantalum	0.3500	
Poisson's ratio	Titanium B 120VCA	0.3000	
Poisson's ratio	Tungsten	0.2800	
Thermal Conductivity	Acetyl	0.2300	W/m*°C
Thermal Conductivity	Acrylic	0.1400	W/m*°C
Thermal Conductivity	Aluminum 2024-T3	190.40	W/m*°C
Thermal Conductivity	Aluminum 3003	233.64	W/m*°C
Thermal Conductivity	Aluminum 6061-T6	155.80	W/m*°C
Thermal Conductivity	Aluminum 7079-T6	121.10	W/m*°C
Thermal Conductivity	Beryllium QMV	147.10	W/m*°C
Thermal Conductivity	Borosilicate glass	1.1300	W/m*°C
Thermal Conductivity	Borosilicate glass (Tempax)	1.1300	W/m*°C
Thermal Conductivity	Concrete (sand & gravel)	1.8000	W/m*°C
Thermal Conductivity	Copper - pure	392.90	W/m*°C
Thermal Conductivity	Diamond	550.00	W/m*°C
Thermal Conductivity	Douglas fir	0.1100	W/m*°C
Thermal Conductivity	Dow Corning 200 (350cSt)	0.1590	W/m*°C
Thermal Conductivity	Dow Corning 739	0.1900	W/m*°C
Thermal Conductivity	Dow Corning 93-500	0.1500	W/m*°C
Thermal Conductivity	Dow Corning Q3-6605	0.8400	W/m*°C
Thermal Conductivity	Epoxy (Epotek 353ND)	0.0490	W/m*°C
Thermal Conductivity	Epoxy (Masterbond 11A0)	1.4400	W/m*°C
Thermal Conductivity	Glass wool	0.0400	W/m*°C
Thermal Conductivity	Gold - pure	297.70	W/m*°C
Thermal Conductivity	Helium	2.7700	W/m*°C
Thermal Conductivity	Ice	2.2000	W/m*°C
Thermal Conductivity	Iron	83.50	W/m*°C
Thermal Conductivity	Lead - pure	37.04	W/m*°C
Thermal Conductivity	Limestone	0.5000	W/m*°C
Thermal Conductivity	Magnesium HK31A-H24	114.20	W/m*°C
Thermal Conductivity	Magnesium AZ31B-H24	95.19	W/m*°C
Thermal Conductivity	Methane	0.3030	W/m*°C
Thermal Conductivity	Molybdenum - wrought	143.60	W/m*°C
Thermal Conductivity	Nickel - pure	91.73	W/m*°C
Thermal Conductivity	Nitrogen	0.1460	W/m*°C
Thermal Conductivity	Nylon	0.2400	W/m*°C
Thermal Conductivity	Platinum	69.23	W/m*°C
Thermal Conductivity	Polycarbonate	0.2000	W/m*°C
Thermal Conductivity	Polypropylene	0.4000	W/m*°C
Thermal Conductivity	Polystyrene foam	0.3600	W/m*°C
Thermal Conductivity	Polyurethane foam	0.0260	W/m*°C
Thermal Conductivity	PTFE	0.2400	W/m*°C
Thermal Conductivity	Quartz	1.3200	W/m*°C
Thermal Conductivity	SiC Alpha	77.50	W/m*°C
Thermal Conductivity	SiC sintered KT	80.00	W/m*°C
Thermal Conductivity	Silastic E	0.1800	W/m*°C
Thermal Conductivity	Silastic L	0.2800	W/m*°C
Thermal Conductivity	Silicone foam (Poron)	0.0600	W/m*°C
Thermal Conductivity	Silver - pure	417.10	W/m*°C

(continued)

Thermal Conductivity	Snow (light)	0.6000	W/m*°C
Thermal Conductivity	Snow (packed)	2.2000	W/m*°C
Thermal Conductivity	Soil (coarse)	0.5200	W/m*°C
Thermal Conductivity	Soil (dry w/stones)	0.5200	W/m*°C
Thermal Conductivity	Soil (dry)	0.2300	W/m*°C
Thermal Conductivity	Soil (w/42% water)	1.1000	W/m*°C
Thermal Conductivity	Steel AISI 304	16.27	W/m*°C
Thermal Conductivity	Steel AISI C1020	46.73	W/m*°C
Thermal Conductivity	Tantalum	53.65	W/m*°C
Thermal Conductivity	Titanium B 120VCA	7.4420	W/m*°C
Thermal Conductivity	Tungsten	164.40	W/m*°C
Thermal Conductivity	Water	0.6030	W/m*°C
Thermal Conductivity	White pine	0.1100	W/m*°C
Thermal expansion coefficient	Aluminum 2024-T3	22.68	μm/m*°C
Thermal expansion coefficient	Aluminum 6061-T6	24.30	μm/m*°C
Thermal expansion coefficient	Aluminum 7079-T6	24.66	μm/m*°C
Thermal expansion coefficient	Beryllium QMV	14.94	μm/m*°C
Thermal expansion coefficient	Borosilicate E6 -30 to +70C	2.8000	μm/m*°C
Thermal expansion coefficient	Copper - pure	16.56	μm/m*°C
Thermal expansion coefficient	Gold - pure	4.39	μm/m*°C
Thermal expansion coefficient	Lead - pure	52.74	μm/m*°C
Thermal expansion coefficient	Magnesium AZ31V-H24	26.10	μm/m*°C
Thermal expansion coefficient	Magnesium HK31A-H24	25.20	μm/m*°C
Thermal expansion coefficient	Molybvdenum - wrought	5.4000	μm/m*°C
Thermal expansion coefficient	Nickel - pure	12.96	μm/m*°C
Thermal expansion coefficient	Platinum	9.0000	μm/m*°C
Thermal expansion coefficient	SiC Alpha	4.0000	μm/m*°C
Thermal expansion coefficient	SiC sintered KT	5.0000	μm/m*°C
Thermal expansion coefficient	Silver - pure	19.80	μm/m*°C
Thermal expansion coefficient	Steel AISI 304	17.82	μm/m*°C
Thermal expansion coefficient	Steel AISI C1020	11.34	μm/m*°C
Thermal expansion coefficient	Tantalum	6.4800	μm/m*°C
Thermal expansion coefficient	Titanium B 120VCA	9.3600	μm/m*°C
Thermal expansion coefficient	Tungsten	4.5000	μm/m*°C
Viscosity	Air (0°C,101 kPa)	1.708E-05	N*s/m^2
Viscosity	Carbon dioxide (0°C,101 kPa)	1.390E-05	N*s/m^2
Viscosity	Helium (0°C,101 kPa)	1.860E-05	N*s/m^2
Viscosity	Hydrogen (0°C,101 kPa)	8.345E-06	N*s/m^2
Viscosity	Methane (0°C,101 kPa)	1.026E-05	N*s/m^2
Viscosity	Nitrogen (0°C,101 kPa)	1.660E-05	N*s/m^2
Viscosity	Oxygen (0°C,101 kPa)	1.919E-05	N*s/m^2
Viscosity	Benzene (0°C)	9.121E-04	N*s/m^2
Viscosity	Carbon tetrachloride (0°C)	1.346E-03	N*s/m^2
Viscosity	Glycerin (0°C)	12.07	N*s/m^2
Viscosity	Kerosene (0°C)	2.959E-03	N*s/m^2
Viscosity	Mercury (0°C)	1.685E-03	N*s/m^2
Viscosity	Oil - light machine (0°C)	0.3534	N*s/m^2
Viscosity	Water (0°C)	1.753E-03	N*s/m^2

APPENDIX H

Emissions Resistivity

Table of Total Emissivity

The following tables are presented for use as a guide when making infrared temperature measurements with the OMEGASCOPE® or other infrared pyrometers. The total emissivity (ε) for metals, non-metals and common building materials are given.

Since the emissivity of a material will vary as a function of temperature and surface finish, the values in these tables should be used only as a guide for relative or delta measurements. The exact emissivity of a material should be determined when absolute measurements are required.

Appendix H

METALS

Material	Temp °F (°C)	ε–Emissivity
Alloys		
20-Ni, 24-CR, 55-FE, Oxid.	392 (200)	.90
20-Ni, 24-CR, 55-FE, Oxid.	932 (500)	.97
60-Ni, 12-CR, 28-FE, Oxid.	518 (270)	.89
60-Ni, 12-CR, 28-FE, Oxid.	1040 (560)	.82
80-Ni, 20-CR, Oxidized	212 (100)	.87
80-Ni, 20-CR, Oxidized	1112 (600)	.87
80-Ni, 20-CR, Oxidized	2372 (1300)	.89
Aluminium		
Unoxidized	77 (25)	.02
Unoxidized	212 (100)	.03
Unoxidized	932 (500)	.06
Oxidized	390 (199)	.11
Oxidized	1110 (599)	.19
Oxidized at 599°C (1110°F)	390 (199)	.11
Oxidized at 599°C (1110°F)	1110 (599)	.19
Heavily Oxidized	200 (93)	.20
Heavily Oxidized	940 (504)	.31
Highly Polished	212 (100)	.09
Roughly Polished	212 (100)	.18
Commercial Sheet	212 (100)	.09
Highly Polished Plate	440 (227)	.04
Highly Polished Plate	1070 (577)	.06
Bright Rolled Plate	338 (170)	.04
Bright Rolled Plate	932 (500)	.05
Alloy A3003, Oxidized	600 (316)	.40
Alloy A3003, Oxidized	900 (482)	.40
Alloy 1100-0	200-800 (93-427)	.05
Alloy 24ST	75 (24)	.09
Alloy 24ST, Polished	75 (24)	.09
Alloy 75ST	75 (24)	.11
Alloy 75ST, Polished	75 (24)	.08
Bismuth, Bright	176 (80)	.34
Bismuth, Unoxidized	77 (25)	.05
Bismuth, Unoxidized	212 (100)	.06
Brass		
73% Cu, 27% Zn, Polished	476 (247)	.03
73% Cu, 27% Zn, Polished	674 (357)	.03
62% Cu, 37% Zn, Polished	494 (257)	.03
62% Cu, 37% Zn, Polished	710 (377)	.04
83% Cu, 17% Zn, Polished	530 (277)	.03
Matte	68 (20)	.07
Burnished to Brown Colour	68 (20)	.40
Cu-Zn, Brass Oxidized	392 (200)	.61
Cu-Zn, Brass Oxidized	752 (400)	.60
Cu-Zn, Brass Oxidized	1112 (600)	.61
Unoxidized	77 (25)	.04
Unoxidized	212 (100)	.04
Cadmium	77 (25)	.02
Carbon		
Lampblack	77 (25)	.95
Unoxidized	77 (25)	.81
Unoxidized	212 (100)	.81
Unoxidized	932 (500)	.79
Candle Soot	250 (121)	.95
Filament	500 (260)	.95
Graphitized	212 (100)	.76
Graphitized	572 (300)	.75
Graphitized	932 (500)	.71
Chromium	100 (38)	.08
Chromium	1000 (538)	.26
Chromium, Polished	302 (150)	.06
Cobalt, Unoxidized	932 (500)	.13
Cobalt, Unoxidized	1832 (1000)	.23
Columbium, Unoxidized	1500 (816)	.19
Columbium, Unoxidized	2000 (1093)	.24
Copper		
Cuprous Oxide	100 (38)	.87
Cuprous Oxide	500 (260)	.83
Cuprous Oxide	1000 (538)	.77
Black, Oxidized	100 (38)	.78
Etched	100 (38)	.09
Matte	100 (38)	.22
Roughly Polished	100 (38)	.07
Polished	100 (38)	.03
Highly Polished	100 (38)	.02
Rolled	100 (38)	.64
Rough	100 (38)	.74
Molten	1000 (538)	.15
Molten	1970 (1077)	.16
Molten	2230 (1221)	.13
Nickel Plated	100-500 (38-260)	.37
Dow Metal	0.4-600 (−18-316)	.15
Gold		
Enamel	212 (100)	.37
Plate (.0001)		
Plate on .0005 Silver	200-750 (93-399)	.11-.14
Plate on .0005 Nickel	200-750 (93-399)	.07-.09
Polished	100-500 (38-260)	.02
Polished	1000-2000 (538-1093)	.03
Haynes Alloy C, Oxidized	600-2000 (316-1093)	.90-.96
Haynes Alloy 25, Oxidized	600-2000 (316-1093)	.86-.89
Haynes Alloy X, Oxidized	600-2000 (316-1093)	.85-.88
Inconel Sheet	1000 (538)	.28
Inconel Sheet	1200 (649)	.42
Inconel Sheet	1400 (760)	.58
Inconel X, Polished	75 (24)	.19
Inconel B, Polished	75 (24)	.21
Iron		
Oxidized	212 (100)	.74
Oxidized	930 (499)	.84
Oxidized	2190 (1199)	.89
Unoxidized	212 (100)	.05
Red Rust	77 (25)	.70
Rusted	77 (25)	.65
Liquid	2760-3220 (1516-1771)	.42-.45
Cast Iron		
Oxidized	390 (199)	.64
Oxidized	1110 (599)	.78
Unoxidized	212 (100)	.21
Strong Oxidation	40 (104)	.95
Strong Oxidation	482 (250)	.95
Liquid	2795 (1535)	.29
Wrought Iron		
Dull	77 (25)	.94
Dull	660 (349)	.94
Smooth	100 (38)	.35
Polished	100 (38)	.28
Lead		
Polished	100-500 (38-260)	.06-.08
Rough	100 (38)	.43
Oxidized	100 (38)	.43
Oxidized at 1100°F	100 (38)	.63
Gray Oxidized	100 (38)	.28
Magnesium	100-500 (38-260)	.07-.13
Magnesium Oxide	1880-3140 (1027-1727)	.16-.20
Mercury	32 (0)	.09
"	77 (25)	.10
"	100 (38)	.10
"	212 (100)	.12
Molybdenum	100 (38)	.06
"	500 (260)	.08
"	1000 (538)	.11
"	2000 (1093)	.18
" Oxidized at 1000°F	600 (316)	.80
" Oxidized at 1000°F	700 (371)	.84
" Oxidized at 1000°F	800 (427)	.84
" Oxidized at 1000°F	900 (482)	.83
" Oxidized at 1000°F	1000 (538)	.82
Monel, Ni-Cu	392 (200)	.41
Monel, Ni-Cu	752 (400)	.44
Monel, Ni-Cu	1112 (600)	.46
Monel, Ni-Cu Oxidized	68 (20)	.43
Monel, Ni-Cu Oxid. at 1110°F	1110 (599)	.46
Nickel		
Polished	100 (38)	.05
Oxidized	100-500 (38-260)	.31-.46
Unoxidized	77 (25)	.05
Unoxidized	212 (100)	.06
Unoxidized	932 (500)	.12
Unoxidized	1832 (1000)	.19
Electrolytic	100 (38)	.04
Electrolytic	500 (260)	.06
Electrolytic	1000 (538)	.10
Electrolytic	2000 (1093)	.16
Nickel Oxide	1000-2000 (538-1093)	.59-.86
Palladium Plate (.00005 on .0005 silver)	200-750 (93-399)	.16-.17
Platinum	100 (38)	.05
"	500 (260)	.05
"	1000 (538)	.10
Platinum, Black	100 (38)	.93
"	500 (260)	.96
"	2000 (1093)	.97
" Oxidized at 1100°F	500 (260)	.07
"	1000 (538)	.11
Rhodium Flash (0.0002 on 0.0005 Ni)	200-700 (93-371)	.10-.18
Silver		
Plate (0.0005 on Ni)	200-700 (93-371)	.06-.07
Polished	100 (38)	.01
"	500 (260)	.02
"	1000 (538)	.03
"	2000 (1093)	.03
Steel		
Cold Rolled	200 (93)	.75-.85
Ground Sheet	1720-2010 (938-1099)	.55-.61
Polished Sheet	100 (38)	.07
	500 (260)	.10
	1000 (538)	.14
Mild Steel, Polished	75 (24)	.10
Mild Steel, Smooth	75 (24)	.12
Mild Steel, Liquid	2910-3270 (1599-1793)	.28
Steel, Unoxidized	212 (100)	.08
Steel, Oxidized	77 (25)	.80
Steel Alloys		
Type 301, Polished	75 (24)	.27
Type 301, Polished	450 (232)	.57
Type 301, Polished	1740 (949)	.55
Type 303, Oxidized	600-2000 (316-1093)	.74-.87
Type 310, Rolled	1500-2100 (816-1149)	.56-.81
Type 316, Polished	75 (24)	.28
Type 316, Polished	450 (232)	.57
Type 316, Polished	1740 (949)	.66
Type 321	200-800 (93-427)	.27-.32
Type 321 Polished	300-1500 (149-815)	.18-.49
Type 321 w/BK Oxide	200-800 (93-427)	.66-.76
Type 347, Oxidized	600-2000 (316-1093)	.87-.91
Type 350	200-800 (93-427)	.18-.27
Type 350 Polished	300-1800 (149-982)	.11-.35
Type 446, Polished	300-1500 (149-815)	.15-.37
Type 17-7 PH	200-600 (93-316)	.44-.51
Type 17-7 PH Polished	300-1500 (149-815)	.09-.16
Type C1020, Oxidized	600-2000 (316-1093)	.87-.91
Type PH-15-7 MO	300-1200 (149-649)	.07-.19
Stellite, Polished	68 (20)	.18
Tantalum, Unoxidized	1340 (727)	.14
"	2000 (1093)	.19
"	3600 (1982)	.26
"	5306 (2930)	.30
Tin, Unoxidized	77 (25)	.04
"	212 (100)	.05
Tinned Iron, Bright	76 (24)	.05
"	212 (100)	.08

Emissions Resistivity

METALS

Material	Temp °F (°C)	ε–Emissivity
Titanium		
Alloy C110M,		
Polished	300-1200 (149-649)	.08-.19
" Oxidized at		
538°C (1000°F)	200-800 (93-427)	.51-.61
Alloy Ti-95A,		
Oxid. at		
538°C (1000°F)	200-800 (93-427)	.35-.48
Anodized onto SS	200-600 (93-316)	.96-.82

Material	Temp °F (°C)	ε–Emissivity
Tungsten		
Unoxidized	77 (25)	.02
Unoxidized	212 (100)	.03
Unoxidized	932 (500)	.07
Unoxidized	1832 (1000)	.15
Unoxidized	2732 (1500)	.23
Unoxidized	3632 (2000)	.28
Filament (Aged)	100 (38)	.03
Filament (Aged)	1000 (538)	.11
Filament (Aged)	5000 (2760)	.35

Material	Temp °F (°C)	ε–Emissivity
Uranium Oxide	1880 (1027)	.79
Zinc		
Bright, Galvanized	100 (38)	.23
Commercial 99.1%	500 (260)	.05
Galvanized	100 (38)	.28
Oxidized	500-1000 (260-538)	.11
Polished	100 (38)	.02
Polished	500 (260)	.03
Polished	1000 (538)	.04
Polished	2000 (1093)	.06

NON-METALS

Material	Temp °F (°C)	ε–Emissivity
Adobe	68 (20)	.90
Asbestos		
Board	100 (38)	.96
Cement	32-392 (0-200)	.96
Cement, Red	2500 (1371)	.67
Cement, White	2500 (1371)	.65
Cloth	199 (93)	.90
Paper	100-700 (38-371)	.93
Slate	68 (20)	.97
Asphalt, pavement	100 (38)	.93
Asphalt, tar paper	68 (20)	.93
Basalt	68 (20)	.72
Brick		
Red, rough	70 (21)	.93
Gault Cream	2500-5000 (1371-2760)	.26-.30
Fire Clay	2500 (1371)	.75
Light Buff	1000 (538)	.80
Lime Clay	2500 (1371)	.43
Fire Brick	1832 (1000)	.75-.80
Magnesite, Refractory	1832 (1000)	.38
Gray Brick	2012 (1100)	.75
Silica, Glazed	2000 (1093)	.88
Silica, Unglazed	2000 (1093)	.80
Sandlime	2500-5000 (1371-2760)	.59-.63
Carborundum	1850 (1010)	.92
Ceramic		
Alumina on Inconel	800-2000 (427-1093)	.69-.45
Earthenware, Glazed	70 (21)	.90
Earthenware, Matte	70 (21)	.93
Greens No. 5210-2C	200-750 (93-399)	.89-.82
Coating No. C20A	200-750 (93-399)	.73-.67
Porcelain	72 (22)	.92
White Al_2O_3	200 (93)	.90
Zirconia on Inconel	800-2000 (427-1093)	.62-.45
Clay	68 (20)	.39
" Fired	158 (70)	.91
" Shale	68 (20)	.69
" Tiles, Light Red	2500-5000 (1371-2760)	.32-.34
" Tiles, Red	2500-5000 (1371-2760)	.40-.51
" Tiles,		
Dark Purple	2500-5000 (1371-2760)	.78
Concrete		
Rough	32-2000 (0-1093)	.94
Tiles, Natural	2500-5000 (1371-2760)	.63-.62
" Brown	2500-5000 (1371-2760)	.87-.83
" Black	2500-5000 (1371-2760)	.94-.91
Cotton Cloth	68 (20)	.77
Dolomite Lime	68 (20)	.41
Emery Corundum	176 (80)	.86
Glass		
Convex D	212 (100)	.80
Convex D	600 (316)	.80
Convex D	932 (500)	.76
Nonex	212 (100)	.82
Nonex	600 (316)	.82
Nonex	932 (500)	.78
Smooth	32-200 (0-93)	.92-.94

Material	Temp °F (°C)	ε–Emissivity
Granite	70 (21)	.45
Gravel	100 (38)	.28
Gypsum	68 (20)	.80-.90
Ice, Smooth	32 (0)	.97
Ice, Rough	32 (0)	.98
Lacquer		
Black	200 (93)	.96
Blue, on Al Foil	100 (38)	.78
Clear, on Al Foil (2 coats)	200 (93)	.08 (.09)
Clear, on Bright Cu	200 (93)	.66
Clear, on Tarnished Cu	200 (93)	.64
Red, on Al Foil (2 coats)	100 (38)	.61 (.74)
White	200 (93)	.95
White, on Al Foil (2 coats)	100 (38)	.69 (.88)
Yellow, on Al Foil (2 coats)	100 (38)	.57 (.79)
Lime Mortar	100-500 (38-260)	.90-.92
Limestone	100 (38)	.95
Marble, White	100 (38)	.95
" Smooth, White	100 (38)	.56
" Polished Gray	100 (38)	.75
Mica	100 (38)	.75
Oil on Nickel		
0.001 Film	72 (22)	.27
0.002 "	72 (22)	.46
0.005 "	72 (22)	.72
Thick "	72 (22)	.82
Oil, Linseed		
On Al Foil, uncoated	250 (121)	.09
On Al Foil, 1 coat	250 (121)	.56
On Al Foil, 2 coats	250 (121)	.51
On Polished Iron, .001 Film	100 (38)	.22
On Polished Iron, .002 Film	100 (38)	.45
On Polished Iron, .004 Film	100 (38)	.65
On Polished Iron, Thick Film	100 (38)	.83
Paints		
Blue, Cu_2O_3	75 (24)	.94
Black, CuO	75 (24)	.96
Green, Cu_2O_3	75 (24)	.92
Red, Fe_2O_3	75 (24)	.91
White, Al_2O_3	75 (24)	.94
White, Y_2O_3	75 (24)	.90
White, ZnO	75 (24)	.95
White, $MgCO_3$	75 (24)	.91
White, ZrO_2	75 (24)	.95
White, ThO_2	75 (24)	.90
White, MgO	75 (24)	.91
White, $PbCO_3$	75 (24)	.93
Yellow, PbO	75 (24)	.90
Yellow, $PbCrO_4$	75 (24)	.93
Paints, Aluminium	100 (38)	.27-.67
10% Al	100 (38)	.52
26% Al	100 (38)	.30
Dow P-31 0	200 (93)	.22
Paints, Bronze	Low	.34-.80
Gum arnish (2 coats)	70 (21)	.53
Gum arnish (3 coats)	70 (21)	.50
Cellulose Binder (2 coats)	70 (21)	.34

Material	Temp °F (°C)	ε–Emissivity
Paints, Oil		
All colors	200 (93)	.92-.96
Black	200 (93)	.92
Black Gloss	70 (21)	.90
Camouflage Green	125 (52)	.85
Flat Black	80 (27)	.88
Flat White	80 (27)	.91
Gray-Green	70 (21)	.95
Green	200 (93)	.95
Lamp Black	209 (98)	.96
Red	200 (93)	.95
White	200 (93)	.94
uartz, Rough, Fuse d	70 (21)	.93
Glass, 1.98 mm	540 (282)	.90
Glass, 1.98 mm	1540 (838)	.41
Glass, 6.88 mm	540 (282)	.93
Glass, 6.88 mm	1540 (838)	.47
Opaque	570 (299)	.92
Opaque	1540 (838)	.68
Red Lead	212 (100)	.93
Rubber, ard	74 (23)	.94
Rubber, Soft, Gray	76 (24)	.86
Sand	68 (20)	.76
Sandstone	100 (38)	.67
Sandstone, Red	100 (38)	.60-.83
Sawdust	68 (20)	.75
Shale	68 (20)	.69
Silica, Glazed	1832 (1000)	.85
Silica, Unglazed	2012 (1100)	.75
Silicon Carbide	300-1200 (149-649)	.83-.96
Silk Cloth	68 (20)	.78
Slate	100 (38)	.67-.80
Snow, Fine Particles	20 (7)	.82
Snow, Granular	18 (8)	.89
Soil		
Surface	100 (38)	.38
Black Loam	68 (20)	.66
Plowed Field	68 (20)	.38
Soot		
Acetylene	75 (24)	.97
Camphor	75 (24)	.94
Candle	250 (121)	.95
Coal	68 (20)	.95
Stonework	100 (38)	.93
Water	100 (38)	.67
Waterglass	68 (20)	.96
Wood	Low	.80-.90
Beech P aned	158 (70)	.94
Oak, Planed	100 (38)	.91
Spruce, Sanded	100 (38)	.89

APPENDIX I

Physical Properties of Some Typical Liquids

Property	Argon Solid	Argon Liquid	Benzene Solid	Benzene Liquid
Density (kg/m^3)		1636		1000
		1407		899
Latent heat of fusion (kJ/mol)		7.86		34.7
Latent heat of evaporation (kJ/mol)	6.69		2.5	
Heat capacity (J/(mol K))		25.9		11.3
		22.6		13.0
Melting point (K)		84.1		278.8
Liquid range (K)	3.5		75	
Isothermal compressibility (1/(N m^2))	1	20	8.1	8.7
Surface tension (mJ/ m^2)	...	13	...	28.9
Viscosity (Poise=0.1kg/(m s))	...	0.003	...	0.009
Self-diffusion coefficient (m^2/s)	10^{-13}	1.6×10^{-9}	10^{-13}	1.7×10^{-9}
Thermal conductivity (J/(m s K))	0.3	0.12	0.27	0.15

Property	Water Solid	Water Liquid	Sodium Solid	Sodium Liquid
Density (kg/m^3)		920		951
		997		927
Latent heat of fusion (kJ/mol)		5.98		109.5
Latent heat of evaporation (kJ/mol)	40.5		107.0	
Heat capacity (J/(mol K))		37.6		28.4
		75.2		32.3
Melting point (K)		273.2		371.1
Liquid range (K)	100		794	
Isothermal compressibility (1/(N m^2))	2	4.9	1.7	1.9
Surface tension (mJ/ m^2)	...	72	...	190
Viscosity (Poise=0.1kg/(m s))	...	0.01	...	0.007
Self-diffusion coefficient (m^2/s)	10^{-14}	2.2×10^{-9}	2×10^{-11}	4.3×10^{-9}
Thermal conductivity (J/(m s K))	2.1	0.58	134	84

APPENDIX J

Speed of Sound in Various Bulk Media

Gases	
Material	v (m/s)
Hydrogen (0°C)	1286
Helium (0°C)	972
Air (20°C)	343
Air (0°C)	331
Liquids at 25°C	
Material	v (m/s)
Glycerol	1904
Sea water	1533
Water	1493
Mercury	1450
Kerosene	1324
Methyl alcohol	1143
Carbon tetrachloride	926
Solids	
Material	v (m/s)
Diamond	12000
Pyrex glass	5640
Iron	5130
Aluminum	5100
Brass	4700
Copper	3560
Gold	3240
Lucite	2680
Lead	1322
Rubber	1600

APPENDIX K

Batteries

Values given for size and weight are typical.

$$\text{Power Density} = \frac{\text{Current} \times \text{Voltage}}{\text{Battery Weight}}$$

Type	Size	Dimensions (in)	Weight (oz)	Capacity (Ah)	Voltage (V)	Energy Density (Wh/kg)
Li-Ion					3.6	100
NiCad						40-60
NiMH					1.25	60-80
Li-Polymer					2.7	150-200
Carbon Zinc	D	1.34 x 2.42	3.07	0.80	1.5	
Zinc Chloride	D	1.34 x 2.42	3.74	2.5	1.5	
Alkaline	D	1.34 x 2.42	4.50	4.8	1.5	
NiCad	D	1.34 x 2.42	5.30	3.5	1.25	
Carbon Zinc	C	1.02 x 1.97	1.59	0.36	1.5	
Zinc Chloride	C	1.02 x 1.97	1.80	1.3	1.5	
Alkaline	C	1.02 x 1.97	2.20	2.4	1.5	
NiCad	C	1.02 x 1.97	2.50	1.6	1.25	
Carbon Zinc	AA	0.57 x 1.99	0.53	0.16	1.5	
Zinc Chloride	AA	0.57 x 1.99	0.71	0.36	1.5	
Alkaline	AA	0.57 x 1.99	0.75	0.75	1.5	
NiCad	AA	0.57 x 1.99	0.85	0.45	1.25	
Zinc Chloride	AAA	0.41 x 1.75	0.32	0.36	1.5	
Alkaline	AAA	0.41 x 1.75	0.40	0.40	1.5	
Zinc Chloride	N	0.47 x 1.18	0.23	0.25	1.5	
Carbon Zinc	9V	1.03 x 1.94 x .69	1.31	0.16	9.0	
Zinc Chloride	9V	1.03 x 1.94 x .69	1.36	0.16	9.0	
Alkaline	9V	1.03 x 1.94 x .69	1.59	0.30	9.0	
NiCad	9V	1.03 x 1.94 x .69	1.25	0.065	9.0	

APPENDIX L

Temperatures

Temperatures (°C) of Some Physical Phenomena

Absolute zero	−273.15
Helium boils	−269
Nitrogen boils	−196
Oxygen boils	−183
Dry ice (CO_2) freezes	−79
Mercury freezes	−39
Water freezes	0
Room temperature	~20
Butter melts	31
Body temperature	~37
Paraffin boils	~54
Alcohol boils	78
Water	100
Saturated salt solution boils	108
Tin melts	232
Lead melts	327
Sulfur boils	445
Aluminum melts	657
NaCl melts	801
Silver melts	961
Gold melts	1063
Copper melts	1083
Glass melts	1000-1400
Steel melts	1300-1400
Iron melts	1530
Lead boils	1620
Platinum melts	1774
Bunsen burner	1870
Iron boils	2450
Tungsten melts	~3410
Oxyacetylene flame	3500
Carbon arc	5500
Surface of the sun	6000
Iron welding arc	6020

Contributor's Biographies

EDITOR-IN-CHIEF

Jon Wilson
Principal Consultant, Jon S. Wilson Consulting, LLC
Chandler, Arizona
Dynamic-Consultant@cox.net

Born and reared in Oklahoma, Jon Wilson earned his BS in Mechanical Engineering at Oklahoma University, a Master's degree in Automotive Engineering at Chrysler Institute of Engineering, and his MSE in Industrial Engineering at Arizona State University. His experience includes assignments as Senior Test Engineer, Laboratory Manager, Applications Engineering Manager, and Marketing Manager. He has worked for Chrysler Corporation, ITT Cannon Electric Co., Motorola Semiconductor Products Division, and Endevco Division of Meggitt Aerospace. Since founding his own company in 1985, he has been a consultant, writer, and educator.

Mr. Wilson's technical experience encompasses all phases of dynamic, environmental, functional, electrical, mechanical, and chemical testing of automotive, industrial and aerospace equipment and components. He was also involved in marketing management and applications engineering of shock, vibration and dynamic pressure measurement instrumentation. He has participated in national and international standards committees and technical divisions of SAE, ISA and IEST. Mr. Wilson provides technical writing of testing standards, specifications, and procedures; he also witnesses and audits test performance, assists in test equipment and services specification and procurement, and consults for and trains engineers, technicians, and non-technical personnel on all phases of dynamic and environmental testing and instrumentation.

He has over 40 years experience writing and presenting training courses and technical papers. He managed and presented Endevco's Shock and Vibration Measurement and Dynamic Pressure Measurement short courses for several years. He has written and/or presented over 30 technical articles and papers.

His recent teaching includes short courses on shock and vibration measurement, testing and calibration; climatic testing; dynamic pressure measurement; and instrumentation for test and measurement. He has presented short courses and tutorials for Endevco, Kistler, Vibrometer, Tustin Technical Institute, Technology Training Incorporated, Institute of Environmental Sciences and Technology, Society of Automotive Engineers, Instrument Society of America, American Society of Testing and Materials, Society of Experimental Mechanics, International Telemetering Conference, Transducer Workshop, Shock and Vibration Symposium and Sensors Expo. He has presented short courses in England, Scotland, Germany, Sweden, Canada, Japan, Singapore, and many parts of the United States.

He is currently an active member of the IEST Working Group, developing a Recommended Practice for vibration transducer selection. Mr. Wilson is also a member of ISA S37 working group and chairman of the S37.11 subcommittee (RP 37.11 deals with vibration transducers).

CONTRIBUTORS

Tom Anderson
Application Development Manager
SpaceAge Control, Inc.
Tom is Application Development Manager at SpaceAge Control, a manufacturer of miniature displacement transducers. He is responsible for application engineering and product marketing efforts and led development efforts for a broad range of displacement sensing solutions, including those for flight data recorders, environmental control systems, actuator controls, and space vehicles. Before working for SpaceAge Control, Tom worked for Hewlett-Packard's engineering workstation division as a product manager. He has a degree in systems science from the University of California at Los Angeles.

Steven Arms
President
MicroStrain, Inc.
Steven Arms is founder and President of MicroStrain, Inc., a manufacturer of precision sensors and wireless sensing instrumentation. Steve founded MicroStrain in 1987 with the mission to create advanced microminiature sensors and wireless instrumentation. Sensors are literally changing our world; we're inspired to work with our customers to introduce advanced sensing technology that will enable the next generation of smarter and safer machines, civil structures, and implanted devices.

Contributor's Biographies

Adolfo Cano Muñoz, Product Manager
Honeywell Sensing and Control
Adolfo Cano Muñoz is the product manager for Honeywell Sensing and Control's speed and position sensing group and has been with the company for five years. Adolfo was born and raised in France and earned his degree in international business, attending ECS Reims University in France and ICADE in Spain. He went on to obtain his MBA from the University of Sheffield in the United Kingdom and is currently based at Honeywell's Newhouse, Scotland facility.

John Carey
Marketing Coordinator
Larson Davis, Inc.
John Carey has been active in the sound and vibration measurement field for 35 years. He holds B.S., M.S. and Ph.D. degrees in Mechanical Engineering from Carnegie-Mellon University and served as an invited professor at the Ecole Polytechnique in Montréal, Canada. He is presently Marketing Coordinator for Larson Davis, Inc. in Provo, Utah.

Anthony Chu
Director of Marketing of Test Instrumentation
ENDEVCO Corporation
Anthony Chu is the Director of Marketing of Test Instrumentation at Endevco - Meggitt PLC. He has been with Endevco for over 21 years in various capacities, from engineering to marketing. His areas of expertise include accelerometer design/manufacturing, and sensor product development. Mr. Chu received his Mechanical Engineering degree from California State Polytechnic University, Pomona; and master's degree in Business Administration from Pepperdine University, Malibu. He has published many papers in various technical societies, and is noted internationally for his contribution in the field of shock measurement. He is an active member of the Institute of Environmental Sciences & Technology (IEST) Pyroshock Committee and Shock and Vibration Committee. Mr. Chu is currently the Working Group chairman of Measurement Transducer Selection, working on the Transducer Selection Recommended Practice for IEST. He is also one of the chapter contributors of the newly revised Shock and Vibration Handbook from McGraw-Hill.

Contributor's Biographies

Paul Engeler
Head of Pressure Sensor Design
Kistler Instrumente

Paul Engeler is currently Head of Pressure Sensor Design at Kistler Instrumente. He has been active in the development of pressure sensors at Kistler for over 30 years. He graduated at the University of Applied Sciences in Zurich in 1972.

John Fontes, Senior Applications Engineer
Honeywell Sensing and Control

John Fontes is a Honeywell senior applications engineer specializing in humidity, thermistors, RTDs and probes. He graduated from Fitchburg State College in Massachusetts, earning bachelor's degrees in both manufacturing and electrical engineering. John joined Honeywell in 1993 and developed his expertise in environmental sensing through various manufacturing and application engineering roles. He currently solves applications for customers of Honeywell Sensing and Control's thermal group. John is based at the company's Pawtucket, Rhode Island facility and lives with his wife and daughter in Uxbridge, Massachusetts.

Timothy J. Geiger
Division Manager, Industrial Sensors Division
PCB Piezotronics, Inc.

Timothy Geiger graduated with a BBA from the University of Notre Dame in 1987 and a MBA from the University of Chicago in 1992. He originally worked in various accounting and financial positions prior to joining PCB Piezotronics, Inc. in the technical community. Currently, Tim holds the role of Division Manager for the Industrial Sensors Division.

Glenn Harman, Global Product Leader
Honeywell Sensing and Control

Glenn Harman is the global product leader for stainless steel pressure transducers at Honeywell Sensing and Control. He has been with Honeywell for over 25 years, working in a variety of sales and marketing positions in the United States and in Europe. His experience includes sensing applications for commercial, automotive and industrial markets. Glenn was born and raised in the United Kingdom where he received a bachelor's degree in industrial engineering. He has been living and working in the United States for the past seven years and is based at Honeywell Sensing and Control's headquarters in Freeport, Illinois.

Contributor's Biographies

William Hennessy
Senior Project Engineer
BMT Scientific Marine Services, Inc.
William Hennessy is a senior design engineer with over 25 years experience in instrumentation system design, fabrication, and testing with particular emphasis on the offshore and marine industries. He has instrumented America's Cup yachts, Arctic icebreakers, offshore drill rigs, and model test basins, to name just a few of the hundreds of projects he has managed and contributed to.

Thomas Kenny, Ph.D.
Assistant Professor, Department of Mechanical Engineering, Design Division
Stanford University
Thomas Kenny has always been interested in the properties of small structures. His PhD research was carried out in the Physics Department at UC Berkeley, where he focused on a measurement of the heat capacity of a monolayer of helium atoms. After graduating, his research at the Jet Propulsion Laboratory focused on the development of a series of microsensors that use tunneling displacement transducers to measure small signals. Currently, at Stanford University, research in Tom's group covers many areas including MEMS devices to detect small forces, studies of gecko adhesion, micromechanical resonators, and heat transfer in microchannels. Tom teaches several courses at Stanford, including Introduction to Sensors. Tom's hobbies include Ultimate Frisbee, hiking, skiing, and an occasional friendly game of poker.

Walt Kester
Analog Devices, Inc.
Walt Kester is currently a Corporate Staff Applications Engineer at Analog Devices. During more than 30 years at ADI, he has designed, developed, and given applications support for high-speed ADCs, DACs, SHAs, op amps, and muxes. Besides writing many papers and articles, he has prepared and edited eleven major application books, which have formed the basis for the Analog Devices world-wide technical seminar series. Prior to joining Analog Devices, he worked at Bell Telephone Laboratories where he was involved in high-speed ADC development for phased array radar receivers. Walt has a BSEE from NC State University (1964) and MSEE from Duke University (1966).

Contributor's Biographies

Mark Kretschmar
Lion Precision
Mark Kretschmar studied electrical engineering at the University of Minnesota and worked as a product design engineer for Lion Precision in St. Paul, Minnesota for thirteen years where he developed several market leading products. He now serves as the company's communications director and is responsible for oversight of product interface design.

Dave Lally
Vice President of Engineering
PCB Piezotronics
Dave is a mechanical engineer who graduated from Bucknell University and then went to the University of Cincinnati to concentrate on research in structural dynamics. Afterwards, he started his professional career as a Sensor Design Engineer at PCB Piezotronics. Subsequently, he has been able to balance his professional experience by working closely with customers as Marketing Manager and then Product Manager of PCB's Vibration Sensor Division. Currently, he holds the position of VP of Engineering and is primarily responsible for research and development of new products.

Dr. Young H. Lee
Professor of Chemical Engineering
Drexel University
Young H. Lee is a professor of Chemical Engineering at Drexel University. He received his Ph.D. from Purdue University in 1977. His primary research interests include immobilized cell processes, development of ultramicro probes for local environment monitoring, biosensors for fermentation process monitoring, bioreactor instrumentation and control, bioenergetics and microbial stoichiometry, and biological wastewater treatment.

Dan Losea
Product Manager
Honeywell
Dan Losea is a product manager for Honeywell's precision thermostats. His expertise comes from working with sensors for 27 years in various engineering, sales, marketing and quality control roles. Born and raised in Providence, Rhode Island, Dan holds a bachelor's degree in secondary education from Rhode Island College and a master's in machine design from the Rhode Island School of Design. He joined Honeywell Automation and Control in 2001 with the company's acquisition of Invensys Sensor Systems and is based at Honeywell's Pawtucket, Rhode Island facility.

Contributor's Biographies

George C. Low
Engineering and Quality Manager
HITEC Corporation

George Low graduated from Massachusetts Maritime Academy in 1993 with a Bachelor of Science in Marine Engineering as a major and Mechanical Engineering as minor. George has been with HITEC Corporation for 17 years, including college years, with the most current position of Engineering and Quality Manager. He has published a paper on CTE Testing and also contributes to ASTM specifications regarding instrumented bolts. George is the HITEC Corporation delegate for the Western Regional Strain Gage Committee.

Dr. Raj Mutharasan
Frank A. Fletcher Professor of Chemical Engineering
Drexel University

Raj Mutharasan is the Frank A. Fletcher Professor of Chemical Engineering at Drexel University. He received his Ph.D. from Drexel in 1973. His primary research interests include development of cultivation strategies for recombinant animal cells, the use of fluorescence in detecting metabolic characteristics with application to cancer detection, consequences of intracellular pH on glycosylation and recombinant protein expression, metabolic response due to hydrodynamic shear, cellular mechanotransduction and ophthalmic applications.

David J. Nagel, Ph.D.
Research Professor, Electrical and Computer Engineering Department
The George Washington University

David J. Nagel received a B.S. degree in Engineering Science from the University of Notre Dame and graduate degrees (M.S. in Physics and Ph.D. in Engineering Materials) from the University of Maryland. He joined the civilian staff of the Naval Research Laboratory in 1964, where he held positions as a Research Physicist, Section Head, Branch Head and, finally, Superintendent of the Condensed Matter and Radiation Sciences Division. In this last position, Nagel was a member of the Senior Executive Service, and managed the experimental and theoretical research and development efforts of 150 government, contractor and other personnel. He has written or co-authored over 150 technical articles, reports, book chapters and encyclopedia articles. Nagel spent 30 years on active and reserve duty for the Navy and retired as a Captain in 1990. He became a Research Professor in the School of Engineering and Applied Science of The George Washington University in 1998. His current interests include applications of MEMS and nano-technologies, as well as low energy nuclear reactions.

Contributor's Biographies

Sharon Smith, Ph.D.
Director of Technology
Lockheed Martin Corporation
Sharon Smith is a Corporate Executive and Director of Technology at Lockheed Martin's Corporate Headquarters in Bethesda, Maryland. Dr. Smith is responsible for research and technology initiatives, including independent research and development projects, university involvement, and various other R&D activities. She is also Chair of the Lockheed Martin's Steering on Nanotechnology. Dr. Smith has over twenty years of experience in management, program management, and engineering at Eli Lilly and Company, IBM Corporation, Loral, and Lockheed Martin Corporation. She has more than twenty technical publications and has given numerous technical presentations in the US and Europe.

Roland Sommer
Head of Technology
Kistler Instrumente
Roland Sommer is currently Head of Technology at Kistler Instrumente and is primary responsible for the evaluation of new technologies and new piezoelectric materials. He graduated in Physics at the Swiss Federal Institute of Technology Lausanne in 1987 and received his PhD degree from the same institute in 1992. From 1994 to 1995, he was a postdoctoral fellow at Lehigh University (Pennsylvania, USA). He joined Kistler in 1996.

Chris Townsend
Vice President
MicroStrain Inc.
Chris Townsend is Vice President of Engineering for MicroStrain, Inc., a manufacturer of precision sensors and wireless sensing instrumentation. Chris's current main focus is on research and development of a new class of ultra low power wireless sensors for industry. Chris joined MicroStrain as its first employee and electrical engineer in 1989. Since then he has been involved in the design of a number of products, including the world's smallest LVDT, inertial orientation sensors, and wireless sensors.

Ken Watkins
Product Manager, Force/Torque Division
PCB Piezotronics, Inc.

Ken earned his electrical engineering degree from Lawrence Technological University in 1987. He spent 11 years working at Eaton Corporation Lebow Products. Then, Ken joined Key Transducers. He has held several positions including Sales Engineer, Regional Sales Manager, and Marketing Manager. Presently Ken holds the position of Product Manager for the Force/Torque Division of PCB Piezotronics in Buffalo, NY.

Scott D. Welsby
Manager
Lion Precision Inductive

Scott D. Welsby, Manager of Lion Precision Inductive in Colorado Springs, Colorado, has more than 20 years experience in design, test, application engineering, and manufacture of precision electromechanical devices and sensor systems. Several products developed under his leadership have won industry awards for innovation and performance. Mr. Welsby holds a degree in mechanical engineering, has published and presented numerous papers on sensors and precision measurement, and holds one U.S. patent.

Contributing Companies

Analog Devices Inc.
One Technology Way
Norwood, MA 02062-9106
www.analog.com
800-262-5643 or 781-461-3333

Analog Devices, Inc. is a world-leading semiconductor company specializing in high-performance analog, mixed-signal and digital signal processing (DSP) integrated circuits (ICs). Since ADI was founded in 1965, its focus has been to solve the engineering challenges associated with signal processing in electronic equipment. ADI's products play a fundamental role in converting real-world phenomena such as temperature, motion, pressure, light and sound into electrical signals to be used in a wide array of applications ranging from industrial process control, factory automation, radar systems and CAT scanners to cellular base stations, broadband modems, wireless telephones, computers, cars and digital cameras. As a MEMS pioneer, Analog Devices has held a leadership position as the industry's largest volume supplier of integrated MEMS accelerometers and gyroscopes. 10 years ago, Analog Devices revolutionized automotive airbag systems with its iMEMS (integrated Micro Electro Mechanical System) inertial sensor technology. Today, low-g, high-g, single axis, dual axis iMEMS accelerometers are used to measure shock, vibration, tilt, position and motion in a wide assortment of applications including sports and medical devices, laptop hard drive protection systems, and keystone correction in digital projectors. With QS-9000 management and control systems in place since 1993, the company's iMEMS® technology has been embraced by the automotive and consumer products industries for their quality and reliability dependent applications.

Contributing Companies

BMT Scientific Marine Services, Inc.
California office
2240 Vineyard Ave.
Escondido, CA 92029
760-737-3505
www.scimar.com

BMT Scientific Marine Services, Inc. (SMS) provides specialized technical consulting services and custom engineered products to the marine, energy and offshore industries applying advanced methods and technologies. The multi-disciplinary nature of the solutions to projects in the marine environment demands expertise in a wide range of fields. Their team of dedicated engineers, naval architects, programmers and technicians possess a combined experience of over 150 years in a broad range of fields including naval architecture and marine engineering, ocean and coastal engineering, electronic, civil, and mechanical engineering. SMS provides technical consulting services in the areas of offshore platforms and subsea systems instrumentation, monitored data reporting and analysis, LNG simulation and risk assessment, ship instrumentation, field trials, coastal and ocean engineering, naval architecture and ship systems repair, maintenance, and design. They also have developed a wide range of products including Integrated Marine Monitoring Systems, Environmental Position Monitoring Systems, Ship Structural Monitoring Systems, and others. SMS has office and electronic fabrication facilities in Escondido, California, and Houston, Texas along with an engineering office in the Washington D.C. area. services, environmental

Endevco Corporation
30700 Rancho Viejo Road
San Juan Capistrano, CA 92675
1-800-982-6732 Toll Free
1-949-493-8181 Telephone
1-949-661-7231 Fax
applications@endevco.com
www.endevco.com

Endevco is a leading designer and manufacturer of dynamic instrumentation for vibration, shock and pressure measurement. The company's comprehensive line of piezoelectric, piezoresistive, ISOTRON® and variable capacitance accelerometers are used to solve measurement problems in a wide variety of industries including aerospace, automotive, defense, medical, industrial and marine. Other products include pressure transducers, microphones, electronic instruments, and calibration systems. Endevco has a factory-direct sales force in the U.S. and is represented in the rest of the world by Brüel & Kjær. Brüel & Kjær has offices in 55 countries and 7 accredited calibration centers worldwide. Endevco is a subsidiary of U.K.-based Meggitt PLC, an international group of companies renowned for their specialized engineering skills.

HITEC Corporation
537 Great Road, Littleton, MA 01460 US
Tel +1 978 742-9032, Fax +1 978 742-9033
www.hitecorp.com

HITEC Corporation was established in 1971 to provide a wide range of specialized engineering services for elevated temperature and extreme environment sensor installations and measurements. Specific advances by the founders of the company include the development of various sensor application techniques that contribute to their success in the sensor industry. They have assisted in improving processes for elevated temperature applications, and offer custom transducerization of components to a wide variety of industries and custom OEM transducers in large production quantities. In 2000 the company added a laboratory facility in Detroit to better serve the automotive industry. The company is represented throughout the continental United States by manufacturers representative organizations that are well-versed in the needs of the sensor and stress analysis industries. HITEC Corporation is part of the First Technology group of companies.

Honeywell Sensing and Control
11 W. Spring St.
Freeport, IL 61032
Domestic: 800.537.6945
International: 815.235.6847
Fax: 815.235.6545
www.honeywell.com/sensing

Honeywell International is a $23 billion diversified technology and manufacturing leader, serving customers worldwide with aerospace products and services; control technologies for buildings, homes and industry; automotive products; turbochargers; and specialty materials. Based in Morris Township, N.J., Honeywell's shares are traded on the New York, London, Chicago and Pacific Stock Exchanges. It is one of the 30 stocks that make up the Dow Jones Industrial Average and is also a component of the Standard & Poor's 500 Index. For additional information, please visit www.honeywell.com. Honeywell Sensing and Control is part of Honeywell's Automation and Control Solutions group, a global leader in providing product and service solutions that improve efficiency and profitability, support regulatory compliance, and maintain safe, comfortable environments in homes, buildings and industry. For more information about Sensing and Control, access www.honeywell.com/sensing.

Contributing Companies

**Kistler Instrument Corp.
75 John Glenn Drive
Amherst, NY 14228-2171 / USA
(1-888-547-8537),
Fax 1-716-691-5226
sales.us@kistler.com**

Kistler is a leading manufacturer of force, pressure, acceleration and strain sensors, as well as instrumentation systems and accessories. Kistler's technology includes a variety of sensing principles including piezoelectric, piezoresistive and variable capacitance. Kistler Instrument Corporation, founded in Western New York over 45 years ago, is a sales, engineering, and manufacturing subsidiary of Kistler Instrumente AB in Switzerland. Kistler's products are well-known worldwide for their quality and reliability. ISO 9001 certification in August 1994 was the first approval granted to a U.S. piezoelectric sensor manufacturer. Their hardware and software products are designed to serve a variety of application markets, including test and measurement, industrial applications, combustion research and engine monitoring, and a broad range of OEM applications.

**Larson Davis, Inc.
1681 W. 820 N.
Provo, UT 84601
801-375-0177, 888-258-3222
www.larsondavis.com**

Larson Davis has as its mission to provide quality and innovative sound and vibration application solutions, through its diverse product line, that empower their customers' success. From their inception in 1981, Larson Davis has brought to market innovative and powerful, yet surprisingly cost effective, acoustic analysis solutions. Their products are segregated into two main product groups: Acoustic Test Products Group for laboratory, research, development, and environmental noise products, and Industrial Hygiene & Safety Products Group for industrial hygiene and worker safety products. Products include hand-held sound level meters, portable and hand-held real-time frequency analyzers, noise dosimeters, noise and vibration analyzers, environmental noise monitoring systems, condenser microphones and accessories. Since 1999, Larson Davis has been a member of the PCB Group of companies, which has over thirty years of experience in consistently delivering Total Customer Satisfaction to the markets it serves.

Lion Precision
Corporate Headquarters, Sales Offices, and Capacitive Products
563 Shoreview Park Rd.
St. Paul, MN 55126
651-484-6544, 800-229-6544

Inductive Products Engineering:
3415 Van Teylingen Dr.
Suite 100
Colorado Springs, CO 80917
Tel: 719-637-3088
Fax: 719-637-3071

info@lionprecision.com
www.lionprecision.com

Founded in 1958 by Dr. Kurt Lion, Lion Precision is a world leader in the design, application, and manufacture of capacitive and inductive noncontact sensors for measuring displacement, dimension, motion, vibration, runout, thickness, and more with nanometer resolutions and 10kHz bandwidth. Lion Precision has been a mainstay of many of the USA's largest companies as well as smaller enterprises. The company has also embraced the global market by becoming ISO9001 certified and forming partnerships with companies in Asia, Europe and throughout the world.

Lockheed Martin
www.lockheedmartin.com

Lockheed Martin, an advanced technology company, was formed in March 1995 with the merger of two of the world's premier technology companies, Lockheed Corporation and Martin Marietta Corporation. Headquartered in Bethesda, Maryland, Lockheed Martin employs about 130,000 people worldwide and is principally engaged in the research, design, development and manufacture of advanced technology systems, products and services.

Contributing Companies

MicroStrain, Inc.
310 Hurricane Lane, Suite 4
Williston, VT 05495
802-862-6629, 800-449-3878
info@microstrain.com
www.microstrain.com

MicroStrain develops and produces innovative, smart, wireless, microminiature displacement, orientation, and force sensors and associated instrumentation. Their products include wireless sensing networks, data logging, data acquisition software, and robotic linear displacement calibration units. Applications for their sensors include position and orientation sensing, high elongation strain measurements, automotive, virtual reality, materials science, civil structures, military, aerospace, and medical. Their sensors have won numerous awards, and the company prides itself on innovation and responsiveness to their customer's unique requirements. The company also offers engineering services such as custom sensors, embedded hardware and firmware, and design. Founded in 1987, MicroStrain is based in Williston, Vermont and is a privately held corporation.

PCB Piezotronics, Inc.
3425 Walden Ave
Depew NY 14043
Phone 716-684-0001
Fax 716-684-0987
Toll Free 1 800 828-8840
Sales@pcb.com
webmaster@pcb.com
www.pcb.com

PCB is a privately owned, high-technology company specializing in the marketing, development and production of piezoelectric, capacitive, and resistive-based instrumentation for measuring pressure, force, torque, load, shock and vibration. The sensors and associated hardware are used for a wide variety of industrial, research, university, aerospace and military applications, ranging from the characterization of underwater blast wave propagation to monitoring micro-g vibrations on the solar panels of orbiting satellites. In essence, any structure that vibrates, pulsates, moves, compresses, surges or makes noise has the potential to be monitored by this instrumentation. PCB Piezotronics was founded in 1967 as a manufacturer of piezoelectric quartz sensors, accelerometers, and associated electronics for the measurement of dynamic pressure, force, and vibration. The unique expertise of the company was the

Contributing Companies

incorporation of microelectronic signal conditioning circuitry within these sensors to make them easier to use and more environmentally compatible. These ICP® sensors gained wide popularity and became the foundation for the company's success. Subsequent growth and steady investment in facilities, machinery, and equipment permitted a constant broadening of the product offering. Measurement capabilities expanded with the addition of piezoceramic, tourmaline, capacitive, piezoresistive, and strain gage sensing technologies. Ensuing products include industrial accelerometers, DC accelerometers, load cells, torque sensors, microphones, pressure transmitters, and calibration equipment. The company's mission statement is Total Customer Satisfaction, and it is backed by products offering great value and by an unconditional warranty.

SpaceAge Control, Inc.
38850 20th Street East
Palmdale, CA 93550 USA
http://spaceagecontrol.com/
email@spaceagecontrol.com
661-273-3000
661-273-4240 (fax)

SpaceAge Control is the innovation leader in displacement measurement and air data measurement. The company specializes in all-environment miniature displacement transducers, also referred to as draw wire transducers, string pots, string encoders, cable extension transducers, and yo yo pots; in fact, they were the originator of the miniature and subminiature string potentiometer. SpaceAge Control also offers air data products that obtain aircraft angle of attack (AOA, alpha), sideslip (AOS, beta), static pressure, total pressure (pitot static), outside air temperature (OAT), and total air temperature (TAT). Their technologies have solved customer problems for over 35 years. The company has 29 representatives in over 39 countries.

Sensor Suppliers

Following is a list of suppliers for sensors discussed in this book, organized alphabetically by type of sensor. This information is presented courtesy of *Sensors* magazine, a publication of Advanstar Communications. To locate other suppliers of similar products, please visit the Sensors Buyer's Guide Online (www.sensorsmag.com/buyersguide), offering a comprehensive listing of companies by product type; and Sensors Express (www.sensorsmag.com/express) a database of recent product announcements that you can search by type.

ACCELERATION

These companies offer a variety of piezoelectric, piezoresistive and variable capacitance accelerometers for a wide range of measurement applications.

Endevco Corporation
30700 Rancho Viejo Rd.
San Juan Capistrano, CA 92675
949-493-8181, 800-982-6732
applications@endevco.com
www.endevco.com
Contact: Bob Arkell, Marketing Manager
ISO 9001 certified.

PCB Piezotronics Inc.
3425 Walden Ave.
Depew, NY 14043
716-684-0001, 800-828-8840
info@pcb.com
www.pcb.com
ISO 9001 certified.

Sensor Suppliers

Kistler Instrument Corporation
75 John Glenn Drive
Amherst, NY 14228
888-547-8537
sales.us@kistler.com
www.kistler.com
ISO 9001 certified.

ACOUSTIC

Companies listed here provide a variety of acoustic sensors and instruments, including sound level meters, high intensity acoustic sensors, noise dosimeters, and microphones.

Larson Davis Labs
1681 W. 820 N.
Provo, UT 84601
801-375-0177, 888-258-3222
sales@larsondavis.com
www.larsondavis.com
Contact: John Carey, Mktg. Coord.

**Measurement Specialties, Inc.
Sensors Group**
1000 Lucas Way
Hampton, VA 23666
757-766-1500, 800-745-8008
sales@msiusa.com
www.msiusa.com
Contact: Angela Watkins,
Global Mktg. Comm. Mgr.
ISO 9001 certified.

Columbia Research Laboratories, Inc.
1925 MacDade Blvd.
Woodlyn, PA 19094
610-872-3900
800-813-8471
Fax 610-872-3882
sales@columbiaresearchlab.com
www.columbiaresearchlab.com
Contact: Robert R. Reymos

UE Systems Inc.
14 Hayes St.
Elmsford, NY 10523
914-592-1220, 800-223-1325
Fax 914-347-2181
info@uesystems.com
www.uesystems.com
Contact: Gary Mohr, VP Sales & Ops.

BIOMEDICAL / BIOSENSORS

The companies listed here offer a variety of analytical and control biomedical sensors and biosensors, for applications that include diagnostics, life science R&D, bio defense, industrial process control, and others.

Honeywell Sensing and Control
1845 57th St.
Sarasota, FL 34243
941-355-8411, 800-446-5762
wayne.garris@honeywell.com
www.honeywell.com
ISO 9001 certified.

Sensor Suppliers

Omron Electronics LLC
One E. Commerce Dr.
Schaumburg, IL 60173
800-556-6766
Fax 847-843-7787
omroninfo@omron.com
www.omron.com/oei
Contact: Mike Frey, Sensor Group, Product Marketing Mgr.
ISO 9001 certified.

Texas Instruments, Analytical Sensors
13536 N. Central Expressway, MS945
PO Box 655012
Dallas, TX 75243
972-995-7914
spreeta@ti.com
www.ti.com/spreeta
Contact: Donna Whitmarsh, Product Mgr.

CHEMICAL / GAS

The companies listed here offer a wide selection of semiconductor, IR, electrochemical, and catalytic bead sensors for measuring combustible gases, toxic gases, oxygen, pH, conductivity, and various chemicals, for applications including medical, automotive, environmental, industrial and many others.

ABB Inc., Instrumentation Div.
125 E. County Line Rd.
Warminster, NY 18974-4995
800-922-2475
instrumentation@us.abb.com
www.abb.com/instrumentation
Contact: Deb Brewer, Cust. Info. Svcs..
ISO 9001 certified.

City Technology Ltd.
City Technology Center, Walton Rd.
Portsmouth, Hampshire, PO6 1SZ
UNITED KINGDOM
44-2392-325511
Fax 44-2392-386611
sensors@citytech.co.uk
www.citytech.com
Contact: Brian Roake, Sales & Mktg. Dir.
ISO 9001 certified.

Figaro USA Inc.
3703 W. Lake Ave., Suite 203
Glenview, IL 60025
847-832-1701
Fax 847-832-1705
figarousa@figarosensor.com
www.figarosensor.com
Contact: Taro Amamoto, VP Mktg.
ISO 9001 certified.

Honeywell Sensing and Control
1845 57th St.
Sarasota, FL 34243
941-355-8411, 800-446-5762
wayne.garris@honeywell.com
www.honeywell.com
ISO 9001 certified.

Rosemount Analytical Inc.
Process Analytic Div.
1201 N. Main St., PO Box 901
Orrville, OH 44667
330-682-9010, 800-433-6076
Fax 330-684-4434
gas.csc@emersonprocess.com
www.processanalytic.com
ISO 9001 certified.

Sensor Suppliers

Thermo Electron Corp.
27 Forge Pkwy.
Franklin, MA 02038
508-520-0430, 866-282-0430
Fax 508-520-1460
thermo@thermoei.com
www.thermo.com

Sierra Monitor Corp.
1991 Tarob Ct.
Milpitas, CA 95035
408-262-6611, 800-727-4377
Fax 408-262-9042
www.sierramonitor.com
Contact: Steve Ferree, VP Mktg.

DISPLACEMENT

The companies listed below offer a large range of both contact and noncontact displacement and position sensors and systems, using various technologies, including string potentiometers, capacitive, inductive, LVDTs, optical, eddy current, potentiometric, and others.

SpaceAge Control Inc.
38850 20th St. E.
Palmdale, CA 93550
661-273-3000
email@spaceagecontrol.com
www.spaceagecontrol.com
ISO 9001 certified.

Capacitive and Inductive Sensors

Baumer Electric Ltd.
122 Spring St., C-6
Southington, CT 06489
800-937-9336
Fax 860-628-6280
sales.us@baumerelectric.com
www.baumerelectric.com

Sensor Suppliers

Lion Precision
563 Shoreview Park Road
St. Paul, MN 55126
651-484-6544, 800-229-6544
www.lionprecision.com
Contact: Bob Benjamin, Product Mgr.
ISO 9001 certified.

Micro-Epsilon
D-94496 Ortenburg
Germany
+49 / 85 42 / 1 68-1 21
www.micro-epsilon.com
ISO 9001 certified.

TURCK Inc.
3000 Campus Drive
Minneapolis, MN 55441-2656
800-544-7769
Fax 763-553-0708
mailbag@turck.com
www.turck.com
Contact: Jeff Ursell, VP Sales & Marketing

Capacitive Sensors Only

ADE Technologies
80 Wilson Way
Westwood, Massachusetts 02090
USA
781-467-3500
www.adetech.com

Capacitec
87 Fitchburg Road
P.O. Box 819
Ayer, MA 01432
USA
978-772-6033

Sensor Suppliers

MTI Instruments
325 Washington Avenue Extension
Albany, New York 12205-5505
USA
518-218-2550
www.mtiinstruments.com

Rechner Electronics Industries Inc.
8651 Buffalo Ave.
Niagara Falls, NY 14304
800-544-4106
Fax 888-283-2127
custserv@rechner.com
www.rechner.com
Contact: Glen MacIntyre, Tech. Sales
ISO 9001 certified.

Inductive Sensors Only

Kaman Sensors
217 Smith Street
Middletown, CT 06457
USA
860-632-4573

Keyence
50 Tice Blvd.
Woodcliff Lake, NJ 07677
USA
201-930-0100

ELECTRICAL PROPERTIES

The companies listed below provide a wide selection of sensors and instruments to measure various electrical properties, including ohmmeters, resistance testers, electrometers, voltmeters, pico- and micro-ammeters, and many others.

AEMC Instruments
200 Foxborough Blvd.
Foxborough, MA 02035
508-698-2115, 800-343-1391
Fax 508-698-2118
sales@aemc.com
www.aemc.com
Contact: Howard Myers, App. Eng.

Conductive Technologies Inc.
935 Borom Rd.
York, PA 17404
717-764-6931, 800-706-0618
lmyers@conductivetech.com
www.conductivetech.com
Contact: Bill Ciabattoni, VP Sales and Mktg.
ISO 9001 certified.

Honeywell Sensing and Control
1845 57th St.
Sarasota, FL 34243
941-355-8411, 800-446-5762
wayne.garris@honeywell.com
www.honeywell.com
ISO 9001 certified.

Keithley Instruments Inc.
28775 Aurora Rd.
Cleveland, OH 44139
440-248-0400, 800-534-8453
Fax 440-248-6168
product_info@keithley.com
www.keithley.com
Contact: Ellen Modock, Mktg. Comm. Mgr.

Sensor Suppliers

Omega Engineering Inc.
One Omega Dr., PO Box 4047
Stamford, CT 06907-0047
203-359-1660, 800-826-6342
info@omego.com
www.omega.com
ISO 9001 certified.

Sypris Test & Measurement
(F.W. Bell)
6120 Hanging Moss Rd.
Orlando, FL 32807
407-678-6900, 800-775-2550
Fax 407-677-5765
marketing.stm@sypris.com
www.fwbell.com
Contact: Jeff Willey, Prod. Line Mgr.
ISO 9001 certified.

FLOW

Below are companies that offer a range of flowmeters, including inline and insertion turbine, positive displacement, ultrasonic, and many other flow measuring technologies and support instrumentation.

ABB Inc.
Analytical Instrumentation
2175 Lockheed Way
Carson City, NV 89706
775-883-4366
Fax 775-883-4373
www.abb.com/instrumentation
Contact: Andy Szeto, Mktg. Mgr.
ISO 9001 certified.

Emerson Process Management
Micro Motion
7070 Winchester Cir.
Boulder, CO 80301
800-760-8119, 800-522-6277
Fax 303-530-8459
micromotion@adtrack.com
www.micromotion.com
Contact: Sarah Danaher, MarCom Spec.

FISO Technologies Inc.
500 Ave. St.-Jean-Baptiste, Office 195
Quebec, QC G2E 5R9
CANADA
418-688-8065
sales@fiso.com
www.fiso.com
Contact: Jean-Francois Meilleur

Flowmetrics Inc.
9201 Independence Ave.
Chatsworth, CA 91311
818-407-3420, 800-356-6387
irfan@flowmetrics.com
www.flowmetrics.com
Contact: Irfan Ahmad, Sales and App. Eng.

GE Panametrics
221 Crescent St.
781-899-2746, 800-833-9438
Fax 781-894-8582
panametrics@ps.ge.com
www.gepower.com/panametrics
Contact: Ralph Sargent, MarCom Mgr.
ISO 9001 certified.

Sensor Suppliers

Omega Engineering Inc.
One Omega Dr., PO Box 4047
Stamford, CT 06907-0047
203-359-1660, 800-826-6342
info@omega.com
www.omega.com
ISO 9001 certified.

FORCE, LOAD AND WEIGHT

Listed below is a small selection of companies that manufacture sensors and instruments for measuring force, load, and weight. Applications for this type of measurement are vast, and include automotive, laboratory, and industrial uses; technologies include piezoelectric, piezoresistive, variable capacitive, and many others.

Honeywell Sensing and Control
1845 57th St.
Sarasota, FL 34243
941-355-8411, 800-446-5762
wayne.garris@honeywell.com
www.honeywell.com
ISO 9001 certified.

PCB Piezotronics Inc.
3425 Walden Ave.
Depew, NY 14043
716-684-0001, 800-828-8840
info@pcb.com
www.pcb.comISO 9001 certified.

Endevco Corp.
30700 Rancho Viejo Rd.
San Juan Capistrano, CA 92675
949-493-8181, 800-982-6732
applications@endevco.com
www.endevco.com
Contact: Bob Arkell, Mktg. Mgr.
ISO 9001 certified.

Vishay Transducers
677 Arrow Grand Cir.
Covina, CA 91722
626-331-0502, 800-722-0820
800-626-2616
Fax 626-332-3418
vt.us@vishaymg.com
www.vishaymg.com
Contact: Jim Hynes, Sales Mgr.
ISO 9001 certified.

HUMIDITY

Companies listed here provide sensors and instruments that measure moisture and humidity in air and other gases. These include relative humidity, absolute humidity, dew point sensors, instruments and controls.

Honeywell Sensing and Control
1845 57th St.
Sarasota, FL 34243
941-355-8411, 800-446-5762
wayne.garris@honeywell.com
www.honeywell.com
ISO 9001 certified.

EdgeTech
Moisture & Humidity Systems Div.
455 Fortune Blvd.
Milford, MA 01757
508-478-9500, 800-276-3729
h2o@edgetech.com
www.edgetech.com
Contact: Jeff Plugis, VP

Sensor Suppliers

GE General Eastern Instruments
Div. of General Electric
500 Research Dr.
Wilmington, MA 01887
800-334-8643
Fax 978-203-1920
generaleastern@ge.com
www.generaleastern.com
Contact: Mark DeNovellis, Dir. Mktg.
ISO 9001 certified.

Ohmic Instruments Co.
508 August St.
Easton, MD 21601
410-820-5111, 800-626-7713
Fax 410-822-9633
Contact: Pat Shorts, Cust. Svc.

Vaisala Inc.
100 Commerce Way
Woburn, MA 01801
781-933-4500, 888-824-7252
Fax 781-933-8029
incsales@vaisala.com
www.vaisala.com
Contact: Gerry Ducharme, Dir. Sales & Mktg.
ISO 9001 certified.

LEVEL

The following companies provide sensor solutions for level measurement applications, including RF/admittance, ultrasonic, capacitance, and many other technologies.

AMETEK Drexelbrook
205 Keith Valley Rd.
Horsham, PA 19044
215-674-1234, 800-969-4641
Fax 215-674-2731
deinfo@drexelbrook.com
www.drexelbrook.com
ISO 9001 certified.

Balluff
8125 Holton Dr.
Florence, KY 41042
859-727-2200, 800-543-8390
balluff@balluff.com
www.balluff.com
Contact: Tom Draper, Mktg. Prog. Mgr.
ISO 9001 certified.

Honeywell Sensing and Control
1845 57th St.
Sarasota, FL 34243
941-355-8411, 800-446-5762
wayne.garris@honeywell.com
www.honeywell.com
ISO 9001 certified.

PCB Piezotronics Inc.
3425 Walden Ave.
Depew, NY 14043
716-684-0001, 800-828-8840
info@pcb.com
www.pcb.com
ISO 9001 certified.

Scientific Technologies Inc.
Automation Sensors
1025 W. 1700 N.
Logan, UT 84321
435-753-7300, 888-525-7300
Fax 435-753-7490
sales@automationsensors.com
www.automationsensors.com
Contact: Jim Vella, GM

OPTICAL

These companies provide optical sensing devices for a wide variety of applications, including laser systems, fiber optics, metrology systems, and many others.

Balluff
8125 Holton Dr.
Florence, KY 41042
859-727-2200, 800-543-8390
balluff@balluff.com
www.balluff.com
Contact: Tom Draper, Mktg. Prog. Mgr.
ISO 9001 certified.

Infrared Solutions, Inc.
3550 Annapolis Lane N, Suite 70
Plymouth, MN 55447
Tel: (763)551-0003
Fax: (763)551-0038
Toll Free: 1-800-760-4523
Tim.Schooler@infraredsolutions.com
www.infraredsolutions.com

Honeywell Sensing and Control
1845 57th St.
Sarasota, FL 34243
941-355-8411, 800-446-5762
wayne.garris@honeywell.com
www.honeywell.com
ISO 9001 certified.

POSITION AND MOTION

Listed below are companies that provide sensors for position and motion measurement applications; technologies include photoelectric, laser, ultrasonic, string potentiometers, optical encoder, inductive, and others.

Banner Engineering Corp.
9714 10th Ave. N.
Minneapolis, MN 55441
763-544-3164, 888-373-6767
Fax 763-544-3213
sensors@bannerengineering.com
www.bannerengineering.com
Contact: Chris Benson, Mgr. Media Mktg.

BEI Technologies Inc.
13100 Telfair Ave.
Sylmar, CA 91342
818-362-1836, 800-959-0506
Fax 818-362-1836
Contact: Gregg Stokely
VP Comm. Mktg.

Endevco Corp.
30700 Rancho Viejo Rd.
San Juan Capistrano, CA 92675
949-493-8181, 800-982-6732
applications@endevco.com
www.endevco.com
Contact: Bob Arkell, Mktg. Mgr.
ISO 9001 certified.

Honeywell Sensing and Control
1845 57th St.
Sarasota, FL 34243
941-355-8411, 800-446-5762
wayne.garris@honeywell.com
www.honeywell.com
ISO 9001 certified.

Sensor Suppliers

Pepperl+Fuchs
1600 Enterprise Pkwy.
Twinsburg, OH 44087
330-425-3555
Fax 330-425-4607
sales@us.pepperl-fuchs.com
www.us.pepperl-fuchs.com
ISO 9001 certified.

SpaceAge Control Inc.
38850 20th St. E.
Palmdale, CA 93550
661-273-3000
email@spaceagecontrol.com
www.spaceagecontrol.com
ISO 9001 certified.

TURCK Inc.
3000 Campus Dr.
Minneapolis, MN 55441-2656
800-544-7769
Fax 763-553-0708
mailbag@turck.com
www.turck.com
Contact: Jeff Ursell, VP Sales & Mktg.
ISO 9001 certified.

PRESSURE

The companies in the list below offer a wide range of pressure sensing technologies, including piezoresistive silicon sensors, ceramic capacitive, dielectric, absolute, differential and combined pressure sensors.

Druck Inc.
Div. of GE Industrial Systems
4 Dunham Dr.
New Fairfield, CT 06812
203-746-0400
Fax 203-746-2494
usa.sales@druck.com
www.pressure.com
Contact: Paul Lupke, App. Eng.
ISO 9001 certified.

Endevco Corp.
30700 Rancho Viejo Rd.
San Juan Capistrano, CA 92675
949-493-8181, 800-982-6732
applications@endevco.com
www.endevco.com
Contact: Bob Arkell, Mktg. Mgr.
ISO 9001 certified.

Kavlico Corp.
A Solectron Company
14501 Los Angeles Ave.
Moorpark, CA 93021
805-523-2000
Fax 805-523-7125
sales@kavlico.com
www.kavlico.com
Contact: Robert Marousek, App. Mgr.
ISO 9001 certified.

Sensor Suppliers

Honeywell Sensing and Control
1845 57th St.
Sarasota, FL 34243
941-355-8411, 800-446-5762
wayne.garris@honeywell.com
www.honeywell.com
ISO 9001 certified.

Kistler Instruments
75 John Glenn Drive
Amherst, NY 14228
888-547-8537
sales.us@kistler.com
www.kistler.com
ISO 9001 certified.

Omega Engineering Inc.
One Omega Dr., PO Box 4047
Stamford, CT 06907-0047
203-359-1660, 800-826-6342
info@omega.com
www.omega.com
ISO 9001 certified.

Pressure Systems Inc.
34 Research Dr.
Hampton, VA 23666
757-865-1243, 800-678-7226
Fax 757-865-8744
sales@pressuresystems.com
www.pressuresystems.com
Contact: Mats Dahland, Sales Dir.
ISO 9001 certified.

SHOCK

Listed are companies that provide a range of sensor solutions for use in static and dynamic shock and vibration measurements; technologies include piezoelectric, piezoresistive, variable capacitance, and others.

Endevco Corp.
30700 Rancho Viejo Rd.
San Juan Capistrano, CA 92675
949-493-8181, 800-982-6732
applications@endevco.com
www.endevco.com
Contact: Bob Arkell, Mktg. Mgr.
ISO 9001 certified.

Honeywell Sensotec
2080 Arlingate Lane
Columbus, OH 43228-4112
614-850-5000, 800-848-6564
sales@sensotec.com
www.sensotec.com

PCB Piezotronics Inc.
3425 Walden Ave.
Depew, NY 14043
716-684-0001, 800-828-8840
info@pcb.com
www.pcb.com
ISO 9001 certified.

Sensor Suppliers

STRAIN

Companies in this section provide a wide range of strain gages and systems, including custom strain gage installations.

HITEC Corporation
537 Great Road,
Littleton, MA 01460 US
Tel +1 978 742-9032, Fax +1 978 742-9033
sales@hitecorp.com
www.hitecorp.com
Contact: Nicole McInnis, Mktg. Mgr.

Honeywell Sensotec
2080 Arlingate Lane
Columbus, OH 43228-4112
614-850-5000, 800-848-6564
sales@sensotec.com
www.sensotec.com

Kistler Instrument Corp.
75 John Glenn Dr.
Amherst, NY 14228
888-547-8537
sales.us@kistler.com
www.kistler.com
Contact: Mike Murphy, Inside Sales Mgr.
ISO 9001 certified.

MicroStrain Inc.
310 Hurricane Lane, Suite 4
Williston, VT 05495
802-862-6629, 800-449-3878
Fax 802-863-4093
info@microstrain.com
www.microstrain.com
Contact: Steven Arms, President

Sensor Suppliers

Omega Engineering Inc.
One Omega Dr.
Stamford, CT 06907-0047
203-359-1660, 800-826-6342
info@omega.com
www.omega.com
ISO 9001 certified.

Vishay Transducers
677 Arrow Grand Cir.
Covina, CA 91722
626-331-0502, 800-722-0820
800-626-2616
Fax 626-332-3418
vt.us@vishaymg.com
www.vishaymg.com
Contact: Jim Hynes, Sales Mgr.
ISO 9001 certified.

TEMPERATURE

Companies listed here offer a variety of temperature sensing and control devices, including thermistors, thermocouples, RTDs, thermostats, heat flux sensors, and strain gage compensating devices.

Airpax, Inc.
550 Highland Street
P.O. Box 500
Frederick, MD USA 21701
Phone: 301-663-5141
Fax: 301-698-0624
www.airpaxtsv.com
www.airpax.net
jim.breuer@airpax.net
ISO 9001 certified.

Sensor Suppliers

GE Thermometrics
General Electric Industrial Systems
967 Windfall Rd.
St. Marys, PA 15857
814-834-9140, 800-246-7019
Fax 814-781-7969
nancy.stouffer@indsys.ge.com
www.thermometrics.com
Contact: Peter Straub, Automotive Strategy Mgr.
ISO 9001 and QS-9000 certified.

Honeywell Sensing and Control
1845 57th St.
Sarasota, FL 34243
941-355-8411, 800-446-5762
wayne.garris@honeywell.com
www.honeywell.com
ISO 9001 certified.

Keystone Automation Inc.
5649 Memorial Ave. N.
Stillwater, MN 55082
651-439-4268
Fax 651-439-4279
info@keystoneautomation.com
www.keystoneautomation.com
Contact: Tim Simonson, VP Sales

RDF Corp.
23 Elm Ave., PO Box 490
Hudson, NH 03051-0490
603-882-5195, 800-445-8367
Fax 603-882-6925
sensor@rdfcorp.com
www.rdrcorp.com
Contact: Irwin Bluestein, Natl. Sales Mgr.
ISO 9001 certified.

Vatell Corp.
PO Box 66
Christiansburg, VA 24068
540-961-3576
Fax 540-953-3010
mkt@vatell.com
www.vatell.com
Contact: Phil Prosser, VP Sales & Mktg.

TORQUE

These companies provide multi-axis torque/force sensors for a variety of applications and load ranges, from outer space to the ocean floor.

AMTI
176 Waltham St.
Watertown, MA 02472
617-926-6700, 800-422-2684
Fax 617-926-5045
sales@amtimail.com
www.amti.biz
Contact: Gary Blanchard, Prod. Mgr.
ISO 9001 certified.

Honeywell Sensotec
2080 Arlingate Lane
Columbus, OH 43228-4112
614-850-5000, 800-848-6564
sales@sensotec.com
www.sensotec.com

Kistler Instrument Corp.
75 John Glenn Dr.
Amherst, NY 14228
888-547-8537
sales.us@kistler.com
www.kistler.com
Contact: Mike Murphy, Inside Sales Mgr.
ISO 9001 certified.

Sensor Suppliers

Lebow Products
1728 Maplelawn Dr.
Troy, MI 48084
248-643-0220, 800-803-1164
Fax 248-643-0259
lebowsales@lebow.com
www.lebow.com
ISO 9001 certified.

Magtrol Inc.
70 Gardenville Pkwy.
Buffalo, NY 14224
716-668-5555
Fax 716-668-8705
magtrol@magtrol.com
www.magtrol.com

S. Himmelstein & Co.
2490 Pembroke Ave.
Hoffman Estates, IL 60195
847-843-3300, 800-632-7873
Fax 847-843-8488
sales@himmelstein.com
www.himmelstein.com
Contact: Steven E. Tveter, Natl. Sales Mgr.

VIBRATION

Listed below are companies that provide vibration measurement sensors and systems, using various technologies, including piezoresistive, piezoelectric, variable capacitance and others; applications include industrial, test and measurement, and military.

Analog Devices Inc.
Micromachined Products Div.
21 Osborn St.
Cambridge, MA 02139
800-742-7024
www.analog.com

Columbia Research Laboratories, Inc.
1925 MacDade Blvd.
Woodlyn, PA 19094
610-872-3900
800-813-8471
Fax 610-872-3882
sales@columbiaresearchlab.com
www.columbiaresearchlab.com
Contact: Robert R. Reymos

Endevco Corporation
30700 Rancho Viejo Rd.
San Juan Capistrano, CA 92675
949-493-8181, 800-982-6732
applications@endevco.com
www.endevco.com
Contact: Bob Arkell, Mktg. Mgr.
ISO 9001 certified.

PCB Piezoelectronics Inc.
3425 Walden Ave.
Depew, NY 14043
716-684-0001, 800-828-8840
info@pcb.com
www.pcb.com
ISO 9001 certified.

Kistler Instrument Corporation
75 John Glenn Dr.
Amherst, NY 14228
888-547-8537
sales.us@kistler.com
www.kistler.com
Contact: Mike Murphy, Inside Sales Mgr.
ISO 9001 certified.

Wilcoxon Research Inc.
21 Firstfield Rd.
Gaithersburg, MD 20878
301-330-8811, 800-945-2696
Fax 301-330-8873
sensors@wilcoxon.com
www.wilcoxon.com
ISO 9001 certified.

Subject Index

Numeric
4-20 mA output, 293

A
absolute humidity, 271
AC bias, 8
accelerometers, 137-158, 291, 396-398
    applicable standards, 153
    calibration, 153
    for shock measurement, 459
    MEMS, 396
    mounting, 155-157
    selecting, 150-152
    with ADC inside, 158
accuracy, 3, 29, 203, 419
acoustic sensor, 481-499
    frequency response, 484
acoustic wave, 485
active sensor, 16
actuator, 1, 18
actuator, electrostatic, 485
ADC, 18, 92-107
AKAS, 154
amperometry, 171
amplifier DC error budget analysis, 60
amplifier selection and use, 25, 45-91
amplifying bridge output, 36
analog-front-end (AFE) integration, 134
analog output sensors, 198
analog-to-digital converter (see ADC)
anemometer, 237
ANSI, 153, 493
antibodies, 169
array microphone, 495
assembly inspection, 209

B
bandgaps, 310-311
bandpass filters, 13
bandwidth, 4, 5, 202, 211
bending beam strain-gage sensor, 264
biased head-on sensing, 337
bi-metal thermostat, 531-532, 539, 553
biophotonic sensors, 178
bioreactor control, 167
bioreceptor, 161, 169
biosensors, 161-180, 569
    applications, 164-168
    design process, 163
    future prospects, 177-178
    nanotechnology applied to, 569
    origin, 168
    transducers for, 171-173
bipolar head-on position sensor, 336
Bluetooth, 581
Bode plot, 115
Bragg cell, 245
bridge circuits, 31-44
bulb and capillary thermostat, 532, 540, 553

C
cable position transducers (CPTs), 370-377
    advantages, 371
cables and connectors, 24
cabling, accelerometer, 301
calibration, microphone, 497-499
CAN technology, 431
capacitance measuring circuits, 8-9, 172
    limitations, 10
capacitive accelerometers, 146-148
capacitive displacement sensors, 194-196, 200

Index

applications, 204, 211
  with inductive sensors, 219
  sensitivity, 195
  target, 195
capacitive RH sensors, 272
  selecting, 276
charge amplifier, 437
charge coupled devices (CCDs), 108, 130-135
charge mode accelerometer, 140-141, 459
charge mode piezoelectric force sensor, 257
charge output sensors, 108, 126-130
CHEMFET, 190
chemical sensor basics, 181-191
  applications, 188-191
chemical sensors, nanotechnology-based, 568
chopper stabilized amplifiers, 84-86
chromatography instruments, 185
closed loop bandwidth, 116
column strain gage sensors, 264
common mode rejection, 58
compression mode accelerometer, 143
condensation, 282
condenser microphone, 483
conductimetric transducers, 172
contact image sensor (CIS), 130
convergent beam scanning, 364
Coriolis meter, 242
correlated double sampling (CDS), 132
cost of sensor ownership, 591-594

D

DAC, 18
dark current, 109, 123, 311
data acquisition, 26
  system on a chip, 101
data sheet, 2
DC open loop gain, 50
DC response accelerometer, 293
deflection/deformation measurement, 207
deuterated tri-glycine sulfide (DTGS), 319
dew point, 271
dielectric constant, table of, 352
  change in, 272
differential pressure flowmeter, 237, 238
diffuse scanning for photoelectric, 362
digital Hall effect sensor, 338
digital-to-analog converter (see DAC)
diode, 308
discharge time constant (DTC), 256
displacement sensors, 193-196
  capacitive, 194-196
  inductive, 197-198
  selecting, 401-408
DNA sensors, 178
Doppler flow sensor, 243
double feed detection (paper), 214
Dunmore type humidity sensor, 273
dynamic range, 2, 5

E

eddy current sensors, 196
effective number of bits (ENOB), 106
electret microphone, 484
electrical runout, 197, 211
electrochemical detection techniques, 186
electromagnetic flow sensor, 242
electromagnetic interference in accelerometers, 473
electromagnetism in sensing, 196, 223-235
electro-mechanical temperature sensors, 538-539
  selecting, 538-539
elevated temperature strain gage installation, 525
embedded sensors, 212
emissivity, 532
environmental monitoring, 167
enzyme thermistor, 173
enzyme-electrode, 168
excitation voltage, 422

F

Faraday's Law of Induction, 225
fatigue rated load cell, 265
feedback, 14
ferromagnetism, 230
FET as transducer (see FET)
fiber optic photoelectric sensors, 365
field calibration, for microphone, 498
filters, 11-13
  bandpass, 13
  for shock sensors, 479
  high pass, 12
  low pass, 11

flash converter, 92
flexural mode accelerometer, 143
flip chips, 565
flow nozzles, 240
flow sensors, 237-249
    calibrating, 248
    installing, 247
    selecting, 246-247
flux gate magnetometer, 232
foil strain gage technology, 262
force balance accelerometer, 149
force sensing, 255-269
force sensor types (quartz)
    general purpose, 259
    impact, 260
    link, 261
    miniature, 260
    multi-component, 261
    penetration, 259
    ring, 260
free field correction, 487
free filament strain gage, 522
frequency modulated continuous wave radar (FMCW), 252-253

G

gas sensor, nanotube based, 568
general purpose load cell, 265
general stress analysis strain gage installation, 522
Giant Magnetoresistive Effect (GMR), 234
guarding technique, 112
guided wave radar (GWR) level sensing, 252

H

Hall effect sensors, 233, 330, 384
    selecting, 333
    used as rotation sensor, 385
heat conduction equation, 559
heat flow, 559
high-frequency industrial accelerometers, 292
high-impedance sensors, 108
    signal conditioning for, 108-135
high-sensitivity microphone, 491
high-temperature pressure sensors, 451
humidity sensors, 271-284
    integrated signal conditioning for, 283
    interfacing to, 280
    selecting and specifying, 275
    standards, 279
hybrid star-mesh network, 579
hydrophones, 128
hydrostatic tank gauging (HTG), 250
hysteresis, 3, 5, 420, 558

I

I/V converter, 121
IC sensors, 412, 541
ICP accelerometer, 289, 291
IEC (International Electrotechnical Commission), 492
IEEE 1451 Smart Transducer Interface, 158, 268, 607-609
IEEE 1451.2, 607
IEEE 1451.4, 607
IEEE 1451.5, 583
IEEE 802.11x, 581
IEEE 802.15.1 and .2 (Bluetooth), 581
IEPE accelerometers, 138-139, 289, 459
impedance change, 273
impedance converter, 439
impedance, 8
    output, 10
inductance measuring circuits, 9
inductive displacement sensors, 196-198
    applications, 204, 216
    compared with capacitive sensors, 203, 219
Inductosyns, 392
industrial accelerometers, selecting, 294
    cable orientation, 298
    calibration interval, 300
    duty, 299
    frequency range, 296
    mounting, 302
    resolution, 297
    size, 297
    temperature range, 297
inertial measuring unit (IMU), 149
infrared (IR) pyrometry, 532, 542
infrared spectra of molecules, 182
input bias current, 49
input offset voltage error, 46

Index

input voltage noise, 118
insert voltage calibration, for microphone, 498
insertion tube flow sensor, 239
installation of sensor system, 27-28
instrumentation amplifiers (in-amps), 70-84
    configurations, 72
    DC error sources, 79
    noise sources, 81
    error budget analysis, 82
    performance tables, 83
    overvoltage protection, 84
integrated differential flowmeter, 249
intrinsic safety enclosure, 301
ISA (Instrumentation, Systems, and Automation Society), 473
ISO, 154
isolation amplifiers, 87

J
JFET vs. bipolar op amps, 129
Johnson noise, 119

K
Kelvin sensing, 41
Kirchhoff's Rules, 6

L
label sensing, 215
laser Doppler anemometer (LDA), 243-245
laser Doppler velocimeter, 460
lead resistance, 9
Lenz's Law, 225
level sensors, 250-254
    selecting, 254
    standards, 254
limit switches, 322-326
    interfacing to, 325
    selecting, 323
    standards, 324
linear output sensors, 199
linearity, 3, 5, 202, 419, 482
linearizing bridge output, 37-38
liquid strain gage, 502
Lissajous pattern, 211
load cell, 263
load sensor, 255

Lorentz force, 225, 228
low-cost accelerometers, 291-292
LVDT, 231, 379-384

M
machinery vibration monitoring, 285-306
magnetic field sensors, 232
magnetic position sensors, 330-340
    interfacing to, 335
magnetic runout, 197
magnetometers, 232
magneto-resistive sensors (MR), 331-340
    array device, 332
    selecting, 332
magnetoresistor, 234
magnetostrictive level sensor, 251-2532
mass flowmeters, 242
mass spectra of molecules, 184
measurement criteria, 29-30
mechanical shock sensors, 457-480
    design techniques, 478
    interfacing, 474
    mounting, 474
    signal conditioning, 477
    standards, 473
membranes in biosensors, 176
mercury cadmium telluride (MCT), 312
mesh network, 578
microbolometer, 317-318
microcontrollers, 19
microdevices, 505
micro-electromechanical systems (MEMS), 249, 396, 480, 563
microfabrication techniques, 186
microphones, test and measurement, 481-499
    condenser, 482-483
        dynamic, 482
        electret, 484
        environmental effects, 491
        free field, 487
        limitations, 490
        piezoelectric, 482
        pressure, 486
        pressure response, 485
        random incidence, 489
        specialized, 494-496
        standards, 492

microsensors, 144, 431
microwave radar level sensing, 252
miniaturization, 455
molecule detectors, 182
multiple channel systems, 206
multiplexing, 100

N
Nanobarcode, 569
nanotechnology, 177
nanotechnology-enabled sensors, 563-574
NIST, 30, 154
noise analysis, 117
noise gain (NG), 50, 116
noise shaping, 104-105
noise suppression, 440
noise, 3, 5
noncontact displacement sensors, 193, 222
noncontact position sensor, 321
nonlinearity, 3
nonrepeating runout (NRR), 212

O
Oersted, 224
offset, in displacement sensor, 201
Ohm's Law, 6, 223
op amp noise, 53-58
open loop gain nonlinearity, 52
operational amplifier, 13-15
optical encoder, 310, 386-387
optical sensors, 308
opto-interruptor, 309
oscillatory shock, 458
oversampling ratio, 103

P
passive sensor, 16
personal area network (PAN), 581
photoconductive mode, 308
photodiode preamplifier, 108
photodiode, 308
    spectral response, 310
photoelectric sensors, 358-367
    interfacing to, 366
    specifying, 359
photometry, 171-172

photosensors, 307
phototransistor, 309, 316-317
photovoltaic mode, 308
piezoelectric accelerometer, 137-143, 459
    for shock measurement, 459
    monitoring machinery vibration with, 285
    structures for, 142
piezoelectric force sensor, 258
piezoelectric pressure sensing, 433-456
    sensor design, 441
piezoelectric sensors, 108, 172, 255-256
piezoelectricity, 434
piezoresistive accelerometers
    for shock measurement, 459
piezoresistive pressure sensing, 411-432
    interfacing, 427
    selecting, 418
    interfacing, 427
piezoresistor, 501-510
piston phone calibrator, 497
Pitot tube, 238
plane wave, 485
Plug and Play, 609
polarized scanning in photoelectric sensors, 359
poling process for ceramics, 141
polycrystalline ceramics, 141
polymer-based sensor, 272
position sensors, 321-408
    selecting, 401-408
position window measurement, 207
positive displacement flow sensor, 241
potentiometers, 327
potentiometry, 171
power supply rejection ratio (PSRR), 59
power supply, 24
    rejection, 58
preamplifier AC design, 115
precision spindle error motion, 211
precision strain gage installation, 524
prepolarized microphone, 484
pressure sensors, 411-456, 516
probe microphone, 496
proximity detector, 309
proximity sensors, 345-358
    capacitive, 346
    inductive, 349

Index

interfacing to, 354-358
selecting, 349
standards, 353
proximity switches, 198
PSRR, 59
PTB, 154
pulsed radar level sensing, 253
PVDF film, 319
pyroelectric effect, 469
pyroelectric plastic, 319

Q

QCM (see quartz crystal microbalance)
quadruple analyzers, 185
quantization noise, 103
quantum detectors, 307, 308-313
quartz crystal microbalance (QCM), 178
quartz sensors, 141, 255

R

rail-to-rail amplifier, 63
Raman spectroscopy, 178
random incidence microphone, 489
ratiometric linear sensors, 334
reciprocity calibration for microphone, 498
relative humidity integrated circuit (RHIC), 281
relative humidity sensors, 273
relative permittivity, 224
relative position measurement, 206
repeatability error, 420
resistive humidity sensor, 273
selecting, 277
resistive position sensors, 327
interfacing to, 329
selecting, 327-329
standards, 329
resistive sensors, 6, 31
resistive strain gage, 511
resolution error, 60-61
resolution, 4, 5, 103, 202
resolvers and synchros, 387-391
retroreflective scanning, 359
RF capacitance level sensor, 251
righthand rule, 224, 225, 263
ROKIDE flame spray process, 527
RTDs, 32, 534, 547-549
self-heating, 560

RTI, 49
RTO, 49
runout, 212

S

seismic transducer, see Accelerometer
semiconductor strain gage, 514
semiconductor technology pressure sensors, 411
sensitivity, 2, 4, 202, 481
sensor application issues, 21-28
sensor basics, 1-20
sensor classification, 17
sensor selection checklist, 23
sensor systems, 16
servo accelerometers, 149
shear mode accelerometers, 142, 288
shear-web load cell, 264
shock sensors, 457-480
environmental effects on, 468-473
failure modes, 464
structural resonance, 468
sigma-delta ADC, 92, 102-107
signal bandwidth, 116
signal conditioning, 15, 17, 22, 31-136, 139, 517
amplifiers for, 45-91
silicon sensors, 412, 532, 540
silicon temperature sensor, 540
single supply op amps, 61-70
input stages, 63-66
output stages, 66-69
ski slope error, 289
slide-by position sensor, 337
smart differential pressure transmitter, 250
smart sensor, 19, 607-609
solid-state sensors, 412
sound intensity microphones, 494
sound level calibrator, 494, 497
span, 2
sputtering, 528
strain gage, 32, 262-269, 460, 501-529
accuracy, 507
applications, 507
flow measure, 516
installation, 522-529
selecting, 265
sensor types, 263

Index

standards, 268
string potentiometer, 370-377
structural loop in displacement measurement, 221
submicron electrometer, 568
successive approximation ADC, 92, 93-100
surface-enhanced Raman scattering (SERS), 178
synchros and resolvers, 387-391

T

target considerations
    for capacitive displacement sensor, 195
    for inductive displacement sensor, 197
TEDS (Transducer Electronic Datasheet), 158, 268, 290, 304, 456, 607-609
    sensor output, 609
Teflon lens, 320
temperature coefficient, 5
temperature sensor, 531-561
    contact, 531
    electro-mechanical, 531
    noncontact, 531
    selecting, 535-538
    standards, 549
thermal anemometer, 237-238
thermal conductivity humidity sensor, 274
    selecting, 278
thermal detectors, 313-317
    sensitivity, 315
thermal errors, 203
thermal expansion measurement, 208
thermal transient response of accelerometers, 471
thermistors, 534, 545
    positive temperature coefficient (PTC), 534
    selecting, 545
thermocouple, 533, 543-545
thermoelectric potentials, 115
thermometric transducer, 172
thermoset polymer-based capacitive RH sensor, 272
thickness measurement, 208-209
thin film metallization, 144
thin-film magnetometer, 235
thin-film strain gage, 504-505
thread detection, 218
three op amp in-amp, 75
through-air radar level sensing, 253
through-scan technique for photoelectric, 360
total indicator reading (TIR), 210
tracking ADC, 92
transducer, 15, 161
transfer function, 2, 4
transverse motion in accelerometers, 472
turbine-based flow sensors, 241
two op amp in-amp, 72

U

ultrasonic flow sensor, 243
ultrasonic level sensor, 251
ultrasonic position sensor, 340-345
    dead zone, 343
    interfacing to, 345
    selecting, 341-342
uncooled detector arrays, 317

V

vector AC induction motor control, 394
vector Hall sensors, 335
velocimeters, 460
velocity shock measurement, 457
Venturi tubes, 240
vibration measurement, 210
voltage mode piezoelectric force sensor, 258
vortex-shedding flow sensors, 241

W

wave, plane, 485
weight sensor, 255
weldable strain gage installation, 529
Western Regional Strain Gage Committee wiring code, 263
Wheatstone bridge, 7, 32, 144, 263, 413, 556
wireless sensors, 304, 575
wireless sensor networks (WSN), 575-589
    applications, 585
    network architecture, 576
    node architecture, 576
    power considerations, 583
    power consumption, 576
    radio layer, 580

Z

Zigbee Alliance, 582

Sensor Technology Index

Acoustic
pp. 481-499

Bimetallic
pp. 531-532, 539, 553

Biosensor
pp. 161-180, 569

Capacitive
pp. 8-9, 146-148, 172, 194-196, 200, 204, 211, 219, 251, 346

Charge-coupled Device
pp. 108, 130-135

Charge Output
pp. 108, 126-130, 140-141, 459

Conductive
pp. 172

Differential Pressure
pp. 237-238, 250

Doppler
pp. 243-245, 460

Electromagnetic
pp. 223-235, 242, 473

Electromechanical
pp. 531, 538-539

Encoder/Resolver/Synchro/LVDT
pp. 231, 379-384, 387-391, 389

Fiber optic
pp. 365

Force Transducer
pp. 149

Giant Magnetoresistive
pp. 234

Hall Effect
pp. 338, 233, 330, 335, 384, 385

Hydrostatic
pp. 250

Inductive
pp. 9, 196-198, 203, 204, 216, 219, 225, 349, 392, 394

Infrared
pp. 182, 317

Laser
pp. 243-245, 460

Load cells
pp. 265, 263, 264

Magnetic
pp. 230-232, 235, 330-340

Sensor Technology Index

Magneto-resistive
pp. 234, 251-252, 331-340

MEMS
pp. 144, 431, 186, 249, 396, 480, 455, 505, 563

Microwave
pp. 252

Nanotechnology
pp. 177, 568, 563-574

Optical
pp. 308-309, 310, 386-387

Photoelectric
pp. 362, 171-172, 308, 358-367

Phototransistor/diode
pp. 108, 307, 308, 309, 310, 316-317

Piezoelectric
pp. 141, 257-258, 482, 434, 433-456

Piezoresistive
pp. 144, 411-432, 459, 501-510

Potentiometric
pp. 171, 327, 370-377

Pyroelectric
pp. 319, 469, 532, 542

Quartz
pp. 141, 178, 259-261

Radar
pp. 252, 243

Resistive
pp. 6, 31, 273, 277, 327-329

RTD
pp. 32, 534, 547-549, 560

Semiconductor/IC
pp. 281, 411, 412, 541, 514, 532, 540

Strain gage
pp. 32, 262, 263, 264, 269, 460, 501-529

Spectroscopy
pp. 178

Thermal/Thermistor/Thermocouple
pp. 115, 172, 173, 203, 208, 274, 278, 313-317, 533, 534, 543-545

Thermometer
pp. 531-561

Ultrasonic
pp. 243, 251, 340-345

Vortex shedding
pp. 241

Wireless
pp. 304, 578, 575-589

OUR READERS ARE DESIGNING THE FUTURE

sensors®
YOUR RESOURCE FOR SENSING, COMMUNICATIONS, AND CONTROL

Launched in January 1984, Sensors celebrated its 20th anniversary in 2004.

"Sensors will completely reshape the information landscape in this decade."

—Paul Saffo
Institute for the Future

Sensors magazine is the premier industry resource for the latest technical information on sensing, communications, and control. The magazine is distributed monthly to a qualified readership of 75,000 buyers, specifiers and recommenders of sensors and sensors-related products. Its in-depth articles and columns focus on the use of sensing devices to increase efficiency, economy and productivity in applications ranging from manufacturing to process control, from aerospace to consumer products.

Sensors • Essential • Everywhere
www.sensorsmag.com

Sign up for a free subscription (U.S. and Canada)!
www.sensorsmag.com/subscribe

 sensorsonline
There's a lot more to Sensors than what you see in print.

Check out **Wireless Sensors**!

A quarterly publication covering wireless sensor networks.

sensors expo
& conference

Founded in 1986, the International Sensors Expo and Conference is the leading industry event for sensing technologies.

www.sensorsexpo.com

Edwards Brothers Malloy
Ann Arbor MI. USA
June 28, 2016